T0138192

The Ornaments of Life

INTERSPECIFIC INTERACTIONS
A Series Edited by John N. Thompson

The Ornaments of Life

Coevolution

and Conservation

in the Tropics

Theodore H. Fleming and W. John Kress

The University of Chicago Press · *Chicago and London*

Theodore H. Fleming is professor emeritus of biology at the University of Miami in Coral Gables, Florida. **W. John Kress** is curator and research scientist as well as director of the Consortium for Understanding and Sustaining a Biodiverse Planet at the Smithsonian Institution.

The University of Chicago Press, Chicago 60637
The University of Chicago Press, Ltd., London
© 2013 by The University of Chicago
All rights reserved. Published 2013.
Printed in China by C&C Offset Printing Co., Ltd.

22 21 20 19 18 17 16 15 14 13   1 2 3 4 5

ISBN-13: 978-0-226-25340-4 (cloth)
ISBN-13: 978-0-226-25341-1 (paper)
ISBN-13: 978-0-226-02332-8 (e-book)

Library of Congress Cataloging-in-Publication Data

Fleming, Theodore H., author.
  The ornaments of life : coevolution and conservation in the tropics / Theodore H. Fleming and W. John Kress.
      pages ; cm — (Interspecific interactions)
  Includes bibliographical references and index.
  ISBN 978-0-226-25340-4 (cloth : alkaline paper) — ISBN 978-0-226-25341-1 (paperback : alkaline paper) — ISBN 978-0-226-02332-8 (e-book) 1. Pollination by animals—Tropics. 2. Seed dispersal by animals—Tropics. 3. Angiosperms—Pollination—Tropics. 4. Vertebrates—Tropics. 5. Animal-plant relationships—Tropics. 6. Mutualism (Biology)—Tropics 7. Coevolution—Tropics. 8. Conservation biology—Tropics. I. Kress, W. John, author. II. Title. III. Series: Interspecific interactions.
  QH549.5.F54 2013
  576.8'75—dc23

                              2013000803

♾ This paper meets the requirements of ANSI/NISO Z39.48-1992 (Permanence of Paper).

We dedicate this book to our

colleagues around the world who are

working hard to understand and conserve

our "Ornaments of Life" for

future generations

Contents

Preface

This book is about the ecological and evolutionary interactions between tropical plant-visiting birds and mammals and their food plants. Our title was inspired by a quote from the late Robert Whittaker in his landmark review of patterns of terrestrial biodiversity: "Birds do pollinate certain flowers, distribute certain seeds, and realize a secondary productivity of ca. 10^{-5} of community net primary productivity. On the whole, however, birds seem to be evolution's ornaments on the tree of community function" (Whittaker 1977, 42). While it is true that tropical plant-visiting birds and, to a lesser extent, their mammalian counterparts, as well as their floral and fruit resources, are "ornaments" in terms of their spectacular beauty, it is certainly not true that they play an insignificant role in community function. To counter Whittaker's conclusion, we have synthesized current knowledge about the impact of tropical vertebrate pollinators and frugivores on the reproductive success, population dynamics, and population genetic structure of their food plants. We have attempted to place these interactions into a historical (phylogenetic) framework by reviewing (1) the effect that these animals and plants have had on each other's diversification and (2) the evolution of plant and animal traits involved in these pollination and frugivory mutualisms. Finally, we provide a discussion of the conservation status of vertebrate pollinators and seed dispersers in tropical habitats around the world.

The motivating questions that prompted us to write this book included:

1. What are the ecological and evolutionary consequences of pollinator and frugivory mutualisms for vertebrates and their food plants? In what ways

have these interactions influenced the taxonomic diversity and morphol-ogy, physiology, and behavior of vertebrate and plant mutualists?
2. What are the demographic and genetic effects of tropical vertebrate mu-tualists on their food plants? What are the potential ecological and evolu-tionary consequences of the extinction of these vertebrates on their food plants and plant communities?
3. What are the historical patterns of these interactions? What are the evo-lutionary pathways that have produced these mutualisms and to what extent have these interactions influenced speciation rates and patterns of diversification of vertebrate plant-visitors and their food plants? What is the extent of convergent evolution in characters of both the plants and animals engaged in these mutualisms?
4. How threatened by and in danger of extinction by human activities are these mutualisms and what can we do to conserve them?

Providing a global overview of tropical vertebrate-plant mutualisms is timely because many of the mutualisms we describe are in danger of go-ing extinct, either directly or indirectly, by the hand of man. Indeed, we know from the archaeological studies of David Steadman and others that many South Pacific islands have already lost substantial fractions of their avian and mammalian faunas, including nectarivores and frugivores (e.g., Steadman 2006). In some cases, these losses have had a negative effect on the reproductive success of native plants. For example, reduction in the size of populations of fruit-eating pteropodid bats on certain South Pacific is-lands has effectively destroyed their functional role as seed dispersers of tropical trees (McConkey and Drake 2006). Likewise, the widespread kill-ing of monkeys and large frugivorous birds in many tropical forests will ultimately reduce tree species richness by changing the demography and dispersion patterns of large-seeded plants (Muller-Landau 2007; Webb and Peart 2001). In general, if nectar- and fruit-eating birds and mammals do play an important functional role in the ecology of tropical habitats, then their disappearance will affect the demography, genetic structure, and, ulti-mately, the evolutionary potential of their food plants. It is our aim in this book to describe the functional relationships between vertebrate mutualists and their food plants, how these relationships have evolved, and why they must be conserved.

We have many agencies and people to thank for financial support and assistance with this project. We thank the following agencies for providing financial support for our research over the years: the US National Science Foundation, National Geographic Society, Fulbright Program, Earthwatch Institute, US Fish and Wildlife Foundation, Ted Turner Foundation, and our home institutions: the Universities of Missouri-St. Louis and Miami (Fleming) and the Smithsonian Institution (Kress). Fleming thanks the College of Arts and Sciences, University of Miami, for generously providing him with funding and release time to work on this project and the Department of Botany, National Museum of Natural History, for its hospitality and access to the academic resources of the Smithsonian Institution. Kress thanks the Smithsonian for its support of independent scholarly research as well as his many colleagues who have contributed to our knowledge of plant-vertebrate interactions. We both thank the Department of Botany, National Museum of Natural History; the College of Arts and Sciences, University of Miami; and Bat Conservation International for providing funds for color publication of this book.

We especially thank Ida Lopez and Alice Tangerini (both in the Department of Botany, National Museum of Natural History) for help with preparing figures and other manuscript details (Lopez) as well as figure 9.2 and cover art (Tangerini). We also thank Yvonne Zipter for her careful copyediting of this book.

For reading and commenting on parts or all of the manuscript, we thank P. Ashton, D. Levey, M. Potts, J. Thompson, and two anonymous reviewers. The following people helped with literature, shared papers and ideas, or helped with data analysis: R. Borges, S. Buchmann, C. Chapman, R. Corlett, C. Dick, E. Dumont, E. Edwards, J. Ganzhorn, F. Gill, T. Givnish, J. Gonzalez, A. Herre, M. Hoffman, C. Janson, C. Jones, S. Knapp, D. Kissling, S. Lomáscolo, D. Levey, C. Mancina, M. Mello, N. Muchhala, R. Nathan, J. Ollerton, D. Pearson, P. Racey, R. Robichaux, A. Sebben, R. Sussman, L. Taylor, E. Temeles, D. Waldien, and D. Wilson.

Neither of us could have completed this intellectual task without the support and inspiration of our wives, Marcia and Lindsay, who encouraged each of us at every step of the way. We are forever grateful.

Finally, it has been a privilege to conduct our research on plant-animal interactions in many tropical and subtropical countries around the world. We could not have written this book without these experiences. We thank the many government agencies, colleagues, and local people for making this research possible. We extend special thanks to Dr. Jin Chen, director of the Xishuangbanna Tropical Botanical Garden in Yunnan Province, China, and the Chinese Academy of Sciences for supporting our visits to that marvelous institution. We hope that this book will increase public awareness of the beauty, ecological significance, and evolutionary importance of the many "ornaments" we describe here and will provide a convincing argument as to why this biological heritage needs to be preserved. A world lacking these ornaments would be a dreary one, indeed.

1   The Scope of This Book

Tropical forests and their marine counterparts, coral reefs, are the most species-rich and colorful ecosystems on earth. Much of this color comes from their animal inhabitants. In coral reefs, the corals themselves as well as anemones, gorgonians, sea fans, sponges, and giant clams, among the invertebrates, and wrasses, damselfish, angelfish, and butterfly fish, among the vertebrates, are major contributors to the color palette. In tropical forests, nectar-feeding birds—hummingbirds, sunbirds, and lorikeets—and their fruit-eating counterparts—tanagers, trogons, manakins, toucans, hornbills, and birds of paradise—are among the most colorful and conspicuous vertebrates (plates 1–5). Less colorful but nonetheless still notable because of their size, conspicuousness, and/or abundance are their mammalian ecological counterparts—phyllostomid and pteropodid bats and primates (plates 6–7). Supplying all of these animals with food is a cornucopia of colorful flowers and fruits produced by understory herbs (e.g., *Heliconia, Musa*), shrubs (e.g., Melastomataceae, Rubiaceae), vines (e.g., Bignoniaceae), epiphytes (e.g., Bromeliaceae, Gesneriaceae, Loranthaceae), and subcanopy, canopy, and emergent trees (e.g., Bignoniaceae, Fabaceae, Lauraceae, Malvaceae [Bombacoideae], Myristicaceae, Myrtaceae) (plates 8–13).

In emphasizing the beauty of tropical nature, the prominent twentieth-century evolutionary biologist Theodosius Dobzhansky wrote:

> Becoming acquainted with tropical nature is, before all else, a great esthetic experience. Plants and animals of temperate lands seem to us somehow easy to live with, and this is not because many of them are long familiar. Their style

is for the most part subdued, delicate, often almost inhibited. Many of them are subtly beautiful; others are plain; few are flamboyant. In contrast, tropical life seems to have flung all restraints to the winds. It is exuberant, luxurious, flashy, often even gaudy, full of daring and abandon, but first and foremost enormously tense and powerful. (Dobzhansky 1950, 209)

While lowland tropical rain forests usually harbor the greatest diversity of plant-visiting vertebrates on earth, some of this diversity and its outstanding beauty spill over into other tropical and subtropical habitats along gradients of rainfall and elevation. Thus, lowland tropical and subtropical dry forests as well as montane forests harbor many colorful plant-visiting vertebrates. The same is true of tropical islands. Although their diversity is usually lower than that of mainland communities, tropical islands nonetheless harbor a substantial number of plant-visiting birds and mammals. Finally, some of these animals migrate to temperate latitudes to breed. Tropical forests routinely export some of their colorful vertebrate mutualists (e.g., hummingbirds, tanagers, and parulid warblers in the New World) to far-flung places on a seasonal basis, thereby temporarily increasing the color palette of temperate habitats.

Why have tropical and subtropical plants evolved pollination and frugivory mutualisms with a wide array of colorful nectar- and fruit-eating birds and mammals? What are the ecological and evolutionary "rules" that govern these interactions and how important has earth history been in formulating these rules? How many variations on the themes of nectarivory and frugivory have evolved in tropical birds and mammals? We seek to answer these questions in this book, which is divided into three major sections. The first section (chaps. 2–4) examines regional and local species diversity patterns as well as patterns of resource availability and the functional (ecological) relationships between plant-visiting birds and mammals and their food plants. These chapters seek to determine (1) how communities of these mutualists are structured in space, (2) the nature of the resource base supporting plant-visiting vertebrates, (3) the extent to which fruit and seed set and seedling recruitment of tropical plants depend on the feeding and foraging behavior of nectarivorous and frugivorous birds and mammals, and (4) the population genetic consequences of this feeding behavior for their food plants. One possible genetic consequence is reproductive isolation and speciation, and the question arises, what impact, if any, have plant-visiting vertebrates had on speciation rates of their food plants? And, conversely, what effect have their food plants had on speciation rates of vertebrate nectar and fruit eaters? We address these questions in chapter 5 of the second section. In addition, we review the phylogeny and biogeography of the major families of animals and plants involved in these two mutualisms (chap. 6) as background for a detailed examination of the large array of morphological, physiological, and behavioral adaptations that have arisen during the evolution of these mutualisms (chaps. 7 and 8). Finally, in the third section we will synthesize the ecological and evolutionary consequences of these mutualisms (chap. 9) before discussing their conservation implications (chap. 10). How vulnerable are tropical vertebrate pollinators and frugivores to natural and human disturbances, and how can the loss of these species via extinction be minimized? In the rest of this introductory chapter, we briefly review the major players in this story and outline the basic ecological and evolutionary features of pollination and frugivory mutualisms.

A Brief Taxonomic Overview of Vertebrate Pollinator and Frugivore Mutualisms

Although fish and reptiles (e.g., tortoises and lizards) eat fruit and disperse seeds in certain contemporary habitats (e.g., fish in the Amazon Basin and lizards on islands; Anderson et al. 2009; Correa et al. 2007; Goulding 1980; Olesen and Valido 2003;) and reptiles probably were important seed dispersers in the Cretaceous (Ridley 1930; van der Pijl 1982; Wing and Tiffney 1987), we will focus on mutualistic interactions between higher vertebrates (birds and mammals) and their food plants in this book. We do this for the simple reason that birds and mammals account for the vast majority of vertebrate pollination and frugivory mutualisms today and undoubtedly have done so throughout the Cenozoic Era (Proctor et al. 1996; Tiffney 2004; van der Pijl 1982; Wing and Tiffney 1987).

Major groups of contemporary nectar- and fruit-eating birds and mammals are listed in table 1.1. Nectar-feeding birds and mammals currently exhibit relatively low taxonomic diversity. Nectarivorous birds occur in three orders and 11 families containing about 870 species. Seven of these families, totaling about 840 species, can be considered to be specialized nectarivores (i.e., species that are morphologically adapted for probing into flowers for nectar; Stiles 1981). Specialized nectar-feeding mammals occur in only two orders and three families containing about 49 species. A number of arboreal mammals, including lemurs, callitrichid and cebid monkeys, and

Table 1.1. Summary of the Major Groups of Nectar- and Fruit-Eating Birds and Mammals Discussed in This Book

Mutualism/Order	Family	Common Name	N Genera	N Species	Distribution	Geological Age (Ma)	Mass Range (g)	Total Length (mm)*	% Threatened Species
Birds:									
Nectarivore (3):									
Apodiformes	**Trochilidae**	**Hummingbirds**	104	331	Neotropical	67	2–20	63–216	9
Psittaciformes	**Psittacidae, Loriinae**	Lorikeets	11	54	SE Asia, Australasia	50	20–240	165–320	29.1
Passeriformes	**Nectariniidae**	**Sunbirds**	16	127	Africa, SE Asia, Papua New Guinea	35	3.6–38.4	95–254	4.9
Passeriformes	**Meliphagidae**	Honeyeaters	44	174	Africa, Australasia	50	6.8–244	102–355	8.2
Passeriformes	**Promeropidae**	**Sugarbirds**	1	2	Africa	39	40		
Passeriformes	**Zosteropidae**	White-eyes	14	95	Palearctic, Africa, SE Asia, Australasia	39	7.9–30	102–39	24
Passeriformes	**Dicaeidae**	**Flowerpeckers**	2	44	Palearctic, SE Asia, Australasia	35	5.2–12.6	76–190	7
Passeriformes	**Drepanididae (part)**†	**Hawaiian honeycreepers**	7	36	Hawaii	5	7.9–36	114–222	91.2
Passeriformes	Emberizidae, Coeribinae	Bananaquits	1	1	Neotropical	12	9.5	90	25
Passeriformes	Philepittidae	False sunbirds	2	4	Madagascar	56	34.4	102–65	7.1
Passeriformes	Irenidae	Fairy bluebirds	1	2	SE Asia	34	22.4–45.8	120–240	7.1
Passeriformes	Icteridae (part)	New World blackbirds	...	several	Neotropical	...	...	...	11.7
Passeriformes	Parulidae (part)	New World warblers	...	several	Neotropical	...	...	...	13.1
Sum	13		203	870					
Frugivore (11):									
Casuariiformes	**Casuariidae**	**Cassowaries**	1	3	Australasia	68	44,000	1,320–1,650	66.7

Tinamiformes	Tinamidae	Tinamous	9	47	Neotropical	97	439–4133	520–990	16.3
Galliformes	**Cracidae**	Guans	11	50	Neotropical	89	49–810	150–840	*31.4*
Columbiformes	**Columbidae (part)**	**Fruit pigeons**	12	126	Australasia	90			*22.9** *
Coliiformes	Coliidae	Mousebirds	2	6	Africa	37	34–62.2	290–355	0
Musophagiformes	**Musophagidae**	Turacos	6	23	Africa	60	198–965	370–710	8.7
Caprimulgiformes	Steatornithidae	Oilbirds	1	1	Neotropical	52	414	430–80	0
Trogoniformes	**Trogonidae**	Trogons	6	39	Neotropical, Africa, SE Asia	55	40–206	228–340	2.5
Coraciiformes	**Bucerotidae**	**Hornbills**	13	49	Africa, SE Asia	55	110–2,385	380–1600	16.1
Piciformes	**Lybiidae**	**African barbets**	6	36	Africa	25	...	...	...
Piciformes	Megalaimidae	Asian barbets	2	26	SE Asia	32	...	...	
Piciformes	**Capitonidae**	**New World barbets**	2	14	Neotropical	13	33–97.5	89–317	2.4
Piciformes	Ramphastidae	Toucans	5	38	Neotropical	13	117–734	305–610	2.4
Passeriformes	Eurylaemidae	Broadbills	9	14	Africa, SE Asia	56	21–82	127–280	20
Passeriformes	**Cotingidae**	**Cotingids**	33	96	Neotropical	39	28–380	89–457	19.5
Passeriformes	**Pipridae**	**Manakins**	13	48	Neotropical	64	7–25.6	83–159	6.8
Passeriformes	**Ptilonorhynchidae**	**Bowerbirds**	8	18	Australasia	61	103–228	228–368	5.3
Passeriformes	Corvidae	Crows	26	106	World	26	41–1240	177–699	11.7
Passeriformes	**Paradisaeidae**	**Birds of paradise**	16	40	Australasia	23	44–313	139–1,016	9.1
Passeriformes	**Meliphagidae**	**Honeyeaters**	44	174	Australasia	50	6.8–244	102–355	8.2
Passeriformes	Oriolidae	Orioles	2	29	Africa, SE Asia, Australasia	33	45.5–135	177–305	10
Passeriformes	Irenidae	Fairy-bluebirds	1	2	SE Asia	34	62.5–81.4	120–240	7.1
Passeriformes	**Turdidae**	**Thrushes**	24	165	Cosmopolitan	27	21.6–140	114–330	
Passeriformes	**Mimidae**	**Mockingbirds**	12	34	Nearctic, Neotropical	29	36.3–100	203–305	17.1
Passeriformes	**Sturnidae**	**Starlings**	25	115	Africa, SE Asia, Australasia	29	39.5–217	177–432	12.2
Passeriformes	**Dicaeidae**	**Flowerpeckers**	2	44	SE Asia	35	5.2–12.6	76–190	
Passeriformes	Pycnonotidae	Bulbuls	22	118	Africa, SE Asia	22	13.6–79.5	139–286	9.8
Passeriformes	**Thraupidae*** **	Tanagers	50	202	Neotropical	15	8–114	76–305	*9.3** *
Passeriformes	Estrildidae	Waxbills	26	130	Africa, SE Asia, Australasia	25	5.9–24.8	95–273	7.2
Sum	29		389	1,793					

Table 1.1. (continued)

Mutualism/Order	Family	Common Name	N Genera	N Species	Distribution	Geological Age (Ma)	Mass Range (g)	Total Length (mm)*	% Threatened Species
Mammals:									
Nectarivore (2):									
Diprotodontia	**Tarsipedidae**	Honey possum	1	1	Australasia	39	10–12	70–85	0
Chiroptera	**Pteropodidae (part)**	Old World fruit bats	6	12	Paleotropical	56	23–82	85–125	40.4
Chiroptera	Phyllostomidae (part)	American leaf-nosed bats	16	36	Neotropical	20	8–30	55–95	20.1
Sum	3		23	49					
Frugivore (10):									
Diprotodontia	Phalangeridae	Brushtail possums	6	27	Australasia	43	1,800–7,000	320–650	19
Dermoptera	Cynocephalidae	Flying lemurs	2	2	SE Asia	?	1,000–1,750	340–420	50
Chiroptera	**Pteropodidae (part)**	Old World fruit bats	36	160	Paleotropical	56	25–1,500	100–400	40.4
Chiroptera	**Phyllostomidae (part)**	American leaf-nosed bats	20	70	Neotropical	22	6–1,000	47–130	20.1
Chiroptera	Mystacinidae	New Zealand short-tailed bats	1	2	New Zealand	45	15–35	70–90	100
Scandentia	Tupaiidae	Tree shrews	4	19	SE Asia	60	60–350	100–220	31.6
Primates	Cheirogaleidae	Mouse lemurs	5	21	Madagascar	37	98–600	125–275	44.4
Primates	**Lemuridae**	Large lemurs	5	19	Madagascar	37	2,500–4,500	280–458	80
Primates	Indriidae	Leaping lemurs	3	11	Madagascar	37	1,200–10,000	400–900	83.3
Primates	Lorisidae	Lorises	5	9	Africa, SE Asia	34	120–2,000	100–460	22.2
Primates	**Cebidae**	Marmosets, capuchins	6	56	Neotropical	25	100–5,000	152–500	33.3
Primates	**Aotiidae**	Night monkeys	1	8	Neotropical	25	950–1,100	350–420	28.6
Primates	**Pithecidae**	Sakis, titis	4	40	Neotropical	25	2,000–3,000	375–500	25.6
Primates	**Atelidae**	Spider monkeys, howlers	5	24	Neotropical	25	8,000–12,125	550–675	40
Primates	**Cercopithecidae (part)**	Old World monkeys	9	57	Paleotropical	29	1,280–54,000	325–1,100	45.8**

Order	Family	Common name			Distribution				
Primates	**Hylobatidae**	Gibbons	4	14	SE Asia	19	8,000–13,000	440–900	58.3
Primates	**Hominidae**	**Great apes and humans**	4	7	Paleotropical	19	38,000–90,000	828–1,500	*100*
Rodentia	Anomaluridae	Scaly-tailed squirrels	3	7	Africa	48	18–1,090	60–432	0
Rodentia	Gliridae (part)	Dormice	1	14	Africa	62	18–30	70–165	
Rodentia	Platacanthomyidae	Spiny dormice	2	2	SE Asia	24	60–80	130–212	
Rodentia	Dinomyidae	Paracana	1	1	Neotropical	20	10,000–15,000	730–90	*100*
Rodentia	Echimyidae	American spiny rats	21	90	Neotropical	7	39–700	105–480	11.8
Rodentia	Dasyproctidae	Agoutis and pacas	2	13	Neotropical	26	1,300–12,000	380–795	23.1
Proboscidea	**Elephantidae**	**Elephants**	2	3	Paleotropical	1. Miocene	5,400,000–7,500,000	640–750	*100*
Carnivora	**Procyonidae**	**Racoons**	6	14	Nearctic, Neotropical	28	800–1,2000	305–670	42.1
Carnivora	**Viverridae (part)**	**Palm civets**	6	8	Paleotropical	40	2,100–5,000	530–760	26.5**
Perissodactyla	Tapiridae	Tapirs	1	4	Neotropical, SE Asia	49	180,000–320,000	180–250	*100*
Perissodactyla	Rhinoceratidae	Rhinoceroses	4	5	Africa, SE Asia	49	1,000,000–3,500,000	200–420	*80*
Artiodactyla	Tragulidae	Chevrotains	3	8	Africa, SE Asia	46	8,000–15,000	750–850	0
Sum	29		172	715					

Sources. Data for taxonomy, diversity, and geographical distributions of birds from Cracraft et al. (2003), Gill (1990), and Stiles (1981) and, for sizes, from Dunning (1993) and Van Tyne and Berger (1976). Data for taxonomy, diversity, and geographic distributions of mammals from Wilson and Reeder (2005) and, for sizes, from Nowak (1991). Estimates of the geological ages of families of birds and mammals come from data in Hedges and Kumar (2009) and are based on time-calibrated molecular phylogenies. The conservation status of families is based on data in Baillie et al. (2004).

Note. Families that are specialized for nectarivory or frugivory are highlighted in bold. Families whose percentage of threatened species is significantly different from all families of birds or mammals are indicated in *italics*. Numbers in parentheses in stub column indicate number of orders represented for that mutualism.

*Total length for mammals equals head-to-body range.

**Includes entire family.

***Classification of this family is in a state of flux; numbers here reflect its traditional classification. Euphonias are now included in family Fringillidae (Zuccon et al. 2012).

†This family has recently been classified within subfamily Carduelinae in family Fringillidae (Zuccon et al. 2012).

procyonids, as well as many fruit-eating phyllostomid and pteropodid bats are occasional nectar feeders, but the number of morphologically specialized mammalian nectarivores is small. Overall, there are about 17 times more species of nectarivorous birds than nectarivorous mammals.

Fruit-eating birds and mammals are much more diverse taxonomically than nectarivores (table 1.1). Fruit-eating birds are widely distributed throughout avian phylogeny and are found in 10 orders and at least 23 families containing nearly 1,800 species. At least 18 of these families, with about 1,400 species, contain specialized frugivores (i.e., species whose diet contains a high percentage of fruit; Corlett 1998; Snow 1981). In mammals, frugivores are found in 10 orders and at least 24 families, containing about 600 species. Members of 12 families, with about 480 species, contain specialized frugivores. In contrast to the high ratio of species of nectarivorous birds to mammals (17:1), this ratio for species of frugivores is about 3:1.

As a final taxonomic point, neither feeding mode (nectarivory or frugivory) is especially common in birds and mammals. Of the 127 families of terrestrial birds listed by Gill (1990), only 11 (8.7%) and 23 (18.1%) contain nectar- or fruit-eating species, respectively. Insectivory is by far the most common feeding mode in birds (140 of 168 families [83.3%]). Similarly, of 119 families of terrestrial mammals listed by Vaughan et al. (2000), only three (2.5%) and 26 (21.8%) contain nectar- or fruit-eating species, respectively. Herbivory and granivory are the most common feeding modes in mammals (Eisenberg 1981; Vaughan et al. 2000).

We will discuss in detail other information contained in table 1.1 in subsequent chapters, but we wish to point out here the differences in body sizes associated with nectarivory and frugivory in birds and mammals. In both groups, nectarivores are much smaller than frugivores (table 1.2). Median body masses in families of nectarivores are 22–30 g, and the size ranges of nectar-feeding birds and mammals overlap broadly with each other. In contrast, median body masses in families of frugivores are 127–7,000 g, with mammalian frugivores exhibiting a vastly larger range of sizes than their avian counterparts because of the greater overall size range of terrestrial mammals (shrews to elephants; cf. hummingbirds to ostriches). In both groups, the size range of frugivores spans virtually the entire size range of all terrestrial species. As we will see, body size has profound consequences for the evolution of pollination and frugivory mutualisms.

Major families of flowering plants that provide nectar and/or fruit for their avian and mammalian mutualists are shown in table 1.3. Here and

Table 1.2. Summary of the Sizes of Plant-Visiting Birds and Mammals by Family

	Birds		Mammals	
Diet	Mass (g)	Length (mm)	Mass (g)	Length (mm)
Nectar	22.0 (9–130) [11]	154.0 (90–243) [10]	30.0 (9–82) [3]	82.0 (80–105) [3]
Fruit	126.9 (9–44,000) [26]	278.5 (121–1,485) [26]	7000.0 (30–5,400,000) [25]	650.0 (80–6,400) [25]

Note. Data include median values plus range (in parentheses) and number of families (in brackets) from table 1.1. Length is head to tail in birds and head to body in mammals.

throughout this book we will use APG III (2009) as our basis for plant classification. This table focuses on many (but certainly not all) of the plant families that interact with birds, bats, and primates, three of the major groups of higher vertebrate nectarivores and frugivores. Plants producing bird- or bat-pollinated flowers occur in at least 30 orders (28 of which contain bird flowers and 16 contain bat flowers) and 62 families (54 with bird flowers and 28 with bat flowers). Plants producing fruits eaten by birds, bats, and primates occur in at least 40 orders (40 with bird fruits, 14 with bat fruits, and 20 with primate fruits) and 86 families (82 with bird fruits, 25 with bat fruits, and 35 with primate fruits). Thirty-four of the 112 families (30%) listed in table 1.3 contain taxa that are involved in both vertebrate pollination and frugivory. As in the case of their animal mutualists, more families of angiosperms are involved in vertebrate frugivory than in nectarivory by a factor of about 1.4. Overall, about 40% of the nonaquatic families listed in APG III contain species that produce vertebrate-pollinated flowers or vertebrate-dispersed fruits.

Brief sketches of the important biological features of the major families of animals and plants discussed in this book can be found in appendixes 1 and 2. A cornucopia of images of these animals and their food plants can be found in plates 1–13.

Basic Features of Pollination and Seed Dispersal Mutualisms

GENERAL CONSIDERATIONS

The two mutualisms examined in this book—pollination and seed dispersal—involve the transport of plant propagules, either pollen or seeds, from their point of origin to a point of deposition, usually someplace else in the environment. While acquiring these propagules, most often passively rather than actively, animals obtain a nutritional reward in the form of nectar, pollen, or fruit pulp. When effective mutualistic interactions occur between plants and their pollinators, animal vectors deposit pollen on conspecific

Table 1.3. Summary of the Major Families of Plants Providing Food for Vertebrate Nectar and Fruit Eaters

Family	Lineage	Order	N Genera	N Species	Geographical Distribution	Geological Age	Bird Flower	Bird Fruit	Bat Flower	Bat Fruit	Primate Fruit
Acanthaceae	Asterid	Lamiales	229	3,500	Pantropical	67	1	0	1	0	0
Agavaceae	Monocot	Asparagales	23	637	Pantropical	23	0	0	1	0	0
Amaryllidaceae	Monocot	Asparagales	59	800	Pantropical	91	1	1	0	0	0
Anacardiaceae	Rosid	Sapindales	70	985	Pantropical	56	0	1	0	1	1
Annonaceae	Basal angiosperm	Magnoliales	129	2,220	Pantropical	82	0	1	0	1	1
Apocynaceae	Asterid	Gentianales	415	4,555	Pantropical	64	1	1	0	1	1
Aquifoliaceae	Rosid	Aquifoliales	1	405	Pantropical	77	0	1	0	0	0
Araceae	Monocot	Alismatales	106	4,025	Pantropical	128	0	1	0	1	1
Araliaceae	Rosid	Apiales	43	1,450	Pantropical	69	1	1	0	0	1
Arecaceae	Monocot	Arecales	189	2361	Pantropical	119	0	1	1	1	1
Asteraceae	Asterid	Asterales	1,620	22,750	Pantropical	51	1	0	1	0	0
Balanophoraceae	Rosid	Santalales	7	50	Pantropical	NA	1	0	1	0	0
Balsaminaceae	Asterid	Ericales	2	1,000	Pantropical	56	1	0	1	0	0
Bignoniaceae	Asterid	Lamiales	110	800	Pantropical	67	1	0	1	0	0
Boraginaceae	Asterid	?	48	2,740	Pantropical	104	1	1	0	1	1
Bromeliaceae	Monocot	Poales	57	1,400	Neotropical	111	1	1	1	0	0
Bruniaceae	Asterid	Bruniales	12	75	S. Africa	109	1	0	0	0	0
Burseraceae	Rosid	Sapindales	18	550	Pantropical	51	0	1	0	0	1
Cactaceae	Caryophyllidae	Caryophyllales	87	2,000	Neotropical	30	1	1	1	1	0
Campanulaceae	Asterid	Asterales	84	2,380	Pantropical	41	1	0	1	0	0
Canellaceae	Basal angiosperm	Canellales	5	13	Neotropical, Africa	99	0	1	0	0	0
Capparaceae	Rosid	Brassicales	16	480	Pantropical	Late Cretaceous?	1	1	0	0	0
Caprifoliaceae	Asterid	Dipsacales	5	220	Pantropical	101	0	1	0	0	0
Celastraceae	Rosid	Celestrales	89	1,300	Neotropical, Asia, Australasia	58	0	1	0	0	1

Family	Clade	Order	Genera	Species	Distribution							
Chloranthaceae	Basal angiosp	Chloranthales	4	75	Neotropical, Asia, Australasia	147	0	0	1	0	0	0
Chrysobalanaceae	Rosid	Malpighiales	17	460	Pantropical	41	0	0	1	1	0	1
Clusiaceae	Rosid	Malpighiales	14	595	Pantropical	45	1	1	1	0	0	1
Combretaceae	Rosid	Myrtales	20	500	Pantropical	79	1	1	1	0	1	0
Connaraceae	Rosid	Oxidales	12	180	Pantropical	NA	0	0	1	1	0	0
Convolulaceae	Asterid	Solanales	57	1,600	Pantropical	88	1	0	1	1	0	1
Cornaceae	Asterid	Cornales	2	85	Pantropical	104	0	0	1	0	0	0
Crassulaceae	Basal eudicot?	Saxifragales	4	1,370	Neotropical, Africa, Australasia	75	1	1	0	0	0	0
Cucurbitaceae	Rosid	Cucurbitales	118	845	Pantropical	65	1	1	1	1	0	0
Cunoniaceae	Rosid	Oxidales	26	280	Pantropical	66	0	0	1	0	0	0
Cyclanthaceae	Monocot	Pandanales	12	225	Neotropical	86	0	0	1	1	1	0
Dilleniaceae	Basal eudicot	Dilleniales	12	300	Pantropical	105	0	0	1	0	0	1
Ebenaceae	Asterid	Ericales	4	548	Pantropical	100	0	0	1	0	0	1
Elaeocarpaceae	Rosid	Oxidales	12	625	Neotropical, Asia, Australasia	66	1	1	1	0	0	0
Ericaceae	Asterid	Ericales	126	3,995	Pantropical	90	1	1	1	0	0	0
Erythroxylaceae	Rosid	Malpighiales	4	240	Pantropical	49	0	0	1	0	0	0
Euphorbiaceae	Rosid	Malpighiales	222	5,970	Pantropical	69	1	1	1	0	0	1
Fabaceae	Rosid	Fabales	730	19,400	Pantropical	79	1	1	1	1	0	1
Fagaceae	Rosid	Fagales	7	670	N. America, Eurasia	61	0	0	0	0	0	1
Gentianaceae	Asterid	Gentianales	87	1,650	Pantropical	64	0	0	0	1	0	0
Geraniaceae	Rosid	Geraniales	7	805	Pantropical	71	1	1	0	0	0	0
Gesneriaceae	Asterid	Lamiales	147	3,870	Pantropical	78	1	1	0	1	0	0
Heliconiaceae	Monocot	Zingiberales	1	ca. 200	Neotropical, IndoPacific	78	1	1	0	1	0	0
Houmiriaceae	Rosid	Malpighiales	8	50	Neotropical, Africa	67	0	0	0	0	0	1
Icacinaceae	Asterid	Garryales?	23	149	Pantropical	118	0	0	1	0	0	1
Iridaceae	Monocot	Asparagales	66	2,025	Pantropical	101	1	1	0	0	0	0
Lacistemataceae	Rosid	Malpighiales	2	14	Neotropical	53	0	0	1	0	0	0

Table 1.3. (continued)

Family	Lineage	Order	N Genera	N Species	Geographical Distribution	Geological Age	Bird		Bat		Primate Fruit
							Flower	Fruit	Flower	Fruit	
Lamiaceae	Asterid	Lamiales	236	7,175	Pantropical	63	1	0	0	0	0
Lauraceae	Basal angiosperm	Laurales	50	2,500	Pantropical	80	0	1	0	0	1
Liliaceae	Monocot	Liliales	66	635	Pantropical	86	1	1	0	0	0
Loganiaceae	Asterid	Gentianales	13	420	Pantropical	66	0	1	0	0	0
Loranthaceae	Basal eudicot	Santalales	68	950	Pantropical	53	1	1	0	0	0
Lythraceae	Rosid	Myrtales	31	620	Pantropical	57	0	0	1	0	0
Magnoliaceae	Basal angiosperm	Magnoliales	2	227	Neotropical, Asia, Australasia	92	0	1	0	0	0
Malpighiaceae	Rosid	Malpighiales	68	1,250	Pantropical	68	0	1	0	0	0
Malvaceae	Rosid	Malvales	243	4,225	Pantropical	54	1	0	1	0	0
Marantaceae	Monocot	Zingiberales	31	550	Pantropical	61	1	1	0	0	0
Marcgraviaceae	Asterid	Ericales	7	130	Neotropical	64	0	1	1	1	0
Melastomataceae	Rosid	Myrtales	188	5,005	Pantropical	79	0	1	0	1	1
Meliaceae	Rosid	Sapindales	52	621	Pantropical	40	1	1	0	1	1
Melianthaceae	Rosid	Geraniales	3	11	Africa	92	1	0	1	0	0
Menispermaceae	Basal eudicot	Ranunculales	70	420	Pantropical	103	0	1	0	0	1
Monimiaceae	Basal angiosperm	Laurales	22	200	Pantropical	85	0	1	0	0	0
Moraceae	Rosid	Rosales	38	1,100	Pantropical	36	0	0	0	1	1
Musaceae	Monocot	Zingiberales	2	35	Africa, Asia, Australasia	78	1	0	1	1	0
Myricaceae	Rosid	Fagales	3	57	Pantropical	36	0	1	0	0	0
Myristicaceae	Basal angiosperm	Magnoliales	20	475	Pantropical	113	0	1	0	0	1
Myrsinaceae	Asterid	Ericales	41	1,435	Pantropical	42	0	1	0	1	0
Myrtaceae	Rosid	Myrtales	131	4,625	Pantropical	68	1	1	1	1	1
Nyctaginaceae	Basal eudicot	Caryophyllales	30	395	Pantropical	21	0	1	0	0	0
Ochnaceae	Rosid	Malpighiales	32	535	Pantropical	29	0	1	0	0	0
Olacaceae	Basal eudicot	Santalales	14	103	Pantropical	85	0	1	1	0	1

Family	Group	Order			Distribution						
Oleaceae	Asterid	Lamiales	24	615	Pantropical	90	0	1	0	0	0
Onagraceae	Rosid	Myrtales	24	650	Pantropical	57	1	0	0	0	0
Orchidaceae	Monocot	Asparagales	880	21,950	Pantropical	117	1	0	1	0	0
Pandanaceae	Monocot	Pandanales	4	885	Africa, Asia, Australasia	86	0	1	1	1	0
Passifloraceae	Rosid	Malpighiales	16	605	Pantropical	51	1	1	1	1	0
Peneaceae	Rosid	Myrtales	9	29	Africa	NA	1	0	0	0	0
Phytolaccaceae	Basal eudicot	Caryophyllales	18	65	Pantropical	21	0	1	0	0	0
Piperaceae	Basal angiosperm	Piperales	5	3,600	Pantropical	96	0	1	0	1	0
Pittosporaceae	Asterid	Apiales	9	200	Africa, Asia, Australasia	66	0	1	0	0	0
Polygonaceae	Basal eudicot	Caryophyllales	43	1,100	Neotropical, Asia, Australasia	37	0	1	0	0	0
Proteaceae	Basal eudicot	Proteales	80	1,600	Pantropical	113	1	1	1	1	0
Ranunculaceae	Basal eudicot	Ranunculales	62	2,525	Pantropical	84	1	0	0	0	0
Rhamnaceae	Rosid	Rosales	50	900	Pantropical	62	0	1	1	1	0
Rhizophoraceae	Rosid	Malpighiales	16	149	Pantropical	49	1	0	0	0	0
Rosaceae	Rosid	Rosales	95	2,830	Pantropical	76	1	1	1	1	1
Rubiaceae	Asterid	Gentianales	600	10,000	Pantropical	78	1	1	1	1	1
Rutaceae	Rosid	Sapindales	161	1,815	Pantropical	47	1	1	0	0	1
Salicaceae	Rosid	Malpighiales	55	1,010	nearly cosmopolitan	NA	1	0	1	1	0
Santalaceae	Basal eudicot	Santalales	44	935	Pantropical	53	0	1	0	0	0
Sapindaceae	Rosid	Sapindales	135	1,580	Pantropical	56	0	1	0	0	1
Sapotaceae	Asterid	Ericales	53	975	Pantropical	102	0	1	1	1	1
Saxifragaceae	Basal eudicot	Saxifragales	29	630	Pantropical	69	0	1	0	0	0
Scrophulariaceae	Asterid	Lamiales	65	1,700	Pantropical	76	1	1	0	0	0
Simaroubaceae	Rosid	Sapindales	19	95	Pantropical	40	0	1	0	0	1
Smilacaceae	Monocot	Liliales	2	315	Pantropical	85	0	1	0	0	0
Solanaceae	Asterid	Solanales	102	2,460	Pantropical	88	1	1	1	1	0
Staphyleaceae	Rosid	Crossosomatales	2	45	Neotropical, Asia	62	0	1	0	0	0
Strelitziaceae	Monocot	Zingiberales	3	7	Neotropical, Africa	70	1	1	1	0	0

Table 1.3. (continued)

Family	Lineage	Order	N Genera	N Species	Geographical Distribution	Geological Age	Bird Flower	Bird Fruit	Bat Flower	Bat Fruit	Primate Fruit
Taccaceae	Monocot	Dioscoriales	1	12	Pantropical	NA	0	1	0	0	0
Theaceae	Asterid	Ericales	7	195	Pantropical	103	1	1	0	0	0
Ulmaceae	Rosid	Rosales	6	35	Pantropical	55	0	1	0	0	1
Urticaceae	Rosid	Rosales	54	2625	Pantropical	42	0	1	0	1	1
Verbenaceae	Asterid	Lamiales	30	1100	Pantropical	63	1	1	0	1	1
Vitaceae	Rosid	Vitales	14	850	Pantropical	117	0	1	0	0	0
Zingiberaceae	Monocot	Zingiberales	50	1200	Pantropical	70	1	1	0	0	0
Zygophyllaceae	Rosid	Zygophallales	26	285	Pantropical	70	0	1	0	0	0
Sum							53	85	27	25	35

Sources. Information on classification, taxonomic diversity, and geographic distributions comes from APG III (2009) and Heywood et al. (2007). Estimates of geological ages come from data in Hedges and Kumar (2009) and are based on time-calibrated molecular phylogenies.
Note. The five right-hand columns indicate the kind of resources (flowers, fruit) each family provides for three groups of vertebrates (birds, bats, and primates) based on the literature. NA = not available.

stigmas, fertilization of ovules ensues, and fruits containing mature seeds are produced. When effective mutualistic interactions occur between plants and their seed dispersers, frugivores deposit intact seeds in places where they will eventually germinate. Effective pollination mutualisms result in the production of a new cohort of seeds, some of which will survive to become members of the next plant generation whenever effective dispersal takes place. Both of these mutualisms can be short-circuited by a variety of invertebrate and vertebrate "cheaters," species that obtain nutritional rewards without providing an effective "pay-off" or service to plants (Bronstein 2001; Ferriere et al. 2002). Thus, many animals, including legitimate pollinators, remove nectar from flowers without effectively pollinating them, and many frugivores destroy seeds in their mouths, in their guts, or by depositing them in inappropriate places, thereby preventing effective dispersal.

General features of these two mutualisms and how they interact ecologically are summarized in figure 1.1, which is adapted from Wang and Smith (2002). The two mutualisms can be viewed as two interlocking cycles. Starting with a population of adult plants, the pollination cycle begins with flower production and ends with fruit production, events that are basically under the control of plants. Pollen acquisition and deposition, in contrast, are often, but not always, under the control of animal pollinators. Once animals have shed their pollen, they have left this mutualism. The seed dispersal cycle begins with a crop of ripe fruit and mature seeds, which frugivores harvest, either as single units (fruits) or in groups (multiple fruits) depending on the size and behavior of each species. Once seeds are spit out, dropped, or defecated, frugivores have left this mutualism. Compared with the pollination cycle, plants have much less control over the different stages of the dispersal cycle. Germination, seedling recruitment, and seedling/sapling survival and growth, for example, often depend on a variety of extrinsic abiotic and biotic factors beyond the direct control of plants (although plants can have indirect control through the physiological and chemical characteristics of their seeds; Vasquez-Yanes and Orozco-Segovia 1993). Extrinsic abiotic factors include the availability of light and soil moisture and nutrients. Extrinsic biotic factors include the availability of soil mycorrhizae (another mutualism) and the impact of competitors, predators (including herbivores), secondary seed dispersers, pathogens, and "death from above" (i.e., death from falling branches and trees; e.g., Clark and Clark 1991). Because many more external processes (factors) are involved in successful plant recruitment from seed dispersal than in successful pollination,

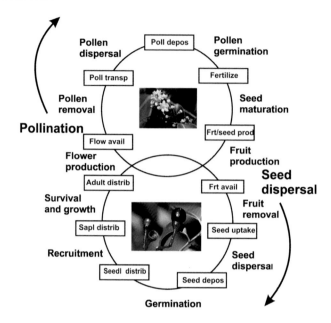

Pollen dispersal — Poll depos — Pollen germination

Poll transp — Fertilize

Pollen removal — Seed maturation

Pollination

Flow avail — Frt/seed prod

Flower production — Fruit production

Adult distrib — Frt avail

Seed dispersal

Survival and growth — Fruit removal

Sapl distrib — Seed uptake

Recruitment — Seed dispersal

Seedl distrib — Seed depos

Germination

Figure 1.1.   A conceptual model of ecological connections between the pollination and seed dispersal mutualisms. Both cycles run in clockwise fashion (as indicated by the *arrows*), starting with flower availability for pollination and fruit availability for seed dispersal. The end point of these two cycles is the production of adult plants. Modified from Wang and Smith (2002). The insets show flowers and fruit of the lauraceous tree *Nectandra coriacea* (photo credit: Pedro Avecedo).

plant-pollinator mutualisms tend to be much more specialized and are subject to potentially stronger plant-animal coevolution, than plant-frugivore mutualisms (Feinsinger 1983; Howe and Westley 1988; Janzen 1983a; Wheelwright and Orians 1982).

With these generalizations in mind, we can now look at the two mutualisms in a bit more detail. Evolutionary aspects of these mutualisms will be thoroughly reviewed in chapters 7 and 8.

THE POLLINATION MUTUALISM

Dispersal propagules in this mutualism are tiny pollen grains, which range in size from about 5 μM to over 200 μM in length or diameter with most kinds averaging 30–40 μM in length (Proctor et al. 1996). The mass of most individual pollen grains is negligible, although their collective mass can be substantial. For example, male flowers of the bat-pollinated cactus *Pachycereus pringlei* produce about 470 mg (nearly 0.5 g) of pollen, with each mg containing about 3,540 pollen grains; individual pollen grains weigh about 2.8×10^{-4} mg (Fleming et al. 1994). Given their small size, pollen grains can be effectively dispersed by a wide variety of agents, ranging from the wind and tiny fig wasps (Agaonidae) to large birds and mammals. Pollen grain size does not seriously constrain the kinds of agents plants use to disperse their pollen.

Although pollen grain size does not constrain the evolution of pollinator choice in plants, other reproductive factors, including individual flower size, nectar reward per flower, and total number of flowers produced per plant per season, do affect pollinator food choice. Flowers, nectar and pollen, and ultimately fruits and seeds can be costly for plants to produce, and, given a limited energy resource budget, selection will produce tradeoffs between these aspects of reproductive effort (Barrett et al. 1996; Cohen and Shmida 1993; Venable 1996). Two classic tradeoffs are those between flower size and number and seed size and number. In both cases, flower or seed size usually decreases as flower or seed number increases. Positive relationships between the size of flowers, fruits, and seeds also exist (Primack 1987). As shown in figure 1.2, large pollinators (e.g., vertebrates) will generally select for large flowers with large nutritional rewards, which selects for large fruits containing a few large or many small seeds. Large fruits and seeds, in turn, can favor the evolution of dispersal by large frugivores. Primack (1987) argues that all aspects of a plant's reproductive strategy as well as many of its vegetative

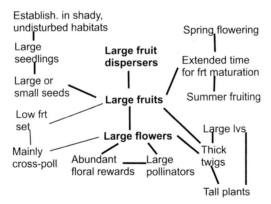

Figure 1.2. Relationships among plant reproductive and vegetative characters during selection for larger sizes of structures by vertebrate pollinators and seed dispersers. A *thick line* designates a strong relationship between pairs of characters; a *thin line* signifies a weaker relationship. Based on Primack (1987).

characteristics (e.g., twig and branch sizes, leaf sizes) are intimately related to its flowering and fruiting biology.

Although wind pollination prevails in non-angiosperm seed plants, it is a relatively uncommon pollination mode in angiosperms. According to Ricklefs and Renner (1994), only 16.9 % of angiosperm families are exclusively or predominantly wind-pollinated (n = 338 families). Wind pollination is especially common in families in the Fagales among dicots (e.g., Betulaceae, Fagaceae) and Alismatales and Poales among monocots (e.g., Graminae, Juncaceae, Cyperaceae). Pollination by unspecialized insects, including beetles, short-tongued wasps, and flies, is probably ancestral in angiosperms (Crepet and Friis 1987; Hu et al. 2008; Proctor et al. 1996). In most contemporary communities, bees are by far the most common animal pollinators (e.g., Bawa 1990). Birds and mammals are currently relatively minor pollinators of angiosperms and have been so throughout the Cenozoic era (Proctor et al. 1996). Nonetheless, as we will see, they are responsible for pollinating a very conspicuous and ecologically (and sometimes economically) important element of tropical and subtropical vegetation.

From both historic and energetic perspectives, it is easy to understand why insects, rather than vertebrates, are the most common pollinators of flowering plants. Historically, insects have been pollinating angiosperm flowers for at least 150 Ma—much longer than vertebrates (Labandeira 2002). Energetically, insects are small and ectothermic and hence are individually much cheaper to feed than large endothermic birds and mammals. The size range of pollinating insects covers over three orders of magnitude (from <1 mg fig wasps to ca. 3 g hawkmoths). Their daily energy budgets are orders of magnitude smaller than those of hummingbirds, the world's smallest nectar-feeding birds (Brown et al. 1978; Winter and Helversen 2001). As a result, most plants can afford to produce "cheap" flowers, at least with respect to their nutritional rewards, when using insects as pollen vectors. Typical nectar volumes of insect-pollinated flowers are a few µL compared with from 100 µL to a few mL for vertebrate-pollinated flowers (Opler 1983; Winter and Helversen 2001). The ability to produce cheap flowers, in turn, has allowed plants to evolve small flowers of a diverse array of shapes to attract different kinds of insect pollinators. Some of these insects (e.g., yucca moths, senita moths, and fig wasps) are actually parasite pollinators that oviposit their eggs in ovaries or seeds (Pellmyr 2002). As a result, they are involved in much more specialized interactions with their host plants than are free-living pollinators such as bees, birds, and mammals (Bronstein et al. 2001).

Although many plants tend to be visited by several species of pollinators and hence have relatively generalized pollination systems (Waser et al. 1996; but see Johnson and Steiner 2003 and Fenster et al. 2004), these species are usually a nonrandom subset of all possible pollinators. Because of their extensive differences in morphology (especially size), physiology (including sensory capabilities), and behavior, different kinds of animals differ enormously in their pollination potential and effectiveness. From a plant's perspective, different kinds of potential animal pollinators (e.g., beetles, flies, short-tongued bees, long-tongued bees, butterflies, etc.) or functional guilds (*sensu* Fenster et al. 2004) therefore represent different pollinator niches or different pollinator adaptive zones. In order to maximize their fruit and seed set, most animal-pollinated plants tend to specialize on particular groups of animal pollinators (i.e., particular pollinator niches) through the evolution of suites of floral characteristics (usually called floral syndromes) that reflect the morphological, physiological, and behavioral characteristics of those animals (table 1.4). In addition to differences among the animals themselves, selective factors that have favored the evolution of floral syndromes include pollen-limited fruit set, which appears to be relatively common in angiosperms (Burd 1994; Larson and Barrett 2000) and which promotes intra- and interspecific competition among plants for pollinators. Competition for pollinators, in turn, promotes the evolution of niche diversification and specialization on particular groups of pollinators (Feinsinger 1983; Pellmyr 2002).

Finally, for most animal pollinators, pollen grains are not a source of nutrition but instead ride passively on their bodies from one flower to another as animals seek a meal of floral nectar. To these species, pollen represents a minor form of ballast. Certain pollinators, including pollen-collecting beetles, flies, bees, heliconiid butterflies, and *Leptonycteris* bats, however, obtain nutrients from pollen grains (or feed them to their larvae) and hence actively seek out pollen for its nutritional reward. These species can act as pollen predators, and their effectiveness as pollinators must be discounted by the amount of pollen they destroy (Barrett 1998, 2003; Porcher and Lande 2005).

THE SEED DISPERSAL MUTUALISM

In contrast to pollen grains, seeds often have substantial mass. Seed mass ranges from the dust- (or pollen-) like seeds of orchids, weighing about 10^{-8} g across 10 orders of magnitude, to seeds of the double coconut (*Lodoicea seychellarum*) weighing 10^4 g (Harper et al. 1970; Grubb 1998). Seed

Table 1.4. Pollination Syndromes of Animal-Pollinated Plants

Animal Group	Time of Opening	Color	Odor	Shape	Nectar
			Flower Characteristics		
Insect pollinated:					
Beetles	Day or night	Dull or white	Fruity or aminoid	Flat or bowl shaped; radial symmetry	Often absent
Carrion or dung flies	Day or night	Brown or greenish	Fetid	Flat or deep; often traps; radial symmetry	Rich in amino acids, if present
Bee flies	Day or night	Variable	Variable	Moderately deep; radial symmetry	Hexose rich
Bees	Day or night	Variable but not pure red	Sweet	Flat or broad tube; radial or bilateral symmetry	Sucrose rich for long-tongued bees; hexose rich for short-tongued bees
Hawkmoths	Night	White or pale green	Sweet	Deep, often with spur; radial symmetry	Ample and sucrose rich
Butterflies	Day or night	Variable; often pink	Sweet	Deep or with spur; radial symmetry	Often sucrose rich
Vertebrate pollinated:					
Birds	Day	Vivid; often red	None	Tube; often hanging; radial or bilateral symmetry	Ample and sucrose rich
Bats	Night	Drab, pale; often green	Musty	Flat "shaving brush" or deep tube; often on branch or trunk; hanging; much pollen; radial symmetry	Ample and hexose rich

Source. Adapted from Howe and Westley (1988).

size is a critical feature of the life history of plants, and its evolution is a complex function of many different selective pressures, including a plant's germination strategy, herbivore and predator pressure, and competition for resources within plants and between seedlings (Grubb 1998; Leishman et al. 2000). While seeds, like pollen, usually represent ballast and are not a source of nutrition to most frugivores, their size (expressed either in linear dimensions or as mass) serves as an important constraint in the plant-frugivore mutualism. Small seeds—those weighing a few milligrams or being a few millimeters long—can be handled by the entire size range of potential seed dispersers (from ants to elephants). As seeds increase in size, however, the size range of their potential animal dispersers quickly decreases. Large seeds (i.e., those weighing ≥ ca. 10 g or ≥ ca. 2.5 cm in length or diameter) can only be handled by relatively large animals. Frugivores harvest fruit and eat fruit pulp and acquire seeds inadvertently during this process. Nonetheless, to the extent that fruit and seed size (or the total mass of small seeds in a fruit) are correlated, seed size has an important bearing on plant-frugivore interactions. Data summarized by Primack (1987) and Niklas (1994), for instance, indicate that fruit and seed size are often positively correlated (fig. 1.2).

Unlike pollination mutualisms, which are dominated by insect-flower interactions, frugivory is basically a vertebrate-plant interaction, mainly because of the seed size constraint. Some contemporary plants, including many woody shrubs in Australia and South Africa and some tropical and temperate understory herbs, have ant-dispersed seeds (Beattie and Hughes 2002; Horvitz et al. 2002). Ants can also serve as important secondary dispersers of small tropical seeds whose primary dispersers are birds or bats (Byrne and Levey 1993; Levey and Byrne 1993). But for most of the history of seed plants, including gymnosperms as well as angiosperms, biotic seed dispersal has involved vertebrates. Herbivorous reptiles (e.g., dinosaurs) were the most likely dispersers of the seeds of Mesozoic and Cretaceous cycads and other seed-bearing plants (Coe et al. 1987; Wing and Tiffney 1987), but, as noted above, birds and mammals have been the most important vertebrate seed dispersers for the last 65 Ma. Biotic dispersal has thus been the rule rather than the exception during the evolution of seed plants (Fleming 1991a; Herrera 1989, 2002: Ricklefs and Renner 1994; Tiffney 2004).

As in the case of pollination mutualisms, fleshy-fruited plants interact with a wide array of potential vertebrate seed dispersers that differ greatly in morphology, physiology, and behavior. These biological differences mean that not all frugivorous vertebrates are equally effective at dispersing particular kinds of seeds (Fleming and Sosa 1994). Variation in disperser effectiveness, in turn, creates an array of potential disperser niches or adaptive zones for plants, and it should come as no surprise that different groups of angiosperms have tended to specialize on different kinds of vertebrates for their seed dispersal. Well-known fruit dispersal syndromes are the result of this specialization (table 1.5). To a greater extent than in pollination syndromes (because of the less-specialized relationships between fleshy-fruited plants and their dispersers), however, these syndromes have many exceptions and should only be viewed as general tendencies in many cases (Heithaus 1982;

Table 1.5. Seed Dispersal Syndromes of Animal-Dispersed Plants

	Fruit Characteristics			
Animal Group	Color	Odor	Form	Nutritional Reward
Primarily vertebrate dispersed:				
Hoarding mammals	Brown	Weak or aromatic	Indehiscent thick-walled nuts	Seed itself
Hoarding birds	Green or brown	None	Rounded seeds or nuts	Seed itself
Arboreal mammals	Yellow, white, green, or brown	Aromatic	Arillate seeds or drupes; often compound and dehiscent	Pulp protein, sugar, or starch
Bats	Pale yellow or green	Musky	Various; often hanging	Pulp starch rich
Terrestrial mammals	Often green or brown	None	Indehiscent nuts, pods, or capsules	Pulp lipid or starch rich
Highly frugivorous birds	Black, blue, red, green, or purple	None	Large drupes or arillate seeds; often dehiscent; seeds >10 mm long	Pulp lipid or starch rich
Partly frugivorous birds	Black, blue, red, orange, or white	None	Small or medium-sized drupes, arillate seeds, or berries; seeds <10 mm long	Pulp often sugar or starch rich
Primarily dispersed by insects:				
Ants	Undistinguished	None to humans	Elaiosome on seed coat; seed <3 mm long	Oil or starch elaiosome with chemical attractant

Source. Adapted from Howe and Westley (1988).

Howe 1986). Frugivores are less likely to place their food plants into discrete "syndromes" than are ecologists.

Summary

In this book we discuss ecological and evolutionary aspects of pollination and seed dispersal mutualisms between tropical plant–visiting vertebrates and their food plants. Whereas tropical plants rely only to a relatively minor extent on nectar-feeding birds and mammals to pollinate their flowers, they rely heavily on fruit-eating vertebrates to disperse their seeds. Plant-pollinator interactions tend to be more specialized, at least in terms of plant and animal morphology, than plant-frugivore interactions, and in both kinds of mutualisms, plants use birds to disperse their pollen and seeds more often than mammals. Nonetheless, these two mutualisms have been responsible for the adaptive radiation of a substantial portion of tropical bird and mammal faunas, and much of the vertebrate biomass in lowland tropical forests is supported by tropical fruits. Plant-visiting birds and mammals are conspicuous elements of most tropical habitats and have played an important role in their ecological and evolutionary dynamics throughout most of the Cenozoic era. Conservation of these colorful and often charismatic animals is therefore of the utmost importance.

2

Patterns of Regional and Community Diversity

Introduction

The pollination and seed dispersal mutualisms we discuss in this book involve interactions between populations of plants and animals exploiting each other for mobility of their propagules and nutritional benefits, respectively. These interactions, and their evolution, occur in a regional and community context. At the regional scale, the composition of floras and faunas depends on earth history and biogeography. Plate tectonics and the dispersal of plants and animals within and between land masses have produced current regional distribution patterns (chap. 6; Lomolino and Heaney 2004; Lomolino et al. 2006). At the habitat or community scale, subsets of this regional flora and fauna co-occur and have the potential to interact ecologically and evolutionarily on a daily basis. Before we discuss these interactions in detail, we will describe patterns of regional and community diversity in this chapter. We will use a rather broad-brush approach in dealing with plants (because there are so many of them!) and a more detailed, quantitative approach in dealing with their vertebrate mutualists.

There are three primary motivating questions. (1) How does the diversity, defined simply as number of species or species richness (S), of these plant and animal mutualists differ among biogeographic regions, including islands, and along gradients of temperature, rainfall, and elevation? (2) What is the diversity and structure of local communities of these plant and animal mutualists? To what extent does community structure differ in different biogeographic regions? (3) Do quantitative community assembly rules exist for these plant and animal mutualists, and, if so, are these rules the same in different biogeographic regions? Broader questions concerning biogeography, phylogeny, and the evolution of regional floras and faunas will be addressed in chapters 5 and 6.

Patterns of Regional Plant Diversity

PATTERNS OF SPECIES RICHNESS

The major tropical regions of the world differ strongly in their overall diversity of plants as well as in their pollinator and seed dispersal mutualists. Of the approximately 250,000 extant species of angiosperms, about 170,000 species occur within the tropics. Half of these (85,000) are found in the Neotropics; 21,000 occur in Africa, 10,000 in Madagascar, and 50,000 in tropical and subtropical Asia, including 36,000 in Malesia and 15,000 on Borneo (Whitmore 1998). Regional differences in tropical angiosperm diversity reflect, in part, the areal extent of tropical forests in different biogeographic regions. South America currently contains about 50% of the world's rain forest area, and Southeast Asia and Africa contain about 33% and 20%, respectively (Morley 2000). It should therefore not be surprising that the Neotropics contains more plant species than other tropical regions. But, as we will see, the Neotropics stands out in more than just number of species. In addition to their outstanding taxonomic diversity, these forests contain an impressive degree of ecological diversity, reflected in their variety of plant life forms—palms, vines, lianas, and epiphytes as well as trees and shrubs— that tend to be less diverse in other tropical regions. This diversity of plant forms provides a multitude of potential food sources for Neotropical birds and mammals.

In the New World, highest plant diversity (of all growth forms) occurs in upper Amazonia near the eastern foothills of the Andes where conditions of high rainfall and relatively fertile soils prevail (Gentry 1988, 1992). Highest plant diversity in Asia occurs in Borneo and Peninsular Malaysia, which were the locations of major refugia during Pleistocene climatic fluctuations (Morley 2000). African forests generally contain much lower plant diversity than elsewhere in the tropics, especially in palms and epiphytes, and highest diversity occurs in wetter areas of west Central Africa (Parmentier et al. 2007). Similarly, highest diversity in Madagascar, which along with New Caledonia and northern Australia represents a major center of endemism for tropical plant families (Morley 2000; Williams et al. 1994), is in the wet northeastern corner. Finally, no forests in Central America, West

Africa, the Indian subcontinent, or tropical Australia contain as many tree species as those in some wet forests in South America and Southeast Asia (Gentry 1992).

While some authors (e.g., Haffer and Prance 2001 and included references) have long argued that high Neotropical plant and animal diversity has arisen in the recent geological past as a result of Pleistocene climatic fluctuations, paleobotanical evidence indicates that diverse tropical forests date from at least the Paleocene in South America (Burnham and Johnson 2004; Graham 2011; Jaramillo et al. 2006; Wing et al. 2009). Studies of fossil plant communities from the Eocene in Patagonia, for example, suggest that high S has prevailed in South America throughout most of the Cenozoic and predates formation of the Andes Mountains (Wilf et al. 2003, 2005). Fossil leaf floras from two moist but nontropical forest sites in Patagonia that date from 48 to 52 Ma are far richer in plant species than any other Eocene flora in the world.

High regional tree diversity reflects the effects of both alpha and beta diversity. The former diversity parameter indicates the within-habitat or community-level component of regional diversity whereas the latter parameter measures the between-habitat component (Chave 2008; Jost 2007; Tuomisto and Ruokolainen 2006; Whittaker 1972; Willig et al. 2003). High overall regional diversity reflects different combinations of high values of alpha and beta diversity. The relative contribution of these two levels of diversity to high Neotropical plant diversity has been extensively discussed recently (e.g., Condit et al. 2002; Kraft et al. 2011; Phillips et al. 2003; Tuomisto et al. 2003a, 2003b; Valencia et al. 2004). Pitman et al. (1999, 2001, 2002, 2008), for example, reported that a few families (Arecaceae, Moraceae, Myristicaceae, and Violaceae) dominate forests over wide areas in lowland Amazonia such that tree species abundance in Ecuadorian plots is correlated with abundance of the same species in Peru. In these forests, common species, which represent only about 15% of all tree species in the Ecuadorian plots, appear to be indifferent to heterogeneity in topography and soil conditions and contribute little to beta diversity. In contrast, rare species (i.e., the majority of tree species in this region) may be more responsive to environmental heterogeneity and contribute more to beta diversity. A similar pattern also emerges from an analysis of western Amazonian palms (Vormisto et al. 2004). Compared to the Amazonian situation, beta diversity in tree species is much higher in Panama, perhaps because climatic and geological gradients are steeper there (Condit et al. 2002). Thus, Panamanian plots 50 km apart share only 1%–15% of their tree species compared to 30%–40% at this distance in Western Amazonia. A detailed analysis of plant communities in central Panama revealed that variation in species composition was related to geographic distance and environmental heterogeneity, particularly rainfall (Chust et al. 2006). In their study of plant species distributions across a large landscape in Peru, Phillips et al. (2003) confirmed that beta diversity is lower than that in Panama, but reported that, contrary to the results of Pitman et al. (1999, 2001), most tree species had significant habitat associations and were not indifferent to soil conditions. In particular, plant associations differed significantly on Pleistocene nutrient-poor and Holocene nutrient-rich soils. Tuomisto et al. (2003a) concluded from their analysis of the distributions of pteridophytes and species of Melastomataceae in Western Amazonia that, despite their relatively uniform appearance, the floristic composition of these forests is significantly influenced by soil edaphic conditions. Recent large-scale work in a 700 km-long transect from the eastern Andes of Ecuador to the Peru-Brazil border indicates that discontinuities in forest tree composition coincide with changes in soil texture from young, nutrient-rich soils of the Andes to older, nutrient-poor soils of the Amazon basin (Pitman et al. 2008).

The contribution of beta diversity to overall regional tree diversity has been less studied in the African and Asian tropics. In Cameroon, Africa, western India, and Borneo, floristic similarity between sampling locations decreases with geographic distance (Davidar et al. 2007; Hardy and Sonke 2004; Slik et al. 2003). The rate of decline in floristic (mature tree) similarity between transects in the Dja Fauna Reserve in Cameroon was comparable to that in Western Amazonia, as reported by Condit et al. (2002), which suggests that similar rates of floral turnover and beta diversity occur in both regions (Hardy and Sonke 2004). In lowland dipterocarp forests of Borneo, Slik et al. (2003, 2009) reported that floristic similarity was more strongly correlated with geographic distance than with annual rainfall. Since distance and rainfall were correlated, however, they felt that "the effect of geographical distance on floristic similarity is non-causal" (1527). The implication here is that floristic turnover is not simply an effect of limited dispersal ability, as postulated by the neutral theory of Hubbell (2001, 2008). Instead, this turnover reflects plant responses to environmental heterogeneity (and niche differentiation), as has been reported for plants in both the Old and New World tropics (e.g., Ashton 1998; Clark et al. 1999; Engelbrecht et al. 2007; John et al. 2007; Kraft et al. 2008; Potts et al. 2002; Pyke et al. 2001). While

Table 2.1. Species Richness of Plants by Growth Habit in the Local Florulas of Five
Neotropical Localities

Growth Habit	Tropical Dry Forest		Tropical Moist Forest		Tropical Wet Forest
	Capeira, Ecuador	Santa Rosa, Costa Rica	Jauneche, Ecuador	Barro Colorado Island, Panama	Rio Palenque, Ecuador
Trees ≥10 cm DBH	69 (15)	142 (21)	108 (20)	291 (22)	154 (15)
Small trees and large shrubs	28 (6)	64 (10)	58 (11)	134 (10)	99 (10)
Herbs and subshrubs	242 (52)	317 (48)	192 (36)	439 (33)	376 (36)
Epiphytes	8 (2)	19 (3)	58 (11)	180 (13)	227 (22)
Parasites	4 (1)	6 (1)	4 (1)	8 (1)	6 (1)
Lianas	46 (10)	52 (8)	54 (10)	149 (11)	87 (8)
Small vines	66 (14)	63 (10)	55 (10)	117 (9)	84 (8)
Total species	463	663	529	1,318	1,033

Source. Based on Gentry and Dodson (1987).
Note. Data include number of species and percentage of total (in parentheses).

still a strong advocate of the utility of his neutral theory for explaining patterns of tree S and relative abundances, Hubbell (2010) agrees that species' identities and ecological characteristics are important for a full understanding of tropical forest dynamics and biogeography.

High tree diversity is only part of the plant diversity story in the tropics. When all vascular plants are systematically sampled in small plots, trees usually represent less than one-third of the species in the Neotropics (Hubbell et al. 2008). Gentry and Dodson (1987), for example, reported that epiphytes represented over 33% of all plant species, followed by terrestrial herbs (13%), shrubs (10%), and nonepiphytic climbers (9%) in three forests in Ecuador. In very wet lowland forest in the Choco Department, Colombia, Galeano et al. (1998) recorded 442 species of vascular plants in a 0.1 ha plot with the following distribution among growth forms: terrestrial herbs, 14%; epiphytes and hemiepiphytes, 27%; palms, 3%; lianas and vines, 7%; small trees and treelets, 46%; and trees ≥10 cm diameter at breast height (DBH), 8%.

Gentry and Dodson (1987) also summarized the distribution of plant S among different growth habits for five local florulas in Central and South America. These data (table 2.1) indicate that overall plant diversity increases along a rainfall gradient and that trees ≥10 cm DBH represent only 15%–22% of all plant species in these floras; herbs and subshrubs represent 33%–52% of all plant species; and epiphytes represent 2%–22% of all species. As in the case of trees, lianas in Western Amazonia show relatively low beta diversity within terra firme forests, but they exhibit a significant habitat effect when floodplain and terra firme forests are compared (Burnham and Johnson 2004). In this region, diversity of vascular epiphytes is correlated with annual rainfall and length of the dry season (Kreft et al. 2004).

These kinds of data are uncommon for Old World tropical forests. Even in Borneo, which appears to harbor the greatest plant diversity in Asia, total vascular plant diversity, as well as the diversity of particular growth forms, in wet forests is likely to be lower than that in New World forests. In evergreen forests in Ghana, West Africa, for instance, epiphytes and palms are uncommon but lianas are common and represent about 30% of vascular plant species (Hall and Swaine 1976).

TAXONOMIC PATTERNS

In addition to major differences in the number of species, different tropical regions differ in the taxonomic composition of their floras. As summarized by Morley (2000), certain families, such as Annonaceae, Apocynaceae, Burseraceae, Clusiaceae, Euphorbiaceae, Fabaceae, Moraceae, Rubiaceae, and Sapindaceae are common in lowland tropical forests throughout the world. But, as shown in figure 2.1, the relative dominance of these families changes from one biogeographic region to another. Thus, species of Fabaceae predominate in local communities in the Neotropics and Africa, but this family is replaced in dominance by the Dipterocarpaceae in various forests in southern Asia. In the wet forests of Queensland, Australia, Lauraceae,

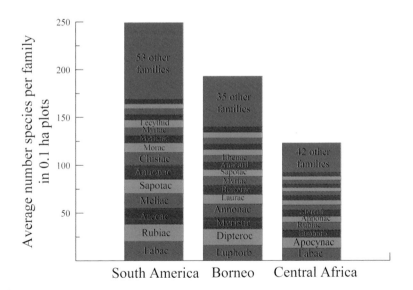

Figure 2.1. Comparison of the average number of tree species by family in 0.1 ha "Gentry plots" in South America (two sites), Borneo (two sites), and central Africa (five sites). Based on Morley (2000).

Elaeocarpaceae, Monimiaceae, Myrtaceae, and Proteaceae are the dominant families. Additional notable regional differences include a high diversity of vines in the Bignoniaceae, palms (Arecaceae), herbaceous understory monocots (e.g., Araceae, Heliconiaceae, Marantaceae), and epiphytes (especially Bromeliaceae and Orchidaceae) in the Neotropics (Gentry 1982; Morley 2000). Distinctive woody elements of the African tropical flora from lowland to montane habitats include canopy and subcanopy trees in the Annonaceae, Euphorbiaceae, Fabaceae, Moraceae, Olacaceae, Rubiaceae, and Sterculiaceae; taxa in the Commelinaceae, Marantaceae, and Zingiberaceae are common as herbaceous and climbing plants in the understory. Distinctive elements of the Southeast Asian flora include Annonaceae, Burseraceae, Clusiaceae, Dipterocarpaceae, Euphorbiaceae, Lauraceae, Myrtaceae, and Rubiaceae among canopy and subcanopy trees and a diversity of herbaceous Gesneriaceae and Zingiberaceae in the understory; bamboo thickets dominate in many areas of monsoonal climates. This flora is also notable for the S and morphological diversity of its palms.

TROPICAL ISLANDS

Compared to most mainland tropical floras, tropical island floras are much less species rich. This trend reflects the well-known relationship between area and S (Rosenzweig 1995). Thus, even though tiny Barro Colorado Island in central Panama contains only a fraction of the flora of Panama (Hubbell 2001), it still contains at least twice as many genera and species of epiphytes, shrubs, and trees in selected plant families than occur on small Caribbean islands such as the Caymans and Bahamas (table 2.2). Larger islands in the Caribbean and elsewhere in the tropics contain more genera and species in these families, but their plant S is still substantially lower than that of floras on adjacent tropical mainlands. For example, despite inhabiting an area about 76 times larger and with much greater topographic complexity than Barro Colorado Island, the flora of Jamaica contains only one to two times more species of plants of the families listed in table 2.2 than Barro Colorado Island.

Patterns of Plant Community Diversity

SPECIES RICHNESS

As summarized by Givnish (1999), a number of trends regarding S occur in tropical plant communities along a variety of ecological gradients. Many

Table 2.2. The Number of Genera and Species in Selected Neotropical Plant Families

Growth Habit/Family	Barro Colorado Island, Panama (15)	Cayman Islands (237)	Bahamas (10,980)	Jamaica (11,400)
Epiphytes or herbs:				
Araceae	14/16	1/1	1/1	4/11
Bromeliaceae	8/20	3/10	4/15	8/61
Heliconiaceae	2/9	0/0	0/0	1/2
Orchidaceae	45/90	11/16	26/50	62/204
Shrubs:				
Melastomataceae	14/35	0/0	1/1	18/70
Piperaceae	3/32	2/3	0/0	3/53
Rubiaceae	37/67	16/19	22/55	46/149
Solanaceae	9/25	5/8	6/21	10/37
Trees or treelets:				
Anacardiaceae	5/7	3/3	4/4	6/9
Annonaceae	7/13	1/3	1/3	3/11
Bombacaeae (= Malvaceae, in part)	7/9	0/0	1/1	2/2
Lauraceae	5/11	2/2	3/3	7/17
Moraceae	15/35	2/3	2/4	7/14
Myrtaceae	5/14	3/4	6/12	6/70
Palmae (= Arecaceae)	12/18	3/3	5/7	7/10
Sapindaceae	6/26	3/4	8/10	11/16
Sapotaceae	2/6	3/4	5/10	6/18

Sources. Numbers recorded from Barro Colorado Island, Panama (Croat 1978), the Cayman Islands (Proctor 1984), the Bahamas (Correll and Cofrrell 1982), and Jamaica (Adams 1972).

Note. Areas (in km^2) are given in parentheses. Numbers in columns represent no. of genera/no. of species.

of these trends reflect well-known general patterns of community richness and include:

- An increase in S with rainfall and a decrease with seasonality in mature lowland forests
- An increase in S with soil fertility in the Neotropics and a decrease at high levels of soil P and Mg in Borneo
- An increase in S with forest stature and time since major physical disturbance
- An increase in S with rate of tree turnover via recruitment and mortality
- A decrease in S with latitude and altitude

A growing number of plant inventories and analyses of plots ranging from 0.1 to 50 ha in size (e.g., Condit 1995; Losos and Leigh 2004; Rees et al. 2001; Ter Steege et al. 2003) indicate that tropical communities can contain at least 300 species of trees per ha. In 50 ha plots located along a dry to wet forest gradient in India, Panama, and Malaysia, total tree S was 71, 303, and 817, respectively (Condit et al. 1996). The late Al Gentry (1988) systematically

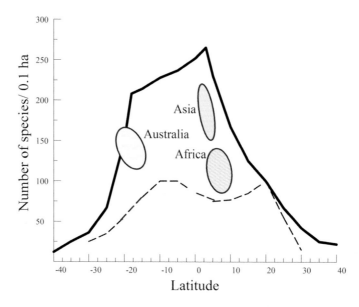

Figure 2.2.   The number of tree species ≥2.5 cm DBH in 0.1 ha "Gentry plots" as a function of latitude in New World moist and wet forests (*solid line*) and dry forests (*dashed line*), including comparable data from Africa, Asia, and Australia (*ellipses*). Modified from Gentry (1988).

sampled 0.1 ha plots for S among tree species ≥2.5 cm DBH in many tropical localities, and his data can be used to illustrate trends in alpha diversity as a function of latitude, rainfall, and biogeographic region. Overall trends are summarized in figure 2.2. In the Neotropics, tree diversity varies with latitude by a factor of 10 (from about 25 to over 250 species per 0.1 ha plot) in forests receiving ≥1600 mm of rain per year. In drier forests, maximum S (about 100 species) is only 40% of that of moist or wet forests, and the latitudinal range is only about fourfold. Wet forest plots contain fewer tree species in Central America than in South America (Condit et al. 2002; Pitman et al. 2001). Available data for wet forests from the Paleotropics fall within the range of Neotropical values, but the richest New World sites contain more tree species (over 250 species) in point samples (plots) than Asian sites (fig. 2.2). Enquist et al. (2002) have pointed out that the number of genera and families recorded in the Gentry plots are negative allometric functions of S in all biogeographic regions. This means that as S increases, more species are added from fewer genera and families such that high-diversity communities become especially rich in close taxonomic relatives.

Hans Ter Steege and colleagues (Ter Steege et al. 2003) have inventoried hundreds of 1 ha plots throughout northern South America and have painted a broad picture of regional variation in the alpha diversity of tree species. As shown in figure 2.3, maximum tree S occurs in a broad swath along the equator from Ecuador, northern Peru, and southern Colombia into central Amazonian Brazil In this region, maximum tree S is negatively correlated ($r^2 = 0.91$) with length of the dry season (i.e., number of months receiving <100 mm of rain) in plots located on terra firme soils. To explain the process behind this pattern, they postulated that "high constant moisture is thought to increase shade tolerance, higher densities, and subsequently a higher number of functional guilds in the understory" (2266). A similar relationship between length of the dry season and S also holds for Amazonian vascular epiphytes (Kreft et al. 2004).

In the Old World tropics, highest tree species diversity occurs in Borneo, a continental island on the Sunda Shelf. Slik et al. (2003) determined the alpha diversity of families and genera of mature trees throughout the lowlands of Borneo and divided this flora into five floristic units dominated by two families, Dipterocarpaceae and Euphorbiaceae. Their analyses indicated

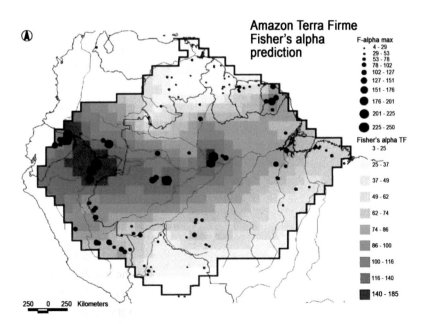

Figure 2.3. Map of the average alpha diversity of rain forest trees in 275 terra firma plots in the Amazon and Guayana Shield of South America. Reproduced with permission from Ter Steege et al. (2003).

that highest family and generic diversity occurs in southeastern Borneo and central Sarawak, and lowest diversity occurs in northeastern and southwestern Borneo. The latter region was probably covered with savanna during the last glacial period and has only been available for colonization by rain forest plants for the last 10,000 years. Total plant S in Borneo, however, shows a somewhat different pattern with highest diversity occurring in the northwest (P. Ashton, pers. comm.).

Understory trees and shrubs are important fruit sources for frugivorous birds and bats, and their diversity in the Neotropics appears to be closely related to rainfall and soil fertility. Gentry and Emmons (1987) documented S of these plants in 500-m long line transects at 13 lowland sites in Central and northern South America. They found that the number of reproductively fertile species per transect ranged from five to about 64 species and that S was lower at sites with lower rainfall and/or low soil fertility. On a seasonal basis, the number of reproductively active species increased twofold from dry to wet seasons at sites with strong dry seasons but changed little throughout the year at aseasonal sites. The top five families in terms of S at most sites were Rubiaceae, Araceae, Piperaceae, Melastomataceae, and Gesneriaceae. Gentry and Emmons (1987) also reported similar increases in S in epiphytes, lianas, and herbs (but not in saplings of canopy trees) along these gradients, although liana abundance generally tends to be negatively correlated with rainfall in tropical forests worldwide (Schnitzer 2005).

Systematic surveys of communities of understory trees and shrubs in the Paleotropics are much fewer in number. In Ghana, West Africa, Swaine and Becker (1999) reported that the number of species of treelets/shrubs and trees increased with rainfall but, unlike Gentry and Emmons's (1987) results, decreased with increasing soil fertility. The prinicipal understory taxa in 155 0.0625 ha mature forest plots were Rubiaceae, Euphorbiaceae, Sapindaceae, and Annonaceae (Hall and Swaine 1981). In dipterocarp-dominated forests of Southeast Asia, understory plant diversity appears to be low, especially from the point of view of vertebrate nectar and fruit eaters (LaFrankie et al. 2006). A high density of dipterocarp seedlings and saplings is a major reason for this (Ashton et al. 2004). Wong (1986) reported that the most common understory families at Pasoh, West Malaysia, were Rubiaceae, Melastomataceae, Myrsinaceae, and Euphorbiaceae. Janzen (1977, 707) has colorfully described the absence of flowering and fruiting plants in the understory of lowland rain forest in Peninsular Malaysia as follows: "I averaged 1.3 plants in flower per kilometer [of trail] in the Malayan rain forest understory [compared with] 21.9 plants in flower per kilometer of Costa Rican rain forest understory . . . if one were to turn loose in Pasoh or Taman Negara [Malaysia] the rain forest understory fauna of flower-visiting hummingbirds, butterflies, moths, and bees found in the Corocovado [National Park, Costa Rica], I predict that they would be dead of starvation in a few days." In the same transects, he found 63 understory individuals of 22 species, or 7.5 individuals per kilometer, in fruit in Malaysia and 345 individuals of 34 species, or 78.4 per kilometer, in fruit in Costa Rica. He wrote, "Again, the fauna of understory birds that frequently eats small fruits in neotropical rain forests would have a very rough go of it in the Malaysian forests" (708). Wong's (1986) more rigorous data from Pasoh confirm that rates of flower and fruit production in the understory of this dipterocarp forest are indeed low.

SUCCESSIONAL PATTERNS

In addition to rainfall and soil fertility, successional age or stage has a strong effect on plant (and animal) community diversity throughout the tropics (e.g., see chapters in Gomez-Pompa et al. 1991). Within the lowland tropics, plant S tends to be lower in secondary (disturbed) forests than in primary (mature) forests in trees but not necessarily in shrubs. In northeastern Costa Rica, for example, Chazdon et al. (2003) recorded 144 tree species and 40 shrub species in old growth wet forests compared with 111 tree species and 53 shrub species in second growth. Letcher and Chazdon (2009) reported that, in this region, aboveground plant biomass and S can return to predisturbance levels in as little as 30 years. At La Selva, Costa Rica, S of *Piper* shrubs, an important food source for fruit-eating bats, is about 63% higher in second-growth forest than in primary forest, and density of *Piper* plants is about three times higher in second-growth than in primary forest (Fleming 2004a). Similarly, Kessler (1999) found that in the Bolivian Andes about 50% of the species of terrestrial herbs and shrubs were restricted to early- to midsuccessional stages on landslides. Lianas are often characterized as being early successional plants, and in central Panama both S and abundance of lianas are higher in early successional stands than in old growth forest (Dewalt et al. 2000; Schnitzer 2005). In Ghana, West Africa, however, only species of small climbers typically occur in secondary forests; species of large climbers are much more common in primary forest than in secondary forest (Hall and Swaine 1981).

Plant species diversity and biomass generally increase with time during tropical secondary succession. Saldarriaga et al. (1988) and Saldarriaga and

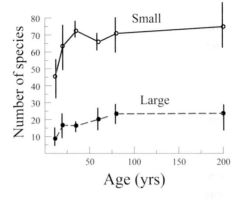

Figure 2.4.   Mean number of species (± SD) of large (DBH ≥10 cm) and small (1 cm < DBH ≤ 10 cm) trees as a function of age since last disturbance in 300 m² plots in Amazonian Venezuela. Based on Saldarriaga et al. (1988).

Uhl (1991) have described successional patterns following abandonment of slash-and-burn agriculture in terra firme tropical moist forests along the Rio Negro of Colombia and Venezuela. Trends in tree S for two size classes of trees (DBH ≥1 cm and DBH ≥10 cm) are shown in figure 2.4. Shortly after land abandonment, plots become dominated by bat-dispersed shrubs of the genus *Vismia* (Clusiaceae). These short-lived plants are replaced by a greater diversity of species as S of small trees increases from about 30 to over 60 species in 300 m² plots during the first 50 years of succession. Later successional trees become established in the shade of pioneer trees and reach a plateau of about 20 species per plot after about 75 years of succession. Basal area and above- and belowground biomass values reach mature forest values in about 190 years.

Ecological implications for wildlife resulting from changes in vegetation structure and composition along successional chronosequences in central Panama have been discussed by DeWalt et al. (2003). Although they documented substantial changes in size distributions among saplings and trees along a 100-year-old sequence, tree S and the importance value (i.e., relative basal area) of animal-dispersed tree species did not change along this sequence. The density of fleshy-fruited understory herbs, shrubs, and palms, however, declined with forest age. Fig trees and species of *Virola*, which are important food sources for vertebrate frugivores, occurred at low densities throughout the sequence while the importance value of palms increased with forest age. DeWalt et al. (2003) concluded that secondary forest contains more fruit resources for small generalist frugivores whereas large frugivorous birds are likely to find suitable food throughout the chronosequence. This pattern is likely to be common throughout the lowland tropics worldwide.

ELEVATIONAL PATTERNS

Throughout the tropics, plant S generally decreases with increasing elevation, although there are several variations on this pattern (e.g., Aiba and Kitayama 1999; Desalegn and Beierkuhnlein 2010; Gentry 1988; Grytnes and Beaman 2006; Lieberman et al. 1996; Ohsawa 1995). Rahbek (1995, 1997), for example, has noted that while S declines monotonically with increasing elevation in many groups, other groups show a hump-shaped pattern with maximum S at midelevations or show fairly constant values of S to midelevations followed by a steady decline thereafter. Two examples of these patterns are shown in figure 2.5. On Mount Kinabalu, Borneo, numbers of tree species, genera, and families in plots ranging from 0.06 ha to 1.0 ha in size decline monotonically with elevation on two different soil types (Aiba and Kitayama 1999; fig. 2.5*A*). In contrast, Kessler (2000, 2001) found hump-shaped patterns in S in Acanthaceae, Araceae (in one transect), epiphytic and terrestrial Bromeliaceae, Cactaceae, and Melastomataceae (in one transect) along two elevational transects in the eastern Bolivian Andes (fig. 2.5*B*). In one transect, three families—Araceae, Arecaceae, and Melastomataceae—showed monotonic declines beyond 1,000–1,500 m. Based on these results, Kessler (2001) concluded that a hump-shaped relationship between S and elevation is the most common pattern in tropical plants. Grytnes and Vetaas (2002) also reported a hump-shaped relationship between plant S and elevation along a 6,000 m gradient in the Himalayas but used computer simulations to demonstrate that such patterns can arise from the statistical effects of "hard boundaries" (i.e., both elevational and latitudinal gradients have finite start and end points) in addition to incomplete sampling along gradients. Finally, in contrast to the above empirical patterns, Lovett (1996) reported that diversity of tree species was not correlated with elevation in the Eastern Arc mountains of Tanzania. In that region, tree diversity was lower in disturbed or seasonal forests irrespective of elevation.

In summary, plant S varies substantially among biogeographic areas within the tropics at both regional and local levels, with the Neotropics containing greater floral diversity than other regions. Trees, which have been the focus of much community-level research recently, represent less than

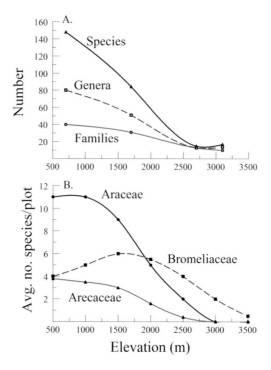

Figure 2.5.   Number of plant taxa along elevational gradients in: *A*, Borneo and *B*, Ecuador. Based on Aiba and Kitayama (1999) and Kessler (2001).

50% of the plant S in most tropical forests. In addition to trees, shrubs, epiphytes, lianas, and vines are important sources of food resources for nectar- and fruit-eating birds and mammals. Within regions, the diversity of these growth forms and their flower and fruit resources vary with time since last habitat disturbance, soil fertility, climatic seasonality, elevation, and latitude. We expect to see similar trends in their animal mutualists.

Patterns of Regional Animal Diversity

Unlike the tens of thousands of vascular plants in different tropical regions, the regional diversity of tropical plant–visiting birds and mammals numbers in the tens to hundreds of species. These much lower levels of diversity allow us to take a closer and more detailed quantitative look at regional and community patterns of animal diversity. Before we do this, however, we will briefly review the major families of plant-visiting birds and mammals that characterize different biogeographic regions. These are listed in table 2.3. More detailed accounts of these families can be found in appendix 1.

As pointed out in chapter 1, taxonomic richness of nectar-eating birds and mammals is much lower than that of fruit-eating birds and mammals. Thus, the number of families of morphologically specialized nectarivorous birds ranges from one to three per region; it is only one per region in mammals, except in Australia where there are two families (Fleming and Muchhala 2008). Hummingbirds (Trochilidae) are the New World nectar specialists. Bananaquits (Coerebidae, one species) and a variety of orioles (Icteridae), tanagers (Thraupinae), and warblers and honeycreepers (Parulinae) also visit flowers in some areas (Stiles 1981). In Africa and Asia, sunbirds (Nectariniidae) are the ecological equivalents of hummingbirds, although they differ significantly from them in morphology and behavior (chap. 7). In the Australasian region, honeyeaters (Meliphagidae) and brush-tongued parrots (Loriinae) are the major avian flower visitors. Among mammals, only certain phyllostomid (New World) and pteropodid (Old World) bats plus the Australian honey possum (Tarsipedidae) are specialized flower visitors. Other animals, including frugivorous phyllostomid bats and a few arboreal

Table 2.3. Major Families or Subfamilies of Vertebrate Mutualists by Biogeographic Region

Mutualism/Taxon	Biogeographic Region			
	Neotropics	Africa	Asia	Australasia
Pollination:				
Birds	Trochilidae	Nectariniidae	Nectariniidae, Zosteropidae	Loriinae, Nectariniidae, Meliphagidae
Mammals	Phyllostomidae	Pteropodidae	Pteropodidae	Pteropodidae
Frugivory:				
Birds	Cracidae, Trogonidae, Ramphastidae, Cotingidae, Pipridae, Turdidae, Thraupinae	Musophagidae, Bucerotidae, Sturnidae, Pycnonotidae	Bucerotidae, Sturnidae, Dicaeidae, Pycnonotidae	Casuariidae, Columbidae, Paradisaeidae, Meliphagidae, Sturnidae
Mammals	Phyllostomidae, Atelidae, Aotidae, Cebidae, Pitheciidae, Procyonidae	Pteropodidae *Africa only:* Cercopithecidae, Hominidae, Tragulidae, Elephantidae *Madagascar only:* Cheirogaleidae, Lemuridae, Indriidae	Pteropodidae, Tupaiidae, Cercopithecidae, Hylobatidae, Hominidae, Elephantidae, Viverridae	Phalangeridae, Pteropodidae

nonvolant mammals (e.g., opossums [Didelphidae] and olingos [Procyonidae]), occasionally visit flowers in the New World, as do frugivorous pteropodid bats and certain other arboreal mammals (e.g., squirrels [Sciuridae] and palm civets [Viverridae]) in the Old World.

Taxonomic diversity of fruit-eating birds and mammals is much greater than that of nectar eaters (table 2.3). Indeed, as Richard Corlett (1998, 440) has noted, "Most forest birds and mammals in the Oriental Region probably eat at least some fruit." The same is probably true in other tropical regions. Nonetheless, certain families of birds and mammals appear to be much more highly frugivorous than others and hence can be considered to be "dedicated" or "obligate" frugivores (Fleming et al. 1987; Snow 1981). Among Neotropical birds, these include trogons (Trogonidae), toucans (Ramphastidae), cotingids (Cotingidae), manakins (Pipridae), and tanagers (Thraupinae). Their mammalian counterparts include certain phyllostomid bats, four families of monkeys, and arboreal procyonids such as kinkajous and olingos. In Africa, turacos (Musophagidae), hornbills (Bucerotidae), certain starlings (Sturnidae), and bulbuls (Pycnonotidae) are highly frugivorous. Fruit-eating African and Madagascan mammals include pteropodid bats, various primates, including lemurs, cercopithecine monkeys, and apes, as well as mouse deer (Tragulidae) and elephants. Major frugivores in tropical Asia include several bird families shared with Africa plus flower peckers (Dicaeidae). Similarly, frugivorous Asian mammals include families shared with Africa plus certain tree shrews (Tupaiidae) and arboreal viverrids. Finally, Australasian frugivorous birds include cassowaries (Casuariidae), fruit pigeons (Columbidae), bowerbirds (Ptilonorhynchidae), and birds of paradise (Paradisaeidae) in addition to certain honeyeaters and starlings. Only two families of mammals, pteropodid bats and marsupial phalangers (Phalangeridae), are dedicated fruit eaters in Australasia.

REGIONAL PATTERNS

Regional patterns of S in plant-visiting birds and mammals generally conform to the expected species-area relationship. Thus, as in plants, the Neotropics is richer in total number of species of nectarivores and frugivores than other tropical regions (fig. 2.6). In both groups of plant visitors, the number of species of birds is far greater than the number of species of mammals. Australia is notable for its low number of frugivorous birds, a situation that is also seen on Madagascar (Fleming et al. 1987; Goodman and Ganzhorn

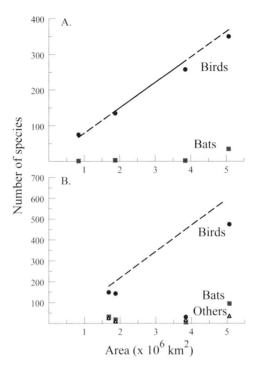

Figure 2.6. Species area relationships in *A*, nectarivorous and *B*, frugivorous birds and mammals in, *from left to right*, African, Asian, Australian, and New World tropical forests. The lines represent least-squares regression curves and indicate total number of species. Sources of data: Davis (1972), Fleming et al. (1987), Mickleburgh et al. (1992), Pizzey (1980), Stiles (1981).

1998; Rakotomanana et al. 2001). A somewhat different picture arises, however, when S is expressed on a per-area basis (fig. 2.7). The species density of nectar-eating birds averages about 75 species per 10^6 km^2 and is relatively constant among regions whereas the species density of nectar-eating bats is somewhat higher in the Neotropics than in the other three regions. Among frugivores, bird species density is relatively similar in the Neotropics, Africa, and Southeast Asia, where it averages about 82 species per 10^6 km^2; it is only half this value in Australia. The species density of frugivorous bats is nearly twice as high in Southeast Asia as in other tropical regions, and primate species density is over twice as high in Africa compared to the Neotropics and Southeast Asia (fig. 2.7). Relatively constant species densities among regions, as seen in birds, suggest that, despite vastly different geological histories and

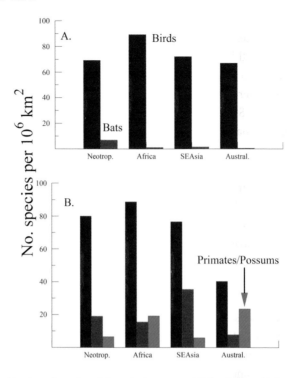

Figure 2.7. Number of species of *A*, nectarivorous and *B*, frugivorous birds and mammals per 10^6 km² in four tropical regions. Based on data in figure 2.6.

topographies, the "carrying capacity" for these animals is rather similar in different tropical regions. Kalmar and Currie's (2007) study of global patterns of bird S supports this conclusion.

As in most other groups of organisms, regional diversity of plant-visiting birds and mammals varies with latitude with diversity generally decreasing north and south of the equator (e.g., Kissling et al. 2011). In the New World, for example, highest S in hummingbirds and nectar-feeding phyllostomid bats occurs north of the equator in montane regions of northwestern South America and adjacent Central America (Fleming 1995; Johnsgard 1997; Rahbek and Graves 2000). At a given latitude in northern South America, generic and S of hummingbirds and nectar bats decreases from west to east. Similar patterns probably also hold for most taxa of Neotropical fruit-eating birds, bats, and primates (e.g., Peres and Janson 1999). In primates, the best predictors of regional S are rainfall and percentage of forest cover; latitude and elevation are less effective predictors (Peres and Janson 1999). Highest primate diversity occurs in the western Amazon basin, as is the case for tree species diversity (Eeley and Lawes 1999).

In Africa, overall land bird S is highest in the wet equatorial zone and decreases sharply to the north but less sharply to the south; holding latitude constant, S is richer in East Africa than in West Africa (Jetz and Rahbek 2001). Changes in S along rainfall gradients in the humid equatorial forest zone are dominated by species' replacements (beta diversity) whereas changes in S in more arid parts of tropical Africa are dominated by richness gradients (i.e., changes in alpha diversity; P. Williams et al. 1999). With respect to particular groups of plant-visiting birds, diversity of sunbirds is higher in open woodlands north and south of the equatorial forests as well as in the mountains of East Africa than in wet forests (Sinclair and Ryan 2003). Similarly, Africa has 23 species of hornbills, of which 13 are savanna dwellers and 10 are forest dwellers (Kemp 1995). Forest species are highly frugivorous whereas terrestrial nonforest species tend to be insectivorous or carnivorous (Kemp 1995). Unlike sunbirds and hornbills, most turacos (Musophagidae) are forest or woodland dwellers and live in central and southern Africa. Eight out of 13 African genera (about 41 of 49 species) of starlings are mostly frugivorous forest dwellers; the other genera are ground-feeding omnivores (Feare and Craig 1999)

Regional patterns of S in African pteropodid bats and primates follow the general diversity pattern of birds, with highest diversity occurring in the wet equatorial region, a sharp drop in diversity north of the equator, and a less steep decline south of the equator. Thus, countries in equatorial Africa contain 10–18 species of pteropodid bats whereas those in the south contain two to five species (Mickleburgh et al. 1992). In anthropoid primates, peak regional diversity in 250,000 km^2 quadrats is 12–16 species in equatorial West Africa compared with only one to three species in South Africa (Eeley and Lawes 1999). In Madagascar, S of lemurs is higher in the wetter eastern side (with up to 13 species per habitat) than in the drier western side (with up to eight species per habitat; Ganzhorn et al. 1999).

We are not aware of any quantitative analyses of regional diversity of plant-visiting birds and mammals in Asia but would expect to see similar trends regarding the relationship between S and latitude as seen in other major tropical regions (see Corlett 2009a; Kissling et al. 2011). Sunbirds (Nectariniidae) are the major nectar-feeding birds in Asia, but their overall diversity is much lower than in Africa (only 40 out of 116 species are Asian; Perrins and Middleton 1985). Nectar-feeding pteropodid bats, however, are slightly more diverse in Asia than in Africa (three vs. one species, respectively). Among frugivores, hornbills are as species rich in Asia as in Africa, but, unlike Africa, all Asian species are forest dwellers where they are the largest avian frugivores (Kemp 1995; Kinnaird and O'Brien 2007). Asian starlings, in contrast, include three omnivorous ground-feeding genera (containing about 13 species) as well as 13 frugivorous forest- or woodland-dwelling genera (containing about 56 species; Feare and Craig 1999). Mainland Asia contains about 21 species of frugivorous pteropodid bats (compared with 32 species in Africa), and an additional 24 species are endemic to the islands of Indonesia (Mickleburgh et al. 1992). Species richness of frugivorous Asian primates (about 11 species) is less than half that of Africa (about 26 species; Fleming et al. 1987). Forest habitats typically contain more species of primates in Southeast Asia than in South Asia (India; Gupta and Chivers 1999).

In the Australasian region, tropical New Guinea is much richer in frugivores, but not necessarily in nectarivores, than tropical/subtropical Australia. Both regions have similar numbers of honeyeaters (Meliphagidae)—65–68 species—but New Guinea has three times as many lorikeets as Australia (21 vs. seven species, respectively). Similarly, New Guinea has about three times as many species of frugivorous birds as Australia (92 vs. 31 species, respectively; Beehler et al. 1986; Pizzey 1980). New Guinea has nearly three times as many pteropodid bats as Australia (34 vs. 13 species, respectively). Of these, five are nectarivores in New Guinea and two are nectar eaters in Australia, although many Australasian species of *Pteropus* are also highly nectarivorous. New Guinea has 12 species of *Pteropus*, whereas Australia has seven (Bonaccorso 1998; Hall and Richards 2000).

ISLAND PATTERNS

As expected, S of plant-visiting birds and mammals is lower on islands than in comparable areas on the mainland. A striking example of this is the difference between S of these animals on Cuba (area = 114,524 km^2) and Costa Rica (area = 50,900 km^2). Despite being over twice as large, Cuba has only three species of hummingbirds compared with 51 species in Costa Rica; it has five fruit-eating birds compared with about 80 in Costa Rica (Raffaele et al 1998; Stiles and Skutch 1989). The Cuban–Costa Rican comparison for bats is similar: three species of nectarivores and three species of frugivores on Cuba compared with 12 and 25 species, respectively, in Costa Rica (Silva Taboada 1979; Wilson 1983).

Species-area relationships for island-dwelling, plant-visiting birds and

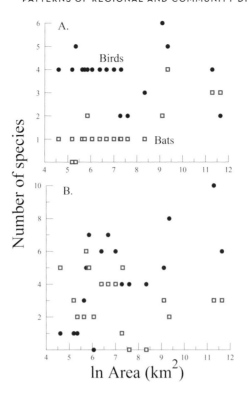

Figure 2.8. Species-area relationships for A, nectarivorous and B, frugivorous birds and bats on 19 Caribbean islands. Sources of data: Raffaele et al. (1998), Rodríguez and Kunz (2001).

mammals are shown in figures 2.8 and 2.9. In the West Indies, the S of nectar-eating birds for 19 islands ranges from two to five per island and is independent of island area ($r^2 = 0.0001$; $P >> 0.50$); richness of nectar bats ranges from zero to four per island and is significantly correlated with island area ($r^2 = 0.77$; $P < 0.01$; fig. 2.8A). The opposite situation holds for West Indian frugivores: S of birds (range = 0–10 species per island) is significantly correlated with island area ($r^2 = 0.43$; $P < 0.01$) but not in bats (range = 0–6; $r^2 = 0.07$; $P >> 0.10$; fig. 2.8B). For six large islands or archipelagos in the Old World tropics (in order of increasing size: Java, the Philippines, Sumatra, Madagascar, Borneo, and New Guinea), S of nectar-eating birds, but not mammals, generally increases with area (with the notable exception of Madagascar), but the correlation is not significant ($r^2 = 0.41$; $P > 0.10$; fig. 2.9A). It generally increases with area in both fruit-eating birds and mammals (again with the exception of Madagascar; fig. 2.9B). The correlation in birds

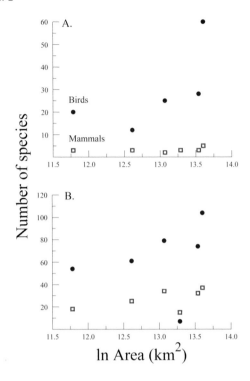

Figure 2.9. Species-area relationships for *A*, nectarivorous and *B*, frugivorous birds and mammals on six large Old World islands or archipelagos (from smallest to largest: Java, the Philippines, Sumatra, Madagascar, Borneo, and New Guinea). Sources of data: Beehler et al. (1986), Bonaccorso (1998), Delacour and Mayr (1946), Langrand (1990), MacKinnon et al. (1993), Mickleburgh et al. (1992), Nowak (1991), Payne et al. (1985), Rowe (1996), Smythies (1968).

approaches significance ($r^2 = 0.68$; $0.10 > P > 0.05$) and is significant for mammals ($r^2 = 0.90$; $P < 0.01$). Overall, absolute S of plant-visiting birds (both nectar and fruit eaters) as well as fruit-eating mammals is generally much higher (by factors of 5.5 to 12 for nectarivores and frugivores, respectively) on the much larger Old World tropical islands that we have analyzed than on the much smaller islands of the West Indies.

ELEVATIONAL PATTERNS

As is true for most groups of organisms, S of tropical plant–visiting birds and mammals generally declines with elevation and probably reflects local climatic gradients (McCain 2007, 2009). Data on birds and bats from an elevational transect in the Andes of eastern Peru illustrate this point (fig. 2.10). Starting from over 300 species of diurnal birds and 89 species of bats in the Amazon lowlands, number of species declines with or without a midelevation hump for birds and bats, respectively. The proportional representation of nectar-feeding birds increases with elevation whereas the proportional representation of frugivorous birds decreases. Proportional representation of these two trophic groups is independent of elevation in bats. The rate of turnover of species with elevation (i.e., beta diversity) is much higher in birds than in bats (Graham 1990). Terborgh (1977) points out that along this transect birds become the major frugivores (in terms of number of species and biomass) at mid- to high elevations; mammalian frugivores (at least in terms of biomass) dominate lowland habitats because most frugivorous primates are lowland species. A similar trend occurs in the Atlantic rainforest of southeastern Brazil (Almeida-Neto et al. 2008). The predominance of avian frugivory at mid- to high elevations in the Neotropics reflects an increase in dominance of bird-dispersed plant families (e.g., Ericaceae, Melastomataceae, and Rubiaceae) in plant communities at

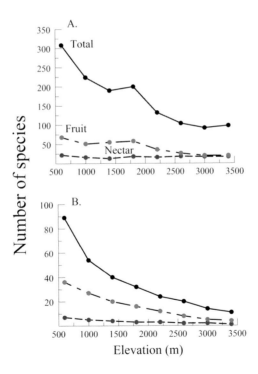

Figure 2.10.   Number of species of A, birds and B, bats along an elevation gradient in the Andes of eastern Peru. Based on Graham (1990).

these elevations. Similarly, an increase in the dominance of hummingbirds and honeycreepers is correlated with an increase of elevation in the representation of bird-pollinated plants (e.g., Bromeliaceae, Ericaceae, Loganiaceae, Loranthaceae, Onagraceae, and Verbenaceae).

Similar trends occur in the Old World tropics. In eastern New Guinea, for instance, lowland forests contain about 150 species of birds, and bird S declines monotonically with a small hump at 1,000–1,500 m to 25 species at 3,600 m (Beehler 1981). As in Peru (and also in Costa Rica; Blake and Loiselle 2000), the proportional representation of avian frugivores decreases slightly with elevation while it increases in nectarivores. Along an elevational gradient on Mount Kinabalu, Borneo, the number of species of frugivorous birds decreases in several families (e.g., barbets, bulbuls, flowerpeckers, and hornbills) and increases in only one family (thrushes; Kimura et al. 2001). Finally, on mountains in the central Philippines, S of pteropodid bats decreases from seven to eight in tall lowland forests to three in short mossy forests at elevations of 1,000–1,500 m (Heaney et al. 1989).

A final pattern that is worth noting is that mean body mass declines with elevation in birds in Peru and New Guinea and in bats in Peru (Beehler 1981; Graham 1990). In eastern New Guinea, median weight of forest birds (of all trophic groups) decreases from about 85 g in the lowlands to about 37 g at or above 2,500 m (Beehler 1981). In both birds and bats, large species tend to be confined to low elevations. This is also true of most primates and other arboreal frugivorous mammals. In hummingbirds, however, body mass tends to be positively correlated with elevation (Stiles 2008). From a botanical perspective, this general trend means that lowland tropical plants likely interact with a wider size range of plant-visiting vertebrates than do highland plants. As a consequence, we expect to see a wider range of fruit sizes, for example, in lowland habitats than in montane habitats. We will examine this prediction in chapter 8.

Patterns of Animal Community Diversity

Communities of tropical plant–visiting birds and mammals represent subsets of their regional species pools and include species living in the same habitat that actually or potentially interact on a daily basis. What is the basic structure of these communities and to what extent does this structure differ among biogeographic regions? In this section, we will focus primarily on two aspects of community structure—S and vertical distributions of species within forest habitats.

PATTERNS OF SPECIES RICHNESS

What is the alpha diversity of nectar- and fruit-eating birds and mammals in lowland tropical forests around the world? Inspired by the research of Robert MacArthur (e.g., MacArthur 1972), this question was initially addressed for birds (of all trophic groups) by James Karr (e.g., Karr 1976) and David Pearson (e.g., Pearson 1977). Results of Pearson's pantropical surveys (Pearson 1982) are summarized in table 2.4. Total number of species of birds per 15 ha plot in primary forest ranged from 114 (in Papua New Guinea) to 232 (in Ecuador). The number of nectarivores/insectivores per plot was less than the number of frugivores and ranged from four to eight species in primary forest and four to nine species in secondary forest. Unlike nectarivores, the number of frugivores in primary forest plots was always greater than the number in secondary forest and ranged from seven to 33 species (table 2.4). In general, New World sites tended to be richer in species of birds than Old World sites, as Beehler (1981) has also noted.

In table 2.5 we expand this survey to include the S of plant-visiting mammals as well as birds in 11 well-studied tropical forest communities. Again, Neotropical forests tend to be conspicuously richer in species of nectarivorous and frugivorous birds and mammals than paleotropical forests. Total number of nectarivores (birds and mammals combined) ranged from five (in Liberia) to 32 species (in Costa Rica); total number of frugivores ranged from 34 (in Liberia) to 117 species (in Peru). At each site, the number of species of frugivores exceeded the number of nectarivores by an average

Table 2.4. Number of Bird Species, Nectarivore/Insectivores, and Highly Frugivorous Species in Tropical Lowland Forests

Site	Total Annual Rainfall (mm)	No. of Months with <150 mm of Rain	Total No. of Species	No. of Nectarivore/ Insectivores	No. of Obligate Frugivores
Ecuador	2,978	0	232	8 (9)	33 (18)
Peru	1,625	4	204	5 (9)	30 (19)
Bolivia	1,995	4	181	4 (6)	27 (12)
Gabon	1,731	7	158	7 (9)	16 (23)
Borneo	2,360	0	142	4 (6)	7 (3)
Papua New Guinea	2,037	4	114	13 (4)	26 (6)

Source. Based on Pearson (1982).
Note. Totals are for six standard census plots of 15 ha. Numbers in parentheses indicate number of species in secondary forest plots.

Table 2.5. Summary of Species Richness in Communities of Nectarivorous and Frugivorous Birds and Mammals

Country/Habitat	Number of Species				Sources
	Nectarivorous Birds	Nectarivorous Mammals	Frugivorous Birds	Frugivorous Mammals	
Costa Rica:					
Tropical dry forest	11 (1)	3 (1)	19 (6)	23 (6)	Fleming (1988), Stiles (1983), Wilson (1983)
Tropical wet forest	22 (1)	7 (1)	53 (6)	26 (7)	Stiles and Levey (1994), Timm (1994)
Premontane moist forest	24 (1)	8 (1)	56 (6)	38 (6)	Fogden (2000), Timm and LaVal (2000)
Panama, tropical moist forest	21 (1)	4 (1)	53 (6)	38 (8)	Glanz (1990), Handley et al. (1991), Karr et al. (1990)
Brazil, tropical moist forest	11 (1)	4 (1)	54 (6)	38 (7)	Bernard (2001), Karr et al. (1990), Malcolm (1990)
Peru, tropical moist forest	13 (1)	5 (1)	68 (6)	49 (7)	Ascorra et al. (1996), Pacheco and Vivar (1996), Servat (1996)
Liberia, tropical moist forest	4 (1)	1 (1)	19 (6)	15 (6)	Karr (1976), Nowak (1991), Rowe (1996), Wolton et al. (1982)
Gabon, tropical moist forest	11 (1)	1 (1)	32 (7)	19 (6)	Brosset (1978), Gautier-Hion et al. (1985), Mickleburgh et al. (1992), Nowak (1991), Pearson (1977, 1982), Rowe (1996)
Borneo, Tropical Moist Forest?	12 (3)	2 (1)	52 (10)	25 (6)	Fogden (1976), Mickelburgh et al. (1992), Nowak (1991)
Malaysia, Tropical Moist Forest	12 (3)	2 (1)	47 (8)	17 (8)	Chivers (1980), Delacour (1947)
Papua New Guinea, Tropical Moist Forest	19 (3)	2 (1)	35 (8)	4 (1)	Bell (1982), Bonaccorso (1998), Nowak (1991)

Note. Parentheses give number of families.

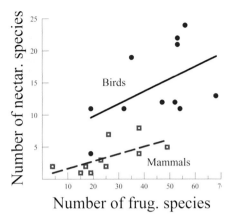

Figure 2.11. Relationship between the number of species of avian and mammalian nectar eaters and fruit eaters in the 11 communities summarized in table 2.5. Lines represent least-squares regression curves.

of 4.4-fold (range: 1.9–6.8), and the number of birds species was always greater than the number of mammal species in both trophic groups. Finally, as shown in figure 2.11, the number of species of nectarivores per site is positively correlated with number of frugivores in mammals ($r^2 = 0.41$, $P < 0.05$) but not in birds ($r^2 = 0.26$, $P > 0.10$). This probably reflects the fact that mammalian nectarivores and frugivores often come from the same families of bats (Phyllostomidae in the New World and Pteropodidae in the Old World), whereas avian nectarivores and frugivores are derived from very different evolutionary clades whose diversities are not necessarily autocorrelated.

Primate communities have attracted substantial attention from ecologists and anthropologists (e.g., Fleagle et al. 1999; Schreier et al. 2009); and hence it is worthwhile to pay special attention to this group (also see Corlett and Primack 2011). According to data presented in Emmons (1999), S of primates (of all trophic groups) at selected sites around the world is highest in Africa (mean = 12.9 ± 2.7 SE species; range = 8–16 species; $n = 7$ communities), followed by Madagascar (mean = 7.7 ± 2.6; range = 4–12; $n = 9$), Asia (mean = 6.5 ± 1.4 species; range = 4–8; $n = 6$), and the Neotropics (mean = 5.7 ± 3.1 species; range = 1–13; $n = 15$). In addition to species richness, primate communities in different parts of the tropics differ in many other characteristics, including average body size, locomotor adaptations, and dietary preferences. Fleagle and Reed (1999) used principal components analysis

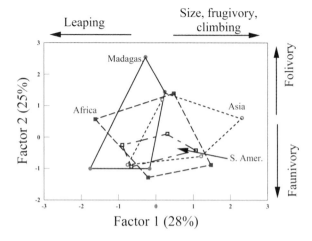

Figure 2.12. Summary of the distribution of species of primates in communities in four tropical regions in two-dimensional ecological space as determined by principal components analysis. PCA factors 1 and 2 account for 53% of the variation found in Fleagle and Reed's (1999) data set. Biological interpretations of factors 1 and 2 are shown along the *top* and *right-hand side* of the figure.

(PCA) to summarize these differences in two-dimensional ecological space (fig. 2.12). In terms of overall ecological space, primate communities in Africa and Madagascar occupy the largest areas, followed by Asia and South America. Only the ecospace of South American communities is completely included within that of the other communities. African communities are unique in the presence of many leaping species and a high degree of animal eating; Madagascan communities are unique in degree of leaping and folivory; and Asian communities are unique in the presence of large species and degree of frugivory (fig. 2.12). Despite these striking intercontinental differences, examples of ecological convergence can also be found in these communities. These include (1) cheirogaleids, lorises, and some platyrrhines (i.e., marmosets and tamarins), which overlap ecologically as small arboreal fruit- and insect-eating quadrupeds; (2) colobines and some indriids, which overlap as medium-sized, arboreal leaping folivore/frugivores; and (3) several platyrrhines (i.e., cebids), cercopithecines, and lemurids, overlap as medium-sized arboreal, quadrupedal frugivores. Niche overlap between potentially competing species varies among biogeographic regions. In Africa and Asia, potentially competing species often differ in their vertical use of forests whereas in the Neotropics and Madagascar, potentially competing species differ most strongly in diet (Schreier et al. 2009). As we will discuss in more detail later, differences in ecological diversity among regions probably reflect differences in the length of time primates have been evolving in each area as well as differences in the constraints on adaptive radiation they have experienced in these regions.

PATTERNS OF VERTICAL STRATIFICATION

A striking feature of the structure of tropical forests is the vertical stratification of plants and their different growth forms (Richards 1952), and a similar vertical structure also characterizes most of the animals living in these habitats. In tropical lowland forests, at least, different sets of species of nectarivores and frugivores forage in different strata of the forest. In lowland Neotropical forests, for example, vertical stratification of hummingbirds tends to reflect a basic dichotomy between the two subfamilies of Trochilidae—hermits (Phaeornithinae) and trochilines or "typical" hummingbirds (Trochilinae). Hermits, which generally have long decurved bills, are primarily understory foragers, whereas trochilines, which usually have short straight bills, forage in the canopy as well as in the understory (Fleming et al. 2005; Stiles 1981). Furthermore, whereas hermits are generally restricted to lowland forests, trochilines occur over a wide elevational range on the mainland and are the only hummingbirds on noncontinental Caribbean islands (Stiles 2004). A similar sharp distinction between understory- and canopy-feeding nectar-eating phyllostomid bats does not appear to exist in lowland Neotropical forests, perhaps because most bat flowers are produced by canopy or subcanopy trees, vines, and epiphytes (Fleming et al. 2005; Kalko and Handley 2001; Tschapka 2004).

Neotropical frugivores can also be roughly classified as "understory" or "canopy" feeders. Understory avian frugivores include manakins (Pipridae), the tyrannid flycatcher *Mionectes oleaginous*, and certain tanagers (e.g., species of *Chlorothraupis*, *Eucometis*, *Habia*, and *Tachyphonus*; Stiles and Skutch 1989) that feed heavily on fruits produced by the Marantaceae, Melastomataceae, and Rubiaceae, among other families. Understory frugivorous phyllostomid bats include species of *Glossophaga* (which also visit flowers), *Carollia*, and *Sturnira*, which feed heavily on fruit produced by shrubs and treelets of the Clusiaceae, Piperaceae, and Solanaceae. Canopy-feeding avian frugivores (and their chiropteran counterparts) are generally larger than understory feeders and include toucans, trogons, cotingids, and tanagers (fig. 2.13). Phyllostomid bats of the subfamily Stenodermatinae (e.g., *Artibeus*, *Chiroderma*,

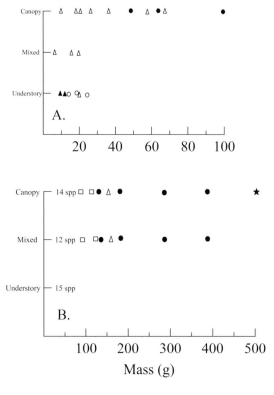

Figure 2.13.   The distribution of sizes of fruit-eating bats (*A*) and birds (*B*) as a function of vertical stratum in a Costa Rican tropical wet forest. Symbols in *A* include phyllostomid bats in subfamilies Carolliinae (*open circles*), Glossophaginae (*solid triangles*), Phyllostominae (*solid circles*), and Sternodermatinae (*open triangles*). Symbols in *B* include Coerebinae (*solid triangles*), Cotingidae (*open triangles*), Cracidae (*solid star*), Pipridae (*solid squares*), Ramphastidae (*solid circles*), Thraupinae (*open circles*), and Trogonidae (*open squares*). Based on Fleming (1988).

Uroderma, and *Vampyrops*) feed heavily on figs and other species of Moraceae and are primarily canopy feeders (Bernard 2001; Bonaccorso 1979; Kalko and Handley 2001). Compared with bats and most birds, frugivorous Neotropical primates are large animals (weighing up to about 10 kg), which restricts them to the canopy and subcanopy for feeding. Only small species such as squirrel monkeys, marmosets, and tamarins, which tend to be highly insectivorous, forage in the understory of lowland forests (Terborgh 1983). All told, over 60% of the biomass of frugivorous birds and mammals in these forests is found in the forest canopy (see chap. 3).

Similar patterns of vertical stratification occur in Old World tropical forests. Understory nectarivores in Africa and Asia include sunbirds (occasionally—they are mostly canopy feeders) and honeyeaters (Australasia) and small pteropodid bats (e.g., *Megaloglossus* in Africa, *Syconycteris* and *Macroglossus* in Australasia; Bell 1982; Karr 1976; Mickleburgh et al. 1992). Karr (1980) has pointed out that, compared to Panama, forests in Liberia, Gabon, and Malaysia contain fewer species of understory frugivorous birds, although all of these forests contain rich sets of canopy frugivores. The principal understory avian frugivores in Ghana and Liberia are bulbuls; they include bulbuls and flowerpeckers in lowland Malaysia, and flowerpeckers and small honeyeaters in Australasia (Bell 1982; Karr 1976; Wong 1986). Old World primates are vertically arrayed in tropical forests (see illustrations in Fleagle and Reed [1999]), again primarily by size, and also include large terrestrial/arboreal species (baboons, macaques, chimpanzees, and gorillas) that are missing from the New World. Also missing from the understory of Neotropical forests are fruit-eating mouse deer (Tragulidae) and elephants.

Old World canopy nectarivores include relatively few species. Principal families of birds include sunbirds in Africa and Asia and lorikeets, sunbirds, and honeyeaters in Australasia. Pteropodid bats (including *Eidolon*, *Rousettus*, and *Epomophorous* in Africa; *Eonycteris* and *Cynopterus* in Asia; and *Pteropus* in Asia and Australasia) are the principal nectar-feeding mammals in the canopy. Compared with their understory counterparts, Old World canopy frugivores are taxonomically more diverse in terms of numbers of species and families and are larger in size. Turacos, hornbills, and most species of starlings and bulbuls are canopy-feeding frugivores in Africa. Asian canopy-feeding birds include hornbills, bulbuls, barbets, and flowerpeckers, whereas Australasian birds include fruit pigeons, hornbills (New Guinea only), birds-of-paradise, bowerbirds, bulbuls (New Guinea only), and certain honeyeaters. Canopy-feeding mammals include cercopithicine monkeys and apes (in Africa and Asia), viverrid carnivores (primarily in Asia), phalangerid marsupials (in Australasia), as well as many pteropodid bats.

A detailed analysis of the frugivorous birds and mammals that feed on *Ficus* fruits in Lambir Hills National Park, Sarawak, East Malaysia, illustrates how species are arrayed vertically in an Asian tropical forest (Shanahan and Compton 2001). At least 80 species of figs are found at this site, and fig fruits occur from the ground (e.g., geocarpic species) to above the canopy in emergent trees (e.g., hemiepiphytic species). Their fruits range in size from 5 mm to >50 mm in diameter. At least 58 species of vertebrates, including 43 birds and 15 mammals, were observed eating figs during diurnal and nocturnal censuses. Shanahan and Compton (2001) recognized three major guilds of figs—one understory guild whose fruit were eaten primarily by bulbuls and a few arboreal mammals and two canopy guilds whose fruit were eaten primarily by fruit pigeons or a mixed group of birds and mammals, respectively. They reported that 32 of 52 (62%) canopy-feeding birds and mammals were restricted to that forest layer whereas only two of 22 (9%) understory species fed exclusively in the understory. Fruits of canopy figs were larger than understory figs, and canopy trees attracted more diverse assemblages of frugivores owing to their large fruit crops.

Similar patterns of stratification were reported for a montane moist forest in Kenya by Schleuning et al. (2011). The diurnal frugivore community here included 83 species of birds, four monkeys, and one squirrel that were grouped into three feeding classes: obligate frugivores ($n = 14$ species), partial frugivores ($n = 28$), and opportunistic frugivores ($n = 46$). Obligate frugivores (e.g., bulbuls, monkeys) fed mostly in the canopy and had relatively generalized fruit diets including many species; opportunistic frugivores (e.g., willow warbler, African yellow white-eye) fed mainly in the understory and had relatively narrow fruit diets. Fruit crops were largest in canopy trees and attracted the greatest number of frugivores.

Community Assembly and Organization

The S and ecological diversity of communities of tropical plant–visiting birds and mammals differ considerably among biogeographic regions and, within these regions, with latitude, elevation, and rainfall, among other factors. Much of this variation parallels variation in the S of their plant communities, and it would be important to know the extent to which variation in animal diversity is correlated with, and is possibly caused by, variation in the diversity of their food plants. The existence of such a correlation would imply that quantitative assembly rules (*sensu* Weiher and Keddy 1999) exist such that knowledge of the number of hummingbird-pollinated plants in a habitat, for example, would allow us to be able to predict how many hummingbird species to expect to find there and vice versa. In addition to knowing whether such correlations exist, we might want to know whether the same quantitative relationships hold for all communities of tropical nectar and fruit eaters, regardless of biogeographic region. If they do, this implies

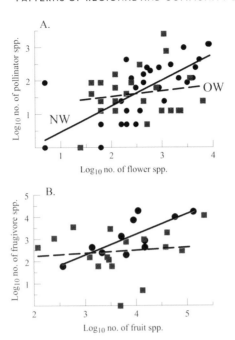

Figure 2.14. Relationship between the species richness of *A*, nectar-eating birds and bats (combined) and *B*, fruit-eating birds and bats (combined) and their food plants in New World (*NW*) and Old World (*OW*) communities. Based on Fleming (2005).

that there is a universal assembly rule in communities of tropical vertebrate mutualists, despite their substantial geological and historical differences. If they do not, then historical contingencies must have played an important role in the assembly of these communities.

To address these questions, Fleming (2005) analyzed data on the number of species of nectar- or fruit-eating birds and/or bats and the number of their flower or fruit species in a total of 87 community studies from around the world. Results indicated that the number of species of plant-visiting birds and bats per community was significantly correlated with number of species of their food plants in both nectarivores and frugivores in the New World but not in the Old World (fig. 2.14). Furthermore, the slopes of the regression equations in the New World were the same for nectarivores and frugivores and produced the following quantitative assembly rule: It takes an average of three species of flowers or fruits to support one species of nectar- or fruit-eating bird or bat (i.e., the 3:1 rule). No such assembly rule appears to exist in the Old World tropics, whose geological history is much more diverse than that of the New World. One implication of this rule is that feeding relationships between plant-visiting birds and bats and their food plants are likely to be more specialized, on average, in the New World than in the Old World. That is, the average Neotropical flower or fruit will likely be visited or eaten by a low and predictable number of mutualist species whereas the number of mutualist species that will visit or eat the average paleotropical flower or fruit cannot easily be predicted. We will discuss why this might be the case and its conservation implications in chapters 9 and 10.

In addition to community assembly rules, we need to know how communities of mutualists are organized in terms of connectance (i.e., how many species does a particular species interact with?) and dependency (i.e., how important is a particular pollinator or frugivore for the reproductive success of a plant and vice versa for animal consumers? [Jordano 1987]). Are communities of tropical plant–visiting vertebrates dominated by feeding generalists characterized by a high degree of connectance with all of their potential food plants or by feeding specialists with a low degree of connectance with these plants? Are particular interspecific dependencies generally strong or weak and are they symmetrical (i.e., equally strong for both plants and animals) or asymmetrical? Based on our current knowledge (Bascompte et al. 2006; Jordano et al. 2003; Thompson 2006), communities of pollination and seed dispersal mutualisms appear to have the following characteristics: (1) they form nonrandom, nested networks of interactions containing a core group of generalists (i.e., animals that feed at many species rather than a few species of flowers or fruits or plants that are pollinated or dispersed by several species rather than by one or two species of animals) that interact with each other and a group of specialists (i.e., species that feed on or are visited by a low number of species) that interact with generalists; and (2) interspecific dependencies are generally weak and highly asymmetrical.

We illustrate these concepts with two Neotropical examples—a frugivore network of bats and plants in Costa Rican tropical dry forest and a pollination network of hummingbirds and plants in tropical moist forest in southeastern Brazil (Buzato et al. 2000; Fleming 1988). A common way of portraying these networks is in the form of a bipartite graph in which species of plants and their mutualists are ordered from most common to least common (in terms of diet breadth for animals and number of visitors or consumers for plants) and lines are drawn showing known feeding relationships; generalists are thus at the top of the graphs and specialists are at the bottom. In the Costa Rican example, three species of bats (*Carollia perspicillata*, *Glossophaga soricina*, and *Artibeus jamaicensis*) dominate the frugivore community, and the three most common fruit sources in bat diets are *Cecropia peltata* and two species of *Piper*. In addition to feeding on the common fruit species, the common bats also feed on uncommon fruit sources (the fruit specialists) whereas the uncommon bats feed nearly exclusively on the common fruit species. The same pattern holds in the Brazilian example: the common hummingbird species (*Ramphodon naevis*, *Thalurania glaucopis*, and *Phaethornis euronome*) feed on both common and uncommon flower species whereas the uncommon hummingbirds feed only on the common flower species.

Instead of portraying these communities as bipartite graphs, however, we feel that a more instructive way to illustrate their structure is via energy-minimization graphs. According to Dr. Marco Mello, who helped us prepare these graphs, energy-minimization graphs are drawn based on the following rule: species with more links to other species (the generalists) are placed in a central position whereas species with fewer links (the specialists) are placed closer to the periphery (Mello et al. 2011a). In addition, when a species is linked with both central and peripheral species, its position is adjusted to occur between the center and periphery of the graph. These graphs are drawn so as to minimize the entropy or random energy in the system, hence the name energy minimization.

Energy-minimization graphs for the two communities are shown in figure 2.15. A glance at the two graphs shows that their structure differs strikingly in a way that we might expect: the community of frugivorous bats is much less compartmentalized (i.e., it is more generalized) than the hummingbird community, which is dominated by clusters of highly connected species. The bat community contains a series of five centrally located generalist species that eat many species of fruit and four peripheral species that also eat fruit eaten by the generalists. In contrast, the hummingbird community contains three obvious clusters, including two centered on a hummingbird species and one centered on a plant. There is one generalist hummingbird and many specialists in this community.

We can get a general feel for degree of interspecific dependency within these networks by plotting the frequency distributions of number of fruit or flower species eaten by each consumer species and number of species of seed dispersers or pollinators that visit each plant species. Low numbers of

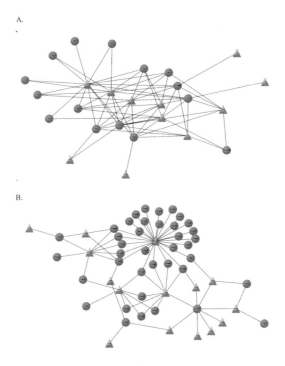

Figure 2.15. Two tropical mutualistic networks portrayed as "energy minimization" graphs (see text and Mello et al. 2011a). *A,* The network of fruit-eating bats (*triangles*) and their food plants (*circles*) in Costa Rican tropical dry forest (Fleming 1988, app. 8). *B,* The network of hummingbirds (*triangles*) and their food plants (*circles*) in a Brazilian tropical moist forest (site Cag in Buzato et al. 2000, app. 1).

food species or visitors per species signify stronger dependencies than high numbers. As we might expect knowing that plant-pollinator interactions generally tend to be more specialized than plant-frugivore interactions (e.g., Feinsinger 1983; Wheelwright and Orians 1982), diets of the Costa Rican fruit-eating bats tend to contain a higher number of species, on average, than diets of the Brazilian hummingbirds (fig. 2.16). The number of fruit species per bat species is multimodal with modes at one, two, six, and nine species (fig. 2.16*A*) whereas the number of flower species per hummingbird species is unimodal with a mode at one species (fig. 2.16*B*). Reflecting the fact that these mutualist networks tend to contain asymmetric interspecific relationships (Vazquéz and Aizen 2004), modal values for the number of animal species visiting each plant species in both networks are strongly unimodal with modal values of one (flowers) or two (fruits) species (fig. 2.16). That is,

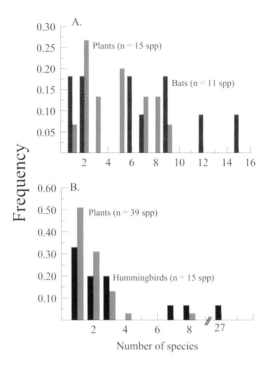

Figure 2.16. The "diet breadths" of species of mutualists and their food plants in the two networks in figure 2.15. A, The frugivorous bat network in a Costa Rican tropical dry forest. B, The hummingbird network in a Brazilian tropical moist forest. For plants, "diet breadth" indicates the number of animal species dispersing or pollinating them.

each plant species is dependent on a smaller number of species of pollinators or frugivores than their consumers.

We will discuss how these communities or networks of mutualists evolve in detail in chapter 9 but note here that this evolution is thought to involve two major processes: complementarity and convergence (Thompson 2006). By complementarity we mean the matching of phenotypic traits between plants and animals (e.g., flower corolla length and snout or bill length in pollinators; fruit hardness and jaw strength in frugivores). By convergence we mean the evolution of a set of similar phenotypic features in unrelated plants or animals (e.g., the evolution of pollination or frugivore syndromes in plants). As in the case of community assembly, it is important to know whether the structure of mutualist networks differs among hemispheres, but addressing this question in detail is beyond the scope of this book (but see

Schleuning et al. [2011] for an African example). This is a promising area for future research because of its important implications for ecology, evolution, and conservation.

Conclusions

Tropical forests contain an overwhelming diversity of vascular plants of many growth forms. They also contain an impressive array of birds and mammals that serve as their pollinators and/or seed dispersers. Many plant families, but few animal families, have pantropical distributions, which gives tropical forests in different biogeographic regions a relatively similar appearance. But, as stressed by Corlett and Primack (2011), differences between these forests, as least in terms of dominance by different groups of plants and animals, often outnumber their similarities, primarily because of different geological and evolutionary histories. Thus, each tropical region tends to have its own unique ecological characteristics.

Regional and community diversity of tropical plants and their vertebrate mutualists tend to vary in predictable fashion along a number of environmental gradients, including latitude, elevation, rainfall, soil fertility, temperature, disturbance regime, and within forests, height above the ground. This environmental variation provides an enormous amount of "adaptive space" for plants and animals in which to evolve and thus countless ways for groups of plants and vertebrate mutualists to interact. The existence of qualitatively different kinds of community assembly rules in different parts of the tropics suggests, however, that some of these interactions are sometimes constrained. Understanding the nature of these constraints—for example, to what extent is animal diversity constrained by plant diversity and vice versa?—is important, and the mutualistic interactions we discuss in this book provide a rich "model system" for beginning to understand in depth the nature of these kinds of evolutionary constraints. In the next several chapters, we will explore this system in detail, beginning with the food resource (energy) base and how it influences the biomass of vertebrate pollinators and seed dispersers in tropical ecosystems.

3   The Resource Base

Communities of tropical plant–visiting birds and mammals are supported by two distinctly different kinds of resources, nectar (and pollen in some cases) and fruit. The nutritional quality and availability of these resources can be highly variable in time and space. In this chapter, we describe the basic nutritional characteristics of the nectar and fruit resource bases and how the availability of these resources varies temporally and spatially. Included here will be quantitative estimates of the biomass of these resources within communities and the biomass of vertebrate nectarivores and frugivores that they support. One fundamental question that these data can address is, To what extent is the biomass of plant-visiting birds and mammals limited by the availability of their plant resources? That is, are populations of these animal mutualists ultimately limited by resource availability in "bottom-up" fashion? Results of the community assembly analysis presented in chapter 2 suggest that the answer to this question may be yes in New World bird and bat communities but possibly no in Old World communities. A positive answer to this question has profound conservation implications because any reduction in plant diversity and biomass in tropical habitats is likely to have a negative effect on populations of nectar- and fruit-eating birds and mammals. If, conversely, the answer to this question is negative, then partial reductions in plant diversity and biomass in these habitats will not necessarily have a negative effect on populations of vertebrate mutualists, although total elimination certainly will. Based on the results of the community assembly analysis, we might expect to find a closer relationship between resource and consumer biomasses in the Neotropics than in the Paleotropics.

To set the stage for our examination of resource-related questions, we will first summarize data on the extent to which tropical plants rely on birds and mammals as their primary or exclusive pollinators and seed dispersers. What proportion of regional or local floras is vertebrate-pollinated or vertebrate-dispersed, and to what extent does this dependency vary among plant growth forms, successional stages, and biogeographic regions? Answers to these questions will indicate the extent to which plants depend on vertebrates for pollination and seed dispersal services and where these food resources are located in tropical habitats.

Proportions of Vertebrate-Pollinated or -Dispersed Plants in Tropical Habitats

In chapter 2 we saw that lowland tropical communities typically contain hundreds of plant species of many growth forms. These plants depend to a much greater extent on vertebrates as seed dispersers than as pollinators. The early pollination systems of angiosperms involved insects, and insect-pollinated plants continue to dominate contemporary tropical floras (e.g., Bawa 1990; Pellmyr 2002; Proctor et al. 1996). Thus, as discussed in chapter 1, it is not surprising that the proportion of plants that rely on birds and mammals for pollination in tropical habitats is generally low. Data for five wet tropical forest sites (table 3.1) illustrate this point and indicate that the frequency of bird or bat pollination ranges are 0–10% and 0–7%, respectively, in these forests. As is true of many islands (Givnish 1998; Traveset 1999), Jamaica lacks bee-pollinated trees and has an especially high frequency of wind pollination compared with the four mainland sites. Compared with these wet forest sites, the frequency of hummingbird and bat pollination (15% and 13%, respectively; $n = 142$ species) is somewhat higher in the shrubby caatinga vegetation of northeastern Brazil (Machado and Lopes 2004).

Pollination syndromes often differ among floral strata and plant growth forms. Thus, in the lowland wet forest at La Selva, Costa Rica, hummingbird pollination is much more frequent in the understory than in the subcanopy and canopy (24% vs. 8% of 276 species, respectively), whereas the opposite situation is true for bat pollination (1.3% vs. 12% of these species, respectively; Kress and Beach 1994). At this site, understory hummingbird-pollinated flowers occur in the Acanthaceae, Costaceae, Gesneriaceae, Heliconiaceae, and Zingiberaceae; canopy and subcanopy bat-pollinated flowers occur

Table 3.1. Summary of the Major Pollination Modes in Five Wet Tropical Localities

Pollination Mode	Percentage of Species at Each Location				
	Kakachi, India	Lambir, Borneo	La Selva, Costa Rica	Jamaica, West Indies	Venezuela
Wind	6	0	1	11	0
Bees	18	33	41	0	13
Other insects	71	50	51	89	77
Birds	0	5.5	4	4	10
Bats/mammals	3	4	7	0	0
Total species	86	127	120	55	38

Source. Based on Devy and Davidar (2003).
Note. Data are for trees and shrubs and exclude epiphytes and vines. "Bees" includes large and small bees; "other insects" includes beetles, moths, butterflies, wasps, flies, thrips, and bugs. Total species indicates the number of species, not a percentage.

primarily in the Malvaceae (Bombacoideae) but also include epiphytes in the Bromeliaceae. In Costa Rican tropical dry forest, Fleming et al. (2005) noted that hummingbirds pollinate at least 16 species of plants (three herbs, two vines, three shrubs, and eight trees), whereas bats pollinate at least nine species (one shrub and eight trees). At another well-studied site in Costa Rica—premontane wet forest at Monteverde—birds and bats, respectively, are thought to pollinate the following percentages of plants: epiphytes—24.8:% and 2.8% (n = 322 species); herbs—13.4% and 0.3% (n = 336); vines and lianas—6.6% and 0.8% (n = 318); shrubs—6.7% and 0.8% (n = 374); and trees—1.4% and 1.6% (n = 706; Murray et al. 2000). In New World tropical forests, therefore, birds generally pollinate plant growth forms other than trees whereas bats generally pollinate trees. In Old World tropical forests, in contrast, both birds and bats pollinate canopy and subcanopy trees (Corlett 2004; Dobat and Peikert-Holle 1985; Fleming and Muchhala 2008; Mickleburgh et al. 1992). In addition, hemiepiphytic mistletoes (Loranthaceae) and understory gingers (Costaceae, Zingiberaceae) are important nectar sources for sunbirds in Bornean lowland forests (Sakai 2000; Yumoto et al. 1997).

It is important to note that estimates of the frequency of different pollination syndromes at sites such as La Selva and Monteverde are based to a large extent on floral morphology and not on detailed pollination studies, including pollinator exclusion experiments, such as those that have been conducted in the Canopy Biology Plot in Lambir Hills National Park, Sarawak, Malaysia (e.g., Momose et al. 1998). While floral morphology is often a reasonably good predictor of the identity of effective pollinators, it is certainly not foolproof (Fenster et al. 2004; Heithaus 1982; Newstrom and Robertson 2005; Ollerton et al. 2009). Consequently, incorrect assumptions can be made about the importance of particular pollinators, and these assumptions, in turn, can sometimes lead to incorrect conservation concerns and decisions. For example, columnar cacti of the Sonoran Desert, which is a subtropical extension of Central American tropical dry forest located in northwestern Mexico and southern Arizona, produce large, light-colored and nocturnally opening flowers that conform to the classic bat pollination syndrome (Valiente-Banuet et al. 1996). Pollinator exclusion experiments indicate, however, that only one of three species (cardon, *Pachycereus pringlei*) relies heavily on the nectar-feeding phyllostomid bat *Leptonycteris yerbabuenae* for fruit set. The important pollinators of the other two species are diurnal birds such as white-winged doves and hummingbirds and bees (Fleming et al. 2001). Because it was incorrectly assumed that certain columnar cacti and paniculate agaves are exclusively bat-pollinated and that populations of *L. yerbabuenae* have declined in recent decades (e.g., Howell and Roth 1981), the US Fish and Wildlife Service declared *L. yerbabuenae* to be a federally endangered species in 1988 and voiced concern about the "collapse of the Sonoran Desert ecosystem" if this bat should go extinct (Shull 1988). The general message from this (extreme?) case of reliance on plant floral characteristics to make sweeping conservation decisions is that one must interpret plant pollinator syndromes cautiously. As mentioned in chapter 1, plants and their animal mutualists do not always conform to ecological expectations.

As expected, the proportion of tropical plants with bird- or mammal-dispersed fruits is much higher than plants pollinated by these vertebrates. Table 3.2 summarizes the frequency of different vertebrate fruit syndromes in five tropical wet forests (Ganesh and Davidar 2001). It should be noted that these frequencies apply only to vertebrate dispersal and ignore other dispersal mechanisms. In tropical habitats, the frequency of vertebrate dispersal ranges from 35% to 40% of all species for a variety of plant growth forms on Mount Kinabalu, Borneo, to 98% for subcanopy trees at La Selva, Costa Rica (Howe and Smallwood 1982). Except for Gabon, Africa, the percentage of species dispersed by birds at these sites is substantially higher than that dispersed by mammals (table 3.2). Willson et al. (1989) reported a similar disparity between the frequencies of bird and mammal fruits in their survey of fruit colors in tropical habitats. In addition, they pointed out that (1) the proportion of fleshy-fruited plants within habitats declines with decreasing rainfall and increasing latitude (also see Howe and Smallwood

Table 3.2. Summary of the Major Fruit-Dispersal Modes in Five Wet Tropical Locations

Forest Type/ Frugivores	Percentage of Species at Each Location				
	Hong Kong, SE Asia (Lowland Wet Forest)	Malawi, Africa (Montane Wet Forest)	Gabon, Africa (Lowland Wet Forest)	La Selva, Costa Rica (Lowland Wet Forest)	Kakachi, India (Midelevation Wet Forest)
Birds	75	94	26	50	60
Mammals	10*	6	48[†]	37	26
Others	15[‡]	0	38[§]	13[‖]	14
Total species	153	52	122	167	82

Source. Based on Ganesh and Davidar (2001).
Note. Total species indicates the number of species, not a percentage.
*Includes monkeys, bats, and civets.
[†]Includes only monkeys.
[‡]Includes other kinds of generalist birds and mammals.
[§]Includes only rodents.
[‖]Includes only bats.

Table 3.3. Percentages of Flowering Plant Species Dispersed in the Cloud Forest of Monteverde, Costa Rica

Dispersal Mode	Epiphytes	Herbs	Vines	Shrubs	Trees
Birds	28	20	43	52	65
Birds and bats	0	1	2	14	1
Bats	1	2	2	14	2
Arboreal mammals	0	0	8	0	9
Terrestrial mammals	0	0	0	1	5
Wind	65*	36	33	15	14

Source. Based on data in Murray (2000, fig. 8.3).
*Many orchids.

1982); (2) understory shrubs and treelets tend to include a higher percentage of fleshy-fruited species than canopy trees in many tropical forests; and (3) in some Neotropical forests, bird dispersal is more common among shrubs and treelets than in canopy trees, whereas dispersal by nonvolant mammals is more common in the canopy than in the forest understory. Data that run counter to the third generalization come from Monteverde, Costa Rica, where dispersal modes have been reported by plant habit (Murray 2000). At this site, birds are the primary animal dispersers for all growth forms and disperse a higher percentage of tree species than shrubs (table 3.3). Finally, early successional plants are often dispersed by bats and birds in many tropical forests (e.g., Charles-Dominique 1993; Foster et al. 1986; Levey 1988). A review of the role of frugivorous bats in tropical succession, however, reveals that New World phyllostomid bats likely play a more important role in early forest succession than do Old World pteropodid bats (Muscarella and Fleming 2007). In the Paleotropics, birds are the most important vertebrate dispersers of early successional seeds (e.g., Gonzales et al. 2009; Ingle 2003). We will return to this topic in chapter 4.

Nutritional Characteristics of Floral and Fruit Resources

NECTAR AND POLLEN

The nutritional reward for most flower-visiting animals is nectar, which is an aqueous solution containing one to three main types of sugar—the monosaccharides glucose and fructose and the disaccharide sucrose—as well as a variety of minor sugars such as maltose, mannose, and lactose (Lotz and Schondube 2006). By dry weight, nectar is about 90% sugar; the remaining 10% includes a diverse array of substances, including amino acids, lipids, antioxidants, mineral ions, and secondary compounds (Adler 2000). Amino acids are often present in nectar but only in concentrations that are about 1% of those of sugars; they tend to occur in higher concentrations in the nectar of butterfly-pollinated flowers (>1 μmol/mL) than in bird- or bat-pollinated flowers (<0.5 μmol/mL; Baker and Baker 1983). As discussed in chapter 7, vertebrate pollinators have alternative sources of proteins and amino acids (e.g., insects, leaves) and hence do not usually rely on nectar for these nutrients.

Both the amount of nectar per flower and its composition vary widely and are generally correlated with flower size and type of pollinator (Baker and Baker 1983; Pacini et al. 2003; Proctor et al. 1996). In Costa Rican tropical dry forest, for example, flowers pollinated by either bats or hawkmoths are substantially larger and contain much more nectar than those pollinated by hummingbirds (fig. 3.1). Detailed studies of hummingbird- and bat-pollinated plants at three sites in the Atlantic rainforest of Brazil indicate

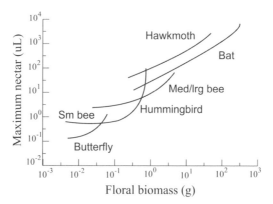

Figure 3.1.   Relationship between flower size and nectar volume in Costa Rican plants belonging to different pollination syndromes. Based on Opler (1983).

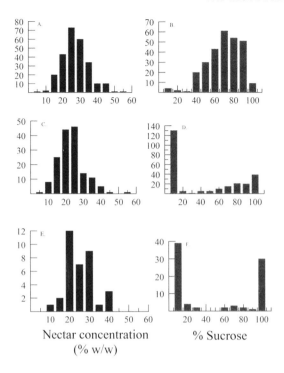

Figure 3.2.   Frequency distributions of nectar concentration (% weight/weight sugar) and sugar composition (% total sugar as sucrose) for plant species pollinated by hummingbirds (*A, B*), sunbirds (*C, D*), and honeyeaters (*E, F*). Based on Nicolson and Fleming (2003).

that these plants are generally similar in three floral parameters—maximum number of flowers per plant at peak flowering, corolla length, and sugar concentration—but differ strikingly in nectar volume per flower (table 3.4). In these plants, sugar concentration is somewhat higher in hummingbird flowers than in bat flowers (median values are 23.2% vs. 17.0%, respectively; also see fig. 3.2), and the range of values in hummingbird plants is twice that of bat plants. In contrast, the nectar volume of bat flowers is nearly 10 times greater than that of hummingbird flowers (median values are 116.6 µL vs. 15.6 µL, respectively; Buzato et al. 2000; Sazima et al. 1999). Data in table 3.4 were pooled over many plant families. When we compare values for just the Bromeliaceae, we again find similar values for hummingbird- and bat-pollinated species regarding maximum number of flowers per plant (means of 6.2 vs. 3.3, respectively; *n* species = 18 and 4; *P* = 0.50 in a two-sample *t*-test) and corolla length (33.0 mm vs. 35.4 mm; *P* = 1.00). Sugar concentration and nectar volume, however, differ significantly between the two

Table 3.4. Summary of Certain Floral Characteristics of Hummingbird- and Bat-Pollinated Plants at Three Sites in Brazilian Atlantic Rain Forest

Characteristic	Hummingbird Plants		Bat Plants	
	Median Values	Range	Median Values	Range
Maximum number of flowers per plant at peak flowering	7.5	2–500	10.0	2–100
Corolla length (mm)	32.0	9.3–79.3	25.1	11.8–63.6
Sugar concentration (% weight/weight)	23.2	11.0–44.2	17.0	10.2–21.6
Nectar volume (μL)	15.6	1.8–150.0	116.6	11.5–408.8

Source. Based on data in Buzato et al. (2000) and Sazima et al. (1999).
Note. For hummingbird plants, n = 58 species in 21 families; for bat plants, n = 15 species in 10 families.

groups. Sugar concentration is higher in hummingbird flowers (29.0% vs. 18.5%; $P < 0.001$), but nectar volume is again markedly higher in bat flowers (131.4 μL vs.19.6 μL; $P < 0.001$). A similar difference in sugar concentration and nectar volume between hummingbird- and bat-pollinated species occurs in tribe Sinningieae of Gesneriaceae (Perret et al. 2001). Overall, New World bat flowers produce, on average, over four times more sugar per flower than hummingbird flowers (Cruden et al. 1983). From a plant's perspective, therefore, glossophagine bats appear to be more expensive pollinators, at least in terms of volume of nectar and sugar produced per flower, than hummingbirds. This cost difference probably reflects, in part, differences in the average size (mass) of these two kinds of pollinators (mean hummingbird mass = 5.1 ± 0.2 (SE) g; mean glossophagine mass = 13.3 ± 0.9 g; table 1.1 and Fleming et al. 2005).

Old World avian and chiropteran nectar feeders are often substantially larger than their New World counterparts (table 1.1 and Fleming and Muchhala 2008), so we might expect their flowers to be larger and to produce greater amounts of nectar and energy. Unfortunately, except for a substantial research effort on the feeding ecology of honeyeaters (Meliphagidae) in Australia (reviewed in Armstrong 1979; Ford et al. 1979; Franklin and Noske 2000; Paton 1986), relatively little work has been done on Old World nectar-feeding birds and mammals. As a result, our knowledge about quantitative aspects of the floral biology of their food plants is spotty. Available data, however, support the generalization that, as in the New World, Old World bird- and bat-pollinated flowers are larger and contain larger nectar or energy rewards compared with other pollination syndromes. For example, Sakai et al. (1999) used a multivariate approach to classify 29 species of gingers (Zingiberaceae and Costaceae) in a lowland Bornean forest into three pollination guilds: a sunbird-pollinated guild containing eight species;

an anthophorid bee guild (11 species); and a halictid bee guild (10 species). Compared with flowers in the other two guilds, sunbird-pollinated flowers had significantly longer corolla tubes (55.6 mm vs. 36.4 mm in anthophorid flowers), differed in color, had slightly lower sugar concentrations in their nectar (26% vs. 27%–29% in the bee guilds), and had much higher sugar mass per inflorescence (149 mg vs. 2.5–6.1 mg). In the same forest, two species of sunbirds pollinate three species of hemiepiphytic mistletoes (Loranthaceae) whose floral characteristics include: corolla length = 45.6–167.8 mm; nectar per flower at 0800–0900 hours = 2.8–149.4 µL; sugar concentration = 8.0%–17.5%; and energy/flower = 3.3–404.2 joules (Yumoto et al. 1997). Interestingly, the large-flowered, large-reward mistletoe (*Trithecanthera xiphostachys*) occurs at much lower densities than the other two species at this site, which supports the hypothesis of Heinrich and Raven (1972) that low-density plants might be expected to provide larger nectar rewards to attract pollinators than high-density plants. Finally, both sunbirds and bats pollinate species of wild banana (*Musa*) in lowland Asian forests that exhibit significant differences in floral presentation, floral structure, and nectar rewards (Itino et al.1991). Regarding floral and nectar characteristics, the bat-pollinated *Musa acuminata* (section *Musa*) has a longer perianth (45 mm vs. 26 mm), higher sugar concentration in its nectar (24% vs. 20%), but a lower rate of nectar production (7.9 µL/h vs. 18 µL/h) than the sunbird-pollinated *M. salaccensis* (section *Callimusa*).

Along with brush-tongued parrots (lorikeets; Loriinae), honeyeaters (Meliphagidae) are the major avian pollinators in forests and shrublands of Australasia (Brown and Hopkins 1996; Ford et al. 1979; Franklin and Noske 2000). Unlike African and Asian sunbirds, which, like hummingbirds, often pollinate flowers with tubular corollas (Fleming and Muchhala 2008), honeyeaters and lorikeets often pollinate flowers that are small and cup-shaped or brushlike and that are produced in substantial numbers per inflorescence (e.g., flowers of Myrtaceae and Proteaceae). These flowers individually may offer small nectar rewards, but collectively they offer substantial rewards per inflorescence and per plant. Data for 17 species of honeyeater-pollinated flowers from southeastern Australia illustrate this trend (table 3.5). These data indicate that trees in the Myrtaceae and Proteaceae can indeed be rich sources of nectar. When in full flower, they are visited by a large and colorful array of honeyeaters and lorikeets (Brown and Hopkins 1996; Franklin and Noske 2000).

Pteropodid bats are the main mammalian pollinators in the Old World.

Table 3.5. Summary of Daily Flower and Nectar Production in Honeyeater-Pollinated Plants from Southeastern Australia

Family (Growth Habits)	No. of Species	Range of Values		
		Number of Flowers/Day	Joules/Flower	kJoules/Plant/Day
Myrtaceae (trees, shrubs)	6	100–11,500	7–250	25–1,400
Proteaceae (trees, shrubs)	3	3–1,000*	14–3,100*	9–51
Loranthaceae (hemiepiphytes)	3	50–500	15–106	0.8–10
Rutaceae (shrubs)	2	14–35	14–15	0.2–0.5
Epacridaceae (shrubs)	3	15–200	2–42	0.2–2.0

Source. Based on data in Paton (1986).
*These values are for inflorescences.

Table 3.6. Summary of the Floral and Nectar Characteristics of Species at Two Asian Sites

Location/Pollinator Species (Family)	Flower Size (diameter in cm)	No. of Flowers/ Cluster or Inflorescence	Maximum No. of Flowers/ Tree/ Night	Nectar Volume/ Flower/ Night (µL)	Sugar Concentration (%)	Total kJ/ Cluster/ Night
Borneo:						
*Durio grandiflorus** (Bombacaceae)	5–7	. . .	. . .	10.6	14	. . .
*D. oblongus**	8–11	. . .	. . .	2,600	11	. . .
D. kutejensis[†]	10–12	. . .	. . .	9,100	9	. . .
Malaysia:						
D. zibethinus[†]	. . .	1–18	>1,000	360		2.6–9.2
Oroxylum indicum[†] (Bignoniaceae)	>10[‡]	1	1–40	1,800	. . .	7.2

Sources. For Borneo—Yumoto (2000); for Malaysia—Gould (1978).
Notes. All of these species are canopy or subcanopy trees.
*Pollinated by sunbirds.
[†]Pollinated by bats.
[‡]Flower size given is length, not diameter.

With body masses that range from about 13 g to over 1,000 g, they are substantially larger than New World glossophagine nectar bats. Reflecting their larger size, pteropodids are much more likely to visit large, nectar-rich flowers produced by canopy or subcanopy trees than are glossophagines (Fleming and Muchhala 2008). Examples of the floral and nectar characteristics of some pteropodid-visited flowers are shown in table 3.6. In addition, Baker and Harris (1957) reported that the large flowers of *Parkia clappertoniana* (Fabaceae) in Ghana produce about 5,000 µL of nectar per night. These flowers produce much larger volumes of nectar than glossophagine flowers studied in Brazil (table 3.4). Overall, these data indicate that Old World bat flowers can be rich energy sources.

In addition to quantitative differences, nectars also differ in their qualitative composition, particularly in their sugars (reviewed by Baker et al. 1998; Lotz and Schondube 2006; Nicolson and Fleming 2003). Herbert and Irene Baker (1982) pointed out that the nectar of bird-pollinated plants tended to be either sucrose- or hexose- (i.e., glucose or fructose) rich, depending on whether they are pollinated by hummingbirds or passerines, respectively. We now know that many passerine-pollinated flowers also have sucrose-rich nectars and that only flowers pollinated by the sturnid-mimid clade of passerines (i.e., starlings and mockingbirds) lack sucrose presumably because these birds lack the digestive enzyme sucrase (Martínez del Rio 1990). Nectar of bat-pollinated plants worldwide tends to be hexose rich, as is the pulp of most vertebrate-dispersed fruits (Baker et al. 1998). Data on sugar concentration and composition for three groups of nectar-feeding birds are summarized in figure 3.2. These data show a striking difference in degree of sucrose dominance in flowers pollinated by hummingbirds, sunbirds, and honeyeaters. As originally proposed by the Bakers, hummingbird nectar is sucrose rich, whereas the nectar of flowers pollinated by the two groups of passerines has a bimodal distribution regarding percent sucrose. One mode occurs at <10% sucrose and the other mode occurs at >90% sucrose; honeyeater flowers are more strongly bimodal than sunbird flowers (Nicolson and Fleming 2003). We will discuss the evolutionary and physiological implications of these trends in chapter 7.

Finally, although we generally consider nectar to have positive nutritional value to attract animal pollinators, the presence of toxins such as nonprotein amino acids, phenolics, and alkaloids in nectar is not uncommon (Adler 2000). Baker (1978), for instance, reported that 50% of the floral nectars from 164 species of plants of all life forms in Costa Rica forests contained phenolics and 12% of 198 species contained alkaloids. Rhoades and Briggs (1981) suggested that toxic substances in nectar may function to reduce its attractiveness to ants, nectar thieves, or ineffective pollinators, but Adler's (2000) review found little support for any of these functions. The clearest demonstration that phenolic compounds can actually serve as a "pollinator filter" comes from a study of the pollination biology of three species of *Aloe* (Aloaceae) in South Africa (Johnson et al. 2006). Unlike sunbird-pollinated *Aloes*, which have long corollas and clear, sweet nectar, these species have short corollas and dark, bitter nectar that deters sunbirds from visiting them but attracts nonspecialist avian pollinators such as bulbuls, white-eyes, and chats. These short-corolla *Aloes* are more effectively pollinated by short-billed bulbuls than by long-billed sunbirds. Hence, toxic nectar in these plants has positive adaptive value by repelling ineffective pollinators.

Another potential source of energy and nutrients for some, but not all, flower-visiting vertebrates is pollen. Among birds, a few lorikeets are known to harvest pollen, and certain glossophagine bats (e.g., *Leptonycteris* species), the New Zealand nectar- and fruit-eating bat *Mystacina tuberculata* (Mystacinidae), and the Australian blossom bat (*Syconycteris australis*), and eastern pygmy possum (*Cercartetus nanus*) also consume and digest pollen (Daniel 1976; Howell 1974; Law 1992b; Richardson and Wooller 1990; van Tets and Hulbert 1999). Compared with nectar, pollen protoplasm is very nutritious. Its gross energy value (11.3 kJ/g) is somewhat lower than that of nectar (16.7 kJ/g), but it is relatively rich in protein and amino acids (16%–30% dry matter) in addition to containing minerals (e.g., P, K, Ca, Na), vitamins, and carbohydrates (Gartrell 2000; Roulston and Cane 2000).

FRUIT

As discussed in detail by van der Pijl (1982), Spujt (1994), and Lorts et al. (2008), the term "fruit" can have a variety of botanical meanings, reflecting the different ways in which plants produce fleshy structures surrounding or containing seeds. Fruit pulp, for example, can come from the ovary wall, seed coats (e.g., in arillate fruits), or floral bracts or receptacles, among other sources (Coombe 1976). Following Fleming (1991a), we will use "fruit" to denote any fleshy diaspore consumed by vertebrates. The most common kinds of diaspores we will consider are drupes, berries, and arils.

In general, fruit pulp—the nutritional reward for frugivores—is high in water and carbohydrates and low in protein and lipids with considerable variation within and between taxa; it also contains vitamins, carotenoids, amino acids, and minerals (Herrera 2002). Data summarized by Jordano (1992, table 5.3) allow us to compare the general nutritional value of fruit with that of insects (an alternate food for many plant-visiting birds and bats and small primates) and leaves (eaten by many frugivorous primates and some bats). These data, expressed as percentage of dry pulp mass, include: protein—fruit, 1.2%–24.5%; insects, 59.9%–75.9%; mature leaves, 7.1%–26.1%; lipids—fruit, 0.7%–63.9%; insects, 9.4%–21.2%; mature leaves, 0.7%–10.7%; and nonstructural carbohydrates—fruit, 5.6%–98.3%; insects, 0.5%–20.0%; mature leaves, 1.9%–14.7%. Because of the low protein content of most fruits, vertebrate frugivores often have to consume animals or leaves to meet their daily protein requirements (e.g., Kunz and Diaz 1995; Kunz and Ingalls 1994). Avian frugivores, for example, seldom feed fruit to their nestlings and instead feed them insects or small vertebrates (e.g., Morton 1973; Wheelwright 1983). The oilbird (*Steatornis caripensis*; Steatornithidae) of northern South America, which feeds the lipid-rich pulp of *Euterpe* and *Jessenia* palm fruits to its nestlings, is an exception (Snow 1962).

Some of the basic design features and nutritional characteristics of bird- or mammal-dispersed fruits are summarized in table 3.7. Compared with bird fruits, mammal fruits are significantly larger (mean fruit length = 29.3 mm vs. 11.9 mm), heavier (5.6 g vs. 1.5 g), and richer in total energy (17.0 kJ vs. 2.2 kJ), but are lower in lipids (8% vs. 15% of dry pulp mass; Jordano 1995). On average, a gram of fruit pulp contains 15–19 kJ of energy. As Jordano (1995) points out, much of the variation in these fruit characteristics can be accounted for by phylogenetic effects (at the family level and above). When the effect of phylogeny is removed from these comparisons, only size differences and their energetic consequences remain significant. Differences in fruit size and energy content again reflect the fact that mammalian frugivores tend to be larger than avian frugivores (table 1.1).

Lipids and proteins are of particular nutritional importance to frugivores, and these fruit traits vary considerably among plant families. Jordano's (1995) multivariate analysis indicates that there is a tradeoff between the lipid and nonstructural carbohydrate content of fruit pulp. Plant families whose fruits are especially lipid-rich include Arecaceae, Lauraceae, Myristicaceae, and some Anacardiaceae and Celestraceae. These families include canopy and subcanopy trees and, except for Arecaceae, most produce fruits eaten by birds rather than mammals (Snow 1981). Families whose fruit are especially rich in nonstructural carbohydrates include Apocynaceae, Caprifoliaceae, Flacourticeae (now included within Salicaceae), Melastomataceae, Myrtaceae, Rhamnaceae, Rubiaceae, Sapindaceae, and Sapotaceae. These families include understory and canopy and subcanopy plants that are dispersed by both birds and mammals. Finally, families with fruits with relatively high protein and mineral contents include trees (Meliaceae, Olacaceae) and understory shrubs, treelets, and herbs (Piperaceae, Zingiberaceae). Fruits of Solanaceae are notable for their high content of nonprotein nitrogen (see below).

Figs are an extremely important fruit source for birds and mammals throughout the tropics (Harrison 2005; Shanahan et al. 2001), and it is worthwhile to examine their nutritional characteristics in some detail. Although figs have sometimes been described as low-quality, sugary fruits with low nutritional value (e.g., Fleming et al. 1977; Herbst 1985; Morrison 1980b), this is not necessarily true. Wendeln et al. (2000), for example, analyzed the

Table 3.7. Summary of the Design and Nutrient Characteristics of Fruits Dispersed by Birds and Mammals

Trait/Characteristic	Bird Fruits		Mammal Fruits		Significant by ANOVA?	Significant after Removal of Phylogenetic Effects?
	Mean ± SE	No. of species	Mean ± SE	No. of species		
Design:						
Fruit length, mm	11.9 ± 0.44	266	29.3 ± 3.36	46	Yes	Yes
Fruit diameter, mm	10.2 ± 0.29	332	23.6 ± 2.7	48	Yes	Yes
Wet mass, g	1.5 ± 3.5		5.6 ± 2.9	3	Yes	No
Dry pulp, g	0.14 ± 0.02	238	0.64 ± 0.19	19	Yes	Yes
Seeds/fruit	8.9 ± 2.3	203	8.3 ± 4.1	24	Yes	No
Dry pulp/wet fruit mass	0.14 ± 0.0	234	0.20 ± 0.03	19	Yes	No
Nutrients:						
Energy/g pulp, kJ	17.9 ± 0.4	221	14.6 ± 0.6	59	Yes	No
Energy/fruit, kJ	2.2 ± 0.4	159	17.0 ± 4.9	11	Yes	Yes
Water in pulp, %	71.8 ± 1.0	259	72.3 ± 2.5	44	No	No
Lipids	0.15 ± 0.01	228	0.08 ± 0.02	60	Yes	Yes
Protein	0.06 ± 0.0	264	0.05 ± 0.0	66	No	No
Nonstructural carbohydrates	0.54 ± 0.02	204	0.60 ± 0.03	61	No	No
Fiber	0.13 ± 0.01	98	0.15 ± 0.02	31	No	No

Source. Based on Jordano (1995).
Note. Values for lipids, protein, nonstructural carbohydrates, and fiber represent proportions of dry mass of pulp.

nutritional content of 14 species of Panamanian bat-dispersed figs and noted that they exhibit considerable interspecific variation. In terms of wet mass, free-standing members of section *Pharmacosycea* produce larger fruits (3.1–37.6 g) than stranglers of section *Urostigma* (1.2–10.1 g; Kalko et al. 1996). Among all species, range of lipids (as percent dry weight of pulp) was 7.0%–11.2%; range of protein was 3.6%–9.6%; range of water-soluble carbohydrates was 9.2%–35.6%; and range of metabolizable energy in fruit pulp was 11.3–15.0 kJ/g. Levels of potassium (9.4–33.0 parts per thousand [ppt]) and calcium (5.5–16.9 ppt) were particularly high in these fruits. Wendeln et al. (2000) concluded that bats could easily meet their daily energy and nutrient requirements with a diet containing two or more species of figs.

Fig species can generally be placed in different disperser guilds based on their morphological and nutritional characteristics. Thus, Kalko et al. (1996) identified two guilds of fig species—bird figs and bat figs—among 17 species in a Panamanian moist tropical forest, and Shanahan and Compton (2001) identified six guilds of *Ficus* among 34 species in lowland Bornean rain forest. The Bornean guilds included three sets of mammals (terrestrial species, bats, and primates) and three sets of birds and mammals (one understory and two canopy/subcanopy guilds). In Panama, bird figs are small (0.3–0.8 g) and sugary when ripe, whereas, as indicated above, bat figs are larger and more nutrient rich. Although nutritional analyses have not yet been done on the Bornean figs, their considerable size range (from 4.6 to >50 mm in diameter; Panamanian figs rarely exceed 25 mm in diameter) suggests that at least some of them can also be excellent sources of energy and nutrients. As is generally the case in tropical fruits (Foster 1986; Foster and Janson 1985; Schaefer et al. 2002), canopy figs at this site are larger than understory figs, and their fruit crops attract a higher diversity of frugivores than do understory species.

As in the case of nectar, the pulp of many fruit species contains toxic substances such as phenolics, alkaloids, saponins, and cyanogenic glycosides, among many others (reviewed in Cipollini and Levey 1997). The presence of these substances in an essentially "attractive" plant structure is paradoxical but may represent an antimicrobial or anti-insect herbivore adaptation (Herrera 2002). That is, animal-dispersed plants need to attract effective mutualists with nutritious fruit pulp but still have to contend with antagonists ("cheaters") that also view ripe fruit pulp as a good food source. The presence of these antagonists forces plants to defend their fruit with chemicals that may reduce its attractiveness to legitimate dispersers. Tewksbury (2002) suggested that the presence of toxic compounds in fruits has played an extremely important, but often overlooked, role in the evolution of fruit-frugivore mutualisms.

Cipollini and Levey (1997) and Cipollini (2000) discuss a series of hypotheses that have been put forth to explain the adaptive significance of actually or potentially toxic secondary compounds in the pulp of ripe fruits. These hypotheses include: (1) secondary compounds attract dispersers with a strong chemical signal; (2) they inhibit premature seed germination; (3) they cause frugivores to leave a plant after ingesting only a few fruit thereby promoting seed dispersal; (4) they increase seed passage rates through frugivores; (5) they deter specific kinds of (ineffective) frugivores; (6) they repel harmful insects (e.g., seed predators and herbivores) and microbes; and, last, (7) they may have a direct nutritional or health benefit by acting as antioxidants or by reducing the effects of microbes and other pathogens in frugivores. Cipollini (2000) points out that most of these functions are not mutually exclusive and that these chemicals act and react with many other chemicals in plant tissues and hence should be viewed in a "holistic" or multifunctional fashion.

It is also possible that toxic secondary compounds in ripe fruits may not have any positive adaptive value (Ehrlen and Eriksson 1993). Given the history of the study of secondary plant substances in other plant structures such as leaves and seeds (e.g., Hulme and Benkman 2002; Strauss and Zangerl 2002), however, it is unlikely that these substances have no adaptive value. Studies of the dispersal ecology of wild chilies (*Capsicum* species, Solanaceae), for example, provide strong support for the directed deterrence hypothesis (number 5 above; e.g., Tewksbury and Nabhan 2001; Levey et al. 2006). The main bioactive secondary compound in chilies is capsaicin, a member of the vanilloid chemical group that causes a burning sensation in the vanilloid receptors in mammals but not in birds (Cordell and Araujo 1993). In both southern Arizona and Bolivia, only birds remove chili fruit from plants, and they defecate seeds in germinable condition. Furthermore, in southern Arizona, curve-billed thrashers deposit chili seeds nonrandomly under other bird-dispersed plants. Chilis gain several advantages from this. Under bird-dispersed plants, the probability of chili seedling establishment is high, and their fruits are less likely to be damaged by insects and are more likely to be removed by birds than fruits on chili plants growing under non-bird-dispersed species (Tewksbury and Nabhan 2001). Finally, compared with many other kinds of vertebrate-dispersed fruits (table 3.7), chili fruits are rich in lipids (24%–35% of dry pulp mass), low in nonstructural carbohydrates (16%–24%), and higher than average in protein (9%–10%) (Levey et al. 2006). In terms of their nutritional composition, chilies clearly are "bird" fruits and not "mammal" fruits, and a potent secondary substance, capsaicin, helps to ensure that effective dispersers consume their fruit. This compound is also a potent antimicrobial agent that helps protect fruit from fungi before they are dispersed (Tewksbury et al. 2008).

Temporal Variation in Resource Availability

In addition to having to contend with considerable nutritional variation in their food supplies, vertebrate nectar feeders and frugivores face food supplies that vary significantly in time and space. Compared with temperate regions, the production of flowers and fruit by animal-pollinated or -dispersed plants is less seasonal in tropical and subtropical habitats (Jordano 1992). Indeed, the year-round availability of nectar and fruit in the tropics has been the resource template on which the adaptive radiation of nectar- and fruit-eating vertebrates (and many other animals) has taken place. Despite the year-round availability of these resources in the tropics, however, significant intra- and interannual fluctuations in the production of flowers and fruit are universal. Here we will summarize the major phenological patterns that have been documented in the tropics with an eye toward addressing two questions: (1) How predictable are flower and fruit resources in space and time (i.e., what is the spatiotemporal predictability (STP) of these resources)? (2) Does STP vary among different biogeographic regions in the tropics?

GENERAL PHENOLOGICAL PATTERNS

Reviews of tropical flowering and fruiting phenology (Newstrom et al. 1994; Sakai 2001; Van Schaik and Pfannes 2005; Van Schaik et al. 1993) indicate that resource seasonality is common throughout the tropics. The climatic factors behind this seasonality are correlated with the annual march of the sun between the Tropics of Cancer and Capricorn (23.50 N and S, respectively) and its effect on the location of the Intertropical Convergence Zone, which is a region of increased cloudiness and high rainfall. In general, months of maximum sunshine or irradiance precede the month when the sun is directly overhead and occur in March–June in the Northern Hemisphere and September–December in the Southern Hemisphere. The onset of the rainy season follows the irradiance peak by about 1 month and can result in

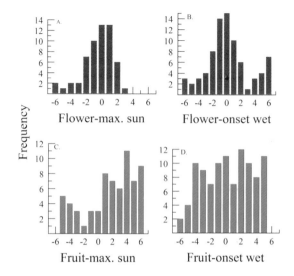

Figure 3.3. Timing of peaks in flowering (A, B), and fruiting (C, D) in tropical woody-plant communities in relation to the timing of maximum irradiance or the onset of the wet season. Positive values indicate that the phenology peak followed the climate event; negative values indicate that the phenology peak preceded the climate event. Based on Van Schaik and Pfannes (2005).

two wet and dry seasons at latitudes just north and south of the equator. The length of the dry season and degree of rainfall seasonality generally increase with latitude but can be highly variable. Van Schaik and Pfannes (2005) suggest that, historically, most of the Neotropics has had stronger dry seasons than the Paleotropics and that equatorial plants in the Neotropics may come from more seasonal stock than most equatorial plants in the Paleotropics. To support this, they report that 83% of 24 sites for which there are detailed phenological data within 10 degrees of the equator in the Neotropics have a single annual rainfall peak compared with 54% of 39 sites in the Paleo-tropics, where two peaks are more common.

Leaf flushing and flowering tend to be more seasonal than fruiting in tropical plants. Figure 3.3 shows that flowering peaks (and peaks of leaf flushing; data not shown) tend to be more closely centered on the onset of wet seasons than with periods of maximum irradiance but that fruiting peaks are less centered on the onset of the wet season than flowering peaks. This is true even in strongly seasonal climates away from the equator (Van Schaik and Pfannes 2005). The time interval between flowering and fruiting peaks, however, does decrease with increasing seasonality and latitude.

The time interval between leaf flushing and fruiting in tropical trees has important implications for the feeding ecology of primates (Terborgh and Van Schaik 1987). The shorter this interval, the less likely it is that primates can shift to a diet of leaves in times of fruit scarcity. When this interval is short, therefore, primate densities are more likely to be limited by fruit availability than by the availability of leaves. Van Schaik and Pfannes (2005) report that this interval varies both seasonally and regionally. The length of this interval decreases as the length of dry seasons increases, and it is shorter in the Neotropics and Madagascar (with an average of about 1 month) than in Africa and Asia (with averages of 2.8 and 3.8 months, respectively). As a result, primate density and biomass tend to be lower in the Neotropics and Madagascar than in Africa (but see below).

The flowering and fruiting rhythms of trees in a submontane rain forest in the central Philippines can be used to illustrate the general pattern of resource availability experienced by tropical vertebrate nectar and fruit eaters (Hamann 2004; fig. 3.4). At this site, annual rainfall is about 4,000 mm, and there is a brief dry period in March and April in most years. An El Niño event that began in November 1997 resulted in very low rainfall early in 1998 but very high rainfall at the end of that year. Flowering phenology showed a single strong peak in the dry season in two years and two peaks in 1 year; flowering activity peaked late in the year in 1998. Fruiting in plants dispersed by birds showed a strong annual peak after the dry season in all

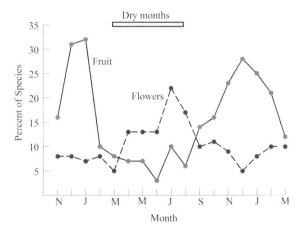

Figure 3.4. Flowering and fruiting phenology in one of four years in a premontane rain forest in the northern Philippines. Flowering is indicated by the *long dashed line* and fruiting by bird- or bat-dispersed species by *solid lines*. Based on Hamann (2004).

4 years and a second peak in midyear in 1998 (not shown; fig. 3.4). Fruiting in bat-dispersed plants, in contrast, was low and more or less continuous in two years before becoming more seasonal in 1998 and 1999. Overall, fruiting in some vertebrate-dispersed plants occurred year round whereas fruiting in wind- or gravity-dispersed plants was concentrated in the dry season. Additional examples of annual fruiting rhythms in different habitat types, including four tropical habitats, can be found in Jordano (1992, fig. 5.2). As expected, fruiting peaks and troughs are much more pronounced in tropical dry forests than in tropical wet forests.

Finally, Stiles's (1978b) detailed study of the flowering phenology of hummingbird-pollinated plants in a Costa Rican lowland wet forest provides a useful overview of phenological rhythms of tropical plants by growth form and habitat. In primary forest at La Selva, flowering in canopy trees, vines, and epiphytes is seasonal with peaks in the brief dry season (February and March); flowering in understory plants peaks in the first half of the wet season. In contrast, flowering in plants in large gaps and in the understory of second growth occurs year round but with peaks in the dry and early wet seasons. Plants in large light gaps and second growth have longer flowering periods than those in the understory of primary forest. In the canopy, vines have shorter flowering seasons than canopy trees, and both groups have shorter flowering periods than understory herbs and shrubs. These data emphasize the spatiotemporal mosaic nature of flower resources used by a particular group of nectar-feeding birds. Vertebrate frugivores undoubtedly encounter a similar spatiotemporal mosaic in their food supplies in this and other kinds of tropical forests (e.g., Levey 1988; Loiselle and Blake 1994). As we discuss below, the spatial and temporal scale of this mosaic varies geographically, and this variation has had an important effect on the evolution of nectarivores and frugivores in different tropical regions.

THE CLASSIFICATION OF PHENOLOGICAL PATTERNS

General flowering and fruiting curves we've just described summarize the overall phenology of entire plant communities or, as in figure 3.4, particular groups of plants. These patterns result from a variety of species-specific flowering or fruiting patterns. In describing the flowering patterns of tropical Bignoniaceae, Gentry (1974) grouped these patterns into four different phenological strategies: (1) steady-state or continuous flowering (few flowers produced per plant per day for extended periods of time), (2) "cornucopia" or mass flowering (many flowers produced over a month or more once a year),

(3) "big bang" (one brief, massive flowering event per year), and (4) "multiple bang" (several brief episodes of flowering per year). More recently, Newstrom et al. (1994) proposed a phenological classification scheme based on flowering or fruiting frequency and regularity in an effort to describe phenological predictability. Their scheme is based on data on the frequency, duration, amplitude, degree of synchrony, and start date of phenological events within species. Based on these variables, Newstrom et al. (1994) recognized four major frequency classes: (1) continual (i.e., Gentry's [1974] steady state), (2) subannual (irregular in frequency with multiple events per year; multiple bang of Gentry), (3) annual (one major event per year; big bang if the event is short or cornucopia if it is long), and (4) supra-annual (events occur on a multiyear frequency). They applied this scheme to 12 years of data from the trees at La Selva, Costa Rica, and found that 7% (n = 254 trees of 211 species) had continual flowering cycles, 55% had subannual cycles, 29% had annual cycles, and 9% had supra-annual cycles. In stark contrast, using this classification scheme, Sakai (2001) reported the following frequencies of flowering strategies in 187 species of trees in lowland wet forest at Lambir Hills National Park, Borneo: continual = 0%, subannual = 3%, annual = 10%, and supra-annual = 61%; one-quarter (26%) of the species did not flower during the 4-year study. As discussed below, these two forests have strikingly different phenological patterns that have profound implications for their nectar- and fruit-eating vertebrates.

Because it has a strong influence on the foraging behavior of pollinators and seed dispersers, another way to classify flowering (and fruiting) phenology is by a two-phase system: extended flowering versus mass flowering (Newstrom et al. 1994). An extended flowering pattern (i.e., continuous or steady-state flowering) tends to promote traplining foraging behavior in which animals visit the same plants in a regular daily circuit that can last for several weeks to months. This pattern is more common in tropical wet forest understories than in the canopy and usually involves widely spaced plants that produce a few flowers (or fruit) per plant per day. It apparently is not common in dry forest plants. Neotropical examples of this kind of foraging pattern include euglossine bees, *Heliconius* butterflies, hermit hummingbirds, and some glossophagine bats (Newstrom et al. 1994).

The antithesis of extended flowering is mass flowering (i.e., cornucopia or big bang flowering) in which species produce large crops of flowers (and fruits) over relatively short periods of time (usually less than 1 month). This pattern is common in canopy trees in tropical wet forests and in tropical dry forest trees in the dry season (Frankie et al. 1974; Newstrom et al. 1994; Stevenson et al. 2008). Mass-flowering plants are often visited by large numbers of opportunistic nectar feeders and pollen collectors. High levels of intra- and interspecific aggression plus avoidance of predators that are attracted to these trees often cause pollinators to spend limited amounts of time in them despite their large energetic rewards.

Although supra-annual flowering and fruiting or "masting" phenological behavior is known to occur in scattered species or families in many tropical areas—for example, certain species of Lecythidaceae in South America, species of Fabaceae (Caesalpinoideae) in South America and Africa, and species of *Chrysophyllum* (Sapotaceae), *Cryptocarya angulata*, and *Beilschmiedia bancroftii* (Lauraceae) in tropical Australia (Connell and Green 2000; Henkel et al. 2005; Mori and Prance 1987; Newbury et al. 1998, 2006; Norden et al. 2007)—it is best known from the aseasonal lowlands of Southeast Asia where it occurs as far west as Sri Lanka (Brearley et al. 2007; Sakai 2002; Van Schaik and Pfannes 2005; Visser et al. 2011). In this pattern, plants flower and fruit at intervals of 3–7 years or more and hence produce nectar and fruit resources that are highly unpredictable in space and time (i.e., low STP). In Southeast Asia, masting behavior is a community-wide event and, in addition to the dipterocarp forest dominants, it includes many other families and growth habits of plants (Sakai 2002; Sakai et al. 2006). Certain understory herbs such as gingers, however, do not exhibit supra-annual flowering and hence provide relatively reliable nectar rewards to their flower visitors (Sakai 2000). Considerable geographic variation occurs in the timing of masting events in Southeast Asia, even over relatively short distances, which further emphasizes the low STP of the region's flower and fruit resources (Numata et al. 2003; Wich and Van Schaik 2000 and included references).

PHENOLOGICAL PATTERNS AT THE GUILD LEVEL

Considerable attention has been paid to the organization of tropical flowering and fruiting patterns at the level of ecological guilds, primarily in the Neotropics (reviewed by Newstrom et al. 1994; Van Schaik et al. 1993). Examples of these guilds include plants pollinated by hummingbirds and plants pollinated or dispersed by bats (e.g., Buzato et al. 2000; Fleming 1985; Heithaus et al. 1975; Sazima et al. 1999; Stiles 1978b; Thies and Kalko 2004). Two phenological patterns might be expected to occur in these guilds: (1) multispecies staggered sequences or (2) multispecies mixed sequences (Newstrom et al. 1994). In the former pattern, temporal overlap in the flowering and/or fruiting seasons of guild members is low, ostensibly to reduce interspecific competition for pollinators or dispersers. Despite the conceptual attractiveness of this hypothesis, however, strong statistical support for it still is lacking (e.g., Van Schaik et al. 1993; Wright and Calderon 1995; but see Aizen and Vazquéz [2006] and Poulin et al. [1999]). For flowering plants, staggered or nonoverlapping flowering periods can also reduce pollen wastage that occurs when pollen is deposited on nonconspecific stigmas. In pattern 2, temporal overlap between species is high, perhaps to increase pollination or seed dispersal efficiency by supporting larger populations of nectar feeders or frugivores. This pattern involves facilitation, which can be defined as the enhancement of a population of one species by the activities of another (Ricklefs 1990), and is the antithesis of competition (see chap. 9).

Neotropical examples of multispecies staggered sequences (though not necessarily statistically significant) include plants pollinated by hermit hummingbirds in tropical wet forest and bats in tropical dry forest, bat-dispersed *Piper* shrubs in tropical moist and dry forests, and manakin-dispersed *Miconia* shrubs in tropical moist forest (Fleming 1985; Heithaus et al. 1975; Poulin et al. 1999; Stiles 1975 1978b; fig. 3.5). Mixed species sequences (pattern 2) occur in plants pollinated by nonhermit or "typical" hummingbirds and bird-dispersed *Psychotria* shrubs and species of Melastomataceae (Poulin et al. 1999; Stiles 1978b; Stiles and Rosselli 1993). Thies and Kalko (2004) described both of these flowering and fruiting patterns in a study of the reproductive phenology of 12 species of *Piper* shrubs and treelets in central Panama. Eight species of forest-dwelling *Piper*s had short and staggered fruiting peaks whereas four species of gap-dwelling *Piper*s had extended and broadly overlapping fruiting periods. Extended flowering and fruiting seasons are typical of early successional and disturbance-adapted plants in both the Neotropics and Paleotropics (e.g., Frankie et al. 1974; Opler et al. 1980; Toriola 1998; Van Schaik 1986).

Staggemeier et al. (2010) documented the flowering and fruiting phenology of 34 co-occurring species of Myrtae (Myrtaceae) in southeastern Brazil to test for patterns 1 and 2 above. These trees and shrubs are pollinated by melaponine bees, and their fleshy fruits are dispersed by birds. Their results indicated that flowering was aggregated among species (pattern 2) but that fruiting was random. They further examined the relationship between phylogeny (using a well-resolved molecular phylogeny), climatic variables, and phenology and found that climate was a better predictor of the timing of flowering and fruiting than phylogeny. Future phenological studies would

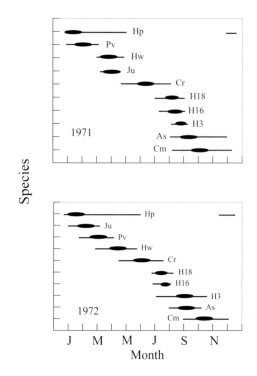

Figure 3.5. Blooming periods of the major food-plants of hermit hummingbirds at La Selva, Costa Rica in 2 years. *Thin lines* = period of major flowering for each species; *thick lines* = period of peak flowering. Abbreviations; As = *Aphelandra sinclairiana*; Cm = *Calathea macrosepala*; Cr = *Costus ruber*; H3, 16, 18 = *Heliconia* spp.; Hp = *Heliconia pogonatha*; Hw = *Heliconia wagneriana*; Ju = *Jacobinia umbrosa*; Pv = *Passiflora vitifolia*. Based on Stiles (1978a).

also benefit from considering both phylogenetic and environmental influences on patterns of flowering and fruiting.

Another important fruiting guild for vertebrates in tropical forests is the fig (*Ficus*) guild. Because of its unique pollination biology (Janzen 1979), we do not expect communities of figs to exhibit staggered flowering and fruiting patterns. Instead, many species of monoecious figs have continuous and overlapping flowering/fruiting schedules at the population level. These schedules are composed of subannual or supra-annual schedules at the individual level (Newstrom et al. 1994). Subannual flowering is individually advantageous in these plants because it produces a large odor plume that is easily detectable by fig wasp pollinators and by fig-eating and seed-dispersing bats (Korine et al. 2000). It also results in high levels of outcrossing whenever different individual trees fruit asynchronously and provides a continuous supply of fig wasps as well as a year-round supply of fruits for many vertebrates (Shanahan et al. 2001).

Detailed phenological studies of guilds of Paleotropical plants pollinated or dispersed by vertebrates are uncommon. The closest comparison to Stiles's (1975, 1978b) work on the flowering phenology of hummingbird-pollinated plants at La Selva, Costa Rica, comes from Sakai's (2000) study of gingers (Costaceae, Zingiberacae) at Lambir National Park, Borneo. At this site, six species of gingers are pollinated by two species of spiderhunters (Nectariniidae), and their phenology involves sporadic flowering, short flowering intervals, and considerable interspecific flowering overlap. In contrast to plants pollinated by Neotropical hermit hummingbirds, there was no evidence of staggered flowering seasons in these plants. Unlike most other plants in this dipterocarp-dominated forest, however, gingers flower annually rather than supra-annually.

GEOGRAPHIC VARIATION IN REPRODUCTIVE PHENOLOGICAL PATTERNS

Comparisons of the flowering and fruiting rhythms at La Selva, Costa Rica, and Lambir, Borneo, reveal a striking difference in the frequency of annual or subannual flowering versus supra-annual flowering: supra-annual flowering is uncommon at La Selva but common at Lambir. In general, how does the frequency of these two patterns (annual/subannual vs. supra-annual reproductive phenology) vary geographically? Because these patterns result in strong differences in the spatiotemporal predictability (STP) of flower and fruit resources, they have important consequences for the ecology and evolution of plant-visiting vertebrates (and many other kinds of animals). How geographically widespread is low resource STP?

Reviews of reproductive phenology of tropical plants (e.g., Corlett and LaFrankie 1998; Newbury et al. 1998; Van Schaik et al. 1993; Van Schaik and Pfannes 2005) indicate that predictable annual seasonal patterns of flowering and fruiting prevail throughout most of the tropics and that community-wide supra-annual masting is most common in the aseasonal lowlands of Southeast Asia. Like the Neotropics and most of Africa, seasonal parts of the Asian tropics have annual phenological cycles and hence relatively high values of STP. In seasonal Asian forests, the Dipterocarpaceae also display annual cycles, but interannual variation in flowering and fruiting can be substantial (Corlett and LaFrankie 1998).

A summary of studies of tropical phenological patterns based on the

Table 3.8. Summary of the Phenological Patterns of Tropical Trees in a Series of Locations

Region/Location	Sample Size (No. of species)	Phenology Pattern	
		Subannual or Annual (% of species with this pattern)	Supra-annual (% of species with this pattern)
Neotropics:			
La Selva, Costa Rica (lowland wet forest)	254*	84	9
Chapada Diamantina N. P., Bahia, Brazil (premontane moist gallery forest)	46	91	9
Central French Guiana (lowland moist tropical forest	48†	54	23
Africa:			
Kibale N.P., Uganda (premontane moist forest)	Many species	75–79 (flowering); 43–58 (fruiting)	21–25 (flowering); 42–57 (fruiting)
Southeast Asia:			
Negros, Philippines (premontane wet forest)	57	60	5
Lambir N.P., Borneo (lowland wet forest)	187	13	61
Papua New Guinea:			
SE of Pt. Moresby (lowland moist forest)	15	73	20
NW of Pt. Moresby (premontane wet forest	93	42	5
Australia:			
NE of Darwin (lowland monsoon forest)	16	75	6
SE of Darwin (lowland mesic savanna)	50	90	4
S of Cairns (lowland wet forest)	57	49	49

Sources. Brown and Hopkins 1996; Chapman et al. 1999; Crome 1975; Franklin and Bacht 2006; Funch et al. 2002; Hamann 2004; Newstrom et al. 1994; Norden et al. 2007; Sakai 2001; P. Williams et al. 1999, Wright 1998.
Note. Phenological patterns use Newstrom et al.'s (1994) categories. Habitat type follows location name. Unless noted, phenology refers to flowering. Difference between sum of percentage of subannual/annual and supra-annual and 100 represents percentage of continuous flowering.
*Trees.
†28 trees, 20 lianas.

Newstrome et al. (1994) classification system also shows that, with a couple of exceptions, annual/subannual patterns prevail throughout the tropics (table 3.8). In addition to lowland Borneo (and elsewhere in Southeast Asia), high frequencies of supra-annual flowering and fruiting are known to occur in central French Guiana, at Kibale National Park, Uganda, and in lowland wet forest in northern Queensland. Supra-annual flowering is uncommon elsewhere in tropical Africa and in the dry Australian tropics (Franklin and Bach 2006; Newbury et al. 1998; Tutin and White 1998). Although they flower annually, however, many Australian eucalypts showed marked year-to-year variations in flowering intensity (e.g., hundredfold differences) and exhibit flowering peaks every 2–7 years (MacNally and McGoldrick 1997). Thus, in addition to mast-flowering species, Australian eucalypts, which are key food resources for nectar-feeding honeyeaters, lorikeets, and pteropodid bats, also exhibit low degrees of STP.

Van Schaik and Pfannes (2005) have attempted to summarize regional differences in resource predictability by calculating coefficients of variation of annual mean number or percentage of flowering and fruiting species at sites with at least 3 years of phenological data. They referred to these coefficients of variation as estimates of interannual variation and compared these values for three biogeographic regions—Southeast Asia plus northern Australasia, Africa plus Madagascar, and the Neotropics. As expected, phenology in the former region was notably more variable than that of the latter two regions, especially in the interannual variation of fruiting (fig. 3.6). Although more long-term studies of plant phenology are needed for firm conclusions, these data suggest that the STP of flower and fruit resources is likely to be lower in Southeast Asia and parts of northern Australasia, including New Guinea and Queensland, Australia, than in other regions of the tropics. If high resource STP promotes feeding specialization, then we might expect to find feeding specialization to be less common among plant-visiting vertebrates in Southeast Asia and northern Australasia than in other tropical regions because of lower resource reliability. Regional differences in many aspects of the biology of nectar-feeding and fruit-eating birds and bats support this prediction (Fleming and Muchhala 2008; Fleming et al. 1987).

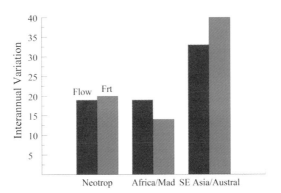

Figure 3.6.   Mean interannual variation (i.e., coefficients of variation of annual mean values) for community-level estimates of flowering (*red bars*) and fruiting (*green bars*) for three biogeographic regions: the Neotropics, Africa/Madagascar, and Southeast Asia/Australasia. Based on Van Schaik and Pfannes (2005).

KEYSTONE AND CORE PLANT RESOURCES

Particular kinds of plant-visiting vertebrates typically feed on nonrandom subsets (guilds) of the flowers and fruits available in their habitats, and the nutritional and spatiotemporal characteristics of these subsets should have the greatest ecological and evolutionary effects on their consumers. Here we discuss two of these critically important kinds of resources: keystone species and core plant taxa.

KEYSTONE SPECIES.—Phenological studies at many tropical sites indicate that the transition period between wet and dry seasons is often a time of low flower and fruit availability for vertebrates (Terborgh 1986; Van Schaik et al. 1993). This period of low food levels is a critical time for plant-visiting birds and mammals, and resources that are available then may be especially important for them. Among other things, resources produced during this period may determine animal carrying capacities (i.e., their maximum density or biomass). Terborgh (1986) called resources that sustain frugivores during the annual 3-month period of fruit scarcity at Cocha Cashu, Peru, "keystone plant resources" (KPRs). For five species of primates, these resources included palm nuts and figs (for two species of *Cebus* monkeys), nectar from *Combretum* vines (for two species of *Saguinus* tamarins), and figs for the squirrel monkey, *Saimiri sciureus*. Terborgh (1986, 339) described these KPRs as "playing prominent roles in sustaining frugivores through general periods of food scarcity" and estimated that about 12 species of plants (out of a total flora of 2,000 species) sustains nearly the entire frugivore community at Cocha Cashu for 3 months each year.

Peres (2000) critically reviewed the KPR concept for tropical habitats and attempted to make it fully operational. His quantitative measure of "keystoneness" is based on four parameters, scaled from 1 to 10: (1) *temporal redundancy*, which is the degree to which a potential KPR overlaps temporally with other plant resources (1 = no overlap; 10 = complete overlap); (2) *consumer specificity*, which is the degree to which a resource is used by different species (1 = very specialized; 10 = very generalized); (3) *resource reliability*, which is the degree of year-to-year variation in its availability (1 = very unreliable; 10 = completely reliable); and (4) *resource abundance*, which is the relative amount of this resource available at a site (1 = very rare; 10 = superabundant). From the perspective of vertebrate frugivores or nectarivores, KPRs are species with low temporal redundancy, high percentage of species' use, and high annual reliability. Based on these criteria, Peres (2000: table 3) examined many species that have been proposed to be KPRs and found that most species had low keystoneness scores, primarily because they are used by only a fraction of their potential consumer community. Given the relatively low overlap between food species used by different kinds of vertebrate nectar or fruit eaters, however, it is not surprising that many KPRs are consumed by only a fraction of their potential consumers. For example, fruit-eating birds and bats eat very different fruits in many tropical habitats (e.g., Gorchov et al. 1995; Hodgkison et al. 2003; Lobova et al. 2009), so it is unreasonable to expect them to use the same KPRs. Westcott et al. (2005b) make this same point for Australian rain forests. Degree of feeding specificity, when defined to include all possible consumers, seems to be less useful as a mark of keystoneness compared to degree of temporal redundancy and reliability. Despite this uncertainty about how to best define KPRs operationally, it seems likely that in the Neotropics, at least, figs are legitimate examples of these kinds of resources (Peres 2000). Whether figs also qualify as important KPRs in the Paleotropics has been questioned (Borges 1993; Gautier-Hion and Michaloud 1989; Patel 1997), although Lambert and Marshall (1991) argued that bird-dispersed figs are keystone resources in lowland Malaysian forests. Low plant population densities, territorial behavior of certain frugivores, and high temporal redundancy reduce the keystoneness of figs at some localities in Africa and Asia. Finally, Peres (2000) and Stevenson (2005) have questioned whether the loss of a putative KPR actually threatens the existence of its consumers. To date, there is no evidence that any species of tropical frugivore or nectarivore has gone extinct locally because of the loss of one or more of its KPRs. Nonetheless, it is likely that the loss of certain critically important plants (e.g., figs in many tropical forests, columnar cacti in arid tropical and subtropical regions) would have a strong negative effect on their mutualists as well as indirect negative effects on many other plants and animals in their habitats.

In addition to predictable annual periods of food scarcity, tropical frugivores (and nectar feeders?) sometimes experience unpredictable years of exceptionally low food availability. These so-called famine years can markedly increase the mortality rates of many species of frugivores and other vertebrates. Such events have been most thoroughly documented for tropical moist forest on Barro Colorado Island, Panama, where famine years have occurred in 1931, 1958, 1970, and 1993; each of these years followed an El Niño event characterized by unusually wet dry seasons after which many plants that typically produce fruits eaten by mammals failed to flower (Foster 1982; Wright et al. 1999). Foster (1982, 208) dramatically described the fate of fruit-eating and other mammals on Barro Colorado Island during the 1970 fruit failure, which began in July: "Dead animals were found increasingly often until late November and December, when one could find at least one dead animal every 300 m along trails well away from the laboratory clearing. The most abundant carcasses were those of coatis, agoutis, peccaries, howler monkeys, opossums, armadillos, and porcupines; there were only occasional dead two-toed sloths, three-toed sloths, white-faced monkeys, and pacas. At times it was difficult to avoid the stench: neither the turkey vultures nor the black vultures seemed able to keep up with the abundance of carcasses."

Famine years do not appear to be universal throughout the tropics. They have not been described for three other well-studied sites—La Selva, Costa Rica; Cocha Cashu, Peru; and Ketambe, Sumatra (Van Schaik et al. 1993). In the aseasonal lowlands of Southeast Asia, famine years for vertebrate frugivores and seed eaters actually appear to be the rule, and years of high food availability (i.e., general flowering or masting years) are the exception (Curran and Leighton 2000; Leighton and Leighton 1983; Wong et al. 2005). The opposite appears to be the case in the Neotropics where years of good food availability for frugivores are the rule and famine years are the exception (Van Schaik and Pfannes 2005).

CORE PLANT TAXA.—Detailed studies of the feeding ecology of virtually all species of nectar- or fruit-eating birds and mammals indicate that they have relatively broad diets containing many species and sometimes several different food types. For example, throughout their geographic ranges, two well-studied frugivorous phyllostomid bats, *Carollia perspicillata* and *Artibeus jamaicensis*, are known to eat at least 37 and 84 species of fruit, respectively; at particular localities, they eat 13–24 and 10–16 species of fruit, respectively (Fleming 1986a). Both of these species also visit flowers and sometimes eat insects, and *A. jamaicensis* also eats leaves (Fleming 1988; Handley et al. 1991; Heithaus et al. 1975; Kunz and Diaz 1995). Because of their large size, primates have particularly broad and diverse diets. At Lope Forest Reserve, Gabon, for example, the diets of eight species of diurnal primates include 20–114 species of fruit; most species also eat significant amounts of seeds, plant pith, leaves, and flowers (Tutin et al. 1997). Among nectarivores, the number of flower species visited by 15 species of hummingbirds ranged from one to 38 per species in forests in southeastern Brazil; six species visited

≥10 flower species (Buzato et al. 2000). Many of these birds also eat insects. As another avian example, the broadly distributed African collared sunbird (*Hedydipna collaris*) visits >40 species of flowers in addition to eating fruit and insects (Cheke and Mann 2001). Similarly, in addition to eating fruit and insects, the nectar-feeding phyllostomid bat *Leptonycteris yerbabuenae* visits >40 species of flowers in Mexico and the southwestern United States (Arizmendi et al. 2002).

Despite their often broad and seasonally changing diets, different groups of tropical plant–visiting vertebrates tend to specialize on or feed selectively on particular groups of tropical plants. In an analysis of feeding specialization in frugivorous phyllostomid bats, Fleming (1986a) called these resources "core" plant taxa and indicated that they share three important characteristics: a high degree of phenological STP, relatively high species richness per habitat, and broad geographic ranges. Examples of core fruit taxa for phyllostomid bats include *Piper* (for *Carollia* species), *Solanum* (for *Sturnira* species), and Moraceae and *Cecropia* (for species of *Artibeus* and related genera). Each of these plant taxa provides a reliable food source on which different groups of tropical vertebrates have specialized. Similarly, although their diets include many species of fruit at a given site, New World spider monkeys (*Ateles* spp.) focus on fruits of *Virola* (Myristicaceae) as their core food resources (Russo et al. 2005). If KPRs are critically important for sustaining vertebrate plant visitors during annual periods of low food availability, then core plant resources are also important because they form the basis of a species' diet for most of the year. From a conservation viewpoint, therefore, it is important to identify and protect both kinds of resources to minimize the chance that plant-visiting tropical vertebrates will go extinct.

In table 3.9 we present a list of core plant taxa, at the level of family and genus, for different groups of nectar- and fruit-eating tropical vertebrates. Although feeding selectivity has not been rigorously demonstrated in most cases (cf. Fleming 1986), these families and genera nonetheless occur frequently in animal diets and therefore represent major components of their diets. In many cases, these families are species rich within and across sites and have broad, often pantropical, distributions. They also tend to be the numerically dominant members of their growth forms in most habitats. Much of the important plant structure in different tropical and subtropical habitats is therefore reflected in these plant families, especially by Fabaceae for many nectar feeders, except in Australasia where Myrtaceae assumes

Table 3.9. Examples of Core Plant Taxa for a Variety of Tropical Nectar- and Fruit-Eating Birds and Mammals

Region/Animal Taxon	Habitat	Core Plant Taxa
Nectar feeders:		
Neotropics:		
Hummingbirds, hermits	Lowland forests	Heliconiaceae (*Heliconia*), Costaceae (*Costus*)
Hummingbirds, nonhermits	Lowland forests and lower montane forests	Bromeliaceae (*Bilbergia, Vriesia*), Fabaceae (*Calliandra, Erythrina*), Rubiaceae (*Hamelia, Palicouria*), Malvaceae (*Abutilon, Malvaviscus*)
Hummingbirds, nonhermits	Montane habitats	Campanulaceae (*Centropogon, Siphocampylus*), Ericaceae (*Gaultheria, Vaccinium*)
Phyllostomid bats	Arid habitats	Agavaceae (*Agave*), Cactaceae (esp. tribes Cereeae and Pachycereeae)
Phyllostomid bats	Lowland forests	Bromeliaceae (*Vriesia*), Fabaceae (*Inga, Mucuna*), Malvaceae (*Ceiba, Pseudobombax*)
Phyllostomid bats	Montane forests	Campanulaceae (*Burmeistera*)
Africa:		
Sunbirds	Forest and nonforest habitats	Aloaceae (*Aloe*), Fabaceae (*Erythrina, Parkia*), Lamiaceae (*Leonotis, Salvia*), Loranthaceae (*Englerina, Phragmanthera*)
Pteropodid bats	Forests and woodlands	Bignoniaceae (*Kigelia*), Fabaceae (*Erythrina, Parkia*), Malvaceae (*Adansonia, Ceiba*)
Madagascar:		
Prosimians	Lowland and upland forests	Fabaceae (*Delonix*), Malvaceae (*Adansonia*), Strelitziaceae (*Ravenala*)
Asia:		
Sunbirds	Lowland and upland forests	Ericaceae (*Rhododendron*), Fabaceae (*Butea, Erythrina*), Loranthaceae (*Amylotheca*), Zingiberaceae (*Amomum, Plagiostachys*)
Pteropodid bats	Lowland forests	Bignoniaceae (*Oroxylum*), Malvaceae (*Ceiba, Durio*), Musaceae (*Musa*)
Australasia:		
Honeyeaters and lorikeets	Lowland forests and woodlands	Myrtaceae (*Eucalyptus, Syzygium*), Proteaceae (*Banksia, Grevillea*)
Pteropodid bats	Lowland forests and woodlands	Myrtaceae (*Eucalyptus, Syzygium*), Proteaceae (*Banksia*), Sonneratiaceae (*Sonneratia*)
Fruit eaters:		
Neotropics:		
Trogons	Lowland and upland forests	Lauraceae (*Beilschmidea, Ocotea*), Moraceae (*Ficus*)
Cotingids	Lowland and upland forests	Araliaceae (*Didymopanax*), Arecaceae (*Euterpe, Iriartia*) , Burseraceae (*Dacroydes, Protium*), Lauraceae (*Beilschmidea, Ocotea*), Moraceae (*Ficus*)
Toucans	Lowland and upland forests	Lauraceae (*Beilschmidea, Ocotea*), Moraceae (*Ficus*), Myristicaceae (*Virola*), Solanaceae (*Cestrum, Solanum*)
Manakins	Lowland and upland forests	Melastomataceae (*Conostegia, Miconia*), Rubiaceae (*Hamelia, Psychotria*), Solanaceae (*Cestrum, Solanum*)
Tanagers	Lowland and upland forests	Loranthaceae (*Phoradendron, Psittacanthus*), Melastomataceae (*Conostegia, Miconia*), Moraceae (*Cecropia*),
Phyllostomid bats	Arid habitats	Cactaceae (esp. tribes Cereeae and Pachycereeae)
Phyllostomid bats	Lowland and upland forests	Clusiaceae (*Vismia*), Moraceae* (*Cecropia, Brosimum, Ficus*), Piperaceae (*Piper*), Solanaceae (*Solanum*)

Region/Animal Taxon	Habitat	Core Plant Taxa
Cebid monkeys (sensu lato)	Lowland forests	Annonaceae (*Oxandra, Xylopia*), Fabaceae (*Inga*), Moraceae (*Brosimum, Ficus*), Myristicaceae (*Virola*), Salicaceae (*Casearia*), Sapotaceae (*Pouteria*)
Africa:		
Hornbills	Lowland forests and woodlands	Ebenaceae (*Diospyros*), Meliaceae (*Trichilia*), Moraceae (*Ficus*)
Pteropodid bats	Lowland and upland forests	Combretaceae (*Terminalia*), Ebenaceae (*Diospyros*), Moraceae (*Ficus, Milecia*)
Cercopithecid monkeys	Lowland and upland forests	Ebenaceae (*Diospyros*), Fabaceae (*Parkia*), Moraceae (*Ficus*), Sapotaceae (*Chrysophyllum, Mimusops*), Ulmaceae (*Celtis*)
Apes	Lowland forests	Burseraceae (*Dacryodes*), Ebenaceae (*Diospyros*), Euphorbiacae (*Uapaca*), Fabaceae (*Dialium, Parkia*), Moraceae (*Ficus*), Sapotaceae (*Gambeya*)
Elephants	Lowland forests	Euphorbiaceae (*Uapaca*), Fabaceae (*Tetrapleura*), Rubicaceae (*Nauclea*), Sapotaceae (*Chrysophyllum*)
Madagascar:		
Prosimians	Lowland and upland forests	Clusiaceae (*Calophyllum, Garcinia*), Euphorbiaceae (*Uapaca*), Moraceae (*Ficus*), Rubiaceae (*Canthium, Genipa*)
Asia:		
Hornbills	Lowland and upland forests	Lauraceae (*Ocotea*), Meliaceae (*Aglaia*), Moraceae (*Ficus*), Myristicaceae (*Myristica*)
Flowerpeckers	Lowland and upland forests	Loranthaceae (*Loranthus, Viscum*)
Pteropodid bats	Lowland forests	Anacardiaceae (*Anacardium, Mangifera*), Moraceae (*Artocarpus, Ficus*), Musaceae (*Musa*)
Cercopithecid monkeys	Lowland forests	Annonaceae (*Polyalthia, Xylopia*), Fabaceae (*Dialium, Pithecellobium*), Meliaceae (*Aglaia, Dysoxylum*), Moraceae (*Antiaris, Ficus*), Sapindaceae (*Nephelium, Paranephelium*)
Apes and gibbons	Lowland forests	Annonaceae (*Polyalthia, Xylopia*), Euphorbiaceae (*Bridelia, Sapium*), Fabaceae (*Dialium, Parkia*), Meliaceae (*Aglaia, Dysoxylum*), Moraceae (*Artocarpus, Ficus*)
Australasia:		
Cassowaries	Lowland forests	Elaeocarpaceae (*Elaeocarpus*), Lauraceae (*Cryptocarya, Endiandra*), Moraceae (*Ficus*), Myrtaceae (*Acmena, Eugenia*)
Fruit pigeons	Lowland forests	Araliaceae (*Tieghemopanax*) Elaeocarpaceae (*Elaeocarpus*), Lauraceae (*Cryptocarya, Litsea*), Moraceae (*Ficus*), Myristicaceae (*Myristica*)
Birds of paradise	Lowland and upland forests	Araliaceae (*Schefflera*), Euphorbiaceae (*Antidesma, Endospermum*), Moraceae (*Ficus*), Meliaceae, (*Dysoxylum, Melia*), Myristicaceae (*Myristica*)
Honeyeaters	Lowland and upland forests	Loranthaceae (*Amylotheca, Loranthus*), Moraceae (*Ficus*), Myrtaceae (*Eugenia, Syzygium*)
Pteropodid bats	Lowland and upland forests	Moraceae (*Ficus*), Musaceae (*Musa*), Myrtaceae (*Syzygium*)

Sources. Arizmendi et al. (2002), Birkinshaw and Colquhoun (2003), Chapman et al. (2002), Cheke and Mann (2001), Chivers (1980), Crome (1975), Dobat and Peikert-Holle (1985), Fleming (1986a), Fleming and Muchhala (2008), Frith and Beehler (1998), Hall and Richards (2000), Kemp (1995), McConkey et al. (2002), Mickelburgh et al. (1992), Russo et al. (2005), Sazima et al. (1999), Snow (1981, 1982), Steadman (1997), Stocker and Irvine (1983), Terborgh (1983), Tutin et al. (1997), Wheelwright et al. (1984), White (1994)
Note. Genera (in parentheses) are common members of their families.
* Includes Cecropiaceae.

great importance. For tropical frugivores worldwide, Moraceae (i.e., *Ficus*) provides critically important food resources that can serve both as KPRs and as core resources. We will reexamine these core plant families from a phylogenetic perspective in chapters 5 and 6.

Community Biomass of Plant-Visiting Vertebrates and Their Food Resources

Nectar and fruit pulp represent important energy and other nutritional resources for nectar-feeding and fruit-eating tropical vertebrates, and we have seen that levels of these resources exhibit considerable spatial and temporal variation at the individual, population, and community or habitat level. But what are the actual amounts of nectar and fruit pulp produced per unit area and time in tropical forests? What is the energy density of these resources and how closely does the biomass of vertebrate consumers match these levels? Two examples will introduce this topic. Detailed observations in the Sonoran Desert indicate that nectar availability for cactus-visiting bees, birds, and bats is strongly seasonal and reaches peak densities of 1,600–2,800 kJ/ha in April and May, depending on year and site (Fleming et al. 2001). Estimates of the collective energy demands of vertebrate visitors to these flowers per hectare, however, are three to nine times lower than this. In this habitat, densities of the nectar-feeding bat *Leptonycteris yerbabuenae* are far below its nectar- and pollen-carrying capacity, and bat density here is not limited by food availability. In contrast, New Holland honeyeaters (*Phylidonyris novaehollandiae*) and bees consume nearly all of the nectar produced by eucalypts and other species in sclerophyll woodlands in southeastern Australia, and bird densities there are limited by nectar availability (Paton 1985). How similar are patterns of energy supply and demand in tropical habitats? How often are the population densities and biomasses of tropical plant–visiting vertebrates limited by their food resources? And to what extent does food limitation vary geographically?

NECTARIVORES AND THEIR FOOD RESOURCES

Compared with fruit pulp, nectar is a very scarce resource in virtually all tropical habitats. As a result, the biomass of nectarivores in tropical forests is generally much lower than that of frugivores. At Cocha Cashu, Peru, for example, the biomass of 11 species of nectar-eating birds is about 0.23 kg/km^2 compared with a biomass of 35.3 kg/km^2 for 25 species of frugivorous birds—more than a hundredfold difference (Terborgh et al. 1990; table 3.10). Similarly, Fleming et al. (2005) reported that the ratio of the abundance of frugivorous phyllostomid bats to nectar feeders based on mist net captures averaged 17:1 at 14 Neotropical sites. Both of these differences reflect the much greater availability of fruit pulp compared to nectar in tropical forests.

Except for dry sclerophyll woodlands and *Banksia* heathlands of Australia, detailed data on the biomass and energy density of nectar are scarce, especially for tropical habitats. Available data (table 3.10) indicate that, as expected, nectar biomass usually varies substantially on a seasonal basis within and between habitats. For example, nectar energy levels of plants visited by nectar-feeding phyllostomid bats vary a hundredfold among seasons in a Costa Rican wet tropical forest, eighty-fold in Australian *Banksia*-dominated heathlands supporting *Syconycteris* pteropodid bats, and about sixty-fold in plants visited by honeyeaters and lorikeets in Australian monsoon forest (Franklin and Noske 1999; Law 1994; Tschapka 2004). At Monteverde, Costa Rica, nectar levels in hummingbird-pollinated plants are generally low and stable year round in undisturbed cloud forest, whereas they vary nearly fourfold in treefall gaps (fig. 3.7A). Petit and Pors (1996) used aerial photography and phenological data to estimate total nectar production by three species of columnar cacti pollinated by phyllostomid bats on the island of Curaçao in the southern Caribbean. Nectar energy levels in these cacti ranged from zero (in January) to 460×10^3 kJ (in May). From these data and energetic calculations, Petit and Pors (1996) concluded that observed numbers of Curaçao's two nectar bats, *Glossophaga longirostris* (about 1,200) and *Leptonycteris curasoae* (about 900), were likely to be at the island's carrying capacity as set by levels of cactus and other kinds of nectar in January and February.

Data from a variety of habitats indicate that the biomass of tropical and subtropical nectar-feeding birds and mammals tends to be low (table 3.10). Maximum reported biomass of nectarivores is about 31 kg/km^2 in Papua New Guinea and southeastern Australia. Compared to other areas, the biomass of hummingbirds at two well-studied sites in Panama and Peru seem especially low (0.23–0.64 kg/km^2). Correlation analyses, as well as experiments in one case, indicate that the density and biomass of these nectarivores are likely to be set by the energy density of nectar in their habitats. Energy limitation has been postulated to occur in glossophagine bats at La Selva, Costa Rica, and on Curaçao; in *Syconycteris* bats in subtropical Australia; in hummingbirds at Monteverde, Costa Rica, and on the Caribbean

Table 3.10. Estimates of Resource and Consumer Energy Densities or Biomass in Various Tropical Habitats

Region/Location	Resource or Consumer Type	Energy Density or Biomass
Plant resources:		
Nectar:		
Neotropics:		
La Selva, Costa Rica (tropical wet forest)	Nectar (18 species of trees, shrubs, vines, and epiphytes)	Variable amount and duration of production: ≤10–1000 kJ/ha/day
Curacao, Netherlands Antilles (arid scrubland)	Nectar (3 species of columnar cacti)	Island-wide estimate: monthly average = 190,558 kJ (range: 0–460,977)
Australasia:		
Northern Territory, Australia (monsoonal woodlands)	Nectar (major trees and shrubs)	Annual range: 12–733 kJ/ha
New South Wales, Australia (*Banksia* heathlands)	Nectar (1 species of *Banksia* shrub)	Nightly standing crop: 50–4,180 kJ/ha
Victoria, Australia (dry eucalypt forest)	Nectar (major trees and shrubs)	Winter and spring: 1,500 kJ/ha; Late spring: 4,000–7,000 kJ/ha; Late summer, autumn: ≤400 kJ/ha
Fruit:		
Neotropics:		
Santa Rosa National Park, Costa Rica (tropical dry forest)	Fruit (16 species of shrubs and trees)	84.6 kg/ha/day at fruiting peak
Central Panama (tropical moist forest)	Fruit (many species of all growth forms)	>2,200 kg/ha/yr*
Cocha Cashu, Peru (tropical moist forest)	Fruit (many species of all growth forms)	>2,000 kg/ha/yr*
Rio Orinoco, Venezuela (tropical moist forest)	Fruit (36 species in a vertical transect)	39.0–92.2 kg/ha (67,000–126,000 kJ/ha) between October and January
Luquillo Experimental Forest, Puerto Rico (tabonuco forest)	Fruit (species of all growth forms)	Mature forest: 332 kg/ha/yr*; secondary forest: 820 kg/ha/yr*
Asia:		
Western Ghats, India (premontane wet forest)	Fruit (53 species of trees, lianas, shrubs, and epiphytes)	544–751 kg/ha/yr*
Animal consumers:		
Nectarivores:		
Neotropics:		
Panama Canal Zone (lowland moist tropical forest)	Nectar-feeding birds (≥8 species)	0.64 kg/km^2
Cocha Cashu, Peru (lowland moist tropical forest)	Nectar-feeding birds (11 species)	0.23 kg/km^2
Asia:		
Sarawak, Borneo (tropical wet forest)	Nectar-feeding birds (>13 species)	≥8 kg/km^2
Australasia:		
Port Moresby, Papua New Guinea (tropical moist forest)	Nectar-feeding birds (≥13 species)	31 kg/km^2

Region/Location	Resource or Consumer Type	Energy Density or Biomass
New South Wales (*Banksia* heathlands)	Nectar bat *Syconycteris australis* (18 g/bat)	1.8–31.5 kg/km^2
Frugivores:		
Neotropics:		
Santa Rosa National Park, Costa Rica (tropical dry forest	Frugivorous mammals (6 species of bats, 2 monkeys)	49 kg/km^2
Panama Canal Zone (lowland moist tropical forest)	Frugivorous birds ($\geq$16 species)	35 kg/km^2
Cocha Cashu, Peru (lowland moist tropical forest)	Frugivorous birds (25 species)	35 kg/km^2
Cocha Cashu, Peru (lowland moist tropical forest)	Frugivorous mammals (species not reported)	1050 kg/km^2
Africa:		
Makande, Gabon (tropical moist forest)	Frugivorous primates (7 species)	204 kg/km^2
Lope, Gabon (tropical wet forest)	Frugivorous mammals	Primates: 250–373 kg/km^2; elephants: 523–5,225 kg/km^2; ungulates: 696–5,496 kg/km^2; squirrels: 4–7 kg/km^2
Asia:		
Sarawak, Borneo (tropical wet forest)	Frugivorous birds (>28 species)	$\geq$30 kg/km^2
Australasia:		
Port Moresby, Papua New Guinea	Frugivorous birds (26 species)	188 kg/km^2

Sources. Bell (1982), Fleming (1988), Fogden (1976), Franklin and Noske (1999), Ganesh and Davidar (2001), Law (1994), Lugo and Frangi (1993), Paton (1985), Petit and Pors (1996), Robinson et al. (2000), Schaefer et al. (2002), Terborgh (1986), Terborgh et al. (1990), Tschapka (2004), White (1994).
Note. Habitat type follows location name.
*Based on fruitfall and hence does not include fruit eaten by animals.

islands of Trinidad and Tobago; in sunbirds in *Protea* woodlands of South Africa; and in honeyeaters in a variety of habitats throughout Australia (DeSwardt 1993; Feinsinger et al. 1985, 1988; Franklin and Noske 1999; Law 1994; Paton 1985; Petit and Pors 1996; Tschapka 2004). Evidence that nectar levels are likely to limit population sizes of hummingbirds on Trinidad and Tobago is shown in figure 3.7B. In disturbed habitats on both islands, energy demand by hummingbirds exceeded nectar supply in most months on Trinidad whereas demand exceeded supply in only 4 of 13 months on Tobago. Finally, by adding supplementary honey water to flowers in study plots in *Banksia* heathlands, Law (1995) was able to significantly increase the abundance of *Syconycteris australis* bats compared to control plots. Increases in abundance in these short-term experiments resulted from immigration of adult bats onto experimental plots. In summary, the biomass of vertebrate nectar feeders is likely to be limited by nectar availability throughout most of the tropics and subtropics.

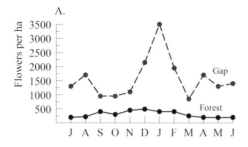

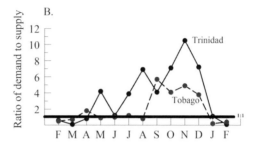

Figure 3.7. *A,* Nectar availability, as indicated by number of flowers per ha, for humming-birds in two habitats at Monteverde, Costa Rica. *B,* Ratio of energy demand by hummingbirds to their energy (nectar) supply in disturbed habitats on two Caribbean islands, Trinidad and Tobago. Based on Feinsinger et al. (1985, 1988).

In addition to influencing the density and biomass of tropical vertebrate nectarivores, nectar levels also influence the species richness of these con-sumers in many habitats. At La Selva, Costa Rica, for example, only two species of nectar-feeding bats are year-round residents. The larger species, *Glossophaga commissarisi* (8.8g), switches to eating fruit when nectar levels are low, whereas the smaller species, *Hylonycteris underwoodi* (7.6 g), re-mains nectarivorous year round by specializing on low-density, steady state flowering species. When nectar levels are high as a result of the flowering of "cornucopia" canopy trees, two other species (*Lonchophylla robusta* and *Lichonycteris obscura*) join this assemblage. *L. robusta* does so by commut-ing long distances daily from higher elevations; *Li. obscura*, in contrast, ap-pears to be nomadic and moves among habitats, tracking flowering peaks (Tschapka 2004). During the annual period of low nectar availability in Sep-tember to early December, three of six species of hummingbirds on Tobago leave that island and probably fly to Trinidad (Feinsinger et al. 1985). Habitat shifts in response to changes in nectar availability are also well known in many other species of hummingbirds as well as in sunbirds, honeyeaters, and pteropodid bats (e.g., Cheke and Mann 2001; Eby 1991; MacNally and McGoldrick 1997; Richards 1995; Schuchmann 1999). Overall, the species composition of assemblages of nectar-feeding birds and bats is highly dynamic as a result of substantial seasonal changes in the availability of nectar.

FRUGIVORES AND THEIR FOOD RESOURCES

As expected, the biomass of fleshy fruits in most tropical habitats far exceeds that of nectar (table 3.10). At well-studied sites such as Barro Colorado Island, Panama, and Cocha Cashu, Peru, annual fruit production, as measured by fruitfall studies, which tend to underestimate total fruit production because they do not include fruit eaten by arboreal or volant frugivores, exceeds 2,000 kg/ha (wet weight of pulp). Data from other sites in the Neotropics indicate that annual fruitfall is positively correlated with annual rainfall (Stevenson 2001 and included references). Whether this correlation holds for other regions is currently unknown but is likely not to hold in regions where masting is common (e.g., Southeast Asia). In regions of resource boom or bust, annual averages are not likely to be relevant to animal consumers that are faced with feast or famine depending on year and habitat. Regardless of region, fruit biomass likely varies substantially within and between habitats (e.g., Levey 1988; Loiselle and Blake 1993). The general pattern is that fruit production is greater in secondary forests than in primary forests and is greater in gaps than in intact forest. Data from La Selva, Costa Rica, illustrate this pattern (fig. 3.8).

As a result of high levels of fruit production, frugivore biomass tends

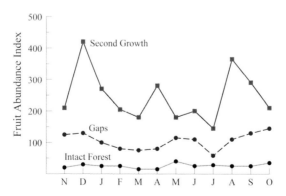

Figure 3.8.   Monthly levels of bird-fruits in three habitats at La Selva, Costa Rica. Based on Levey (1988).

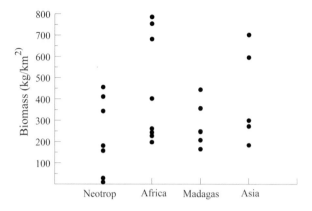

Figure 3.9. Estimates of the biomass of frugivorous primates in four tropical regions. Each point represents a different locality. Based on Janson and Chapman (1999).

to be substantial in most tropical forests (table 3.10). This is particularly true in tropical forests of equatorial Africa where fruit-eating elephants and ungulates are (or were) common. Compared to many areas in the Paleotropics, the Neotropics is depauperate in terrestrial mammalian frugivores, and hence total frugivore biomass tends to be higher in many areas of the Paleotropics than in the Neotropics (Cristoffer and Peres 2003; Primack and Corlett 2005). Large terrestrial birds (e.g., cassowaries, 18–59 kg) are currently restricted to Australasia, where they are by far the largest frugivores in their habitats. Among arboreal birds, biomass estimates range from about 30 kg/km² in Sarawak to nearly 200 kg/km² in Papua New Guinea (table 3.10). Among frugivorous arboreal mammals, primates account for the greatest biomass in lowland tropical forests (e.g., Terborgh 1983). Biomass estimates of frugivorous primates in four tropical regions are summarized in figure 3.9. These estimates range from about 10 kg/km² in the northern Neotropics to nearly 800 kg/km² in parts of equatorial Africa. Although there is a trend for frugivorous primate biomass to be lower in the Neotropics than in Africa and Asia, mean values for these regions do not differ significantly (ANOVA, $F_{3,22} = 1.83$, $P = 0.17$). Including leaf-eating primates in these data would substantially increase estimates of primate biomasses in African tropical forests (Janson and Chapman 1999; Terborgh and Van Schaik 1987).

As is the case for vertebrate nectar feeders, food limitation appears to be common in frugivorous birds and mammals throughout the tropics. At Cocha Cashu, Peru, for example, fruit production appears to exceed frugivore demand (of about 4–8 kg/ha/day) for 8–9 months of the year before falling below levels of demand during the annual 2–3 month lean season. Resource levels during the lean season determine the carrying capacity of frugivores at this site (Terborgh 1986). During this season, frugivorous birds tend to migrate to other habitats, and primates switch to alternate food resources (see above). Based on long-term observations on Barro Colorado Island, Panama, Wright et al. (1999) postulated that fruit levels during irregularly occurring "famine" years play an important role in determining the population densities and biomass of vertebrate frugivores there.

Statistical support for the hypothesis that frugivore population density is controlled by fruit availability comes from a variety of primate studies. The density of orangutans in dryland forests in Sumatra is positively correlated with the density of strangling figs (Wich et al. 2004, 2006). Similarly, the density of common chimpanzees at Kibale National Park, Uganda, is positively correlated with the density of fruiting trees (Balcomb et al. 2000). Brugiere et al. (2002) indicated that primate density tends to be low in forests in Gabon dominated by masting caesalpinaceous legume trees, in South American forests dominated by masting Lecythidacaeae, and in Asian forests dominated by masting Dipterocarpaceae. Stevenson (2001) examined the relationship between primate biomass and the availability of different kinds of food resources (e.g., fruit, seeds, figs, seeds of Lethycidaceae eaten by pitheciid monkeys) at 30 Neotropical sites and found that the biomass of frugivorous species (but not folivores) was most strongly correlated with annual fruit production. He concluded that frugivorous primate biomass was controlled by fruit production at these sites. Similarly, Peres (2008) examined the relationship between soil fertility and primate biomass, at both the community and individual species level, at 60 undisturbed sites in the Amazon Basin and Guianan Shield and found a positive correlation (r^2 = 43%). He concluded that primate biomass is regulated in bottom-up fashion (i.e., by plant productivity that increases with soil fertility).

Many recent studies have reported a positive correlation between the abundance of fruit-eating birds and fruit abundance. Kinnaird et al. (1996) noted that the density of hornbills, the largest nonterrestrial frugivorous birds in African and Asian forests, was positively correlated with the density of fig trees in northern Sumatra as well as in East Africa and Southeast Asia. Similar results have been reported for various frugivorous birds in Amazonian Brazil and northwestern Argentina (Malizia 2001; Wunderle et al. 2006), for tanagers in Puerto Rico (Carlo et al. 2004; Saracco et al. 2004), and for hornbills in Cameroon (Holbrook et al. 2002). Experimental evidence that these correlations have a causal basis comes from studies in Jamaica and Brazil. Brown and Sherry (2006) added fruit (oranges) to five plots and compared the abundance of banded birds on experimental and control plots over two dry seasons in Jamaica. Bird abundance and recapture probabilities were higher on the fruit-supplemented plots. In Brazil, Moegenburg and Levey (2003) removed *Euterpe* palm fruits from experimental plots and monitored the abundance and activity of frugivorous birds and mammals (coatis, squirrels, and monkeys). Compared with control plots, the number of frugivorous birds and mammals and their activity levels were lower on the fruit-removal plots. In summary, both correlative and experimental evidence appear to support the hypothesis that frugivore population levels are determined by fruit availability. Nonetheless, results of long-term studies of fruit production and population dynamics of several species fruit-eating mammals on Barro Colorado Island suggest that we need to be cautious in reaching this conclusion (Milton et al. 2005).

Additional evidence that levels of fruit production determine population densities of vertebrate frugivores comes from between-habitat studies at different research sites. At La Selva, Costa Rica, for example, densities of fruit-eating birds and bats are higher in second-growth habitats than in primary forest. Fleming (1991b) reported that the abundance of *Piper*-eating *Carollia* bats was over twice as high in second-growth as in intact forest, although the opposite situation prevailed on Barro Colorado Island. At La Selva, density and species richness of *Piper* plants in second-growth is two to three times higher than that in primary forest (Fleming 2004a). Similarly, Levey (1988) and Loiselle and Blake (1993) found that within and between habitats at La Selva, the abundance of fruit-eating birds was correlated with the abundance of their food plants and that bird and fruit densities were much higher in second-growth than in primary forest (fig. 3.8). Similar results have been documented for birds in Brazilian terra firme forest and subtropical Bolivian forests (Woltmann 2003; Wunderle et al. 2006)

In addition to influencing the biomass of frugivores, seasonal changes in the availability of fruit also affect the species richness of assemblages of frugivorous vertebrates in many tropical habitats. These changes reflect movements of frugivores among habitats and along elevational and latitudinal gradients. In lowland mixed dipterocarp forests in Malaysia, for instance, the resident assemblage of fruit-eating pteropodid bats includes three species that feed primarily on five species of steady-state fruiters.

When 16 species of big-bang trees are fruiting, however, this assemblage increases to eight species (Hodgkison et al. 2004a). Similar changes occur in assemblages of hornbills in Borneo and fruit pigeons in tropical Australia (Crome 1975; Leighton 1986). At La Selva, Costa Rica, both latitudinal and elevational migrants increase the abundance and species richness of fruit-eating birds when fruit levels are high in October to January (Levey 1988). Elevational movements in response to changes in fruit availability have also been described for birds elsewhere in Central America and Africa (e.g., Blake and Loiselle 2001; Burgess and Mlingwa 2000; Chaves-Campo 2004; Chaves-Campo et al. 2003; Powell and Bjork 2004, among others). These movements occur in charismatic birds such as trogons, cotingids, and toucans in the Neotropics.

In summary, there is substantial (mostly correlative but some experimental) evidence that total fruit production or fruit production by important groups of plants (e.g., figs) plays an important role in determining the biomass of fruit-eating tropical birds and mammals throughout the tropics. Experimentally testing this hypothesis on large spatial scales (i. e., on more than a few hectares) is logistically difficult and ethically unacceptable but is nonetheless occurring on a daily basis throughout the tropics as a result of selective logging and deforestation. Based on our current knowledge, these "experiments," unfortunately, will have easily predictable results: both the biomass and species diversity of frugivores (and nectarivores) will decline as their food supplies are fragmented, reduced, or eliminated. Given the large spatial scales at which many species of vertebrate nectar and fruit eaters operate, in large part in response to spatiotemporal variation in their food supplies (e.g., Chaves-Campos 2004; Fleming and Nassar 2002; Powell and Bjork 2004), deforestation and habitat fragmentation will likely have a strong negative effect on these mutualists. We will discuss this topic in more detail in chapter 10.

Conclusions

The plant-visiting birds and mammals we discuss in this book feed on two distinctly different resources, nectar and fruit. Nectar is a low density, energy rich resource that is well-advertised by conspicuous flowers. Fruit, in contrast, often occurs at rather high densities but is less digestible than nectar. It, too, can be relatively easy to find via visual and olfactory cues. Considerable spatiotemporal heterogeneity characterizes both kinds of resources in all tropical habitats. Temporal heterogeneity results from seasonal flowering and fruiting that is ubiquitous, even in ever-wet lowland tropical forests. Spatial heterogeneity has both a vertical and horizontal (i.e., between-habitat) component. Flowering and fruiting is more seasonal in canopy and subcanopy trees than in understory herbs, shrubs, and treelets; epiphytes also tend to have more extended flowering and fruiting seasons than their host plants. The biomass of nectar and fruit also tends to be greater in forest gaps and in second-growth than in intact forests, especially in the understory. Masting flowering and fruiting behavior produces boom-or-bust levels of nectar and fruit and is most common in the lowland forests of Southeast Asia, where this phenological behavior occurs on a nearly community-wide basis. These forests produce the least predictable resources for nectar- and fruit-eating birds and mammals.

Available evidence strongly supports the hypothesis that population densities, biomass, and, often, species richness of vertebrate nectarivores and frugivores are strongly correlated with levels of their food supplies throughout the tropics. This evidence does not support our initial hypothesis in this chapter, based on hemispheric differences in the relationship between animal species richness and the species richness of their food plants, that food limitation is more likely to occur in Neotropical than in Paleotropical habitats. In many cases this correlation likely has a causal basis so that many populations of tropical plant–visiting birds and mammals are food-limited. Limitation often occurs during lean times of the year or in lean years (common in Southeast Asia but uncommon in the Neotropics) when many animals are forced to migrate or switch to alternate foods. Certain plant families (e.g., Fabaceae and Myrtaceae for nectarivores, Moraceae for frugivores) are especially important in providing food for many species of vertebrate plant visitors. Loss of species in these families would have an especially strong negative effect on these animals in many tropical habitats. In the end, under natural conditions (i.e., in the absence of anthropogenic effects) the fate of many tropical nectar- and fruit-eating birds and mammals appears to be in the hands of their food supplies. To what extent is the fate (i.e., reproductive and recruitment success) of their food plants in the hands (or beaks, jaws, wings, guts, etc.) of these birds and mammals? We address this question in the next chapter.

4

Pollen and Seed Dispersal and Their Ecological and Genetic Consequences

Many tropical birds and mammals eat nectar, pollen, and fruit to obtain energy and nutrients. In so doing, they unwittingly provide important ecological services for their food plants. When nectar and pollen eaters feed, flowers get pollinated; when fruit eaters feed, seeds get dispersed. But, in the words of Howe and Westley (1988), these acts of feeding represent "uneasy partnerships" because in virtually all cases involving vertebrates, pollination and seed dispersal are passive acts. Unlike pollinator-parasites such as yucca moths, fig wasps, and senita moths, vertebrate pollinators do not deliberately place pollen on conspecific stigmas to ensure that their seed-eating offspring will have a food supply. Similarly, to vertebrate frugivores, seeds generally represent extra weight or ballast that needs to be voided, usually in rapid fashion. Where seeds are voided does not concern them (at least in the short term), but it can have very important consequences for their food plants, just as patterns of pollen deposition can affect the reproductive success of the food plants of nectarivores. The feeding and foraging behavior of tropical vertebrates thus has important ecological consequences. And, because pollen and seeds are the two ways by which plants transmit their genes to the next generation, their dispersal also affects the genetic structure of plant populations. Pollen and seed dispersal therefore also have important evolutionary consequences.

In this chapter we will describe general patterns of pollen and seed dispersal by tropical birds and mammals and will discuss their ecological and genetic consequences for plants. The central questions we address here are: (1) How important are nectar- and fruit-eating birds and mammals in the

economy of tropical habitats? To what extent does the reproductive success of their food plants depend on the foraging behavior of plant-visiting
vertebrates? (2) How do nectar-feeding vertebrates affect fruit and seed set,
mating patterns, and gene flow in their food plants, and how do frugivores
influence the density, dispersion, and genetic structure of their food plants?
And (3), if these mutualists were to disappear, how much would this affect
the population and community dynamics of tropical plants? As in previous chapters, we will take a broad biogeographic perspective in addressing
these questions by asking, To what extent do vertebrate pollination and seed
dispersal mutualisms differ among regions? For example, are Paleotropical
sunbirds and honeyeaters as effective pollinators as Neotropical hummingbirds? Are Neotropical and Paleotropical primates functionally equivalent
in terms of their seed dispersal "services"? What is the relative importance
of birds and bats for forest regeneration in different biogeographic regions?
And finally, does the genetic structure of plant populations whose pollen
and seeds are dispersed by vertebrates differ from that of populations with
insect pollination or abiotic seed dispersal?

Although our primary focus in this book is on tropical plant–visiting
birds and mammals and their food plants, we will relax our geographic
boundaries somewhat in this and subsequent chapters in discussing the
ecological functions of these animals. We do this because some of the most
detailed and instructive work of this kind has been done in "near-tropical"
habitats such as the Sonoran Desert of northern Mexico and southern Arizona, Mediterranean shrublands of southern Europe, the flower-rich fynbos
of South Africa, and sclerophyll woodlands and heathlands of eastern Australia. Except for Mediterranean shrublands, many of the plants and animals
in these habitats are closely related to tropical organisms. Studies in these
habitats represent models of how similar studies might be done in the tropics in the future.

Ecological and Genetic Consequences of Plant-Pollinator Interactions

POLLEN LIMITATION

A fundamental question in pollination biology is the extent to which fruit
and seed set is limited by pollen dispersal versus a plant's internal resources.
Pollinators and their behavior influence only one of these two forms of
limitation. Pollen limitation (PL) occurs when levels of fruit and seed set

fall below those set by the availability of plant resources. The usual way to demonstrate PL is by pollen supplementation experiments in which extra conspecific pollen is added to stigmas and the resulting fruit and seed set of experimental flowers is compared with that of unsupplemented control flowers. Although Zimmerman and Pyke (1988) suggested that these experiments need to be done at the level of whole plants (i.e., all flowers on a plant should either be supplemented or not to control for the possibility that plants will selectively reallocate resources to pollen-supplemented flowers), this is rarely done, primarily for practical reasons. For large, many-flowered tropical trees and many other kinds of plants, it is simply impossible to hand-pollinate an entire flower crop or to repeat these experiments over several life cycles. Thus, most pollen supplementation studies only treat subsamples of a plant's flower crop. Despite this potential bias, these studies can still give us valid insights into questions concerning pollen limitation, although Aizen and Harder (2007) warn that such studies can give biased estimates of degree of pollen limitation if they do not control for pollen quality as well as pollen quantity.

Although sexual selection theory predicts that seed set (a female function) should be limited by resources and not by pollen (Bateman 1948; Willson and Burley 1983), empirical data indicate that pollen limitation is common in plants (reviewed by Ashman et al. 2004; Burd 1994; Knight et al. 2005; Larson and Barrett 1999). In their review of PL, Knight et al. (2005) reported that significant PL occurred in 63% of 482 studies and that fruit set (the most common response variable in PL studies) averaged 75% higher in pollen-supplemented flowers than in control flowers. Significant sources of variation in PL among plants included taxonomic family, timing of flower production within species (plants flowering at peak blooming time are less likely to experience PL than plants flowering before or after the peak), and mating system (self-incompatible plants are more likely to experience PL than self-compatible plants). Pollen limitation was not related to size, shape (e.g., actinomorphic vs. zygomorphic—a measure of pollinator specialization), or longevity of flowers or to the number of pollinator species per plant, nor was it related to plant longevity (e.g., monocarpic vs. polycarpic) or degree of asexual reproduction.

Most studies of PL have been conducted on temperate plants in the Northern Hemisphere, and we currently do not know whether PL occurs more frequently in temperate compared with tropical or subtropical plants. Results of a meta-analysis of pollen supplementation studies, however, indicate that extent of PL increases with increasing regional plant species richness, perhaps as a result of increased competition for pollinators in high-diversity areas (Vamosi et al. 2006). These results imply that PL should be significantly more common in tropical plants than in temperate plants. Among subtropical and tropical plants, PL has been reported in three species of bat- and bird-pollinated Sonoran Desert columnar cacti but not in tropical bat-pollinated columnar cacti (Fleming et al. 2001; Molina-Freaner et al. 2004; Nassar et al. 1997; Valiente-Banuet et al. 1997); bird- and bat-pollinated paniculate *Agaves* (Molina-Freaner and Eguiarte 2003; Slauson 2000); the tree *Caryocar brasiliense* (Gribel and Hay 1993); the understory tropical palm *Calypterogyne sarapiquensis* (Cunningham 1996); a tropical weed (*Triumfetta semitriloba*; Tiliaceae)(Collevatti et al. 1997); and in South African wildflowers and Australian bird-pollinated *Banksia* and *Telopea* shrubs (Proteaceae [Dalgleish 1999; Johnson and Bond 1997; Whelan and Goldingay 1989]). In the case of *Banksia aemula*, pollinator exclusion experiments revealed that PL occurred only when flowers were pollinated exclusively by honeyeaters rather than by both bees and honeyeaters (Dalgleish 1999).

Although PL obviously affects individual seed production (and fitness), it will only have an effect on population growth rates, λ, when seed production significantly affects the seed-to-seedling transition probability in a plant's life cycle (fig. 4.1; Ashman et al. 2004). In the terminology of population matrix projection analysis (e.g., Caswell 1989; Howe and Mitiri 2004; Tuljapurkar et al. 2003), PL will have a significant effect on λ when the elasticity (i.e., a factor's relative effect on λ compared with other factors) of the seed-to-seedling transition is large relative to other transition probabilities in the life cycle. If its elasticity is low, then this transition and the extent to which PL affects it will have little effect on a population's λ. Ashman et al. (2004) predicted that PL and its concomitant reduction in seed production will have a greater effect on plants with growing populations (e.g., early successional plants?) than those with stable populations. They also point out that PL is unlikely to have a strong effect on λ in plants whose seed-to-seedling transition is limited by factors other than seed production. Two of those factors include seed dispersal limitation and recruitment limitation, which we discuss later in this chapter.

To our knowledge, studies determining the relative importance of PL for population growth rates in tropical plants are almost nonexistent. This is not surprising given the large amount of work these kinds of studies entail (Godinez-Alvarez and Jordano 2007; Silvertown et al. 1993). Godinez-Alvarez

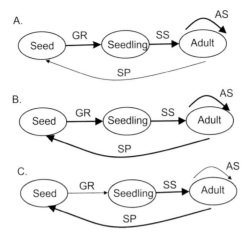

Figure 4.1. Plant life-cycle diagrams indicating how pollen limitation might affect population growth rate, λ. Width of arrows indicates the strength of particular processes on population growth. *A*, Pollen limitation limits seed production (the "natural" condition). *B*, Pollen supplementation increases seed production (SP) which may or may not affect λ. *C*, Pollen supplementation increases seed production but other processes (e.g., seedling survival [*SS*]) limit its effect on λ. *AS* = adult survival and *GR* = germination rate. Based on Ashman et al. (2004).

and Valiente-Banuet (2004) conducted a preliminary study of the demography of the Mexican bat-pollinated columnar cactus *Neobuxbaumia macrocephala* and concluded, without presenting detailed data, that low seed production may account for its low density and restricted distribution compared with the much more common cactus *N. tetetzo*. However, elasticity analyses for both species indicated that survival has a much greater influence on λ than reproduction in these slow-growing, long-lived plants. As discussed below, a more detailed analysis of the demography of *N. tetetzo* was also able to determine the relative contribution of different species of fruit-eating birds and bats to this species' λ (Godinez-Alvarez et al. 2002).

Heithaus et al. (1982) had a more modest demographic goal in determining the relative importance of pollination for seed production, the first stage in the seed-to-seedling transition, in the bat-pollinated legume shrub *Bauhinia ungulata*. Working in Costa Rican tropical dry forest, they used key factor analysis to determine the relative contribution of different mortality factors to overall seed mortality. In addition to lack of pollination, these factors included herbivory on flowers and seed pods, incomplete ovule development, and predispersal seed predation. Results of their analysis showed that a low rate of pollination by the phyllostomid bat *Glossophaga soricina*, along with incomplete ovule development and seed predation, was an important mortality source. This study highlights the importance of considering multiple sources of seed mortality in assessing the relative contribution of PL to plant demography.

Anderson et al. (2011) studied the effects of reduced nectar-feeding bird densities on fruit and seed set in the understory shrub *Rhabdothamnus solandri* (Gesneriaceae) in two regions in New Zealand: the mainland of the upper half of the North Island and several islands 4–20 km away from the mainland. The three native avian pollinators (two meliphagids and a species of Notiomystidae) have been functionally extinct on the mainland since about 1870 but are still common on offshore islands. Anderson et al. reported that pollination limitation was strong on the mainland but not on the islands; fruit set on the mainland was reduced by 69% below maximum levels whereas it was reduced only 15% on the islands. Similarly, mainland seed production per flower was reduced 84% compared with island seed production. They further reported that seedling densities around adult plants were much lower on the mainland than on the islands. Seed addition experiments indicated that the low mainland seedling densities were the result of dispersal limitation. Thus, actual or virtual extirpation of avian nectarivores on the mainland of New Zealand has had a negative effect on both pollination success and recruitment success in this plant. Other similar studies are likely to produce similar results, indicating that pollinators can have significant demographic effects on populations of their food plants.

POLLINATOR EFFECTIVENESS

The evolutionary rationale for the existence of pollinator syndromes centers on the idea that different kinds of pollinators differ in their ability to successfully pollinate flowers. Even within particular pollinator guilds, not all species are likely to be equally effective in contributing to seed and fruit set. For example, both hummingbirds and glossophagine bats visit flowers of *Burmeistera* (Campanulaceae) in the mountains of Central and South America. But Muchhala's (2003, 2006a) experimental studies indicate that bats deposit much more pollen on *Burmeistera* stigmas than do hummingbirds. All else being equal, bats should have a stronger effect on the reproductive success of these plants than hummingbirds, and bats should produce stronger selective effects on their flowers. In discussing whether plants should specialize on particular kinds of pollinators, Stebbins (1970) proposed the

"most effective pollinator principle," which states that plants should specialize on the most abundant and/or most effective pollinator(s) whenever its
spatiotemporal reliability is high. Whenever the spatiotemporal reliability of
the most effective pollinator is low, then plants should not specialize on that
pollinator(s). Applied to the *Burmeistera* case, this principle predicts that
these plants should evolve flowers that conform to the bat pollination syndrome whenever bats are reliable visitors; otherwise, they should specialize
on hummingbirds or have more generalized pollination systems that attract
both bats and hummingbirds (Muchhala 2006a).

   Ne'eman et al. (2010) have reviewed the literature on pollinator efficiency
and effectiveness and recommend that these two terms be avoided in at-
tempts to quantify the relative importance of particular pollinators for the
reproductive success of their food plants. In assessing the performance of
pollinators, they instead recommend addressing two questions: What is
each species' proportional contribution to the pollen deposited on a plant's
stigma? And what is the contribution of this deposition to a plant's reproductive success? While in theory these two questions are key to understanding plant-pollinator interactions, in practice they are difficult to answer because they involve quantifying the results of single pollinator visits in terms
of seed set. Conducting these kinds of experiments in species-rich pollinator environments is likely to be prohibitive in most instances. Nonetheless, many recent studies have examined the relative importance of different
kinds of pollinators for plant reproductive success (e.g., Fumero-Caban and
Melendez-Ackerman 2007; Martén-Rodríguez and Fenster 2010; Ortega-
Baes et al. 2011; Schmid et al. 2011), and some of these have used a single-visit
approach to determine each species' contribution to total seed set. In the
Brazilian bromeliad *Aechmea nudicaulis*, for instance, single-visit experiments revealed that the hummingbird *Thalurania glaucopis* was the most
effective pollinator but that because of their high visitation frequency, several kinds of bees were also effective pollinators (Schmid et al. 2011). These
authors concluded that this bromeliad has a bimodal pollination system involving both hummingbirds and bees.

VERTEBRATE CONTRIBUTIONS
TO FRUIT AND SEED SET

Beyond the questions of pollen limitation and pollinator effectiveness of
particular species, one can ask: To what extent do birds or mammals actually
contribute to fruit and seed set in ostensibly vertebrate-pollinated plants?

Despite the existence of distinctive bird and mammal pollination syndromes (chap. 1), in reality vertebrate and insect pollinators sometimes (often?) visit and pollinate flowers belonging to other syndromes, especially when flowers lack morphological features that strongly limit access to their pollen and nectar (e.g., Dalsgaard et al. 2009). When this happens, it becomes important to know in detail how different kinds of pollinators affect a plant's reproductive output. One reason for this is its conservation implications in terms of ecological redundancy. Ecological redundancy occurs when different species are equally effective (and substitutable *sensu* Tilman [1982]) as pollinators or seed dispersers so that the loss of one or more species does not result in reduced reproductive success of its food plants; the remaining species can compensate for the missing species. In nonredundant or complementary systems, loss of one or more species does result in reduced plant reproductive success; the remaining species cannot compensate for the missing species.

Columnar cacti, whose species diversity and population densities tend to be highest in arid Neotropical habitats (e.g., in the Tehuacan Valley of south-central Mexico), are excellent plants for demonstrating how reliance on different kinds of vertebrate pollinators can vary geographically. Many columnar cacti in tribes Pachycereeae and Cereeae of subfamily Cactoideae produce flowers that conform to the classic bat pollination syndrome (Fleming 2002; Valiente-Banuet et al. 1996). Their flowers tend to be large, white in color, open at or just after sunset, and produce copious amounts of nectar and pollen. Pollinator exclusion experiments in Peru, Venezuela, Mexico, and Arizona indicate that degree of PL and the contribution of bats to fruit and seed set in these plants varies in a predictable geographic pattern. Pollen limitation increases and the importance of bats as the sole pollinators of these plants decreases at the northern and southern range limits of these tribes (Fleming 2002; Fleming et al. 2001). Thus, in the Sonoran Desert, hummingbirds, white-winged doves, and bees are more important than bats as pollinators of saguaro (*Carnegiea gigantea*) and organ pipe (*Stenocereus thurberi*; but not in the primarily bat-pollinated cardon, *Pachycereus pringlei*). In this system, organ pipe and cardon have redundant pollination systems involving both nocturnal and diurnal pollinators whereas organ pipe has a nonredundant pollination system (Fleming et al. 2001). Similarly, in addition to the phyllostomid bat *Platalina genovensium*, hummingbirds are important pollinators of *Weberbauerocereus weberbaueri* in the mountains of southern Peru. In contrast, bats are the most important or exclusive pollinators of columnar cacti in south-central Mexico, northern Venezuela, and

Curaçao; these plants have nonredundant pollination systems. An exception to this is the red-flowered cactus *Marginatocereus marginatus*, which is pollinated by hummingbirds and bats in south-central Mexico (Dar et al. 2006). Unlike the strictly bat-pollinated cacti in this region, fruit set in this species is severely pollen limited, even with an expanded set of pollinators. Dar et al. (2006) suggest that competition for bat pollinators from more common cacti may be responsible for this. A similar trend away from pure bat pollination to a mixture of bird, insect, and bat pollination in a south to north direction in Mexico also occurs in paniculate *Agave*s (Agavaceae; Molina-Freaner and Eguiarte 2003; Rocha et al. 2005; Slauson 2000).

Pollinator exclusion experiments aimed at determining the relative importance of different kinds of pollinators are uncommon for other kinds of tropical plants. Crome and Irvine (1986) studied the relative contributions of four species of honeyeaters and two species of nectar-feeding pteropodid bats to fruit set in the Australian subcanopy rain forest tree *Syzygium cormiflorum* (Myrtaceae). Despite substantially higher flower visitation rates, honeyeaters contributed less to fruit set than did the less common flower bats; fruit set attributed to these two groups was 8%–22% and 32%–48% among experimental trees, respectively. Subsequent research with this system (Law and Lean 1999) found that individual bats carried about 10 times more pollen than honeyeaters and that they were more likely to transport *Syzygium* pollen among forest fragments than birds. Also in tropical Australia, Boulter et al. (2005) used pollinator exclusion experiments to determine that in *Syzygium sayeri* trees about 50% of fruit set is affected by two or three species of nocturnal pteropodid bats and 50% by five species of diurnal honeyeaters. As a final example, hummingbirds, phyllostomid bats, and moths visit flowers of *Aphelandra acanthus* (Acanthaceae) in the Andes of Ecuador. Observations of visitation frequencies and the results of pollen deposition experiments indicate that bats likely contribute to about 70% of pollination (Muchhala et al. 2009). Nonetheless, it is unlikely that these plants are experiencing strong selection pressure to specialize on bats because they deposit many heterospecific pollen grains on *Aphelandra* stigmas; this diminishes their overall effectiveness as pollinators of this plant.

THE SEXUAL AND MATING SYSTEMS OF TROPICAL PLANTS

Along with pollinator foraging behavior, plant sexual and mating systems play an important role in determining levels and distribution of genetic diversity within and among populations. Sexual systems include hermaphrodite (individual flowers have both male and female function), monoecious (separate male and female flowers on a plant), and dioecious (separate male and female individuals) systems. Mating systems include self-compatible, which results in selfing; self-incompatible, which results in outcrossing; and "mixed" systems, which include both selfing and outcrossing. Hermaphrodites and monoecious species can be either self-compatible or self-incompatible and hence can undergo either self-fertilization (if self-compatible) or outcrossing (if self-incompatible); dioecious species can only outcross. It is widely believed that both monoecious and dioecious sexual systems have evolved to reduce selfing rates and the costs associated with inbreeding (Charlesworth and Charlesworth 1987; Charlesworth 1999), but Barrett (2003) has suggested that the main advantage of moneocy is that it allows plants to allocate resources differentially to male and female function depending on resource availability.

Data summarized in table 4.1 indicate that about 62% (range: 47%–84%) of species of tropical trees are hermaphrodites; about 16% (range: 9%–33%) are monoecious; and about 25% (range: 16%–40%) are dioecious. Australian tropical forests are notable for their high percentage of monoecious species (Gross 2005). The frequency of dioecy is lower in tropical shrubs and herbs than in canopy trees (Bawa 1992). Data illustrating this point for tropical wet forest at La Selva, Costa Rica, are shown in table 4.2.

Biologists once thought that many tropical trees must have self-compatible breeding systems and are inbred because most species are rare

Table 4.1. Summary of the Sexual Systems of Tropical Trees at 11 Localities in the New and Old World

Location/Forest Type	N Species	Percentage of Species That Is:		
		Hermaphrodite	Monoecious	Dioecious
Mexico, tropical dry forest	188	58	16	26
Costa Rica:				
Tropical wet forest	333	65.5	11.4	23.1
Tropical dry forest	ca. 130	68	10	22
Central America, tropical dry forest	141	64	15	21
Colombia, tropical wet forest	52	54	17	29
West Indies (Puerto Rico, Virgin Islands), tropical dry forest	463	73.4	8.9	17.7
Nigeria, tropical wet forest	ca. 220	47	13	40
Borneo, tropical wet forest	711	60	14	26
New Caledonia, tropical wet forest	123	84	NA	16
Australia:				
Tropical wet forest (upland)	145	53.8	26.2	20
Tropical dry forest	61	49.2	32.8	18

Source. Based on Gross (2005).

Table 4.2. Sexual Systems and Mating Systems of Flowering Plants in the
Lowland Rain Forest at La Selva, Costa Rica

Sexual System or Mating System	Percent of Species and Forest Stratum			
	Understory (225)	Subcanopy (184)	Canopy (98)	All Strata (507)
Sexual system:				
Hermaphrodite	74.7	67.4	65.3	70.2
Monoecious	15.5	9.8	10.2	12.4
Dioecious	9.8	22.8	24.5	17.4
Mating system:				
Self-compatible	65.8	9.1	25.0	49.1
Self-incompatible	34.2	90.9	75.0	50.9

Source. Based on Kress and Beach (1994).
Note. Number of species is shown in parentheses.

and occur at low densities (e.g., Baker 1959; Federov 1966). This supposition itself resulted in a controversy as to the factors responsible for the great number of species found in tropical forests. If self-compatibility is prevalent, Fedorov (1966) suggested that genetic drift should play a predominant role in the evolutionary divergence of tropical trees, leading to high levels of speciation even within continuous low-density populations. It is now known that most tropical trees are self-incompatible (Bawa 1992; Bawa and Opler 1975; Loveless 2002; Murawski 1995; Ward et al. 2005), and accordingly Ashton (1998) concluded that selection in tropical plants must favor high rates of outcrossing and substantial levels of genetic variation over the greater reproductive assurance, but lower genetic variation, associated with self-compatibility. Given that most tropical trees are rare, widely spaced, and self-incompatible, successful reproduction and the avoidance of strong PL requires mobile pollen dispersers. This, in turn, provides powerful selection for relatively large, mobile pollinators that can carry substantial pollen loads on their bodies and that have the cognitive abilities to locate (and remember the locations of) widely spaced food plants. Volant vertebrates—birds and bats—would seem to fit this bill.

The frequency of self-incompatibility is lower in understory plants than in canopy trees (Bawa 1992; Kress and Beach 1994). At La Selva, for instance, the frequency of self-incompatibility in canopy and subcanopy trees ranges from 75% to 91%, whereas it is only 34% in understory plants (table 4.2). Thus, most tropical trees are likely to be outcrossers whereas selfing is likely to be common in understory plants. We will discuss the genetic consequences of this below.

Estimates of actual rates of outcrossing and selfing in plants became possible with the advent of allozyme studies in the 1970s and 1980s. When Loveless and Hamrick (1984) first reviewed electrophoretic estimates of genetic diversity in plants, very few data were available for tropical trees, but the situation has changed dramatically since then (Hamrick et al. 1992; Hamrick and Godt 1996; Ward et al. 2005). Beginning in the 1990s, dominant nuclear DNA markers (e.g., RAPDS, AFLPs) and microsatellites are now being used to estimate levels of genetic diversity and outcrossing rates in plants. Fortunately, results from studies using allozymes and DNA markers are usually quite similar (Nybom 2004) so that the earlier data still provide important information about the mating systems and the genetic structure of plant populations. In addition to estimating outcrossing rates, allozyme and other genetic studies can be used to determine rates of biparental inbreeding (i.e., the frequency of matings between close relatives) as well as the number of sires that father the seeds in a single fruit (e.g., Irwin et al. 2003; Pardini et al. 2007). These studies now provide us with a very detailed picture of the breeding structure of plants.

Allozyme- and microsatellite-based estimates of outcrossing (OC) rates in a series of Neotropical trees are summarized in figure 4.2A (Ward et al. 2005). These data are broken down by pollinator type in an attempt to see if vertebrate pollinators (hummingbirds and phyllostomid bats) produce higher OC rates than a variety of insect pollinators. Based on their size and potential mobility, we might expect birds and bats to produce higher OC rates than insects, but these data do not support this hypothesis. If anything, trees pollinated by small insects and bees appear to have higher OC rates than those pollinated by vertebrates. One reason why OC rates are low in bat-pollinated trees is that some of these species belong to the Malvaceae-Bombacoideae clade, in which mixed mating systems are well-known (Quesada et al. 2003; Ward et al. 2005). Selfing rates in bat-pollinated *Ceiba pentandra*, for example, range from 0% to 98% in different populations owing to asynchronous flowering among individuals and nonrandom foraging by bats (see below). Selfing rates >50% are also known to occur in some Neotropical Fabaceae (which are commonly insect pollinated) and in Asian Dipterocarpaceae (which are thrip pollinated; Ward et al. 2005).

Detailed studies of the mating systems of bird- and bat-pollinated tropical plants are still in their infancy. Examples of OC rates and other details about mating systems of several Neotropical plants pollinated by hummingbirds and glossophagine bats are summarized in table 4.3. Similar data are not yet available for paleotropical plants pollinated by sunbirds, honeyeaters, or pteropodid bats, so interhemispheric comparisons are not yet possible. Available data show that OC rates can vary quite substantially within

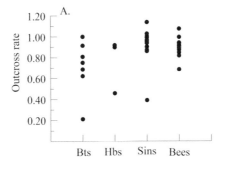

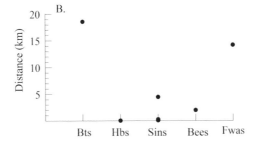

Figure 4.2.   The effects of different pollinators on outcrossing rates *A*, and maximum pollen movement distances *B*, in species of Neotropical trees. Abbreviations: *Bts* = bats, *Hbs* = hummingbirds, *Sins* = small insects, *Fwas* = fig wasps. Based on data in Ward et al. (2005).

species depending on habitat. Plants in low-density populations or that occur in forest fragments or as isolated individuals tend to have lower OC rates than plants in continuous forest. Biparental inbreeding, which usually reflects short-distance pollen movements between related plants, occurs in both hummingbird- and bat-pollinated species. The seed crops of fruits of isolated plants also tend to have fewer sires than fruits produced by trees in continuous forest. Overall, results of these studies suggest that the foraging behavior of tropical hummingbirds and bats is affected by habitat type (e.g., intact vs. fragmented forest; canopy vs. understory) and that this behavior in turn can affect the mating systems of their food plants. Plants in continuous forest or intact habitats likely have higher OC rates and hence larger effective population sizes and larger genetic neighborhoods than plants in fragmented habitats.

One example of the impact of vertebrate foraging behavior on inbreeding, reproductive success, and plant genetic population structure is the large herbaceous, hummingbird-pollinated genus *Heliconia*. Heliconias

Table 4.3. Examples of the Outcrossing Rates (tm) of Neotropical Woody Plant Species Pollinated by Hummingbirds or Glossophagine Bats

Species and Family	Growth Form	Mating System	Study Site	Outcrossing Rate (tm) and Other Genetic Data	Comments
Hummingbird-pollinated plants:					
Helicteres brevispina (Sterculiaceae)	Pioneer shrub/ treelet	Self-compatible	Habitat = transition between savanna and forest; São Paulo State, Brazil	tm = 0.48 (low-density areas) to 0.61 (high-density areas); no evidence of inbreeding	Birds were trapliners in low-density areas and territorial in high-density areas; territorial birds did not produce more selfing than trapliners
Symphonia globifera (Clusiaceae)	Tree	Self-compatible	Tropical moist forest; French Guiana	tm = 0.92 but significant biparental inbreeding	Pollinators also include butterflies and passerine birds at some sites
Bat-pollinated plants:					
Pachira quinata (Malvaceae, Bombacoideae)	Tree	Self-incompatible	Tropical dry forest; Costa Rica	tm = 0.78 (isolated individuals) to 0.92 (continuous forest); lower seed relatedness/fruit and more sires/fruit in continuous forest	Asynchronous flowering results in lower tm and higher seed relatedness/fruit
Ceiba pentandra (Malvaceae, Bombacoideae)	Tree	Self-incompatible	Tropical dry and wet forests; Costa Rica	tm = 0.40 (wet forest) to 0.90 (dry forest); higher seed relatedness/fruit and fewer sires/fruit in wet forest	Two bat species visited flowers in dry forest; no bats visited flowers in wet forest
Ceiba aescufolia and C. grandiflora (Malvaceae, Bombacoideae)	Trees	Self-incompatible	Tropical dry forest; Mexico	In both species, tm ≥ 0.90 in forest fragments and continuous forest	Flowers visited by 2–3 species of bats; species differed in effect of forest cover on visit rates
Caryocar brasiliense (Caryocaraceae)	Tree	Self-incompatible	Cerrado vegetation; Brazil	tm = 1.0; biparental inbreeding found in 1 of 4 populations; all fruits have multiple sires	Two bat species pollinate the flowers
Hymenaea courbaril (Fabaceae)	Tree	Self-incompatible	Tropical dry forest; Puerto Rico	tm = 1.0; no biparental inbreeding; multiple sires/fruit	Extensive gene flow among patches of trees
Hymenaea stigonocarpa	Tree	Self-compatible	Tropical savanna fragments and isolated trees; Brazil	tm = 0.857 (isolated trees) –0.873 (in populations); biparental mating more common in populations than in isolated trees	Extensive gene flow among patches of trees

Sources. Data are from Collevatti et al. (2001); Degen and Roubik (2004); Dunphy et al. (2004); Franceschinelli and Bawa (2000); Fuchs et al. (2003); Lobo et al. (2005); Moraes et al. (2007); Quesada et al. (2004).

Note. tm is a multilocus estimator of the outcrossing rate, which ranges from 0 (complete selfing) to 1.0 (complete outcrossing: Ritland 1989).

are common and conspicuous members of the understory of many Neo-tropical forests (Kress 1983a, 1983b, 1985; Linhart 1973; McDade 1983; Stiles 1975, 1978a, 1979). Hummingbirds are the primary and probably exclusive pollinators of Neotropical *Heliconia*, which typically possess brightly colored inflorescences with long, tubular, diurnal, odorless, hermaphroditic flowers producing considerable amounts of nectar (Stiles 1975, 1979). Species of *Heliconia* fall into two ecological groups, based on the type of hummingbird pollinator (Linhart 1973; Stiles 1975, 1979). The first and largest group consists of species that form small clumps in closed forests, have long curved floral tubes, and produce few flowers each day with abundant or concentrated nectar. Members of this group are pollinated by hermit hummingbirds (Phaethorninae), which do not defend territories but forage over relatively long distances between widely spaced plants. The second smaller group consists of species that grow in large, monoclonal stands at forest edges or in open habitats and that have short, straight perianth tubes and produce many flowers containing relatively small amounts of dilute nectar. This localized and long-lasting nectar source provides a dependable food supply for pollinating nonhermit hummingbirds (Trochilinae), which establish and defend territories around these plants (Stiles 1975; also see below).

Physiological self-incompatibility is rare in Central American *Heliconia*. In an investigation of 19 species, self-compatibility and self-pollination ranged from total self-rejection in one species to full self-compatibility in the majority of taxa tested (Kress 1983a). Autogamy, however, is uncommon in most species; most plants require a hummingbird visitor for pollination and seed set. Spatial separation of the stigma and anthers within a flower (i.e., dichogamy) is less than 2 mm and probably has little effect on outcrossing. Overall, physiological and morphological specializations for outcrossing are few in *Heliconia*.

Observations of the interactions between plants and pollinators suggest that daily phenological patterns of flower production and the foraging patterns of the hummingbirds significantly influence the level of inbreeding and seed set that occurs in a population. In species visited by nonhermit territorial birds, pollen movement between plants appears to be limited. In contrast, pollen flow, and presumably degree of outcrossing, is significantly greater in species that are visited by traplining birds (Linhart 1973).

Data on genetic variation in populations of *Heliconia* pollinated by long-distance and territorial foraging birds are limited. However, a study of six species of Costa Rican *Heliconia* using allozymes showed a wide range of genetic variation associated with hummingbird foraging type (J. Kress, unpubl. data). In general, heliconias pollinated by territorial hummingbirds showed low levels of diversity (percentage of polymorphic loci [P] = 10.7–48.6; expected heterozygosity [H_e] = 0.029–0.143) while traplined species showed considerably higher levels of gene diversity (P = 50.0–66.6; H_e = 0.148–0.207). These results suggest that in the absence of self-incompatibility vertebrate pollinator foraging behavior can have a significant impact on the genetic structure of plant populations (Kress and Beach 1994). We will discuss the genetic consequences of vertebrate foraging behavior in more detail below.

Pollinators do not always deposit pure loads of conspecific pollen on stigmas. Instead, they frequently deposit mixed loads of conspecific and heterospecific pollen on stigmas as a result of generalized flower visitation behavior. Heithaus et al. (1975), for example, reported that 16%–78% of individuals of seven species of Costa Rican flower-visiting bats were carrying two or more types of pollen on their bodies when captured. Similarly, the phyllostomid bat *Glossophaga soricina* was frequently carrying pollen of *Crescentia* spp. trees when visiting flowers of the shrub *Bauhinia pauletia* in western Costa Rica (Heithaus et al. 1974). "Mixed" pollen loads are likely to be common in other kinds of pollinating vertebrates (e.g., in hummingbirds; Feinsinger et al. 1987; Borgella et al. 2001). Deposition of heterospecific pollen on plant stigmas can have both ecological and evolutionary consequences for plants. Ecologically, it can lead to interspecific competition for pollinators with a variety of possible outcomes (Palmer et al. 2003). Among other things, this competition can select for reduced flowering overlap (as discussed in chap. 3) and/or for different floral reward characteristics and morphologies (i.e., floral character displacement; Murcia and Feinsinger 1996). Heterospecific pollination can also result in reduced seed set via stigma clogging (e.g., Brown and Mitchell 2001; Nemeth and Smith-Huerta 2002; but see Waites and Agren 2004). It can also lead to the production of infertile hybrids, which will provide strong selection for reducing this cost (Waser 1983).

Although a combination of high plant resource diversity and frequent visits to different flower species by tropical pollinators would seem to result in the frequent deposition of foreign pollen on plant stigmas, we currently have little information about its ecological and evolutionary consequences. An extreme example of its potentially negative consequences comes from the Sonoran Desert where three species of columnar cacti compete for visits from nocturnal phyllostomid bats and diurnal birds and bees in April and May. In the self-incompatible bat- and hummingbird-pollinated organ pipe cactus (*Stenocereus thurberi*), early-flowering individuals often receive pollen from cardon cacti, whose flower densities are 50–100 times higher than those of organ pipe (Fleming 2006). Surprisingly, however, controlled pollination experiments revealed that organ pipe flowers receiving cardon pollen had nearly as high a probability of setting fruit (74%) as those receiving conspecific pollen (84%); this probability was zero in cardon flowers receiving organ pipe pollen. Instead of being aborted, organ pipe fruits from flowers receiving cardon pollen grew to maturity and contained normal-looking seeds. Close examination of these seeds, however, showed that they lacked embryos; plants apparently gained no fitness advantage by failing to abort these flowers. A comparison of the growth rates of fruits receiving heterospecific or conspecific pollen with those of open-pollinated control fruits indicated that early in its flowering season, most open-pollinated fruits of organ pipe were the product of heterospecific pollination. Fleming (2006) concluded that early flowering in organ pipe should be strongly selected against when it co-occurs with cardon. When it does not co-occur with competitors, however, early flowering is advantageous because it maximizes exposure of its flowers to migrating nectar-feeding bats and hummingbirds. In this example, receiving the wrong pollen does not necessarily reduce fruit set (and fruit pulp available to vertebrate frugivores), but it certainly reduces the fitness of those seeds to zero.

Peter Feinsinger and colleagues (Feinsinger et al. 1986, 1988, 1991) have conducted elegant field and lab experiments on the consequences of floral neighborhood on pollen deposition and seed set in hummingbird-pollinated plants. These studies indicate that low conspecific plant density and/or high heterospecific density does not always have predictable effects. In the shrubs *Palicourea lasiorrachis* and *Besleria triflora* (Rubiaceae and Gesneriaceae, respectively), for example, increasing conspecific flower density resulted in receipt of more pollen grains per stigma and more seeds per fruit in the former species but not in the latter. With increased heterospecific flower density, flowers of *P. lasiorrachis* received fewer conspecific pollen grains and had lower seed set per fruit; only seed set per fruit was affected in *B. triflora*. Results of these studies suggest that plant-vertebrate pollinator interactions are likely to be complex in species-rich tropical forests.

POLLEN DISPERSAL DISTANCES PRODUCED
BY VERTEBRATE FORAGING BEHAVIOR

Data presented above indicate that pollination by tropical birds and bats can produce high rates of outcrossing, a modest amount of biparental inbreeding, and multiple sires per fruit in their food plants. They also indicate that pollinators can be "promiscuous" and visit several flower species in a foraging bout. These results are the consequences of vertebrate food choice and foraging behavior, which in turn depend strongly on factors such as social systems and roosting and ranging patterns. Here we examine another important question based on pollinator foraging behavior: How far do pollinators carry pollen from where they pick it up to where they deposit it? Pollen transport distances can have a profound effect on plant effective population sizes and neighborhood areas. As we discuss below, these two parameters have important evolutionary implications.

Critical information needed to assess the long-distance pollen dispersal potential of birds and bats is how long pollen resides on these animals. In the case of phyllostomid bats, for example, pollen is sometimes a food (Herrera and Martínez del Rio 1998; Howell 1974), and hence these bats often groom themselves while digesting a stomach-full of nectar in night roosts. Whenever this happens, pollen residence times will be relatively short (i.e., <1 hour). Hummingbirds and sunbirds, in contrast, do not consume pollen but instead steadily lose previously acquired pollen and pick up new pollen as they visit flowers. Pollen "carryover" ("the deposition of pollen beyond the first flower after the one where it originated"; Kearns and Inouye 1993, 487) tends to decline rapidly with subsequent flower visits after pollen is picked up, but it still increases pollen dispersal distances in wide-ranging pollinators (Cresswell 2003). An extreme example of pollen retention by a hummingbird was reported by Singer and Sazima (2000), who noted the presence of an orchid pollinium on the beak of a Brazilian bird about 6.5 hours after it had been picked up. For most plants pollen is not presented in pollinia, however, and long pollen resident times on hummingbirds are unlikely to be very common.

In the 1970s and 1980s, the foraging behavior of animals was often discussed in the context of optimal foraging theory (Pyke 1984; Schoener 1971; Stephens and Krebs 1986). Optimal foraging theory suggests that to maximize their net rate of energy gain, foragers should minimize the time and energy spent searching for and moving among food sources. All else being equal, this suggests that foragers should usually be area-restricted foragers:

they should feed in as small an area as possible that meets their energetic and nutritional needs. When food plants occur at low densities, as is often the case in tropical forests, these "small areas" are not necessarily small in an absolute sense. But not all things are equal for nectar-feeding birds and mammals, and as a result, they exhibit a wide array of foraging strategies that produce an equally wide array of pollen dispersal distances. One extreme example of a species that grossly violates the (naïve) predictions of optimal foraging theory is the phyllostomid bat *Leptonycteris yerbabuenae.* When feeding on the nectar and pollen from columnar cacti in the Sonoran Desert, this bat forages over areas that range from 100 ha to 240 ha in size (Horner et al. 1998). Within these areas, bats visit 100–150 flowers (of up to three species) each night to gain 40–80 kJ of energy, depending on reproductive status. If their foraging behavior corresponded to expectations of optimal foraging theory, individuals could easily find enough energy to meet their daily energy requirements in <1 ha. By foraging widely, these bats sometimes carry pollen several kilometers between plants instead of moving pollen short distances (<100 m) between flowers. For whatever reason (see the discussion in Horner et al. 1998), these bats are rather long-distance pollen dispersers and, as a consequence, their food plants have relatively large genetic neighborhoods and low levels of population subdivision (Hamrick et al. 2002).

In summarizing pollen dispersal distances produced by nectar-feeding birds and mammals, we find it useful to group species into four classes that reflect increasingly large potential dispersal distances: territorial foragers, trapliners, nonterritorial foragers, and migrants.

TERRITORIAL FORAGERS.—Most vertebrate nectar feeders are birds (chap. 1), and most passerine birds have (socially) monogamous territorial social systems during the breeding season (Bennett and Owens 2002; Lack 1968). Hence, we would generally expect these kinds of birds to provide short-distance pollen movements for their food plants. Many nonhermit hummingbirds (which are not monogamous), sunbirds, and honeyeaters are known to defend breeding and/or feeding territories aggressively for at least some time each year (Carpenter and MacMillen 1976; Cheke and Mann 2001; Gill and Wolf 1975a, 1975b; Paton 1986; Schuchmann 1999; Stiles 1975; Yumoto et al. 1997). Socially subordinate individuals in all of these birds regularly "raid" territories and likely provide greater mobility for pollen than territory holders. Territory size in hummingbirds, sunbirds, and honeyeaters is often sensitive to floral densities. Elegant experimental field studies (e.g., Carpenter et al. 1983) have shown that hummingbirds will adjust the size of their territories to match energy levels of their flower resources. Hummingbirds, sunbirds, and honeyeaters also abandon territorial behavior when resource levels or intruder pressure are high (e.g., Carpenter and MacMillen 1976; Gill and Wolf 1975a; Wolf 1975).

In contrast to many nectar-feeding birds, most phyllostomid bats do not routinely defend feeding territories (Heithaus et al. 1974; Horner et al. 1998; Muchhala and Jarrin-V. 2002) and hence are potential long-distance pollen dispersers. Exceptions to this include *Glossophaga soricina* when it defends *Agave* inflorescences or *Caryocar brasiliense* trees and *Anoura caudifer* when it defends *Abutilon* flowers (Gribel and Hay 1993; Lemke 1984; L. Lopes, pers. comm.). Pteropodid nectar bats tend to be more aggressive than their phyllostomid counterparts and hence might be expected to provide restricted pollen movement for their food plants. Law (1995, 1996), for example, reported that resident adult males of the Australian blossom bat *Syconycteris australis* apparently defend patches of flowering *Banksia integrifolia* shrubs against intrusions by adult females and juveniles. In this situation, males are likely to be short-distance (i.e., within-patch) pollen dispersers whereas females and juveniles are likely to move pollen greater distances (i.e., between patches). In a spatially more restricted fashion, Australian *Pteropus* bats sometimes establish feeding territories in the canopies of flowering or fruiting trees. These bats defend their small territories against intruders in a "resident and raiders" system (Hall and Richards 2000). For both pollen and fruit, it is likely that intruders ("raiders") provide greater mobility for plant propagules than residents. Similar social behavior was not reported in two species of Malagasy pteropodid bats when they visited flowers of *Adansonia* species and *Ceiba pentandra* (Andriafidison et al. 2006). These bats made brief visits to flowers in a tree before moving on. Similarly, other pteropodid bats make brief visits to flowers of *Ceiba pentandra* in India and to flowers of *Oroxylum indicum*, *Parkia speciosa*, *Musa acuminata*, and *Durio zibethinus* in Malaysia (Gould 1978; Singaravelan and Marimuthu 2004).

Estimates of pollen deposition distances of territorial and other kinds of foragers are summarized in table 4.4. As expected, the few quantitative data for territorial hummingbirds indicate that these birds move pollen very short distances (generally <20 m). Occasional forays away from territories, however, are not uncommon in these birds and represent longer potential pollen movements.

TRAPLINING FORAGERS.—Hermit hummingbirds are quintessential tra-pliners in the understories of Neotropical wet tropical forests. In these habitats, birds typically visit the flowers of widely spaced *Heliconia* clumps or *Costus* gingers and, as indicated in table 4.4, they often move pollen much greater distances, on average, than territorial hummingbirds. Direct comparison of pollen movement by territorial and nonterritorial hummingbirds in a montane Costa Rican forest, for example, indicated that territorial birds moved pollen ≤10 m, which resulted in a relatively high index of inbreeding (F_{is}), whereas traplining birds moved pollen much longer distances, which resulted in no inbreeding (Linhart et al. 1987). Some phyllostomid and pteropodid nectar bats are also likely to be traplining foragers (e.g., Gould 1978; Tschapka 2004), but estimates of their pollen dispersal distances are not yet available.

NONTERRITORIAL FORAGERS.—When the movements of pollinators are not constrained by territorial boundaries, they are potentially free to move pollen long distances, but whether they do so or not is highly variable. Radio-tracking studies of the bat *Pteropus poliocephalus* feeding on *Melaleuca* and *Eucalyptus* blossoms in eastern Australia, for example, indicate that these bats are highly mobile and visit several widely spaced trees within a habitat patch before moving several kilometers to another patch (Eby 1991). In northern Australia, 95% of the intertree movements by *Pteropus scapulatus* feeding on eucalypt blossoms were >50 m whereas only 20% of honeyeater movements were this long (McCoy 1990). *Pteropus* bats in Australia (and presumably elsewhere) also carry much larger loads of viable pollen on their bodies than do nectar-feeding birds and hence are potentially very important long-distance pollinators in eucalypt forests and woodlands (Richards 1995), just as *Leptonycteris* bats are for paniculate *Agaves* and columnar cacti in Neotropical arid habitats (Fleming and Nassar 2002).

Estimates of pollen movement distances by nonterritorial hummingbirds and bats are shown in table 4.4. These data suggest that hummingbirds usually move pollen shorter distances than phyllostomid and pteropodid bats do, which often move several kilometers among forest patches in disturbed landscapes.

MIGRATING FORAGERS.—Many nectarivores migrate seasonally among habitats along both elevational and latitudinal gradients (Fleming 1992; Levey and Stiles 1992) and hence have the potential to be truly long-distance

Table 4.4. Examples of Minimum Pollen Movement Distances by Tropical Nectar-Feeding Birds and Bats

Pollinator and Plant	Habitat and Location	Pollen Movement Distances	Comments
Territorial pollinators:			
Nonhermit hummingbirds at *Heliconia* sp. giant herbs (Heliconiaceae)	Montane and lowland forests; Costa Rica	Most pollen deposited ≤20 m from source; some pollen moves away from territory	Used fluorescent dyes to study pollen movement
Lampornis hummingbirds at *Razisea* sp. and *Hansteinia* sp. herbs (Acanthaceae)	Montane forest; Costa Rica	Most pollen moved ≤10 m	Used fluorescent dyes to study pollen movement. High inbreeding coefficient ($F_{is} = 0.22$) from territorial pollination
Traplining pollinators:			
Phaethornis hummingbirds at various understory plants (mostly *Heliconia* sp. [Heliconiaceae])	Tropical wet forest; Costa Rica	Foraging distances from 2 leks were 1,000–1,500 m; mean distances were about 400 m from leks	Birds feed along routes between leks and maximum foraging distances
Hermit hummingbirds at various plants (mostly *Heliconia* [Heliconiaceae])	Montane and lowland forests; Costa Rica	Pollen moved up to 180 m from source; much moved >>20 m	Used fluorescent dyes to study pollen movement
Phaethornis hummingbirds at *Razisea* sp. and *Hansteinia* sp. herbs (Acanthaceae)	Montane forest; Costa Rica	More pollen moved ≥25 m than by territorial *Lampornis*	Used fluorescent dyes to study pollen movement. Low inbreeding coefficient ($F_{is} = -0.075$) from trap-lining pollination

Nonterritorial pollinators:

Amazilia hummingbirds at *Malvaviscus* sp. shrubs (Malvaceae)	Riparian forest; western Costa Rica	Most moved ≤40 m; maximum moves of 220 m	Fluorescent dyes used to study pollen movement. Pollen from one plant moved to 5–24 other plants in one day.
Hummingbirds and passerine birds at *Symphonia globulifera* trees (Clusiaceae)	Tropical moist forest; French Guiana	27–53 m	Movements estimated by indirect genetic means
Phyllostomid bats at *Hymenaea courbaril* trees (Fabaceae)	Pastures and tropical dry forest; Puerto Rico	Some pollen moved 600–800 m from source to deposition	Estimates based on genetic paternity analyses
Phyllostomid bats at *Hymenaea stigonocarpa* (Fabaceae)	Forest patches in savanna; Brazil	Pollen moved an average of 860 m between plants; some moved 5,229 m, on average	Estimates based on genetic paternity analyses
Phyllostomid bats at *Ceiba* spp. trees (Malvaceae, Bombacoideae)	Isolated trees and continuous tropical dry forest; Mexico	Pollen moves between forest trees and isolated trees	Inferences based on genetic analyses
Syconycteris pteropodid bats at *Syzygium cormiflorum* trees (Myrtaceae)	Patchy upland wet forest; Australia	Most pollen moves <200 m but some up to 5.8 km	Movements based on radio-telemetry and pollen loads on bats

Sources. Data are from Degen and Roubik (2004); Dunphy et al. (2004); Law and Lean (1999); Linhart (1973); Linhart et al. (1987); Moraes and Sebenn (2011); Quesada et al. (2004); Stiles and Wolf (1979); Webb and Bawa (1983).

Note. Pollen movement distances will generally underestimate actual distances because they do not take into account pollen carryover.

pollen movers. As an example, Horner et al. (1998) reported that in their Mexican study area, several radio-tagged individuals of the bat *Leptonycteris yerbabuenae* fed at cactus flowers early in the evening before disappearing. This behavior was not seen in tagged resident bats, and they concluded that the "lost" bats were still in transit to roost sites farther north. Because these bats often carry large pollen loads on their heads and necks and can easily fly >30 km in 1 hour, they have considerable potential for carrying pollen substantial distances in a night. How often other migrant birds and bats carry pollen long distances is currently unknown.

Finally, Ward et al. (2005) reported (maximum) pollen transport distances, as estimated by genetic analyses, by insects, hummingbirds, and bats for 45 species of Neotropical trees. These data are summarized in figure 4.2B. Within vertebrates, bats clearly appear to be longer distance pollen carriers than hummingbirds. Like bats, maximum transport distances for small insects and bees can be several kilometers, and fig wasps can potentially carry pollen enormous distances relative to their tiny size (160 km; Ahmed et al. 2009). Likewise, Degen and Roubik (2004) reported foraging distances of 1.6–8 km for species of African bees. These data attest to the power of genetic analyses for estimating actual pollen transport distances—estimates that would otherwise be very difficult to make. They also suggest that other kinds of tropical pollinators, not just birds and bats, are capable of carrying pollen substantial distances in intact and fragmented tropical habitats.

To summarize this section on pollinator-plant interactions, we note that pollen limitation appears to be common in plants in general and is likely to be common in tropical plants, although relatively little data are yet available for the tropics. Whether PL is an important factor in the demography of long-lived tropical plants, however, is not yet clear. In general, it is less likely to have a significant effect on the demography of long-lived plants than on short-lived plants. Although many tropical plants are pollinated either by birds or mammals, others depend on both groups for pollination, but the relative importance of birds and mammals tends to vary geographically, as illustrated by Neotropical columnar cacti. Most tropical trees are self-incompatible hermaphrodites, but moneocy and dioecy are also common breeding systems. Self-compatibility is more common in understory plants than in canopy trees. Outcrossing rates tend to be high in tropical trees regardless of type of pollinator. Forest fragmentation sometimes results in increased selfing rates in isolated trees, but bats and large bees can counteract this by moving pollen among fragments. Vertebrate pollinators sometimes (often?) deposit mixed-species pollen loads on stigmas, and this can lead to reduced seed set within fruits and reduced fruit set, suggesting that plant-pollinator interactions in high-diversity habitats are likely to be complex. The distance that nectar-feeding vertebrates move pollen between flowers is highly variable and often depends on pollinator social systems. Territorial species (e.g., many hummingbirds and sunbirds) move pollen short distances compared with trapliners (e.g., hermit hummingbirds). Nonterritorial species and migrant species potentially move pollen relatively long distances (many kilometers), with bats moving pollen longer distances than birds. Insects can also move pollen several kilometers in tropical habitats. Finally, outcrossing rates and long-distance pollen movement have important evolutionary consequences regarding effective plant population sizes and the sizes of plant genetic neighborhoods.

Ecological Consequences of Plant-Frugivore Interactions

Pollen movement is just one way in which plants can disperse their genes. The other way is by seed dispersal. Because most seeds are diploid whereas pollen is haploid, seed dispersal can potentially be twice as effective at moving plant genes as pollen movement. But, as we will see, this is not likely to be true in tropical forests, perhaps because vertebrate-dispersed seeds are always substantially larger than pollen grains. Because of its small size and because genes in pollen move twice—once in pollen and again in seeds—pollen-dispersed genes should theoretically have much greater mobility than genes dispersed via animal-ingested seeds, and because of this mobility, we might expect pollen dispersal to have a greater effect on the genetic structure of tropical plants than seed dispersal (Hamrick et al. 1993). Despite differences in mobility, there are many ecological parallels between pollen and seed dispersal, including the concepts of pollen limitation and dispersal limitation. But whereas pollen limitation has the immediate effect of influencing fruit and seed set in individual plants and their populations via its effect on fecundity, seed dispersal limitation has important immediate effects for individuals, populations, and communities (e.g., Muller-Landau et al. 2002). Seed dispersal, along with speciation, for example, plays a major role in Hubbell's (2001) neutral theory of biogeography (Alonso et al. 2006). Seed dispersal is ultimately how most species persist in communities and how they enter new plant communities.

A dramatic example of the importance of seed dispersal in determining the structure of tropical forests comes from Seidler and Plotkin's (2006) analysis of spatial aggregation patterns of 561 species in a lowland Malaysian dipterocarp forest. Using data from a mapped 50 ha plot, these researchers determined patterns of intraspecific spatial aggregation in species classified into seven dispersal modes (ballistic, gravity, gyration [the dispersal mode of winged dipterocarp seeds], wind, animals ingesting seeds <2 cm in diameter, animals ingesting seeds 2–5 cm in diameter, and animals ingesting seeds >5 cm in diameter). For each species, degree of spatial aggregation was estimated by calculating the parameter σ, which represents its mean cluster size and which increases as degree of spatial aggregation decreases. Results of their analysis indicate that σ differs significantly among dispersal modes and that it is substantially higher in animal-dispersed species than in abiotically dispersed species (fig. 4.3). Within the former group, σ increases with seed size, which suggests that mean dispersal distances of large-seeded plants dispersed by large frugivores (e.g., hornbills and primates) are greater than those of species dispersed by smaller frugivores (e.g., barbets and small pteropodid bats). Large-seeded, animal-dispersed plants are thus less spatially aggregated than plants with other dispersal modes. Seidler and Plotkin (2006) also analyzed data from the 50 ha plot on Barro Colorado Island, Panama, and obtained similar results. These results clearly indicate that

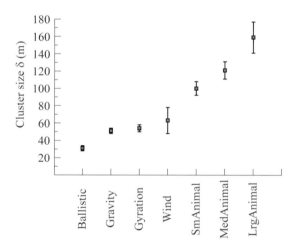

Figure 4.3. Relationship between dispersal type and degree of spatial aggregation, expressed as cluster size (σ), in trees belonging to seven dispersal syndromes in a Malaysian rain forest. Degree of spatial aggregation decreases with an increase in cluster size. Based on Seidler and Plotkin (2006).

dispersal mode helps determine the spatial dispersion patterns of tropical trees. These patterns, in turn, help to determine patterns of intra- and inter-specific interactions among plants that ultimately affect plant community composition.

DISPERSAL AND RECRUITMENT LIMITATION

All tropical plants experience either dispersal limitation or recruitment limitation (or both) to some degree and at some spatial scale. These limitations are due to the fact that plants do not occur everywhere within a local habitat (e.g., most tropical tree species are rare), and they usually do not occur in all habitats within a region. Schupp et al. (2002) recognized three basic kinds of recruitment limitation (*sensu lato*): seed limitation, dispersal limitation, and recruitment limitation (*sensu stricto*). Seed limitation occurs when plants produce too few seeds (perhaps as a result of PL) to saturate all possible re-cruitment sites; it reflects plant fecundity. Dispersal limitation occurs when dispersal agents fail to disperse available seeds to all possible recruitment sites. This can result from low numbers of disperser visits to fruiting plants, distance-restricted dispersal with most seeds falling near parent plants, or spatially contagious seed dispersal in which a few sites are saturated with seeds and many sites receive no seeds as a result of frugivore foraging and/or social behavior. Recruitment limitation occurs when seeds arrive at potential recruitment sites but fail to produce surviving seedlings and juveniles for biotic (e.g., competition, seed predators, pathogens, and herbivores) or abiotic (e.g., light, soil water and nutrients) reasons.

Of these three forms of recruitment limitation, frugivorous animals most strongly affect dispersal limitation as a result of their foraging and other kinds of behavior. As in the case of vertebrate pollinators, tropical frugivores exhibit a diverse array of foraging behaviors, and most of them result in spatially restricted seed dispersal. These behaviors include prolonged feeding in a fruiting tree (e.g., many fruit-eating birds, primates, and territorial pteropodid bats); eating fruit in feeding roosts (e.g., many phyllostomid and small pteropodid bats), depositing seeds at display or nesting sites (e.g., lek mating manakins, cotingids, and birds of paradise; and tree hole-nesting trogons and hornbills), depositing seeds in latrines (e.g., tapirs and rhinoceroses) or in caches (e.g., agoutis and other rodents), depositing seeds at resting or sleeping sites (e.g., cassowaries, primates, and artiodactyls), and depositing seeds at other conspecific or heterospecific fruiting trees (e.g., most frugivorous birds and mammals; Schupp et al. 2002). The net effect of these behaviors is that most vertebrate-dispersed seeds end up either un- der fruiting trees (both conspecifics and heterospecifics) or in clumps away from these trees with a few sites receiving many seeds and many potential recruitment sites receiving no seeds. As nonrandom and idiosyncratic as this pattern appears to be, it still has important ecological and genetic con- sequences for tropical plants. To paraphrase Howe and Mitiri (2004), seed dispersal "matters" in species-rich tropical forests.

Russo et al. (2006) provide an especially elegant example of how the foraging behavior of an important Neotropical frugivore, the spider mon- key *Ateles paniscus*, influences seed dispersal of the tree *Virola calophylla* (Myristicaceae) in Manu National Park, Peru. This tree produces relatively large, aril-covered seeds (length = 1.7 cm; mass = 1.4 g) that are dispersed primarily by primates but also by 17 species of large birds. Based on exten- sive field observations, the deposition fate of all fruit produced by 19 trees was determined. These fates included deposition under conspecific fruiting trees (i.e., nondispersal), in transit while monkeys moved between fruit- ing trees or to resting or sleeping sites, and under resting or sleeping sites. An overall summary of the monkey-produced seed shadows of these trees (fig. 4.4) indicates that the seed dispersal curve of this species is multimodal with a strong mode under a *Virola* canopy, a second mode at about 200 m, and a broad third mode about 800 m from the parent tree. Mean dispersal distance for all seeds was 151 m; it was 245 m for all dispersed seeds. By de- positing many seeds in clumps at restricted locations as well as widely scat- tering single seeds, this monkey creates highly heterogeneous seed shadows and strong seed dispersal limitation.

What is the fate of these monkey-dispersed seeds and does the initial seed dispersal template created by *A. paniscus* have any influence on the density and dispersion patterns of seedlings, saplings, and adults of *V. calophylla*? In addition to dispersal limitation, does recruitment limitation also affect adult distributions? Russo and Augspurger (2004) determined the fate of dispersed and nondispersed seeds of *V. calophylla* and documented the dis- tribution patterns of seedlings, saplings, and adults. They found that 99% of all seeds had failed to establish as seedlings 13–15 months after deposition, primarily as a result of seed predation by spiny rats (*Proechimys*; Echimy- idae) and scotylid beetles. This predation is an intense form of recruitment limitation and occurs in many large-seeded tropical plants. Although per capita seed survival varied inversely with seed density and positively with distance to the nearest fruiting *Virola* tree, highest seedling and sapling

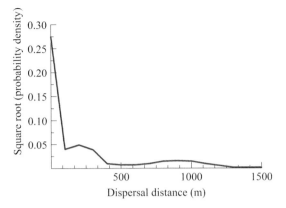

Figure 4.4.   Probability density function of dispersal distances created by the spider monkey *Ateles paniscus* for seeds of *Virola calophylla* in a Peruvian tropical moist forest. Based on Russo et al. (2006).

densities occurred under female *Virola* plants and under sleeping roosts (i.e, in places where seed deposition was greatest). As a result, the dispersion patterns of seedlings, saplings, and adults were clumped at different spatial scales, and the initial template of clumped seeds produced by monkey dispersal persisted from seedlings up to adults. Although strong recruitment limitation occurs in this system, the sheer number of seeds deposited in nonrandom fashion by spider monkeys dictates where adults will eventually become established.

Strong concordance between frugivore seed deposition behavior and seedling recruitment, as seen in the *V. calophylla* study, is likely to be the exception rather than the rule in many vertebrate seed dispersal systems. The African canopy tree *Monodora myristica* (Annonaceae), for example, produces large (16 cm diameter) fruits with a thick woody husk that contain 175–750 1.9-cm-long seeds. At Kibale National Park, Uganda, only three species of large primates—chimpanzees (*Pan troglodytes*), a baboon (*Papio anubis*), and a mangabey (*Lophocebus albigena*)—are strong enough to open fruits and gain access to fruit pulp and seeds (Balcomb and Chapman 2003). Once the fruits are open, smaller monkeys and two large birds (a hornbill and a turaco) can remove seeds. Terrestrial or scansorial frugivores at this site include elephants and civets. Of all of these species, the three large primates are responsible for moving >90% of all seeds away from the canopies of fruiting trees where, like spider monkeys, they defecate them in clumps. Small monkeys, in contrast, remove single seeds from fruits and spit them out, usually within 10 m of the canopy. Experimental studies examining the fate of dispersed seeds indicated that 86%–94% of seedling recruitment came from seeds dispersed in clumps, that seed predation by rodents and curculionid beetles provided the strongest limit to seedling recruitment, and that the density and dispersion of pole-sized individuals of *M. myristica* did not reflect the original template of seed deposition. Postdispersal events thus had a strong effect on seedling growth and survival and, ultimately, on the density and dispersion patterns of adult plants. A similar lack of concordance between initial seed deposition templates and seed survival and seedling recruitment patterns has also been documented in other studies (e.g., Gross-Camp and Kaplin 2005; Herrera et al. 1994; Wenny 2000a).

Dispersal limitation resulting from the failure of seeds to move away from the canopies of parent trees is common in all dispersal modes. For example, Clark et al. (2005) measured seed deposition patterns in nine species of trees (three each that were wind, bird, or monkey dispersed) in an intact forest in Cameroon, Africa. Seed traps were placed in 60-m-long transects from the base of fruiting trees, a design guaranteed to underestimate long-distance dispersal by hornbills and monkeys. At least 60% of all seeds were deposited under the canopies of fruiting trees (nondispersal) in each dispersal mode, and the proportion of seeds deposited >60 m from fruiting trees ranged from about 2% in wind-dispersed species to about 14% in monkey-dispersed species (fig. 4.5). Mean dispersal distances among the dispersal modes did not differ appreciably, but maximum observed dispersal distances for bird-dispersed species (473 m) and monkey-dispersed species (100 m) were substantial. Other studies in this forest indicate that hornbills and monkeys can disperse seeds up to 6.5 km and 2 km, respectively, away from fruiting trees (Holbrook and Smith 2000; Poulsen et al. 2001). Despite the mobility that large fruit-eating vertebrates occasionally provide, however, the fate of most seeds is nondispersal, indicating that dispersal limitation is inevitable in tropical trees (e.g., Terborgh et al. 2011).

Additional examples of dispersal limitation abound in the plant ecology literature (e.g., Svenning et al. 2006 and included references), and we will describe only a few representative studies here. The clearest way to demonstrate dispersal and recruitment limitation is to conduct seed addition experiments (Munzbergova and Herben 2005; Nathan and Muller-Landau 2000; Turnbull et al. 2000; but see Clark et al. (2007) for a critique of current methods). Dispersal limitation occurs when more seedlings are produced in seed addition plots than in matched control plots. Recruitment limitation

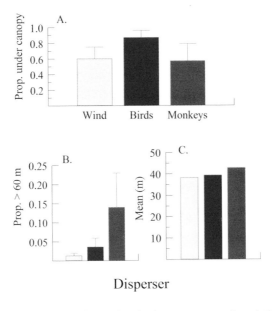

Figure 4.5.   Effect of different dispersal methods on proportion of seeds that were undispersed. *A*, proportion dispersed >60 m. *B*, average dispersal distance. *C*, for nine species of trees (three in each dispersal category) in a Cameroon tropical moist forest. Error bars in *A* and *B* are SDs. Based on data in Clark et al. (2005).

occurs when the addition of seeds to plots does not produce more seedlings than in matched control plots. Svenning and Wright (2005) conducted a seed addition experiment using 32 species of shade tolerant plants of several different growth forms on Barro Colorado Island, Panama. They found that seedling recruitment was significantly higher (by factors of 60–300 after 2 years) in the addition plots compared with control plots in 31 of 32 species after 1 year and in 26 of 32 species after 2 years. Species were initially separated into three classes based on seed size in these experiments, but seed size did not affect the number of seedlings that became established in experimental plots. Svenning and Wright (2005) suggested that predation on large seeds by large vertebrates equalized seedling recruitment rates among seed size classes. Overall, their results support the hypothesis that dispersal limitation strongly affects the density of understory plants in this forest. Dalling et al. (2002) reached a similar conclusion for shade intolerant plants in the same forest by measuring natural seed rain and seedling recruitment of 13 species of pioneer plants in forest gaps. They found that seed rain and seedling recruitment declined with distance to nearest source tree in all 13 species and that seedling abundances in gaps were positively correlated with abundances predicted from statistical models of seed dispersal for these species. They noted that recruitment into gaps in species that were dispersed by birds or mammals was more variable than that of wind- or ballistically dispersed species, perhaps because of the difficulty in measuring long-distance seed dispersal in animal-dispersed taxa. In contrast to the results of studies on Barro Colorado Island, Webb and Peart (2001) reported that rates of seedling recruitment in animal-dispersed plants were substantial in a Bornean dipterocarp forest. They suggested that the presence of a largely intact disperser fauna, including large primates, was the reason for this and predicted, based on simulation models, that removal of these animals would increase dispersal limitation and ultimately reduce local species richness of understory seedlings by 60%.

Like studies conducted in intact forest, studies of forest regeneration in highly disturbed sites often conclude that dispersal limitation is common during its early stages with small-seeded pioneer species having higher probabilities of moving significant distances from primary forest edges than larger-seeded species (e.g., Chazdon et al. 2003; Lawrence 2004; Mesquita et al. 2001; Uhl 1987). In the New World tropics, bat-dispersed pioneer species often have higher recruitment rates in large gaps or abandoned lands than do bird-dispersed species whereas the opposite situation appears to hold in the Old World tropics (reviewed in Muscarella and Fleming 2007). Thus, the early stages of forest regeneration in abandoned fields in the New World are often dominated by bat-dispersed shrubs and trees of the genera *Solanum* (Solanaceae), *Cecropia*, (Cecropiaceae), *Piper* (Piperaceae), and *Vismia* (Clusiaceae), whereas bird-dispersed species of *Macaranga* (Euphorbiaceae) are common in disturbed sites in Asia and Australasia (e.g., Franklin 2003; Lawrence and Mogea 1996).

Recruitment limitation occurs after seeds are dispersed by frugivores, and its study has had a long history. Janzen (1970) and Connell (1971) proposed that natural enemies acting in a density- and/or distance-dependent fashion often prevented seedlings from recruiting near parent plants. Much subsequent research has supported the idea that density-dependent mortality, often caused by insect and vertebrate seed predators, limits recruitment of tropical trees (e.g., Balcomb and Chapman 2003; Hammond and Brown 1998; Harms et al. 2000; Howe 1980; Nathan and Casagrandi 2004; Norghauer et al. 2006; Terborgh and Nuñez-Iturri 2006; Wenny 2000b; Wyatt and Silman 2004). By contrast, Hyatt et al. (2005) conducted a meta-analysis of papers dealing with the effects of distance from parent plants on the survival of seeds and seedlings in temperate and tropical plants and found no support for the hypothesis that dispersal of seeds away from parent plants, within either the same or a different habitat, enhances seed survival. They did find a distance effect for seedlings, and there was a trend toward a distance effect for tropical seeds but not for temperate seeds. In their comprehensive review of over 50 studies testing the Janzen-Connell hypothesis, Carson et al. (2008) concluded that few studies have been conducted for long enough and at a large enough community-level scale to provide a suitable test of its truth. Rigorous tests of this popular hypothesis will remain logistically daunting for the foreseeable future (but see Bagchi et al. 2011). We will return to this topic when we discuss the benefits of seed dispersal below.

Although dispersal limitation and/or recruitment limitation of some kind is likely to be ubiquitous in tropical plants, the question remains: To what extent do these kinds of limitation affect plant population growth rates (λ)? Howe and Mitiri (2004) speculated that, in long-lived plants with relatively stable population sizes, high variance in seedling recruitment and survival and low elasticity in the seed-to-seedling transition likely have little effect on λ compared with adult survival rates. In growing populations, however, this transition is likely to be more important. But, as in the case of the effect of PL on λ, we currently have few data with which to address this question. Godinez-Alvarez et al. (2002) conducted a detailed analysis of the effects of different fruit-eating vertebrates on λ in the Mexican columnar cactus *Neobuxbaumia tetetzo*, which recruits only under three species of legume "nurse plants." Their demographic calculations indicated that the collective effect of nine species of frugivorous birds and bats yielded a λ of 1.0 (a stable population) whereas λ was <1.0 in the absence of these seed dispersers. Of the nine frugivores, only the bat *Leptonycteris yerbabuenae* reliably dispersed large numbers of germinable seeds to recruitment sites and hence was the major contributor to population stability in *N. tetetzo*. This bat is also the most important pollinator of its flowers (Valiente-Banuet et al. 1996), so its role in the population ecology of this cactus is critically important. Esparza-Olguin et al. (2005) also conducted a demographic analysis of *N. tetetzo* and two other species of *Neobuxbaumia* (*N. macrocephala* and *N. mezcalaensis*) in south-central Mexico and found that, compared with the effects of juvenile and adult survival, seed production and seedling establishment contributed little to current population dynamics, as expected in these long-lived plants. In an absolute sense, of course, seed production and seedling establishment are critically important processes for these and other plants. Thus, to the extent that seed dispersal affects seedling establishment probabilities, it is a critical process in plant population dynamics (Bruna 2003).

DISPERSER EFFECTIVENESS

Compared with their nectar-feeding counterparts, frugivorous vertebrates exhibit much greater species richness, a much greater range of body sizes, and a much greater range of foraging behaviors. As a result, frugivores interact with their food plants in a much more heterogeneous way than do nectar feeders, and, to paraphrase Dan Janzen (1983a), the fruit-frugivore mutualism is much "sloppier"—that is, ecologically less efficient—than the flower-pollinator mutualism (also see Wheelwright and Orians 1982). As summarized by Schupp (1993), disperser effectiveness has both quantitative and qualitative components (table 4.5). Quantitative components include a frugivore's abundance, its reliability of visitation to a particular plant species, and the number of seeds it removes and disperses; qualitative components include how it treats seeds in its mouth or gut and where it deposits seeds. From a plant's perspective, its seed dispersers are likely to vary substantially in their effectiveness (reviewed in Schupp 1993 and Schupp et al. 2010; also see Howe and Miriti 2004), and the effectiveness of any given frugivore is likely to vary for different fruit species in its diet. For example, chestnut-mandibled toucans (*Ramphastos swainsonii*) are three to 30 times more effective at dispersing seeds of *Virola nobilis* trees than are other birds and the monkey *Ateles geofroyii* on Barro Colorado Island, Panama (Howe 1993a). Similarly, in southeastern Brazil, the muriqui, *Brachyteles arachnoides*, is a better seed disperser than the howler monkey, *Alouatta guariba*,

Table 4.5. Schupp's (1993) Concept of Disperser Effectiveness

I. Quantity of Seed Dispersal		II. Quality of Seed Dispersal	
A. Number of Visits	B. Number of Seeds Dispersed per Visit	A. Quality of Treatment	B. Quality of Deposition
1. Abundance of disperser	1. Number of seeds handled per visit	1. Destroy or pass seeds intact	1. Movement patterns: (*a*) habitat and microhabitat selection; (*b*) rate and directionality of movement
2. Disperser diet breadth	2. Probability of dispersing a handled seed	2. Alter percentage or rate of germination	2. Deposition patterns: (*a*) rate and pattern of deposition; (*b*) seed (diet) mixing
3. Reliability of visitation			

because it disperses a higher proportion of seeds away from parent plants, defecates more frequently, deposits seeds in more locations, and is more likely to scatter seeds than defecate them in clumps (Martins 2006). Finally, the Central American howler monkey (*Alouatta palliata*) is an effective disperser of the large seeds of *Poulsenia armata* and *Brosimum alicastrum* but destroys small seeds or fruits (e.g., *Cecropia peltata, Muntingia calabura, Vismia baccifera*) in its gut (Chapman 1989; Estrada and Coates-Estrada 1986).

Detailed accounts of the seed dispersal effectiveness of particular species of birds and mammals come from studies of the bat *Carollia perspicillata*, the avian dispersers of an Australian mistletoe, and the gibbon *Hylobates mulleri × agilis.* The 18 g phyllostomid bat *C. perspicillata* is one of the most common chiropteran seed dispersers in the understory of the lowland Neotropics. During the course of a 10-year study, Fleming (1988) examined many aspects of this species' disperser effectiveness in tropical dry forest in western Costa Rica. Although its diet in this habitat includes 18 species of fruit, its most-preferred fruits are produced by five species of *Piper* (Piperaceae) shrubs. Individuals are highly reliable visitors to patches of *Piper* plants and remove nearly all ripe fruit the first night they are available (also see Thies and Kalko 2004). This species carries *Piper* and other fruit to feeding roosts usually located within 50 m of a resource patch where, like the spider monkey *Ateles paniscus* in its sleeping roosts, it defecates many seeds. Bats also defecate seeds while in flight and in conspecific and heterospecific resource patches (Fleming and Heithaus 1981). Seed germination experiments indicated that defecated seeds have high germination rates (e.g., nearly 100% for *Piper* seeds) but that seeds of light-demanding taxa such as *Piper* rarely germinate when deposited under shady night roosts (Fleming 1981). Despite depositing many seeds in unfavorable germination sites, the sheer number of seeds each bat handles each night, which ranges from hundreds to thousands depending on fruit species, plus its abundance (hundreds of individuals per day roost) means that this bat is a highly effective dispersal agent for understory and early successional trees in this and other habitats. Rates of colonization of disturbed habitats by many species of early successional plants would plummet if *C. perspicillata* were to disappear from Neotropical forests (Muscarella and Fleming 2007).

A second detailed example of seed disperser effectiveness involves a comparison of the effects of two Australian birds, the spiny-cheeked honeyeater *Acanthagenys rufigularis* (41–47 g) and the mistletoebird *Dicaeum hirundinaceum* (8–10 g), on seedling recruitment in the mistletoe *Amy-*

ema guandang (Loranthaceae), which parasitizes *Acacia papyrocarpa* trees (Reid 1989). Although the mistletoebird deposited four times more seeds on branches that were appropriate for mistletoe recruitment, honeyeater dispersal actually accounted for more mistletoe recruitment in *Acacia* crowns. In terms of dispersal quality, both species had similar positive effects on mistletoe seed germination, although seeds were treated more gently in the specialized gut of the mistletoebird. While the mistletoebird is a more reliable and specialized visitor to *A. guandang*, the more abundant and larger honeyeater was its most effective seed disperser. Like the *C. perspicillata* study, this example emphasizes the importance of determining both quantitative and qualitative aspects of a seed disperser's effectiveness to assess its overall effect on recruitment in its food species.

A final example of species-specific dispersal effectiveness is the gibbon *Hylobates mulleri* × *agilis* studied in lowland dipterocarp forest in central Borneo by McConkey (2000). This 5–6.4 kg primate lives in family groups in relatively small territories of about 45 ha. Over the course of a year, two groups consumed fruits of 160 species and dispersed the seeds of at least 32 species of trees and 21 species of lianas via defecation. Seeds were treated gently in their guts, were defecated an average of 27 hours after ingestion, and had high germination rates. The fate of ingested seeds included: 28% defecated under feeding trees; 21% dispersed 1–10 m from feeding trees; and 51% defecated >10 m from feeding trees (median distance = 15 m; maximum distance = 1,250 m). Only 1.2% of seeds were defecated under sleeping trees (cf. *Ateles paniscus*). Owing to heavy predation by rodents, only 8% of dispersed seeds recruited as seedlings, but this value is quite high compared to other systems (e.g., see *Ocotea endresiana* below). Hence, it seems likely that this primate provides both high-quantity and high-quality seed dispersal for many of its food plants, but more information on the fates of dispersed seeds and seedlings is needed before we can reach a firm conclusion about this.

QUANTITATIVE ASPECTS OF DISPERSER EFFECTIVENESS.—This aspect deals with the abundance, reliability, and visitation rates of fruit-eating birds and mammals to fruiting trees as well as the number of seeds they disperse per fruit crop. Early in the modern era of studies of fruit-frugivore interactions, the tendency was to classify frugivores as either specialists or opportunists based on their putative seed disperser effectiveness (McKey 1975; Snow 1971). Specialists (e.g., oilbirds, certain cotingids and trogons, toucans, hornbills, etc.) were thought to be more reliable visitors to particular fruiting species, to handle seeds "gently," and to deposit seeds in good recruitment sites. Opportunists (e.g., the majority of fruit-eating birds, including tanagers, turdids, mimids, bulbuls, starlings, etc.), in contrast, were thought to provide lower-quality seed dispersal services. By the mid-1980s, the veracity of this paradigm was under question with the realization that many so-called specialists have very broad diets, are not always reliable visitors to fruit crops, and often deposit seeds in poor recruitment sites (e.g., under the canopies of fruiting trees; see, e.g., Wheelwright 1983). Nonetheless, as discussed by Howe (1993b), by the early 1990s strong tests of this paradigm were scarce enough yet that it still had considerable heuristic value.

The recent frugivory literature contains an ever-increasing number of studies showing that many species of tropical fruit-eating birds and mammals are reliable visitors to fruit crops and move large numbers of seeds some distance away from the canopies of fruiting trees (e.g., Balcomb and Chapman 2003; Davidar and Morton 1986; Fleming and Williams 1990; Fleming et al. 1985; Foster 1990; Graham et al. 2002; Howe 1993a; Ortiz-Pulido et al. 2000; Reid 1989; Russo et al. 2006; Wenny 2000a, 2000b; reviewed by Schupp et al. 2010). These plants include early successional species (e.g., *Cecropia peltata*, *Maesa lanceolata*, and *Muntingia calabura*) and epiphytes or parasites (e.g., Loranthaceae) as well as canopy trees (e.g., *Beilschmiedia pendula*, *Monodora myristica*, *Ocotea endresiana*, *Virola nobilis*, and *V. calophylla*), including figs. Large-bodied frugivores (e.g., toucans, hornbills, and many primates) seem to be especially effective at finding fruit crops of trees and moving seeds away from parent plants, but small species of birds and bats can also be quantitatively effective dispersers (e.g., Fleming 1988; Holbrook and Smith 2000; Kinnaird 1998; Kinnaird et al 1996; Murray 1988; Poulsen et al. 2001; Reid 1989; Chapman and Russo 2007). Nonetheless, the ultimate measure of a species' or group of species' quantitative dispersal effectiveness is the number of new plant recruits they produce and so it becomes crucial to determine the fate of seeds once they have been spit out, dropped, or defecated by frugivores.

QUALITATIVE ASPECTS OF DISPERSER EFFECTIVENESS.—This aspect deals with how frugivores treat seeds and where they deposit them. Many frugivores, including birds, bats, and primates, provide relatively gentle treatment of seeds in their mouths and guts (Chapman and Russo 2007; de Figueredo and Perin 1995; Fleming 1988; McConkey 2000; Naranjo et al. 2003; Reid 1989; Tang et al., 2007; Verdú and Travaset 2004; Whitney et al.

1998). In terms of seed handling behavior, two basic modes have evolved independently in birds, bats, and primates. In New World fruit-eating birds, many species (e.g., manakins, Pipridae) are fruit/seed "gulpers" whereas others (e.g., tanagers, Thraupinae; emberizine finches, Emberizinae) are "mashers" (Levey 1987; Moermond and Denslow 1985; Stiles and Rosselli 1993). Gulpers ingest whole fruits and their seeds and either regurgitate or defecate seeds intact. Most Neotropical avian frugivores are gulpers. A few kinds of birds are mashers, which mandibulate fruits before swallowing fruit pulp and juices and spitting out seeds and fiber. Mashers and regurgitating gulpers generally move seeds shorter distances from fruiting trees than defecating gulpers before releasing seeds (Stiles and Rosselli 1993; Wenny 2000a). Interestingly, a similar fruit-handling dichotomy has evolved in Neotropical phyllostomid bats. Nonstenodermatine phyllostomids such as species of *Glossophaga* and *Carollia* are gulpers, which swallow fruit pulp and seeds and defecate seeds, whereas bats of subfamily Stenodermatinae (e.g., *Sturnira, Artibeus, Chiroderma*) are mashers, which press soft fruits against their palate with their tongue, swallow fruit juices, and spit out the fibrous parts of fruits and many seeds (Bonaccorso and Gush 1987; Dumont 2003). In these bats, gulpers are much faster feeders than mashers.

In the Old World, these two feeding methods apparently occur only in mammals and not in birds. Whereas New World primates usually swallow and defecate seeds, Old World primates either swallow and defecate seeds (apes and some cercopithecines) or collect seeds in their cheek pouches and spit them out without swallowing them (many cercopithecines; Chapman and Russo 2007). Primates that swallow seeds generally deposit them much farther from parent plants than spitters (table 4.6). In addition to swallowing many fruits and seeds, chimpanzees use their lips to form wedges of fruit fiber and seeds that they spit out for certain species (e.g., the soft fruits of *Syzygium guineense*; Gross-Camp and Kaplin 2005). Old World fruit-eating pteropodid bats also form wedges and spit out the fiber and seeds of many of the fruit they eat (Dumont 2003; Utzurrum 1995). Wedges are often deposited under the canopies of fruiting trees, where the seeds they contain are sometimes killed by fungal infections (McConkey and Drake 2006; Utzurrum 1995).

The different means of handling seeds before they are dispersed have three important consequences for the fate of seeds: (1) they determine how far seeds will be carried from fruiting plants before they are voided (short distances for mashers, regurgitators, and spitters; longer distances for defecators); (2) they determine whether seeds will germinate (lower probabilities because of fungal infections for seeds that end up in bat wadges but not necessarily in wadges produced by chimpanzees; see Gross-Camp and Kaplin 2005); and (3) they determine whether seeds will be deposited singly or in clumps (singly by regurgitators and spitters; in clumps by defecators and mashers). Clump sizes vary tremendously among different vertebrate frugivores owing to differences in body sizes. Thus, small birds and bats produce small fecal clumps containing a few to hundreds of seeds whereas cassowaries produce large clumps also containing a few to hundreds of seeds depending on seed size (Fleming and Heithaus 1981; Loiselle 1990; Mack 1995; Westcott and Graham 2000). In Cameroon, lowland gorillas and chimpanzees produce fecal clumps containing an average of 18 and 41 seeds measuring >2 cm long, respectively, whereas clumps produced by four species of smaller monkeys contained averages of one to two large seeds (Poulsen et al. 2001). In theory (Howe 1989), clump size and species composition matters because it can influence the intensity of intra- and interspecific competition among seedlings as well as the susceptibility of seeds and seedlings to predators, pathogens, and herbivores. In reality, however, clump size often doesn't matter whenever all seeds, whether they are dispersed singly or in clumps, are destroyed by predators, as often happens in tropical forests (e.g., Howe 1993b; Lambert and Chapman 2005; Wenny 2000b).

THE THREE MAJOR BENEFITS OF SEED DISPERSAL

Howe and Smallwood (1982) proposed that seed dispersal by fruit-eating vertebrates provides three non–mutually exclusive benefits to their food plants: they allow seeds to escape from density-responsive seed predators or other natural enemies; they allow plants to colonize new habitats; and they provide spatially nonrandom, directed dispersal to particularly favorable recruitment sites. These benefits ultimately determine the dispersal effectiveness of these vertebrates. Below, we review evidence for these benefits for vertebrate-dispersed seeds.

THE ESCAPE BENEFIT.—Although seedlings can sometimes recruit under the canopies of fruiting trees (e.g., *Beilschmiedia pendula*, *Tetragastris panamensis*, *Virola calophylla*; Howe 1980; Russo et al. Augspurger 2006; Wenny 2000a), the vast majority of these "undispersed" seeds die before becoming established seedlings. The escape hypothesis proposes that frugivores provide a benefit to plants by moving some seeds away from the "killing zone"

Table 4.6. Estimates of the Seed Dispersal Distances Provided by Selected Species of Tropical Fruit-Eating Birds and Mammals

Disperser Taxon (N species)	Hemisphere, Country	Plant Taxon	Dispersal Method	Mean Distance (m)	Range or Maximum Distance (m)
Birds:					
Various (4)	New World, Costa Rica	*Beilschmiedia pendula*	Defecated	39.6	70
			Regurgitated	14.3	…
Various (5)	New World, Costa Rica	*Ocotea endreasiana*	Regurgitated/ defecated	≤10	70
Various (3)	New World, Costa Rica	3 early successional spp.	Defecated	≤50*	150–300
Cassowary	Old World, Australia and Papua New Guinea (PNG)	Various in Australia; *Aglaia aff. flavida* in PNG	Defecated	201–338 in Austral.*; 388 in PNG	950–1,500
Hornbills, monkeys	Old World, Borneo	*Ficus*	Spit/defecated	>60*	…
Hornbills, turacos, monkeys	Old World, Cameroon	Various	Spit/defecated		Bird max = 473; monkey max = 100
Hornbills (2)	Old World, Cameroon	Various	Defecated	Small seeds = 1,127–1,422; large seeds = 1,620–1,947	3,558 to >6,919
Turacos (3)	Old World, Rwanda	Various	Defecated	"Fast" seeds = 117–232*; "slow" seeds = 143–292*	…

				≤50*	
Bats:					
Carollia	New World	Various	Defecated	...	ca. 1,500
Primates:					
Saguinus spp.	New World	Various	Defecated	...	34–513
Cebus (3)	New World	Various	Defecated	299	20–1,000
Alouatta (2)	New World	Various	Defecated	225	10–811
Ateles (3)	New World	Various	Defecated	217	11–1,119
Lagothrix (3)	New World	Various	Defecated	300	0–989
Cercopithecus (1)	Old World	Various	Defecated	...	1,178
Cercopithecus (2)	Old World	Various	Spit	2	0–100
Hylobates (1)	Old World	Various	Defecated	220	0–1,250
Pan (1)	Old World	Various	Defecated	...	3,000
Pan (1)	Old World	Various	Spit	4	0–20
Other mammals:					
Loxodonta cyclotis	Central Africa	Various	Defecated	82% of large seeds dispersed >1 km	Some large seeds dispersed up to 57 km

Sources. Primate data come from Chapman and Russo (2007). Other data are from Blake et al. (2009); Clark et al. (2005); Fleming (1988); Holbrook and Smith (2000); Laman (1996); Mack (1995); Murray (1988); Sun et al. (1997); Wenny (2000a, 2000b); Westcott et al. (2005).
*Median distance (m).

under canopies to places where they have a better chance of becoming established seedlings (Connell 1971; Janzen 1970). A variety of studies indicate that seeds do not have to move very far from the canopies of fruiting trees to gain a survival advantage. For example, Howe et al.'s (1985) classic experimental study of seed dispersal in *Virola nobilis* on Barro Colorado Island, Panama, showed that seeds gain a twenty- to fortyfold survival advantage to the age of 3 months when they are carried 45 m away from fruiting plants. Further analysis, however, showed that this simple pattern did not persist, primarily as a result of intense seed and seedling predation by mammals (Howe 1993a). In the end, fruit crop size, rather than initial patterns of seed deposition and mortality from insect and mammal seed predators, best predicted the survival of seeds and seedlings. Plants with large crop sizes produced more surviving seedlings at all distances up to 45 m away than plants with small crop sizes. Nonetheless, for most plants some dispersal was better than none, which led Howe (1993a) to conclude that seed dispersal matters for *V. nobilis*.

Wenny's (2000a, 2000b) studies of two species of Lauraceae at Monteverde, Costa Rica, provide an interesting contrast in dispersal effectiveness, as measured by seedling recruitment, and "escape" in a set of ostensibly specialized frugivorous birds—emerald toucanets, resplendent quetzals, three-wattled bellbirds, and black guans. The first tree species, *Beilschmiedia pendula*, produces large seeds (49 mm long, 13 g) most of which are dispersed (if at all) <20 m from the canopy of fruiting trees. Toucanets, quetzals, and bellbirds typically eat several fruits and regurgitate seeds either beneath fruiting plants or a short distance away, whereas guans swallow seeds and defecate them up to at least 70 m from fruiting plants. Survival of all seeds (dispersed or undispersed) exceeded 80%, and seedlings recruited under as well as away from parent trees. Sixteen months later, seedling survival was greatest 20 m from fruiting plants and was higher in undispersed seeds than in seeds dispersed >20 m from parent plants. The highly frugivorous birds in this system thus provided rather inefficient dispersal, but dispersed seeds were still more likely to recruit to the sapling stage than were undispersed seeds. Modest dispersal provided more escape from the fungal pathogens and herbivores that kill seedlings of *B. pendula* than no dispersal (Wenny 2000a).

The second species Wenny (2000b) studied at Monteverde was *Ocotea endresiana*, which produces smaller seeds than *B. pendula* (15 mm long, 0.75 g). Its disperser coterie contains the above four highly frugivorous birds plus the mountain robin, which is a seed regurgitator. Like *B. pendula*, many seeds of *O. endresiana* were not dispersed (45% one year and 73% the next), and most dispersed seeds were deposited within 10 m of parent trees in closed canopy forest. Unlike *B. pendula*, some seeds dispersed by male bellbirds were deposited in gaps below the birds' singing perches. Mortality from rodent predation killed nearly all seeds, dispersed or not, but the distribution pattern of surviving seedlings had two peaks—a small one close to parent plants (as a result of seed regurgitation) and a somewhat higher one in gaps below bellbird singing sites. This pattern is the reverse of the initial seed deposition pattern in which the highest peak occurs near parent plants. Wenny (2000b) concluded that recruitment in *O. endresiana* depends on the combined effects of seed dispersal, seed predation (which strongly limits recruitment), and seedling mortality. Nonrandom (directed) dispersal (i.e., escape) likely plays an important role in recruitment in this species.

Whereas it can be relatively easy to find the dispersed seeds of large-seeded tree species and to follow their fates individually in detail, such is not the case with small-seeded species. As a result, we currently have much less detailed information about the dispersal effectiveness of vertebrates dispersing small seeds. In an early study, Fleming and Heithaus (1981) documented the seed rain produced around fruiting small-seeded plants (e.g., *Ficus* spp., *Cecropia peltata*, *Muntingia calabura*) by phyllostomid bats in a Costa Rican tropical dry forest. They found that the density of clumps of seeds declined rapidly as a function of distance from a fruiting tree and that, in addition to seeds of the "target" species, this seed rain contained up to seven additional species, most of which were not produced close to the target species. The presence of "foreign" seeds around fruiting trees is clear evidence for relatively long-distance seed dispersal and escape from parent plants. An analysis of the dispersion patterns of saplings and adults of bat-dispersed trees and shrubs in this forest showed that clumped dispersion patterns prevailed within species and that mixed-species clumps of bat-dispersed plants were not uncommon. From this, Fleming and Heithaus (1981) concluded that adult distributions of these species are the result of interactions between the diverse seed shadows created by frugivorous bats and the nonrandom distribution patterns of seed germination and seedling establishment sites.

More recently, Laman (1996) studied the fruit crops of two Bornean hemiepiphytic figs, *Ficus stupenda* and *F. subtecta*, and used seed traps in 60-m-long transects to document their seed shadows. These plants produce enormous crops of up to 40,000 fruits containing up to 6.8 million seeds per crop. In both species, >50% of the fruits and seeds fell below fruiting plants, but plants with the largest fruit crops, which attracted the greatest number of frugivorous birds and primates, dispersed up to 46% of their seeds beyond 60 m. In six of the seven study trees, an inverse power function of the form $y = as^{-b}$ in which y is the density of seed rain, s is the distance from the source, and a and b are constants, provided a better fit to the seed rain data than a negative exponential function of the form $y = ae^{-bs}$, which provides a good fit to data from wind-dispersed species (Willson 1993). From this, Laman (1996, 352) reasoned that "this suggests that vertebrate dispersal may produce seed shadows with shapes fundamentally different from those produced by wind dispersal." Specifically, inverse power functions have longer tails than negative exponential functions and hence should provide more long-distance escape for vertebrate-dispersed seeds. Murray's (1988) results for seed dispersal of three species of pioneer shrubs by birds at Monteverde, Costa Rica, as well as Clark et al.'s (2001) findings for the seeds dispersed by birds and monkeys in a tropical moist forest in Cameroon, support this view.

THE COLONIZATION BENEFIT.—This benefit deals with the dynamic nature of all habitats, which experience disturbances at a wide variety of intensities and spatial and temporal scales (e.g., Platt and Connell 2003; Whitmore and Burslem 1998). Early successional species in particular need seed mobility to maximize their contact with high sunlight recruitment sites ranging from small treefall gaps to larger landslides and new beaches exposed by river meanders in tropical forests. In addition, colonization of new habitats is the way most plants expand their geographic ranges (Clark et al. 1998). Each of these colonization scenarios usually requires moderate- to long-distance seed dispersal. As a result, many woody tropical plants rely on frugivorous vertebrates to maximize their colonization abilities.

Two examples will illustrate the role played by vertebrate dispersal in tropical primary succession. In the New World, plant succession on newly exposed beaches created by meandering rivers in the Amazon basin involves many plants dispersed by fruit-eating birds and bats. Along the Rio Manu in Peru, for example, bats disperse 10 of the 38 species of initial colonizers (Foster et al. 1986). In terms of biomass, two bat-dispersed species (*Cecropia* "*tessmannii*" and *Ficus insipida*) dominate the early stages of floodplain succession. As succession progresses, bird- and nonvolant mammal–dispersed species become more common, but even after 500 years of forest development, two of the four species of forest giants (*Dipteryx alata* and *Poulsenia armata*) are primarily bat-dispersed. Foster et al. (1986, 365) concluded that "wind and bat-dispersal is the most effective way for large or fast-growing trees to reach areas with long-term lack of competition and thus achieve individual or collective dominance. . . . These mechanisms are much more likely than birds or nonflying mammals to get seeds out of the forest and onto the beach edge, or into the river with later deposition downstream."

A classic example of primary succession in the Old World is the reforestation of the islands of Krakatau after they were sterilized by volcanic eruptions in 1883. As described by Whittaker and Jones (1994), the earliest animal-dispersed plants included four species of *Ficus* (Moraceae), *Macaranga tanarius* (Euphorbiaceae), and *Melastoma affine* (Melastomataceae). Of these, either birds or bats could have brought in seeds of *Ficus* and *Melastoma*; only birds disperse seeds of *Macaranga*. By 1992, 124 plant species dispersed either by bats or birds occurred on Krakatau. Of these, 43 species could have been brought in by either group, 31 species were brought in by birds and subsequently dispersed within the islands by bats, and 50 species were solely bird-dispersed. Overall, except for species of *Ficus*, a majority of the animal-dispersed plants on Krakatau were brought there by birds. Therefore, unlike the situation in Amazonia, primary succession on Krakatau was more bird-dependent than bat-dependent, although, as Shilton and Whittaker (2009) point out, individuals of early colonizing figs brought to Krakatau by vagrant bats prior to 1900 could have served as "recruitment foci" for other species of bat- or bird-dispersed plants.

Studies of secondary plant succession also highlight the importance of bird- or bat-dispersal for tropical forest regeneration. As in the case of primary succession, bats appear to play a more important role in secondary succession in the Neotropics than in the Paleotropics (Muscarella and Fleming 2007). As mentioned above, abandoned fields and other large disturbances in the Neotropics are usually first colonized by bat-dispersed taxa such as *Piper*, *Vismia*, *Solanum*, and *Cecropia* (e.g., Guariguata 2000; Saldarriaga et al. 1988; Toledo and Salick 2006; Uhl 1987; Uhl et al. 1981), and bat-dispersed species tend to dominate the seed rain falling into abandoned fields or around isolated fruiting trees in pastures (e.g., Galindo-Gonzalez et al. 2000; Medellin and Gaona 1999). In contrast, bird-dispersed species (e.g., *Musanga* [Cecropiaceae] in Africa and *Macaranga* throughout the Paleotropics) are common in early successional habitats in the Paleotropics (e.g., Corlett 2002; Grubb 1998; Cleary and Pridjati 2005; Lwanga 2003; Turner and Corlett 1996; White et al. 2004), and bird-dispersed species dominate the seed rain around remnant trees in southern Cameroon and in the Philippines (Carriere et al. 2002; Ingle 2003). In summary, bats and birds are important dispersers of the small seeds of many pioneer plants throughout the tropics, and interesting biogeographic differences exist regarding the relative importance of these two vertebrate groups in tropical forest regeneration.

THE DIRECTED DISPERSAL BENEFIT.—This benefit involves the nonrandom dispersal of seeds to particularly favorable recruitment sites. As discussed in detail by Wenny (2001), less attention has been paid to this aspect of seed dispersal than to escape and colonization aspects, but current evidence suggests that it might be more common than was once thought. Convincing evidence of the importance of directed dispersal for recruitment in some plants comes from studies of hemiepiphytic figs dispersed by the phyllostomid bat *Artibeus jamaicensis* (August 1981); mistletoes dispersed by flowerpeckers and mistletoebirds in Australasia, by euphonias and tanagers in montane regions of the Neotropics, by mockingbirds in southern South America, and by waxwings and phainopeplas in southwestern United States (Larson 1996; Martínez del Rio et al. 1995; Reid 1989, 1990; Sargent 2000; Walsberg 1975); dispersal of seeds of *Ocotea endresiana* to forest gaps by singing male three-wattled bellbirds (Wenny 2000b; Wenny and Levey 1998); dispersal of palm seeds to tapir latrines and of seeds of *Trewia nudiflora* to rhinoceros latrines (Dinerstein and Wemmer 1988; Fragoso 1997); and dispersal of seeds of columnar cacti and *Capsicum* species under "nurse plants" (usually nonconspecific shrubs or trees) in desert habitats in North America and Mexico (Sosa and Fleming 2002; Tewksbury et al. 1999; Valiente-Banuet et al. 1991). In addition to these specialized cases, directed dispersal occurs more generally whenever birds and bats deposit seeds under and around isolated fruiting trees in abandoned pastures (e.g., Carriere et al. 2002; Galindo-Gonzalez et al. 2000; Guevara et al. 2004; Slocum 2001). This situation has important conservation implications because it can be a means of rapidly reseeding disturbed lands (Zahwai and Augspurger 2006).

In summary, substantial evidence indicates that fruit-eating birds and mammals provide a variety of benefits to their food plants when they disperse their seeds. Frugivore foraging behavior allows the seeds of many plants to escape from killing zones under fruiting trees, to colonize disturbed lands and new habitats, and, in some cases, to be deposited in especially favorable recruitment sites. As is frequently pointed out, initial seed dispersal patterns, sometimes modified by secondary dispersal agents such as ants, dung beetles, and seed caching mammals (reviewed in Forget et al. 2005), provide the template for later stages in the life histories of plants. Adult distribution patterns may or may not be concordant with the initial seed dispersal template, but without some dispersal, it is likely that the plant populations and communities would be very different than they are in terms of density, distribution, and S.

SEED DISPERSAL DISTANCES PRODUCED BY FRUGIVOROUS VERTEBRATES

Vertebrate frugivores clearly provide mobility for some of the seeds they handle, but obtaining quantitative estimates of median and, especially, maximum dispersal distances can be very difficult (Muller-Landau et al. 2008; Nathan 2006). Nonetheless, such estimates are important because they ultimately determine the sizes of genetic neighborhoods and degree of genetic subdivision within populations of vertebrate-dispersed plants. Although all seed dispersal methods generally produce leptokurtic dispersal kernels (i.e., probability density functions of dispersal distances; Levine and Murrell 2003; Levin et al. 2003) with a peak under or close to the canopy of fruiting plants, different dispersal systems are likely to differ in the size of the tail of their dispersal kernels and hence in median dispersal distances. But measuring this long-distance tail can be logistically challenging.

Both direct and indirect methods have been used to measure the dispersal distances of vertebrate-dispersed seeds. Direct methods, which are much easier to use with large seeds (e.g., those >1 cm long) than small seeds, involve observing or following frugivores to determine where they deposit seeds or associating dispersed seeds with the behavior of their putative dispersers (e.g., Gross-Camp and Kaplin 2005; Reid 1989; Russo et al. 2006; Wenny 2000a, 2000b). For terrestrial frugivores such as cassowaries and agoutis, seeds falling under the canopies of fruiting trees can be labeled and then relocated once they have been dispersed (e.g., Forget and Wenny 2005; Mack 1995). Indirect methods involve modeling the behavior of frugivores based on radio-tracking data and knowledge of seed retention times (e.g., Fleming 1981, 1988; Holbrook and Smith 2000; Levey et al. 2008; Murray 1988; Uriarte et al. 2011; Westcott et al. 2005a). Genetic methods can also be used to identify long-distance dispersal events (e.g., Jones and Muller-Landau 2008; Jordano et al. 2007; Wang et al. 2007).

Estimates of median and maximum seed dispersal distances produced by tropical birds and mammals are summarized in table 4.6. Because these estimates were obtained by a variety of different methods, they are not directly comparable *inter se* (e.g., to determine whether birds disperse seeds farther, on average, than bats). Instead, we present them to indicate the range of dispersal distances that have been reported in the literature. As expected given their wide range of sizes and foraging behaviors, birds exhibit a wide variety of dispersal distances. Large terrestrial cassowaries are moderate- to long-distance seed dispersers; they routinely disperse large numbers of seeds in clumps. An interesting dichotomy in dispersal distances exists among non-terrestrial birds. Certain New World canopy-feeding species (e.g., toucanets, quetzals, and noncourting bellbirds) regurgitate seeds very close to fruiting plants, if they disperse them at all, and are short-distance seed dispersers. Guans, in contrast, defecate seeds of the same species greater distances. At the same site in Costa Rica, small understory birds such as turdids, barbets, and silky flycatchers disperse the small seeds of early successional plants longer distances than the larger canopy species (Murray 1988). In Africa, canopy-feeding turacos disperse seeds 100–300 m, depending on how long seeds are retained in their digestive tracts. Large African and Asian hornbills are remarkably long-distance seed dispersers and routinely seem to disperse seeds farther than Neotropical canopy frugivores (table 4.6).

Except for species such as *Carollia perspicillata*, we lack detailed information about how far fruit-eating bats disperse seeds (table 4.6), but radio-tracking studies provide us with an indication of the range of dispersal distances these bats are likely to provide for their food plants. Most frugivorous bats are refuging species (Hamilton and Watt 1970) that spend the day roosting gregariously in caves, in hollow trees, or in the foliage of canopy or understory trees and commute from their day roosts to one or more feeding areas at night. Commute distances can be ≤1 km for some species (e.g., *Balionycteris maculata, C. perspicillata, Nyctimene robinsoni, Syconycteris australis*; Fleming 1988; Hodgkison et al. 2003; Spencer and Fleming 1989; Winkelmann et al. 2000) or as far as 8–10 km in other species (e.g., *Artibeus jamaicensis, Phyllostomus hastatus, Pteropus poliocephalus*; Eby 1991; McCracken and Bradbury 1981; Morrison 1978a). Except for species of *Pteropus*, which sometimes defend small feeding territories in fruiting trees (McConkey and Drake 2006; Richards 1995), frugivorous bats remove single fruits or parts of larger fruits from plants and fly some distance to a feeding roost to eat. *Carollia* bats often night roost within 50 m of their food plants, but other phyllostomids (e.g., *A. jamaicensis, A. lituratus,* and *Vampyrodes caraccioli*) sometimes fly 100–250 m from fruiting trees to feed (Charles-Dominique 1986; Morrison 1978b, 1980a). Bats usually change feeding areas at least once during a night, and feeding areas can be a kilometer or more apart. All of these movements (commutes, shuttles from fruiting plants to feeding roosts, and changes in feeding areas) provide extensive movement for seeds because gut retention times are short (usually about 30 minutes), and, unlike birds, bats often defecate in flight. Additional evidence for long-distance dispersal in pteropodid bats comes from the recolonization of Krakatau by bat-dispersed plants, by observations of interisland movements of radio-tagged bats in the Krakatau group, and by exceptionally long (up to 19 hours) seed retention times in some species (Shilton and Whittaker 2009; Shilton et al. 1999). Island-dwelling pteropodids in general often fly among islands while foraging and are clearly important long-distance seed dispersers in these systems (McConkey and Drake 2002, 2006; Rainey et al. 1995).

Primates are much larger than bats, often have large home ranges, and have long seed retention times (range = 3.3–35 hours; Chapman and Russo 2007), but their seed dispersal distances aren't necessarily exceptional (table 4.6). Seed handling mode has a dramatic effect on distances that primates move seeds; spitters move seeds about one-tenth as far, on average, as defecators. Average or median seed dispersal distances for most defecators are 200–300 m and maximum dispersal distances are 1–2 km—values that are likely to be comparable to those of many species of canopy-feeding bats. Based on our current knowledge, African and Asian hornbills appear to be longer-distance seed dispersers than their primate counterparts.

Finally, African forest elephants (*Loxodonta cyclotis*) are likely to be the long-distance seed dispersal champions among terrestrial vertebrates. Blake et al. (2009) report that they sometimes move seeds as far as 57 km from parent plants in three days (table 4.6).

FORESTS WITHOUT FRUGIVORES

The best (but ethically unacceptable) way to demonstrate that seed dispersal by vertebrate frugivores plays an important role in the abundance and distribution patterns of their food plants is to do a removal experiment. If a plant population's λ is reduced and its dispersion pattern becomes more clumped postremoval, then, all else being unchanged, we can conclude that vertebrate seed dispersers have an important impact on the demography and distribution of their food plants. Otherwise, we cannot. Even without doing a removal experiment, however, it is easy to predict how the absence of vertebrate frugivores might affect the seed shadows of their food plants. Virtually all seeds will be deposited under or very near the canopies of parent plants where they will suffer overwhelming (but not total?) mortality from the usual biotic processes (competition and predation) and agents (conspecifics, seed predators, pathogens, and herbivores). Surviving seeds/seedlings will produce strongly clumped adult dispersion patterns. Because of their limited disperser coteries, we might expect that populations of large-seeded plants with hard-to-disperse seeds will decline at a faster rate than populations of small-seeded plants. Ultimately, tropical plant communities likely will have lower species richness in the absence of their frugivores (Muller-Landau 2007; Webb and Peart 2001).

Although it is unethical to conduct frugivore removal experiments, such experiments exist for both biogeographical and anthropogenic reasons. An example of a biogeographical "removal experiment" comes from a comparison of the dispersal ecology of two species of *Commiphora* (Burseraceae) trees—*C. harveyi* in South Africa and *C. guillaumini* in Madagascar (Bleher and Böhnin-Gaese 2001). Higher species richness of frugivorous birds (15 vs. six species) in the African site resulted in a nearly tenfold increase in proportion of seed crop dispersed (71% vs. 8%), a greater median distance between established seedlings and fruiting plants (21 m vs. 0.9 m), and a less-clumped adult dispersion pattern in *C. harveyi* compared with *C. guillaumini*. After ruling out other potential causes of these differences, Bleher and Böhnin-Gaese (2001) concluded that differences in both the diversity of frugivores and their seed handling behavior (ingestion in Africa vs. non-ingestion in Madagascar) were responsible for the striking differences in seed dispersal and seedling establishment in the two species.

Anthropogenic effects on frugivore diversity involve hunting pressure and deforestation and habitat fragmentation. Kent Redford (1992) described the effects of subsistence and commercial hunting on the bird and mammal faunas of the Amazon Basin. He pointed out that in the 1980s, hunting resulted in the deaths of at least 23 million of these animals and that about 70% and 84% of this biomass came from species of frugivorous birds and mammals, respectively. In heavily hunted areas, primate density and biomass were reduced by 81% and 94% compared with nonhunted areas. Comparable values for birds were reductions of 74% and 95% for density and biomass. With these reductions, ecological functions such as seed dispersal, seed predation, and herbivory were likely to be significantly affected. He concluded: "We must not let a forest full of trees fool us into believing that all is well. Many of these forests are 'living dead' . . . , and, although satellites passing overhead may reassuringly register them as forest, they are empty of much of the faunal richness valued by humans. An empty forest is a doomed forest" (421).

As noted by Redford (1992), a species does not have to go extinct for its role as a seed disperser to be diminished when its numbers drop to low levels. For example, in areas where population levels of the Pacific island bat *Pteropus tonganus* are low, few seeds of their fruit trees are dispersed away from parent plants (McConkey and Drake 2006). Above a certain abundance threshold, however, the proportion of dispersed seeds steadily increases. At low bat densities, the canopies of fruiting trees are not saturated with feeding territories, territory holders drop seeds beneath tree crowns, and there are no socially subordinate individuals to carry seeds away to feeding roosts (see above). At greater than the threshold, fruiting canopies are saturated with bat territories, and more seeds are removed from trees and dropped by subordinate bats. Thus, this bat acts as a nondisperser at low densities but becomes a seed disperser above a threshold level of abundance.

Since 1992, we have learned much more about how the loss of frugivores and their seed dispersal services might affect plant populations and communities. Island systems continue to be instructive on this issue. Numerous islands in the South Pacific have lost many of their large fruit-eating pigeons and bats as a result of overhunting, to the potential detriment of pollination and seed dispersal services (Cox et al. 1991; Steadman 2006). Rainey et al. (1995), for example, compared the proportion of seeds dispersed away from fruiting trees by bats in transects on American Samoa, where two species of *Pteropus* bats are still relatively common, and on Guam where only one species exists at very low numbers. These proportions averaged 0.37 and 0.002, respectively. Similarly, on islands in the Tonga group, at least three species of canopy trees produce large fruits that were likely dispersed by a now-extinct large species of *Caloenas* fruit pigeon. McConkey and Drake (2002) ask, Who now disperses these species? Moreover, where *Pteropus* bats are uncommon in the Tonga group, the proportions of seeds dispersed more than 5 m from the canopies of fruiting trees are low (0.0–0.07) (McConkey and Drake 2002). Collectively, these results indicate that nondispersal is the usual fate of seeds when frugivores are either extinct or occur in low numbers. Nonetheless, until their recruitment success is studied in detail, we will not know the long-term consequences of apparent nondispersal for forest plants on Pacific islands.

Unsustainable hunting of tropical frugivorous birds and mammals clearly is not restricted to oceanic islands. As discussed in detail by Corlett (2007a), Fa and Peres (2001), and Peres and Palacio (2007), it is rampant in all of the world's tropical forests (see chap. 10). Bushmeat hunting of many species of large birds and mammals has caused local extirpations of some species (e.g., gorillas, chimps, and duikers in Cameroon and Uganda) and has drastically reduced population levels of many other species in many areas of otherwise intact forest throughout the tropics. But the demographic consequences of this hunting pressure are likely to be complex, primarily because hunting selectively removes some but not all species of large frugivores and it also removes seed predators. Loss of these species undoubtedly increases dispersal limitation but it can also reduce recruitment limitation (Muller-Landau 2007). Long-term demographic studies are needed to determine the net botanical consequences for plants whose seeds are dispersed or preyed on by targets of the tropical bushmeat trade.

Increased dispersal limitation, especially of large-seeded plants whose fruits are eaten by large birds and mammals, has either been postulated or documented to occur in many tropical areas (e.g., Africa: Chapman and Chapman 1996; Chapman and Onderdonk 1998; Vanthomme et al. 2010; Wang et al. 2007; the Neotropics: Nuñez-Iturri et al. 2008; Nuñez-Iturri and Howe 2007, Peres and Palacio 2007; Peres and Roosmalen 2002; S. Wright et al. 2007). Reflecting this, studies of the density and species richness of seedlings in hunted and unhunted areas sometimes report a reduced density and diversity of large-seeded species, usually with a concomitant increase in the density and diversity of seedlings of wind-dispersed lianas and small-seeded plants dispersed by small birds and bats, in hunted areas (e.g., Chapman and Onderdonk 1998; Corlett 2007a; Nuñez-Iturri et al. 2008; Nuñez-Iturri and Howe 2007, S. Wright et al. 2007). But the picture can become complicated whenever hunting also removes large seed-eating birds and mammals. Muller-Landau (2007), for example, points out that removal of seed predators can give large-seeded plants a twofold recruitment advantage over small-seeded plants in closed canopy forests: (1) they suffer reduced seed predation and (2) they have enhanced competitive ability as seedlings. Dirzo and Mendoza (2007) note that the removal of large seed-eating and browsing mammals such as peccaries, tapirs, and deer by poaching has resulted in higher seedling densities of large-seeded plants in tropical wet forest at Los Tuxtlas, Mexico. Chapman and Chapman (1996) reported that, in the absence of seed predators, undispersed seeds in two of the four species they studied in Kibale National Park, Uganda, recruited as well as or better under parent plants than dispersed seeds and proposed that there may be an intraspecific tradeoff between small, easily dispersed seeds that recruit poorly under parent plants and large, harder to disperse seeds that can recruit under parent plants (also see Wenny 2000b).

Hunters usually do not target small species of birds and mammals for the bushmeat trade (but see Redford 1992), so their removal from intact forests by hunting is not necessarily a conservation problem. Loss of these species, however, could also have significant ecological consequences for tropical forests. Loiselle and Blake (2002), for example, modeled the seed dispersal consequences of the loss of one or more species of birds that disperse the seeds of four species of understory shrubs in a Costa Rican wet tropical forest. Their results suggested that, depending on which bird species they "removed," changes in the shapes and distributions of seed shadows were substantial or minimal. They noted that the long-term effects on plant population dynamics of these changes were not obvious. Likewise, Henry and Jouard (2007) conducted a short-term (1 month in each of 2 years) bat "removal" experiment in primary forest in French Guiana and monitored seed rain into seed traps on one experimental and one control plot. In this experiment, "removal" involved scaring bats away from the experimental plot with a series of mist nets that cause bats to avoid the area temporarily. They reported that seed traps in the experimental plot received fewer seeds and a lower diversity of seeds than traps in the control plot. Species with fruits containing few seeds per fruit (e.g., species of *Solanum* and *Vismia*) were affected more strongly than species with a high number of seeds per fruit (e.g., epiphytic species of Araceae and Cyclanthaceae). But, again, these results do not speak to the issue of plant recruitment limitation.

Finally, it is sometimes suggested that large species of frugivorous birds or mammals that are not hunted for bushmeat can provide "compensatory" or redundant dispersal services for species eaten by target species. Available evidence does not support this view. In the forests of central Cameroon and Gabon, for instance, the diets of hornbills are quite distinct from those of primates (Gautier-Hion et al. 1985; Holbrook and Smith 2000; Poulson et al. 2001). Hornbills, which receive less hunting pressure than primates in Africa (but are still harvested; Trail 2007), cannot substitute as effective dispersers of primate-dispersed fruits. Similarly, Peres and Roosmalen (2002) point out that only large primates such as spider monkeys (*Ateles*) or woolly monkeys (*Lagothrix*) are effective dispersers of seeds >25 mm long in Amazonian forests. Large primates such as gibbons and macaques, and to a lesser extent hornbills, are also much more effective dispersers of seeds of *Prunus javanica* (Rosaceae) than small birds and squirrels in Thailand (McConkey and Brockelman 2011). Ecological redundancy among frugivores thus is low for these plants. Whenever large frugivores disappear from these forests, large-seeded plants lose their dispersers. Many of these trees have considerable economic value, so a dollar value can be placed on the loss of these primates (Nuñez-Iturri and Howe 2007; Peres and Roosmalen 2002). We will return to this topic in chapter 10.

The Genetic Structure of Tropical Plants

Our final topic in this chapter deals with the effect that vertebrate pollinators and seed dispersers have on the genetic structure, specifically the degree of population subdivision as indicated by Wright's (1965) fixation index F_{st} or its multilocus or molecular analogues of G_{st} and Φ_{st}, of tropical plants. Nectar-feeding and fruit-eating birds and mammals are constantly moving the genes of their food plants around within and between habitats. To what extent do these movements influence the genetic neighborhood size and degree of genetic subdivision of these plants? Do vertebrates provide greater mobility for plant propagules and consequently produce larger neighborhood sizes and less subdivision than other pollination or dispersal methods? Which of the two mutualisms we discuss in this book has the greatest effect on plant genetic structure?

Before attempting to answer these questions, we note that many factors are responsible for determining the genetic structure of plant populations. At the beginning of the era of allozyme studies of tropical plant populations, Loveless and Hamrick (1984) summarized many of these factors and predicted their genetic effects (table 4.7). Their list of factors included a plant's breeding system, floral type, pollination mechanism, seed dispersal mechanism, and flowering phenology. Of these, breeding system was predicted to have a major effect on genetic structure with self-compatible species containing less genetic variation and more population subdivision than self-incompatible species. This prediction has been supported by numerous subsequent studies (e.g., Duminil et al. 2007; Hamrick et al. 1992). They noted that pollen flow from parent plants is generally leptokurtic and varies substantially with type of pollen dispersal (e.g., wind vs. animal pollination) and, within animal-pollinated species, varies with type and size of pollinators. Like pollen dispersal, seed dispersal is generally leptokurtic and varies widely with dispersal method. Because diploid seeds carry two sets of alleles instead of only one as in haploid pollen, effective gene flow via seeds has potentially twice the value of pollen dispersal with respect to gene dispersal; because genes carried in pollen are effectively dispersed twice in seeds, the effect of pollen dispersal is also likely to be more variable in its genetic consequences than seed dispersal. Finally, they predicted that early successional species should have smaller effective population sizes, stronger founder effects (i.e., more genetic bottlenecks), and stronger genetic subdivision than late successional species.

Results of the Loveless and Hamrick (1984) review indicated that, as expected, plants pollinated by birds or bats had lower genetic subdivision (mean G_{st} = 0.158, n = 2 studies) than other pollination systems (range of mean G_{st} values = 0.224–0.322) but that genetic subdivision in plants with seed dispersal via animal ingestion was not particularly low (mean G_{st} = 0.332 [n = 14] vs. means of 0.079–0.446 for other dispersal methods). Regarding the effects of seed dispersal method on population structure, Hamrick and Loveless (1986) suggested that the relatively high G_{st} values in animal-ingested species indicated that (1) most seeds were dispersed locally, (2) if seeds were dispersed long distances, they suffered bottleneck events that increased the level of differentiation between new populations and source populations, or (3) there is little gene movement via seeds among populations.

Subsequent summaries of allozyme-based genetic studies by the Hamrick group (Hamrick and Godt 1996; Hamrick et al. 1992, 1993) using a substantially larger database indicated that outcrossing plants with animal pollination had a mean G_{st} value of 0.099 (n = 37 species) compared with 0.077 for outcrossing wind-pollinated species (n = 146 species). In terms of seed dispersal, animal-ingested species had a mean G_{st} value of 0.051 (n = 14 species) compared with mean G_{st} values for other dispersal mechanisms, which ranged from 0.065 to 0.131. Multivariate analysis revealed that high values of G_{st} (i.e., relatively strong subdivision) in woody plants are associated with tropical (cf. high latitude) distributions, animal pollination, and seed dispersal by gravity (Hamrick et al. 1992). Based on a two-trait analysis of these data, Hamrick and Godt (1996) concluded that outcrossed, animal-dispersed species had somewhat lower G_{st} values (0.223) than animal-dispersed species with mixed mating systems (0.269) but higher values of G_{st} than other dispersal modes, which ranged from 0.101 to 0.189. Finally, in a meta-analysis of 164 studies of genetic structure in plants (biased toward

Table 4.7. Summary of selected ecological factors that can influence the genetic structure of plants

Ecological Factor	Genetic Variation within Populations	Genetic Structure	
		Among Populations	Within Populations
Breeding system:			
Autogamous or selfing	Lower; low heterozygosity	Increased divergence owing to drift and reduced gene flow	Reduced heterozygosity and within-family genetic diversity; increased between-family genetic variation
Mixed mating	More variability	Potential for differentiation, depending on levels of selfing	Potentially subdivided, depending on levels of selfing
Predominantly outcrossing	Higher: high heterozygosity	Reduced divergence owing to increased gene flow	Increased N_e and neighborhood size, reduced subdivision
Floral type:			
Hermaphroditic	Moderate levels, if mixed mating; lower if selfing	Depends on breeding system	Potential for subdivision; depends on mating system and pollen movement
Monoecious	Potentially high, if highly outcrossed	Increased outbreeding and pollen flow reduce differentiation	Depends on mating system and pollinators
Dioecious or heterostylous	High	Enforced outcrossing and pollen movement reduce differentiation	Enforced outbreeding reduces subdivision
Pollination mechanism:			
Small bee	Reduced amounts of variability	Limited pollen movement and local foraging increase differentiation	Limited, leptokurtic distribution, or nearest neighbor pollen movement reduces N_e and promotes subdivision; animals with high variance in pollen carryover and delivery will increase N_e
Large bee			
Butterfly/moth			
Bird/bat	⋯	Rare long-distance pollen dispersal, long-distance traplining prevent differentiation	Large, vagile vectors will visit more plants, reduce subdivision, and produce large N_es and large neighborhood sizes

Seed dispersal mechanism:			
Gravity	Intermediate	Limited dispersal promotes differentiation	Limited seed movement reduces N_e, promotes family structure and inbreeding, increased homozygosity, and subdivision
Animal ingested	Intermediate	Regular long-distance transport reduces subdivision	Dispersal by animals may reduce clumping and family structure
Animal attached	Low		
Phenology:			
Population asynchronous	No prediction	Prevents gene exchange, promotes divergence	Restricts mating, reduces N_e, and promotes subdivision
Population seasonal and synchronous	No prediction	Potential for extensive gene flow, reduces probability of divergence	Large potential N_e; may be restricted by pollinator behavior or family structure, but potentially homogeneous
Extended, steady-state flowering	No prediction	Long-distance pollinator movement prevents divergence	Reduces selfing, increases pollen flow, increases N_e, and prevents subdivision

Source. Based on Loveless and Hamrick (1984).

higher latitudes) based on nuclear and cytoplasmic markers, Duminil et al. (2007) concluded that except for mating system and gravity dispersal, most life history traits, including pollination mode, explain little variation in genetic structure in plants.

Although these analyses of large data sets appear to indicate that animal pollination and seed dispersal generally play minor roles in determining the genetic structure of plant populations, it is still instructive to examine data for specific kinds of tropical plants to see how these two factors affect particular species or groups of species. Differences in genetic subdivision in animal-pollinated arid zone tropical and subtropical cacti, for example, are correlated with particular kinds of pollinators. Bat-pollinated (and bat-dispersed) species tend to have lower values of G_{st} (i.e., more between-population gene flow) than two of three species of insect- or territorial hummingbird–pollinated species (fig. 4.6). Of the two species with high G_{st} values, *Lophocereus schottii* (now *Pachycereus schottii*) is self-incompatible and *Melocactus curvispinus* has a mixed mating system (Fleming and Holland 1998; Nassar et al. 2001).

Additional examples of genetic subdivision in vertebrate-dispersed tropical plants are shown in table 4.8. These data are very heterogeneous regarding experimental design and intent and only provide us with a glimpse of the

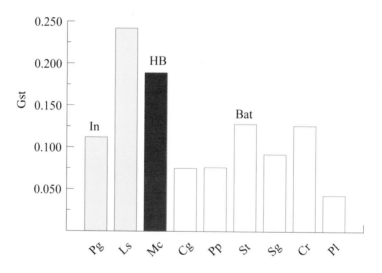

Figure 4.6. Genetic subdivision (G_{st}) in nine species of New World columnar cacti. Pollination systems include insect (*In*), hummingbird (*HB*), and bat. Based on data in Hamrick et al. (2002).

Table 4.8. Examples of the genetic structure of vertebrate-dispersed tropical plants

Species and Family	Location and Habitat	Habitat Status	Growth Habit	Mating System	Pollinators	Seed Dispersers	F_{st} or Its Analogue	Source
Piper amalago (Piperaceae)	Costa Rica, TDF	F	S	SI	Small bees	*Carollia* bats	0.103	Heywood and Fleming (1986)
Piper hispidinervum	Brazil, TMF	F	T (small)	SI?	Insects	*Carollia* bats	0.280	Wadt and Kageyama (2004)
Piper cernuum	Brazil, TWF	F	S	SI?	Insects	*Carollia* bats	0.380	Mariot et al. (2002)
Erythroxylum havanense (Erythroxylaceae)	Mexico, TDF	F	S	SC	Insects	Birds	0.131	Dominguez et al. (2005)
Psychotria faxlucens (Rubiaceae)	Mexico, TWF	C	S	SI	Moths	Birds	0.026	Perez-Nasser et al. (1993)
Psychotria officianalis (Rubiaceae)	Costa Rica, TWF	C	S	SI	Insects	Birds	0.095	Loiselle et al. (1995)
Cecropia obtusifolia (Cecropiaceae)	Mexico, TWF	F	T	SI	Wind	Birds and mammals	0.029	Alvarez-Bullya and Garay (1994)
Syzygium nervosum (Myrtaceae)	Australia, TWF	F	T	SC	Insects, birds, bats	*Ducula* pigeons and *Pteropus* bats	0.118	Shapcott (1999)
Carpentaria acuminata (Arecaceae)	Australia, TMF	F	P (large)	SC	Insects, *Pteropus* bats, birds	*Ducula* pigeons and *Pteropus* bats	0.379	Shapcott (1998)
5 *Pinanga* spp (Arecaceae)	Borneo, TWF	C	P (small)	SC	Insects	Birds	0.099–0.261	Shapcott (1999)

Note. Genetic structure is indicated by Wright's fixation index F_{st} or its multilocus analogue G_{st} or Φ_{st}. Extent of population subdivision ranges from 0 to 1 and increases as this index increases. TDF = tropical dry forest; TMF = tropical moist forest; TWF = tropical wet forest; C = continuous; F = fragmented; P = palm, S = shrub, T = tree; SC = self-compatible, SI = self-incompatible.

diversity in structure that likely characterizes tropical plants. Nonetheless, they allow us to reach several very tentative conclusions. First, as expected, populations in continuous forest tend to exhibit less genetic subdivision than those in fragmented habitats. Most values of F_{st} are below 0.100 in the former group whereas they are greater than 0.100 in the latter group. A potentially confounding factor in this comparison, however, is distance between study plots or populations. This distance tends to be shorter in studies in continuous habitats than in fragmented habitats, which should result in less genetic differentiation by distance in continuous habitats. Second, relatively substantial levels of subdivision (up to $F_{st} = 0.380$) occur in vertebrate-dispersed species that occur in fragmented habitats. Although vertebrates such as *Pteropus* bats can be very mobile, they appear to provide relatively little long-distance dispersal for some kinds of seeds. The Australian rain forest tree *Syzygium nervosum* is interesting in this regard because it is both pollinated and dispersed by these bats. Shapcott (1998, 1999) reported that individual seed crops of this mass-flowering species exhibit high levels of homozygosity as a result of very short pollen movements and that most of the gene flow between isolated populations likely results from bat-mediated seed dispersal. Third, contrary to the prediction of Loveless and Hamrick (1984), at least one early successional species, *Cecropia obtusifolia*, a dioecious wind-pollinated and vertebrate-dispersed tree, does not exhibit strong genetic structure and small genetic neighborhoods. Gene flow (mostly via wind pollination?) is extensive between isolated populations, even ones that are 100 km apart (Kaufman et al. 1998)! Finally, genetic isolation by distance is the expected pattern in most populations of plants and animals, but this pattern does not appear to exist in several of the species included in table 4.8. An isolation-by-distance pattern means that the genetic distance or dissimilarity between pairs of populations increases with distance between them as a result of spatially restricted gene flow. This pattern does not exist in *Carpentaria acuminata, Cecropia obtusifolia, Erythroxylum havanense,* and *Syzygium nervosum*. It is likely that gene flow via seed dispersal rather than by pollen dispersal accounts for the absence of isolation by distance in three of the four species. We need many more data on the genetic structure of tropical plant populations, however, before we will know whether vertebrate seed dispersal regularly reduces genetic differentiation in fragmented populations of their food plants. Especially informative would be studies of the genetic structure of cohorts of recently established seedlings compared with adult populations in different fragments. What proportion of these cohorts was produced by parents living outside the fragment where recruitment took place (e.g., Jordano et al. 2007)?

In the end, we would like to know whether dispersal of pollen and seeds by tropical vertebrates has a different effect on the genetic structure of plant populations than other methods of gene dispersal. Only in the case of seed dispersal can we now reach a firm conclusion about this. Seeds ingested by birds and mammals do appear to move greater distances than seeds that are explosively dispersed or dispersed by gravity, and these differences appear to result in larger genetic neighborhoods and less population subdivision in vertebrate-ingested species. By contrast, it may not be true that vertebrate pollination results in larger genetic neighborhoods compared with other pollination mechanisms. The example of *C. obtusifolia* indicates that wind pollination, which is a relatively uncommon pollination mechanism in tropical plants (chap. 3), can sometimes create very large genetic neighborhoods. This is a common occurrence in wind-pollinated high latitude pines (Hamrick et al. 1992). Even more impressive are tiny fig wasps, which use the wind to move tens of kilometers and whose host plants have the largest genetic neighborhoods (encompassing 108–630 km^2) of any tropical plants (Nason et al. 1998). Less dramatically, recent genetic studies indicate that tropical bees and other insects can move pollen several kilometers between plants (reviewed in Ward et al. 2005) and are likely to be longer distance pollen carriers than many species of nectar-feeding birds and mammals. Nonetheless, about 500 genera of tropical plants clearly "view" nectar-feeding birds as desirable pollinators as do about 250 genera of bat-pollinated plants (Fleming et al. 2009; Sekercioglu 2006), even though long-distance pollen movement is not unique to large, strong-flying vertebrates. There must be other reasons why birds and mammals might be desirable dispersers of pollen in tropical habitats. We will explore these reasons in upcoming chapters.

Conclusions

By pollinating flowers and dispersing seeds, plant-visiting birds and mammals clearly affect the reproductive success of a wide variety of tropical plants, ranging from canopy trees and their epiphytes to understory shrubs and herbs. They are also important pollinators of columnar cacti and agaves in arid tropical and subtropical habitats in the New World. Pollinators affect plant reproductive success by determining the extent to which seed and fruit set is pollen limited and by determining mating patterns among conspecifics. Dispersers affect plant reproductive success by determining the degree to which seedling recruitment is limited by the arrival of seeds at favorable recruitment sites and by the extent that dispersal allows seeds to escape from natural enemies and/or to colonize new habitats. Without at least some seed dispersal, many animal-ingested tropical plants would likely go extinct. Given that biotic dispersal is commonplace in the tropics and subtropics, vertebrate frugivores are indispensible for plant reproductive success.

Pollen limitation (PL) is probably common in tropical plants, although relatively few data are yet available on this. Whether PL is an important factor in the demography of long-lived tropical plants, however, is not yet clear. In general, it is less likely to have a significant effect on the demography of long-lived plants than on short-lived plants. Most tropical trees are self-incompatible hermaphrodites, but monoecy and dioecy are also common breeding systems. Self-compatibility is more common in understory plants than in canopy trees, and as a result, it is likely that understory plants have smaller genetic neighborhood sizes than canopy trees. Outcrossing rates tend to be high in tropical trees regardless of type of pollinator. Forest fragmentation sometimes results in increased selfing rates in isolated trees, but bats and large bees can counteract this by moving pollen among fragments. Vertebrate pollinators sometimes (often?) deposit mixed-species pollen loads on stigmas, and this can lead to reduced seed set within fruits and reduced fruit set, suggesting that plant-pollinator interactions in high-diversity habitats are likely to be complex. The distance that nectar-feeding vertebrates move pollen between flowers is highly variable and often depends on pollinator social systems. Territorial species (e.g., many hummingbirds and sunbirds) move pollen shorter distances than trapliners (e.g., hermit hummingbirds). Nonterritorial species and migrant species potentially move pollen relatively long distances (many kilometers), with bats moving pollen longer distances than birds. Insects can also move pollen several kilometers in tropical habitats. Although vertebrate pollinators may not carry pollen greater distances between conspecific plants than insects, they do offer plants an advantage over insects because of the relatively large pollen loads they carry, often containing many plant genotypes. Bats are especially valuable in this regard, carrying particularly large pollen loads as they do (Muchhala and Thomson 2010).

Dispersal limitation is also common in tropical plants, and the seed shadows created by foraging frugivores are often highly heterogeneous. The net effect of these foraging behaviors is that most vertebrate-dispersed seeds end up under fruiting trees (both conspecifics and heterospecifics) or in clumps away from these trees, with a few sites receiving many seeds and many potential recruitment sites receiving no seeds. Median and maximum seed dispersal distances differ strongly among different kinds of tropical frugivores as a result of different fruit- and seed-handling behaviors as well as different foraging behaviors. For example, seed-spitting primates disperse seeds much closer to parent trees than primates that ingest seeds. Neotropical canopy fruit-eating birds appear to disperse seeds shorter distances, on average, than their paleotropical counterparts, and New World bats are more important dispersers of early successional seeds compared with Old World bats. Throughout the tropics, large frugivorous birds and primates are especially important for dispersing large seeds. Unsustainable hunting of large vertebrate frugivores and seed predators is common throughout the tropics and will undoubtedly affect the future structure and composition of tropical forests. In many areas, large-seeded plant species will lose their dispersers and will be replaced by plants dispersed by small birds, bats, or the wind unless this hunting pressure is reduced.

Pollination and seed dispersal services provided by tropical plant–visiting vertebrates affect the genetic structure of their food plants in a variety of ways. Pollinators determine outcrossing rates and rates of inbreeding and ultimately affect the size of genetic neighborhoods. They also determine rates of gene flow between isolated or fragmented populations. In some species (e.g., columnar cacti), pollen flow mediated by bats is the major component of gene flow whereas in others (e.g., understory plants dispersed by birds or bats), seed dispersal is the major component. Seed ingestion by birds and mammals clearly results in higher rates of between-population gene flow than dispersal by gravity or by ballistic methods, but vertebrate pollinators do not necessarily produce higher rates of gene flow than wind or insect pollination. Nonetheless, the disappearance of vertebrate pollinators and seed dispersers from tropical habitats would severely disrupt their ecological and evolutionary dynamics. Fully functional tropical habitats need intact populations of their vertebrate mutualists.

Plate 1. Avian pollinators, pt. 1: *A*, white-whiskered hermit hummingbird (*Phaethornis yaruqui*); *B*, sword-billed hummingbird (*Ensifera ensifera*); *C*, rufous-tailed hummingbird (*Amazilia tzacatl*); *D*, violet-necked lory (*Eos squamata*); *E*, booted racket-tailed hummingbird (*Ocreatus underwoodii*). Photo credits: Glenn Bartley (*A–C, E*), Doug Janson (*D*).

Plate 2. Avian pollinators, pt. 2: *A*, blue-faced honeyeater (*Entomyzon cyanotis*); *B*, brown-throated sunbird (*Anthreptes malacensis*); *C*, crimson sunbird (*Aethopyga siparaja*); *D*, red-legged honeycreepers (*Cyanerpes cyaneus*). Photo credits: Doug Janson (*A–D*).

Plate 3. Avian frugivores, pt. 1: *A*, double-wattled cassowary (*Casuarius casuarius*); *B*, crested guan (*Penelope purpurascens*); *C*, Hartlaub's turaco (*Tauraco hartlaubi*); *D*, pinon imperial pigeon (*Ducula pinon*); *E*, great hornbill (*Buceros bicornis*); *F*, double-toothed barbet (*Lybius bidentatus*). Photo credits: Glenn Bartley (*B*), Doug Janson (A, C–F).

Plate 4. Avian frugivores, pt. 2: *A*, red-headed barbet (*Eubucco bourcierii*); *B*, pale-mandibuled aracari (*Pteroglossus erythropygius*); *C*, chestnut-mandibuled toucan (*Ramphastos swainsonii*); *D*, resplendent quetzal (*Pharomachrus mocinno*); *E*, collared trogon (*Trogon collaris*). Photo credits: Glenn Bartley (*A–E*).

Plate 5. Avian frugivores, pt. 3: *A*, green broadbill (*Calyptomena viridis*); *B*, spangled cotinga (*Cotinga cayana*); *C*, golden-headed manakin (*Pipra erythrocephala*); *D*, black-headed bulbul (*Pycnonotus atriceps*); *E*, lesser bird of paradise (*Paradisaea minor*); *F*, golden-hooded tanager (*Tangara larvata*). Photo credits: Glenn Bartley (*C*, *F*), Doug Janson (*A*, *B*, *D*, *E*).

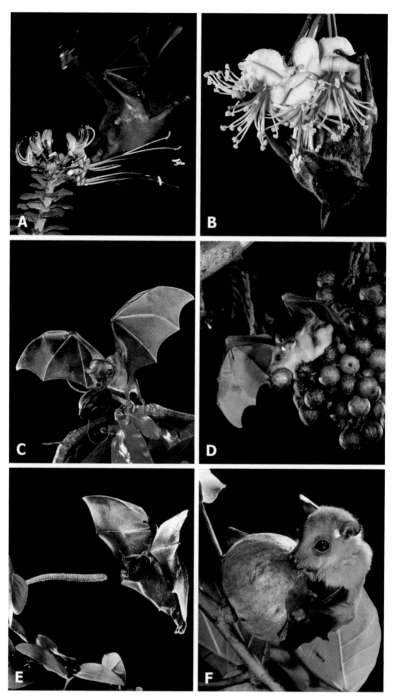

Plate 6. Plant-visiting bats—Phyllostomidae (*A*, *C*, *E*) and Pteropodidae (*B*, *D*, *F*): *A*, *Anoura fistulata* and *Cleome*; *B*, *Eonycteris spelaea* and *Durio*; *C*, *Artibeus jamaicensis* and *Terminalia*; *D*, *Epomophorus gambianus* and *Ficus*; *E*, *Carollia perspicillata* and *Piper*; *F*, *Micropteropus pusillus* and *Psidium*. Photo credits: Nathan Muchhala (*A*), Merlin Tuttle, Bat Conservation International (*B–F*).

Plate 7. Frugivorous primates: *A*, crowned lemur (*Eulemur coronatus*); *B*, red-faced black spider monkey (*Ateles paniscus*); *C*, green monkey (*Chlorocebus sabaeus*); *D*, white-throated capuchins (*Cebus capucinus*); *E*, agile gibbons (*Hylobates agilis*); *F*, Bornean orangutans (*Pongo pygmaeus*). Photo credits: Nicolas Cegalerba (*A*), Mike Lane (*B*), Steve Bloom (*C*, *E*), Michel Gunther (*D*), Antoine Boureau (*F*); all photos from Biosphoto/SteveBloom.com

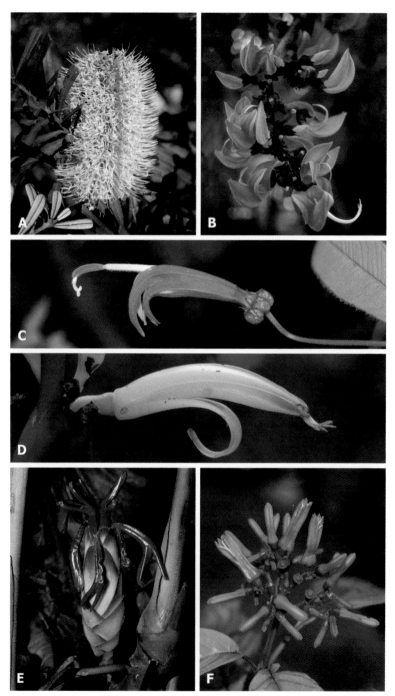

Plate 8. Bird flowers: *A, Banksia* sp. (Proteaceae); *B, Butea monosperma* (Lam.) Taub. (Fabaceae); *C, Centropogon* sp. (Campanulaceae); *D, Heliconia vaginalis* Benth. (Heliconiaceae); *E, Hornstedtia scotiana* K. Schum. (Zingiberaceae); *F, Hamelia patens* Jacq. (Rubiaceae). Photo credits: W. John Kress.

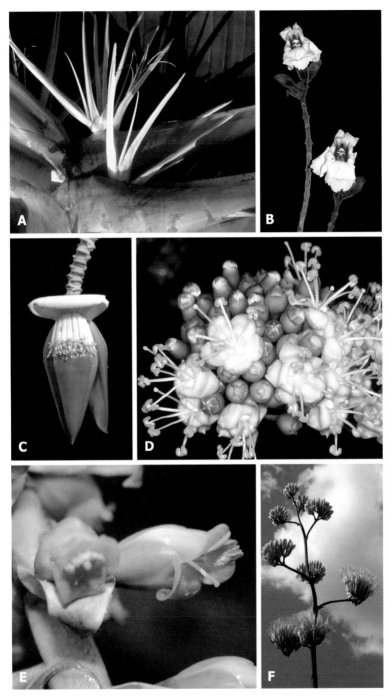

Plate 9.   Mammal flowers: A, *Ravenala madagascariensis* J.F. Gmel. (Strelitziaceae); B, *Markhamia stipulata* Seem. (Bignoniaceae); C, *Musa acuminata* Colla (Musaceae); D, *Ceiba pentandra* Gaertn. (Malvaceae); E, *Vriesia* sp. (Bromeliaceae); F, *Agave chrysantha* Peebles (Agavaceae). Photo credits: W. John Kress (A–C, E); Scott Mori (D); Ted Fleming (F).

Plate 10. Bird fruits, pt. 1: *A, Ocotea* sp. (Lauraceae); *B, Lycianthes sanctae-clarae* (Greenm.) D'Arcy (Solanaceae); *C, Myristica fragrans* Houtt. (Myristicaceae); *D, Gnetum* sp. (Gnetaceae). Photo credits: W. John Kress.

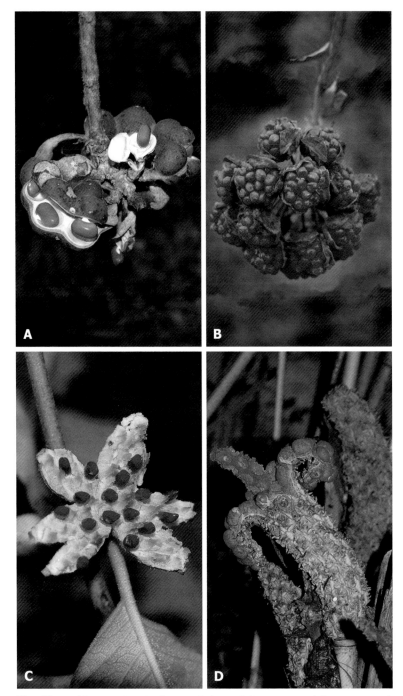

Plate 11. Bird fruits, pt. 2: *A, Erythrina* sp. (Fabaceae); *B, Bomarea* sp. (Alstroemeriaceae); *C, Siparuna* sp. (Monimiaceae); *D, Carludovica palmata* Ruiz & Pav. (Cyclanthaceae). Photo credits: W. John Kress (*A, B*); Carlos Garcia-Robledo (*C, D*).

Plate 12. Mammal fruits, pt. 1: *A*, *Aframomum sanguineum* K. Schum. (Zingiberaceae); *B*, *Musa nagensium* Prain (Musaceae); *C*, *Annona amazonica* R.E. Fr. (Annonaceae); *D*, *Sapranthus palanga* R.E. Fr. (Annonaceae). Photo credits: W. John Kress (*A–C*); Carlos Garcia-Robledo (*D*).

Plate 13. Mammal fruits, pt. 2: *A, Artocarpus heterophyllus* Lam. (Moraceae); *B, Lecythis* sp. (Lecythidaceae); *C, Ficus auriculata* Lour. (Moraceae); *D, Piper sancti-felicis* Trel. (Piperaceae). Photo credits: W. John Kress.

5

Macroevolutionary Consequences of Pollen and Seed Dispersal

As discussed in detail in chapter 4, tropical avian and mammalian pollinators and seed dispersers constantly move around the genes of their food plants within and between habitats—movements that help determine the size of genetic neighborhoods, rates and patterns of gene flow, and degree of genetic subdivision between populations of plants. Current evidence suggests that the distances that these animals move genes via pollen and seeds is very heterogeneous and that, except for seed dispersal by gravity, which creates strong genetic subdivision in plants, different pollination and seed dispersal modes do not have particularly strong effects on the genetic structure of plants. Pollination and dispersal by birds and mammals does not appear to create quantitatively different genetic structures in their food plants compared with most other pollination and seed dispersal modes. Nonetheless, these vertebrates likely create "fat-tailed" pollen and seed dispersal kernels resulting from occasional long-distance movements, and these movements can have important evolutionary consequences. When they involve pollen, long-distance movements can prevent isolated populations from diverging genetically via drift or selection. This can slow down rates of speciation. Long-distance movements of seeds may or may not have a similar effect. When it occurs frequently, long-distance seed dispersal can prevent isolated populations from diverging genetically; when such events are infrequent, they can promote the divergence of isolated populations, especially when this dispersal allows plants to colonize new habitats with different abiotic and biotic selective regimes.

In this chapter we will continue our focus on genetic consequences of these mutualisms by discussing the macroevolutionary consequences of interactions between plant-visiting tropical vertebrates and their food plants. Our major questions here include: (1) To what extent have tropical birds and mammals influenced rates of speciation and diversification in their food plants? Do plant clades that are pollinated or dispersed by tropical birds and mammals contain more species than sister clades that lack vertebrate pollination or dispersal? (2) How do these mutualists influence speciation in their food plants and to what extent is coevolution involved in the speciation process? And (3) how have these interactions influenced speciation and diversification of tropical plant–visiting birds and mammals? To what extent has the diversification of tropical plants influenced rates and patterns of speciation in vertebrate nectar feeders and frugivores? As discussed in chapter 3, the morphological and ecological space occupied by tropical flowers and fruits—the adaptive zones for vertebrate consumers—is extensive. How has this resource diversity influenced the diversification of its consumers?

Throughout this discussion we will assume that the primary mode of speciation in both plants and animals is allopatric speciation, as concluded by Coyne and Orr (2004) and Albert and Schluter (2005), among many others. Speciation via hybridization or polyploidy also occurs in plants, but these two modes are less common than allopatric speciation (Rieseberg and Willis 2007). A variation on the allopatric theme—parapatric speciation or geographic isolation in populations at the margins of species' ranges—is also likely to be an important mode of speciation in tropical plants and animals, particularly in regions of high topographic relief (e.g., Hughes and Eastwood 2006; Knapp 2002). Furthermore, evidence is now accumulating to suggest that speciation in tropical vertebrates can occur along habitat ecotones such as forest-savanna edges via morphological and genetic divergence even in the face of gene flow (Moritz et al. 2000; Smith et al. 2005). The general message here is that total geographic isolation is not necessarily needed for evolution to produce new species of plants or animals. Morphological and genetic divergence can occur whenever environmental gradients produce strong selection gradients. As Johnson (1996) points out, these gradients can involve either abiotic environmental factors (e.g., different soils or moisture regimes) or biotic factors (e.g., different kinds of pollinators) or both.

As an introduction to this topic, consider questions raised by data in figure 5.1. These data come from studies of fruit-frugivore interactions in primary rainforest in central French Guiana. Based on these studies,

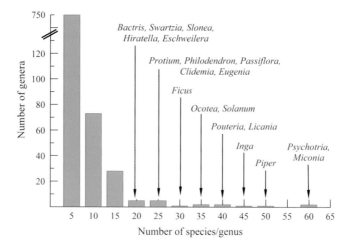

Figure 5.1.   Species richness in genera of woody plants in central French Guiana and their association with specialized vertebrate frugivores (i.e., the italicized genera). Based on Charles-Dominique (1993).

Charles-Dominique (1993) placed avian and mammalian frugivores into two general classes: specialized frugivores that consume a restricted range of fruit types and generalized frugivores that eat many different fruit types. Examples of specialized frugivores include birds such as manakins, cock-of-the-rock, and trogons as well as certain phyllostomid bats (e.g., species of *Artibeus* and *Carollia*). Nonspecialized frugivores include mammals such as opossums, kinkajous, and some primates. Figure 5.1 shows that many of the genera of fruit eaten by specialized frugivores are among the most species rich in the flora of French Guiana. The three most species-rich genera, for example, are core food genera for manakins (*Miconia, Psychotria*) and *Carollia* bats (*Piper*; see table 3.9). Questions raised by these data include: (1) Is there a cause-effect relationship between vertebrate feeding specialization and the proliferation of species within particular genera of fleshy-fruited plants? As suggested by Charles-Dominique (1993), has the close association between particular plant types (genera) and groups of frugivores promoted parallel evolution between them and particularly extensive speciation? Or has speciation in these plants occurred independently of the effects of specialized frugivores, with frugivores responding opportunistically to particularly species-rich groups of plants? (2) Have species-rich groups of tropical plants promoted the diversification of their principal seed dispersers? That is, are genera of specialized vertebrate frugivores more species rich than genera of generalized frugivores? We will attempt to answer these questions in this chapter.

An Overview of Diversification and Speciation in Tropical Angiosperms

Angiosperms are the numerically dominant and taxonomically most diverse plants in most terrestrial habitats. This is particularly true in the tropics, which harbor about two-thirds of all angiosperm species. The rise to dominance of angiosperms, whose evolution began about 130 million years ago (Ma), has long been associated with biotic modes of pollination and dispersal (e.g., Burger 1981; Regal 1977; Tiffney 1984). According to Bawa (1992), animal pollinators can affect rates of speciation in angiosperms in several ways: (1) founder plant populations are likely to interact with new pollinators, which can lead to reproductive isolation from source populations; (2) floral variants in different parts of a species range can interact with different pollinators, again leading to reproductive isolation; (3) sexual selection in biotically pollinated plants can select for different floral variants, which then attract different pollinators; (4) sexual selection within species of pollinators (e.g., in orchid bees) can lead to reduced gene flow within plant species and genetic differentiation (Kiester et al. 1984); and (5) genetic drift, in conjunction with plant-pollinator coevolution and sexual selection, can lead to rapid speciation.

We will use the following definition of "species" in this book: "Species are groups of interbreeding natural populations that are reproductively isolated from other such groups" (Coyne and Orr [2004, 30], quoted from Mayr [1995]). In addition to influencing degree of reproductive isolation among populations, animal pollinators can reduce rates of plant extinction by carrying pollen between widely spaced individuals in low-density populations or between conspecifics in fragmented populations (chap. 4).

Biotic dispersal can also lead to higher speciation rates and lower extinction rates in fleshy-fruited plants. Long-distance seed dispersal is an important benefit of vertebrate frugivory, and it is an important way in which plants colonize new habitats (chap. 4). Although most colonization events are unlikely to lead to geographic isolation and speciation, some do, and over long periods of geological time, new species are likely to accumulate in groups that regularly experience long-distance seed dispersal (Levin 2006). Recent evidence using molecular phylogenies and dating techniques have suggested that, over evolutionary time, long-distance dispersal has played a more important role as an explanation for broad geographic distributions than vicariance events, which was hypothesized to be the driving force behind these patterns for many years (e.g., amphi-Atlantic distributions; see Lavin et al. 2004; Renner 2004; Sarkinen et al. 2007). In ecological time, vertebrate frugivory promotes plant reproductive success by moving seeds away from "killing zones" under and near parent trees and hence reduces extinction probabilities. Vertebrate frugivory also favors the evolution of large seeds in closed canopy habitats, which contributes significantly to a plant's competitive ability and reduces its chances of extinction (Foster 1986; Muller-Landau 2007; Tiffney and Mazer 1995).

Although it has long been argued that biotic pollination and dispersal are key features in the evolutionary success of angiosperms, rigorous tests of this idea are relatively recent. Two early studies by Herrera (1989) and Fleming (1991) asked whether dispersal mode affected species richness in angiosperm families. These studies did not use sister-group comparisons but instead simply classified plant families by major dispersal mode (fleshy fruits, dry fruits, or "mixed" families, in which both fleshy and dry fruits occur) and compared the number of species per family in each of these modes. Results of their analyses indicated that mixed families contained more species, on average, than those producing either fleshy or dry fruits. This result was supported by a more detailed but nonphylogenetically controlled analysis of species richness of angiosperm families by Ricklefs and Renner (1994). In addition to dispersal mode, these authors examined the effect of geographic distribution, growth form, and pollination mode on family S. Their analyses indicated that these four classification variables accounted for about 41% of variation in species richness among families but that animal dispersal and pollination accounted for relatively little of this variation. Instead, families with mixed growth habits (i.e., those containing both herbs and woody species) and mixed pollination and dispersal modes had higher values of species richness than families containing a single growth habit or single modes of pollination and dispersal.

With the advent of more extensive and better resolved molecular phylogenies (e.g., Davies et al. 2004; Soltis et al. 2005), recent studies have asked whether biotic pollination and dispersal promote higher rates of diversification (i.e., the net effect of speciation and extinction) than abiotic means in angiosperms. The results of several sister-group analyses support the hypothesis that biotic pollination, but not necessarily biotic seed dispersal, is

Table 5.1. The Effect of Different Modes of Pollination, Dispersal, and Growth Form on Species Richness in Sister Clades of Angiosperms

Trait/Hypothesis	Number of Sister Groups	Relative Species Richness
Pollination: biotic > abiotic	22	2.36*
Dispersal:		
Biotic > abiotic	55	0.98
Biotic < abiotic, in herbs only	8	0.39
Biotic > abiotic, in woody plants only	30	1.40
Growth form: Herbs only > no herbs	33	1.65*
Pollination: Two modes > biotic only	19	1.36
Dispersal: Two modes > biotic only	45	2.07***
Growth form: Two forms > herbs only	35	2.03*

Source. Based on independent contrasts in Dodd et al. (1999).
Note. Relative species richness is the ratio of species numbers in sister groups with contrasting traits. The last three rows deal with comparisons of groups with more than one kind of pollination or dispersal mode or growth form vs groups with just one kind.
* $0.05 > P > 0.01$.
*** $P < 0.001$.

associated with high species richness in angiosperm families. In an analysis of 299 families in the context of a resolved molecular phylogeny, Dodd et al. (1999) found that pollination mode and growth habit, but not dispersal mode, influenced species richness in the expected directions. As summarized in table 5.1, biotically pollinated clades contain about 2.4 times more species than abiotically pollinated clades; this ratio was only about 1.0 for biotically dispersed clades. Growth form also had a significant effect on clade species richness, with herbaceous clades containing about 1.7 times more species than nonherbaceous clades. Sargent (2004) examined the effect of floral asymmetry on S in angiosperm sister clades. Her results indicated that clades characterized by zygomorphic (asymmetrical) flowers contained more species than clades with actinomorphic (symmetrical) flowers in 15 of 19 comparisons. This result supports the idea that zygomorphic flowers promote more precise relationships between flowers and their pollinators and a greater chance of reproductive isolation between plant populations. Vamosi and Vamosi (2010) reached the same conclusion from their analysis of factors correlated with species richness in 409 angiosperm families. Kay et al. (2006) reanalyzed the data of Dodd et al. (1999) and Sargent (2004) using a more recent phylogeny and more appropriate statistical methods and basically reached the same conclusions as these authors. Clades characterized by biotic pollination and specialized floral morphology tend to contain more species than other clades. To paraphrase Howe and Mitiri (2004), there can be no doubt that biotic pollination matters in the diversification of angiosperms.

While there is strong support for the importance of biotic pollination as

Table 5.2. The effect of different modes of dispersal on species richness in families of eudicots and monocots

Group/Comparison	Number of Families Dispersed		Relative Species Richness
	Biotically	Abiotically	
Eudicots:			
Woody plant families	88	64	3.30***
Mixed growth form families	6	20	1.10
Herbaceous families	10	56	0.025*
Monocots			
Woody plant families	3	1	24.6
Herbaceous families	9	40	0.29

Source. Based on Tiffney and Mazer (1995).
Note. Relative species richness is the ratio of the mean of families that are biotically dispersed to the mean of families that are abiotically dispersed.
* $0.05 > P > 0.01$.
*** $P < 0.001$.

a significant factor promoting high species richness in angiosperm families, the importance of biotic dispersal as a diversifying factor is still being debated. Two studies (Bolmgren and Eriksson 2005; Tiffney and Mazer 1995) found that biotic dispersal is associated with high species richness in families of woody plants but not in herbaceous families, and Eriksson and Bremer (1991) found that high species richness occurs in genera of animal-dispersed shrubs in the Rubiaceae (e.g., in *Psychotria*; fig. 5.1). Tiffney and Mazer's (1995) results are summarized in table 5.2, which shows that, within woody dicots, families that are biotically dispersed contain about 3.3 times as many species as families that are abiotically dispersed. A similar pattern occurs in monocots, but the small number of woody monocot families reduces the statistical power needed to demonstrate this. Families of abiotically dispersed herbs, in contrast, contain many more species than biotically dispersed families (table 5.2). Thus, to a greater extent than pollination modes, the effect of biotic dispersal on plant diversification appears to be contingent on plant growth form (de Queiroz 2002). From an analysis of contemporary European temperate plant families, Bolmgren and Eriksson (2005) reported that fleshy fruits were significantly associated with woody plants, particularly those growing in closed forests and in those living in disturbed habitats. In both of these situations, vertebrate dispersal might be more advantageous than abiotic dispersal for at least two reasons: (1) large seeds are favored in closed forests because they provide more reserves of energy and material for seedlings and (2) vertebrate dispersal permits the evolution of larger seeds (Foster 1986; Tiffney 1984, 1986). Seed mobility provided by vertebrates is also advantageous in early successional plants (Muscarella and Fleming 2007). Finally, Levin (2006) has argued that plant lineages with

highly dispersible propagules (by whatever means) should have high speciation rates because this gains them access to new habitats with opportunities
for ecological and geographic isolation. Broad distributions resulting from
high dispersability also reduce extinction rates.

Additional evidence for the importance of dispersal as a significant factor
in angiosperm diversification comes from an analysis of speciation in Hawaiian angiosperms. In this archipelago, species richness of different clades
is related to mode of dispersal in the following manner: species richness of
clades with external bird dispersal is greater than that of clades with internal
bird dispersal or with abiotic dispersal (Price and Wagner 2004). Price and
Wagner summarize their results in a verbal model that likely has general
applicability (fig. 5.2). This model posits that Hawaiian plant lineages with
an "intermediate" level of dispersability (e.g., via external bird dispersal)
should have higher species richness than lineages with lower (e.g., abiotic
dispersal) or higher (e.g., via internal bird dispersal) levels of dispersability. Low dispersability, in combination with habitat specialization, should
produce lineages with restricted geographic or ecological ranges and low
speciation potential. In contrast, lingeages with high dispersability and generalized habitat requirements should have large geographic or ecological

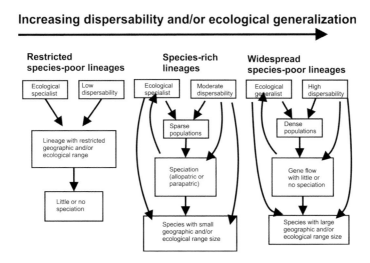

Figure 5.2.   The effect of seed dispersability and ecological generalization on speciation in
Hawaiian plants. In this scheme, plant groups with "intermediate" dispersability should contain more species than those with less or greater dispersability. Based on Price and Wagner
(2004).

ranges and should experience high rates of gene flow between populations that reduce geographic isolation. Lineages with a combination of moderate dispersability and ecological specialization should have the greatest potential for speciation. We will apply this model to mainland tropical dispersal systems below.

Up to this point, we have considered the effects of biotic pollination and dispersal on angiosperm species richness from a very broad perspective. Studies such as those of Ricklefs and Renner (1994), Dodd et al. (1999), and Kay et al. (2006) deal with angiosperms as a whole and do not explicitly examine the effects of biotic pollination and dispersal by tropical vertebrates on species richness of their food plants. Data in figure 5.1 give us a hint that dispersal by vertebrates may help to promote high species richness in their food plants. More generally, we can ask two questions: (1) Are the "core families" (table 3.9) of tropical nectar-feeding and fruit-eating birds and mammals more species rich than the average angiosperm family? And (2), are these families more species rich than sister clades? To address these questions, we used the molecular phylogeny of Davies et al. (2004) to identify sister clades and data from their figure 2 for number of species per family. Data summarized in figure 5.3 bear on the first question. Those data indicate that, throughout the angiosperms, nearly all core families contain many more species than median family size; this trend holds for both flower and fruit core families. In birds but not in bats, core flower families tend to be more species rich than core fruit families. Not shown in figure 5.3 is the fact that several core families contain both flowers and fruit used by tropical vertebrates. These include three of 19 core bird families (16%) and two of 15 core mammal families (13%; table 3.9). None of the 11 primate core fruit families contains primate-pollinated flowers.

To address question 2, we tallied the number of core families occurring in two classes of S: families that contained more species than their sister clades and families that contained the same or fewer species. We eliminated five of the 45 core families from this analysis because the number of species in their sister clades was unavailable. We initially treated dicots and monocots as well as pollinator and disperser mutualisms separately. Chi-square analyses revealed that there was no heterogeneity caused by phylogeny or mutualism ($Ps \gg 0.50$), so data were collapsed into two species richness classes. Twenty-seven of 40 core families (68%) contained more species than their sister clades, but this difference is not significantly different from random expectations (Fisher's exact probability = 0.173). From this, we tentatively

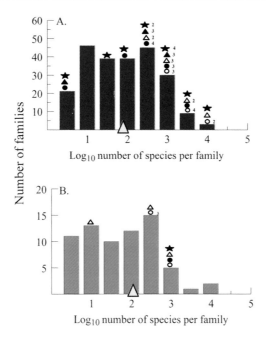

Figure 5.3.   Species richness in families of eudicots (*A*) and monocots (*B*), and the occurrence of core families of vertebrate-pollinated (*open symbols*) or -dispersed (*closed symbols*) plants. Animal taxa: *circles*—birds; *triangles*—bats; *stars*—primates. Numbers next to the symbols indicate number of families (= 1 unless indicated otherwise). The *large triangles* along the ordinate indicate median values of species richness.

conclude that, although species richness is high in the core families providing flower and/or fruit resources for tropical birds and mammals, it is not necessarily higher than that of their sister clades.

Returning to the Price-Wagner model of speciation in Hawaiian angiosperms (fig. 5.2), we expect to see a hump-shaped relationship between clade species richness and dispersability. Does this model apply to other tropical fruit and vertebrate frugivore systems? Several predictions follow from this model. First, plants that are dispersed by large, strong-flying vertebrates should have lower speciation rates and lower species richness than plants dispersed by small, sedentary dispersers. Large canopy foraging birds such as hornbills and toucans should provide greater overall dispersability for the seeds they ingest than small canopy foragers such as barbets, bulbuls, and tanagers. All else being equal, families of their food plants should contain fewer species than those of more sedentary species. Second, plants dispersed by small sedentary understory birds (e.g., manakins) should have higher speciation rates and more species than related canopy plants dispersed by larger, more mobile birds. Third, a similar difference should hold for canopy and understory plants dispersed by bats. Canopy plants dispersed by large phyllostomid and pteropodid bats (e.g., *Artibeus*, *Pteropus*, and their relatives) should have lower speciation rates and fewer species than understory plants dispersed by New World *Carollia* and *Sturnira* and Old World *Cynopterus* and *Haplonycteris*. Finally, plants dispersed by seed-spitting primates (e.g., many African cercopithecines) should have higher speciation rates than those whose seeds are ingested by primates (e.g., *Ateles* in the New World, chimps and gibbons in the Old World).

Testing these predictions is not easy because of the potential confounding effects of other variables such as pollination mode, seed size, successional status, growth form, and degree of habitat specialization in different plant lineages. Nonetheless, a critical examination of the hypothesis that plant speciation rates are influenced by the dispersability of their seeds as mediated by tropical vertebrate frugivores is important for both theoretical and conservation reasons. From a theoretical perspective, it would be important to know if a general rule holds for the relationship between clade richness and dispersibility. For example, is it true that vertebrate-dispersed understory shrubs and treelets are more likely to experience short-distance dispersal distances and higher speciation rates than vertebrate-dispersed canopy and subcanopy trees? If such a rule existed, then results from a dispersal study in one part of the world could be used to make predictions about how a dispersal system in another part of the world was likely to operate. From a conservation perspective, it is important to know how the elimination of particular groups of frugivores (e.g., large birds and primates) will affect the dispersability and speciation potential of their food plants. How much dispersability and speciation potential has been lost in a "forest without frugivores" (chap. 4)?

An analysis of patterns of species richness within tropical understory plants provides some support for the Price-Wagner model of speciation. Inspired by the observation that within Gesneriaceae, subfamily Gesnerioideae, fleshy-fruited genera (e.g., *Columnea*, *Besleria*) contain more species than dry-fruited genera, Smith (2001) compared the species richness of 14 clades of fleshy-fruited understory plants with that of their dry-fruited sister clades. Except for bat-dispersed *Pipers*, each of the fleshy-fruited clades is bird-dispersed. Results showed that fleshy-fruited clades contained more species than their sister clades in 11 of 14 comparisons. The median ratio of fleshy-fruited to dry-fruited species was 5.6 (range, 0.5–337). Smith postulated that clades of fleshy-fruited understory plants have higher values of species richness than dry-fruited clades because, although their bird and bat dispersers generally provide limited dispersal, they occasionally provide long-distance dispersal, which promotes geographic isolation and speciation.

Data summarized in figure 5.1 also provide some support for the Price-Wagner model. Six of the top 12 genera of plants dispersed by specialized frugivores are understory shrubs or epiphytes dispersed by sedentary birds or bats. These include the top three genera (*Miconia*, *Psychotria*, and *Piper*) as well as *Solanum*, *Clidemia*, and *Philodendron*. While we cannot rule out pollinators (which are insects in all cases) as having as strong an effect on genetic isolation in these plants as dispersers, we nonetheless suspect that limited but occasional long-distance seed dispersal by birds and bats has played an important role in speciation in these plants.

Finally, Givnish et al. (unpubl. ms.) have argued that occasional long-distance dispersal by birds eating the fleshy fruits of epiphytic bromeliads in montane regions of the Andes and southeast Brazil has been an important factor in the high net diversification rates of these plants. This dispersal (as well as hummingbird pollination; see below) has led to a tenfold increase in the species richness and thirtyfold increase in the geographic distribution of this important Neotropical family.

Coevolution and Cospeciation

Early theoretical discussions of evolutionary interactions between fruits and frugivores (e.g., Howe and Estabrook 1977; McKey 1975; Snow 1971) suggested that so-called specialized frugivores were more likely to be co-evolved with their food plants than so-called opportunistic or generalized frugivores. Proposed consequences of this process ranged from patterns of fruiting phenology to the nutritional composition of fruit pulp in plants and patterns of foraging constancy in animals and the treatment of seeds in their guts. Similarly, it has long been argued that flowering plants are also often coevolved with their pollinators (reviewed in Feinsinger 1983, 1987; Pellmyr 2002). Prior to Janzen (1980), however, the term "coevolution" was not rigorously defined and meant different things to different authors (Futuyma and Slatkin 1983). Janzen (1980) proposed that coevolution between two species occurs when a trait of one species evolves in response to a trait of another species, which in turn has evolved in response to the trait of the first species. In other words, coevolution is reciprocal evolutionary change in the traits of two interacting species. This definition, of course, implies pairwise evolution between species and does not address the interactions between groups of plant species and their multiple pollinators or seed dispersers. A commonly used term to describe these more generalized types of interactions has been "diffuse coevolution," which was defined as "the evolution of a particular trait in one or more species in response to a trait or suite of traits in several other species" by Futuyma and Slatkin (1983, 2). Although this concept of diffuse coevolution provides a means to describe these multispecies interactions as well as a general mechanism for understanding the evolution of these communities of plants, pollinators, and frugivores, the term often masks the complexity of the interactions and does not provide a rigorous set of testable predictions or hypotheses. In many cases these "diffuse relationships" are now understood to be well-defined and to involve coevolved interactions among multiple species within community networks (Thompson 1994, 2005; see chap. 9).

The conceptual transition from regarding multispecies interactions as a diffuse assemblage of interactions to recognizing them as involving coevolution within mutualistic networks has been a significant contribution to understanding plant-vertebrate ecology and evolution. The recognition of these networks led Thompson (2005, 289) to propose that "reciprocal selection on mutualisms between free-living species favors genetically variable, multispecific networks in which species converge and specialize on a core set of mutualistic traits rather than directly on other species." His hypothesis of convergence in a set of traits within these networks provided a new perspective on what had previously been dismissed as diffuse coevolution. From this viewpoint, coevolution is the foundation of these networks and eventually results in the development of generalist interactions among an expanding set of species that share the same combination of adaptive traits. The recognition of pollination syndromes as defined by Faegri and van der Pijl (1979) is one of the best examples of the widespread convergence of traits between flowering plants and their pollinators. We discuss this concept in further detail in chapter 9.

Contrasting with mutualistic networks of interactions at the community level is the macroevolutionary outcome of pairwise coevolution between a lineage of plants and a lineage of animals. In this case, cospeciation or parallel cladogenesis may occur in which a plant or host phylogeny is precisely mirrored by the phylogeny of its interacting antagonists (e.g., herbivores, parasites) or mutualists (fig. 5.4*a*). Each pairwise interaction is then associated with reproductive isolation and the formation of a new species. Pairwise antagonistic interactions are not likely to have a direct effect on reproductive isolation and rates of speciation, but the same is not true for pairwise mutualistic interactions, particularly those involving pollinators. In some plant-pollinator systems, such as the interaction between the South American Andean clade of evergreen shrubs in the genus *Chuquiraga* (Asteraceae) and their *Oreotrochilus* hummingbird pollinators, coradiation may have occurred during the uplift of the Andes, beginning in the mid-Miocene (Ezcurra 2002). Nonetheless, although the potential for cospeciation exists in highly specialized interactions, such as between *Chuquiraga* and *Oreotrochilus* or figs and their fig wasps and yuccas and yucca moths, current evidence suggests that strict cospeciation has not actually occurred in these systems (Herre et al. 2008; Machado et al. 2001; Marussich and Machado 2007; Pellmyr 2003). Therefore, given the diet breadths of most species of nectar-feeding and fruit-eating birds and mammals (chap. 3), it is highly unlikely that parallel cladogenesis is a common feature of their evolution.

In addition to parallel cladogenesis, Labandeira (2002) identified three other forms of potential evolution and coevolution between plants and herbivorous animals that might be applicable to tropical plants and their vertebrate mutualists (fig. 5.4). In sequential evolution (fig. 5.4*b*), a group of plants radiates and is then "colonized" by a group of animals which then radiates in parallel with the plants. This pattern does not initially involve plant-animal coevolution during the early radiation of the plants, although coevolution is likely to occur as animals radiate onto their new hosts. While we are not aware of any formal analyses of such a process in tropical birds and mammals, it is highly likely that examples of this process will be found. In table 5.3 we list some potential examples of this process.

Among avian pollinators, hermit hummingbirds and Andean trochiline hummingbirds of the Coquette and Brilliant groups (Bleiweiss 1998a; McGuire et al. 2007; Stiles 2008) are good candidates to have radiated and coevolved with lowland/middle elevation *Heliconia* and montane Ericaceae and Asteraceae, respectively. Similar to the hummingbirds, montane- and forest-dwelling sunbirds are associated with specific groups of plants with which they may have radiated, but details of such a phylogenetic association are not yet available (Irwin 1999). In tropical Asia, the association of

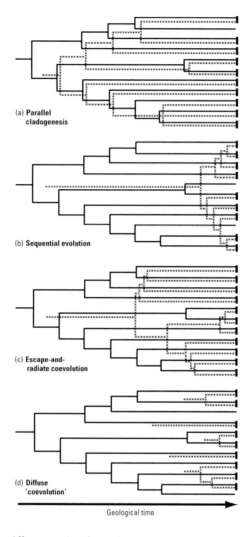

Figure 5.4.   Four different modes of coevolution between plants and their animal antagonists or mutualists. From Labandeira (2002) with permission.

flowerpeckers with the Loranthaceae and of nectar-feeding Australasian meliphagids and lories with certain groups of Myrtaceae and Proteaceae suggests possible coradiations (Ford et al. 1979; Hopper and Burbidge 1986; Paton and Ford 1977). Finally, the diversity of Hawaiian honeycreepers and their associated lobelioid members of the Campanulaceae deserve more intensive phylogenetic study as coradiations (Tarr and Fleischer 1995; Ziegler 2002).

Among mammalian pollinators, phyllostomid bats in the genus *Leptonycteris* are good candidates to have radiated with New World columnar cacti and paniculate agaves, and other phyllostomid bats have radiated with certain Bignoniaceae and the Bombacoideae (Simmons and Wetterer 2002; table 5.3). In southern Asia, *Eonycteris* bats are associated with flowers of *Oroxylum* trees (Bignoniaceae) in a rare example of close coevolution between a plant and a pteropodid bat (Gould 1978; Heithaus 1982). Similarly, the (nontropical) Australian nectar-feeding marsupial *Tarsipes rostratus* (Tarsipedidae) has evolved in response to *Banksia* inflorescences (Rourke and Wiens 1977; Wiens et al. 1979).

While we expect to see weaker evolutionary associations between vertebrate frugivores and their food plants than is the case in vertebrate nectar feeders, a number of such associations are notable (table 5.3). Among New World fruit-eating birds, trogons have likely evolved in response to fruits of the Lauraceae; toucans with fruits of Myristicaceae; oilbirds with palm

Table 5.3. Potential Examples of Sequential Radiation in Tropical Vertebrates in Response to Their Food Plants

Mutualism	Plant Taxa	Animal Taxa
Bird pollination	NW *Heliconia* herbs (Heliconiaceae)	Hermit hummingbirds (Phaethorinae)
	NW Andean Ericaceae	Coquette and Brilliant hummingbirds (Trochilinae)
	Asian Loranthaceae	Flowerpeckers (Dicaeidae)
	Australasian Myrtaceae (eucalypts)	Honeyeaters (Meliphagidae) and lorikeets (Psittacidae, Loriinae)
	Hawaiian Campanulaceae (lobeliods)	Hawaiian honeycreepers (Drepaniidae)
Mammal pollination	NW columnar cacti (Cactaceae) and agaves (Agavaceae)	*Leptonycteris* phyllostomid bats
	NW Bignoniaceae	Glossophagine phyllostomid bats
	OW Bignoniaceae	*Eonycteris* pteropodid bats
	NW and OW Bombacoideae	Phyllostomid and pteropodid bats
	Australian *Banksia* (Proteaceae)	*Tarsipes* marsupials (Tarsipedidae)
Bird frugivory	NW Lauraceae	Trogons (Trogonidae)
	NW Myristicaceae	Toucans (Ramphastidae)
	NW Rubiaceae and Melastomataceae	Manakins (Pipridae)
	NW palms (Arecaceae)	Oilbirds (Steatornithidae)
	Australas Lauraceae	Fruit pigeons (Columbidae)
	SE Asian mistletoes (Loranthaceae)	Flowerpeckers (Dicaeidae)
	SE Asian Meliaceae and Myristicaceae	Hornbills (Bucerotidae) and birds of paradise (Paradiseidae)
Mammal frugivory	NW figs (Moraceae)	Stenodermatine phyllostomid bats
	NW Piperaceae	*Carollia* phyllostomid bats
	NW Araceae and Cyclanthaceae	*Rhinophylla* phyllostomid bats
	NW Solanaceae	*Sturnira* phyllostomid bats
	NW Myristicaceae	Atelid primates (Atelidae)
	African and Asian figs (Moraceae)	Cercopithecine monkeys, gibbons, and orangutans (Cercopithecidae, Hylobatidae, and Hominidae)

Note. NW = New World; OW = Old World.

fruits; and manakins with understory fruits of the Melastomataceae and Rubiaceae. In the Paleotropics, Australasian fruit pigeons appear to be strongly associated with fruits of Lauraceae as well as figs; Southeast Asian and Australasian mistletoebirds with fruits of the Loranthaceae; and hornbills and birds of paradise with fruits of Meliaceae and Myristicaceae.

A particularly clear example of the radiation of a group of mammalian frugivores in response to a group of plants is phyllostomid bats of subfamily Stenodermatinae. This advanced clade of phyllostomids contains about 67 species classified in 18 genera (Simmons 2005). Figs and other Moraceae are the core foods of these bats, and bat species sort themselves out among these fruits on the basis of body size (Korine et al. 2000; Wendeln et al. 2000). Additional examples of New World frugivorous phyllostomids that are strongly associated with specific kinds of plants include *Carollia* with *Piper*, *Sturnira* with *Solanum*, and *Rhinophylla* with epiphytic aroids and Cyclanthaceae (Fleming 1986a, 2004a; Gorchov et al. 1995; Henry and Kalko 2007; table 5.3). Also in the New World, atelid monkeys appear to be strongly associated with fruits in the Myristicaceae (Russo et al. 2005). Finally, like stenodermatine phyllostomids, many pteropodid bats are strongly associated with figs as are many species of New and Old World primates (Shanahan et al. 2001). These associations, however, tend to involve animals that are feeding generalists and thus are less likely to have evolved in parallel with their food plants than the interactions included in table 5.3.

While we currently lack rigorous demonstrations that these plant-animal associations have resulted in parallel evolutionary radiations, circumstantial but strong evidence based primarily on morphology suggests that they have. Whenever one clade of animal mutualists feeds preferentially on a particular clade of plants, then opportunity for parallel radiation would seem to be high. This should be especially true if, as in the case of hermit hummingbirds and stenodermatine bats, for example, morphological and/or behavioral adaptations to their preferred plants are strong (see chaps. 7 and 8). Such adaptations reflect much more than a casual relationship between the interactors; they reflect more than just "diffuse coevolution" as this phrase has often been used (e.g., Feinsinger 1983; Herrera 1985; Morris et al. 2007; Thompson 1999; Tiffney 2004). Instead, they clearly reflect coevolution, not necessarily in the narrow one-on-one specialized sense (although examples of this exist in certain Neotropical plants and their vertebrate pollinators), but in a slightly broader sense involving single clades of plants and animals. We will designate this type of clade-specific, but not species-specific, evolutionary interaction as "generalized coevolution" to differentiate it from mutualistic networks of plants and pollinators that may be derived from multiple lineages of plants and animals. It should be emphasized (as is further detailed below) that hypotheses about generalized coevolution, species-specific coevolution, and the coevolution of mutualistic networks are testable with well-supported phylogenies of the interacting taxa (Donatti et al. 2011; Thompson 2005).

In their seminal paper on coevolution, Ehrlich and Raven (1964) hypothesized that the adaptive radiation of plants and their herbivores involved escape-and-radiate coevolution (fig. 5.4c). In this process, plants are thought to evolve novel antiherbivore chemical defenses that allow them to temporarily escape into enemy-free space and radiate until herbivores breech their defenses and radiate onto them, resulting in another round of escape and radiation. This kind of coevolution clearly does not apply to plants and their mutualists because there is no need for plants to escape from their animal mutualists. In fact, we would expect plants to evolve features in their flowers and fruit that serve to increase, not decrease, the strength of interactions with their mutualists (chap. 1 and beyond).

The coevolution of mutualistic networks in communities of plants and their vertebrate pollinators and seed dispersers is the antithesis of parallel cladogenesis and involves different clades of animals coevolving independently with different clades of plants in the absence of extensive phylogenetic congruity (Guimaraes et al. 2011; fig. 5.4d). This form of coevolution is the one most often associated with interactions between tropical plants and their vertebrate pollinators and seed dispersers. In a phylogenetic context, mutualistic networks often include multiple independent events of convergent evolution in traits associated with pollination or frugivory within a particular plant clade or within the entire angiosperm phylogeny in general (chap. 6). As mentioned earlier, the concepts of pollination and dispersal "syndromes" (e.g., Faegri and van der Pijl 1979; van der Pijl 1982) are examples of mutualistic networks in which groups of unrelated species have converged on a set of adaptive traits. Free-living mutualisms, such as found in pollination or dispersal systems, inherently form multispecies networks and proliferate through diversification of species possessing complementary traits and through convergence of traits in unrelated species. Many of the interactions discussed in this book are described by these pollination and dispersal syndromes and are the result of coevolution between plants and their various groups of vertebrate mutualists. In summary, coevolution in pollination and seed dispersal systems may occur between single or closely related plant species and their counterpart animal species (i.e., specialized coevolution), between multiple members of a single plant lineage and multiple species in a single animal lineage (i.e., generalized coevolution), or between unrelated plant species in diverse lineages and a series of unrelated species in various animal lineages (i.e., coevolution through mutualistic networks). Of these three processes, coevolution through mutualistic networks would appear to be the most common and specialized coevolution the least common of the interactions discussed in this book.

The Geographic Mosaic of Coevolution

A major advance in the study of coevolution has been the recognition that the structure of interactions between two species will vary across a geographic landscape (Thompson 1997, 1999, 2005). This variation in space, and in most cases time as well, is the foundation of Thompson's concept of the geographic mosaic of coevolution (GMC; Thompson 1999, 2005). The GMC is based on three evolutionary hypotheses (Thompson 1997, 2005): (1) natural selection on interspecific interactions varies among populations owing to geographic differences in how the fitness of one species depends on the distribution of genotypes in the other; (2) interspecific interactions are subject to reciprocal selection only in some local communities, which results in coevolutionary hotspots embedded within a matrix of coevolutionary coldspots where selection is nonreciprocal; and (3) the range of coevolving traits undergoes continual population remixing as a result of the selection mosaic, coevolutionary hotspots and coldspots, gene flow, genetic drift, and local extinction. The recognition of and subsequent support for the GMC has resulted in a paradigm shift in our understanding of coevolution between parasites and their hosts (Lively 1999), pathogens and plants (Burdon and Thrall 1999), dispersers and seeds (Benkman 1999), predators and prey (Brodie et al. 2002), and pollinators and plants (Thompson and Cunningham 2002; Thompson and Pellmyr 1992).

To date, analyses of the GMC with respect to mutualisms have been limited, particularly those dealing with interactions between plants and their pollinators (e.g., Anderson and Johnson 2008; Gomez et al. 2009; Thompson and Cunningham 2002). Because the geographic ranges of cointeracting species seldom completely overlap (Thompson 2005), studies of plant-pollinator interactions across the range of both partners provide insights into how one partner is replaced by a new party to the interaction, as well as the mechanisms that result in convergences of traits among unrelated species. When combined with phylogenetic histories, such studies will shed light on how cointeracting species are genetically constrained in the traits and kinds of interactions in which they are involved.

One coevolutionary system involving plants and their vertebrate partners that has received significant study with respect to the GMC is the pollination system between hummingbirds and heliconias found across the archipelago of the eastern Caribbean (Temeles and Kress 2003, 2010; Thompson 2005). This system involves a hummingbird, the sexually dimorphic purple-throated carib (*Eulampis jugularis*), and its two principal food plants, *Heliconia caribaea*, which is endemic to the Lesser Antilles, and *H. bihai*, which occurs from northern South America to the Greater Antilles. It has been shown that a strong association exists between the body sizes and bill morphologies of males and females of purple-throated caribs and the energy rewards and flower morphologies of the two species of *Heliconia* that they visit (Temeles and Kress 2003; Temeles et al. 2000; fig. 5.5). In the most specialized interaction, which occurs on the island of Dominica, the small-bodied females exclusively visit the flowers of *H. bihai*, which are long and curved and fit the long and curved bills of the females. The larger-bodied males, in contrast, defend territories around dense patches of flowers of *H. caribaea*, which has shorter and straighter flowers more adapted to the shorter and straighter bills of the males (Temeles et al. 2009). Although females may also visit the flowers of *H. caribaea*, their visits are controlled and regulated by territorial males (Temeles and Kress 2010).

Evidence important for testing the GMC suggests that some of the plant traits involved in the interaction, and perhaps the hummingbird traits as well, vary from population to population in a predictable way depending on which sex is the primary visitor to the plant species (Temeles and Kress 2003; Temeles et al. 2000). This variation provides support for local matching between plant and pollinator. Results of on-going research indicate that the plant-pollinator interactions differ from population to population and island to island in the Caribbean and that particular islands are potential hotspots and coldspots in the coevolutionary mosaic of this system. For example, *H. bihai* relies on female purple-throated caribs as its sole pollinator on Dominica whereas at its most southerly distribution on the island of Trinidad, six species of hummingbirds are known to visit its flowers. Most interesting for the GMC is that the flowers of this species significantly

Figure 5.5. Coevolution between the purple-throated carib hummingbird and two species of Antillean *Heliconias*. Bills of a female and a male are shown in *A* and *B*, respectively. *Heliconia* flowers or inflorescences are shown as follows: flowers of *H. bihai* (*C*), and *H. caribaea* (*D*), on Dominica; *H. bihai* green inflorescence morph on St. Lucia (*E*); *H. bahai* red-green morph on St. Lucia (*F*); *H. caribaea* on St. Lucia (*G*); *H. bihai* red-and-yellow-striped inflorescence on Dominica (*H*); *H. caribaea* red inflorescence on Dominica (*I*); and *H. caribaea* yellow inflorescence on Dominica (*J*). From Temeles and Kress (2003) with permission.

decrease in both curvature and length from the highly specialized interaction with a single pollinator on Dominica in the north to the generalized plant-pollinator interaction with numerous hummingbird visitors on Trinidad and Tobago in the south (Kress and Temeles, unpublished; also see Martén-Rodríguez et al. 2011). The extreme specialization between the hummingbird and heliconia on Dominica may be an exception to most coevolutionary interactions because, in contrast to host-parasite relationships, free-living mutualists interact with many different species and individuals during their lifetimes. These kinds of mutualistic systems are not expected to favor such a high degree of specialization (Thompson 2005). Future investigations of additional plant-pollinator systems, especially plant-vertebrate mutualisms, will most likely provide further examples supporting a geographic mosaic of coevolution.

Phylogenetic Consequences for Tropical Plants from Their Interactions with Vertebrate Pollinators and Seed Dispersers

As discussed in detail below, there can be no doubt that vertebrate-plant interactions have resulted in the formation of new plant species. But what are the higher-level taxonomic or phylogenetic consequences of these interactions? How many plant families, subfamilies, or tribes have evolved from these interactions? To what extent has the taxonomic diversity of tropical angiosperms been increased as a result of their interactions with nectar-feeding and fruit-eating vertebrates? To address these questions, we extracted data from APG III (2009) supplemented with data from Mabberley's (1997) review of the taxonomy of angiosperm plant families, looking for families, subfamilies, and so on with particularly strong associations with vertebrate pollinators and seed dispersers. Results of this survey are summarized in table 5.4.

INTERACTIONS WITH VERTEBRATE POLLINATORS

Seven families of tropical angiosperms (out of a total of approximately 217 nonaquatic tropical families) appear to be exclusively (or nearly so) associated with vertebrate pollination (table 5.4). None of these families is particularly large. Five families (Caryocaraceae, Heliconiaceae, Musaceae, Sonneratiaceae [= Lythraceae in part in APG], and Strelitziaceae) contain one to six genera and seven to 200 species. The remaining two families (Bombacaceae

Table 5.4. Higher Order Plant Taxa That Are Associated Primarily with Tropical Vertebrates for Pollination or Seed Dispersal

Families Strongly Associated with Vertebrate Mutualists	
Pollination	Seed Dispersal
Bombacaeae (Pantropical, 26/250)—bats	Burseraceae (Pantropical, 17/540)—birds and mammals
Caryocaraceae (New World, 2/25)—bats	Cecropiaceae (New World & Africa, 6/180)—birds and mammals
Loranthaceae (Pantropical, 68/900)—birds	Cyclanthaceae (New World, 12/200)—bats
Musaceae (Pantropical, 6/200)—birds and mammals	Davidsoniaceae (trop. Australia, 1/2–3)—birds
Sonneratiaceae (Old World, 2/8)—bats	Ebenaceae (Pantropical, 2/485)—mammals
Heliconiaceae (primarily New World, 1/200+)—birds and bats	Lauraceae (Pantropical, 52/2850)—birds
Strelitziaceae (Pantropical, 3/7)—birds, bats, lemurs	Moraceae (Pantropical, 38/1100)—birds and mammals
	Myristicaceae (Pantropical, 19/400)—birds and mammals
	Piperaceae (Pantropical, 8/3000)—birds and bats
	Rubiaceae (Pantropical, 630/10,200)—birds
	Sapotaceae (Pantropical, 53/975)—mammals
Families with Subfamilies or Tribes Strongly Associated with Vertebrate Mutualists	
Pollination	Seed Dispersal
Acanthaceae (Pantropical, 250/2500): Acanthoideae—hummingbirds	Anacardiaceae (Pantropical, 70/875): 2 of 5 tribes—mammals
Agavaceae (New World, 18/600): Agavoideae—bats	Annonaceae (Pantropical, 112/2150): Annoideae—mammals
Bignoniaceae (Pantropical, 110/650): 2 of 7 tribes—bats	Euphorbiaceae (Pantropical, 313/8100): Acalyphoideae—birds and mammals
Cactaceae (New World, 121/1500): Cactoideae, 2+ tribes—bats	Loganiaceae (Pantropical, 29/570): Potalieae—birds and mammals
Campanulaceae (Pantropical, 70/2000): Lobelioideae—hummingbirds and bats	Melastomataceae (Pantropical, 188/4950): Melastomatoideae—birds
Ericaceae (cosmopolitan, 107/3400): Vaccinioideae—hummingbirds	Meliaceae (Pantropical, 51/565): Meliodideae, 4 tribes—birds and mammals
Fabaceae (cosmopolitan, 730/19,400): Mimosoideae, 2 tribes—birds and bats	Myrsinaceae (Pantropical, 33/1225): Myrsinoideae—birds
Malvaceae (cosmopolitan, 111/1800): 3 of 5 tribes—hummingbirds	Myrtaceae (Pantropical, 129/4620): Myrtoideae—birds and mammals
Myrtaceae (Pantropical, 140/3000): Leptospermoideae—birds and bats	Sapindaceae (Pantropical, 131/1450): Sapindoideae—mammals
Pandanaceae (Old World, 3/875): Freycinetiodeae—bats	Solanaceae (Pantropical, 94/2950): Solanoideae—birds and bats
Proteaceae (± Old World, 80/2000): Grevilleoideae—birds and bats	

Sources. Data come from APG III (2009) and Mabberley (1997).

Note. Family characteristics include geographic distribution, number of genera/number of species, and vertebrate group. Two of the traditional families we are recognizing here are treated either as a subfamily (Malvaceae, Bombacoideae) or have been included within a larger family (Sonneratiaceae in Lythraceae) in APG. We do this to emphasize the close association of these plant taxa with vertebrate pollinators. Sonneratiaceae (Old World, 2/8) is not likely to be a monophyletic clade according to recent molecular work (APG III 2009).

[= Malvaceae in part in APG] and Loranthaceae) contain 26–68 genera and up to 900 species. Species in these two families are conspicuous canopy trees or forest parasites, respectively, in many lowland tropical forests. Three families are pantropical in distribution; one occurs only in the New World; and one occurs only in the Old World. One family is associated with avian pollinators; three are linked with bats; and one contains species pollinated by both birds and bats.

At least 11 families contain subfamilies or tribes that are strongly associated with vertebrate pollination (table 5.4). Except for Pandanaceae, which has only three genera (but 875 species), all of these families are relatively large and contain a minimum of 18 genera and at least 600 species. Seven of the 11 families are pantropical in distribution; two each occur only in the New or Old World tropics. Three of the subfamilies or tribes are strongly associated with hummingbirds; one with hummingbirds and phyllostomid bats; four with bats; and three with both birds and bats.

INTERACTIONS WITH VERTEBRATE FRUGIVORES

Not surprisingly, given the widespread occurrence of fleshy fruits in tropical plants (chap. 3), a larger number of angiosperm families appear to be strongly associated with vertebrate frugivores than with vertebrate nectar feeders (table 5.4). Thus, most species in at least 11 families are vertebrate dispersed. Four of these families (Cecropiaceae, Cyclanthaceae, Davidsoniaceae, and Ebenaceae) are small in terms of number of genera or species; the others are relatively large. One family (Rubiaceae) is very large and contains both fleshy and dry fruits. Except for Cyclanthaceae (New World) and Davidsoniaceae (tropical Australia), all of these families have broad tropical distributions. Three families are mostly bird dispersed; one is bat dispersed; two are dispersed by mammals other than bats; and five are dispersed by different combinations of birds, bats, and primates.

At least 10 families contain subfamilies or tribes that are strongly associated with vertebrate dispersal (table 5.4). All of these families are relatively large and have broad tropical distributions. Two families contain subfamilies or tribes that are mostly bird dispersed; three are mostly mammal dispersed; and the other five are dispersed by combinations of birds, bats, and other mammals.

In summary, at least 16 angiosperm families have evolved to be either pollinated or dispersed by tropical vertebrates, and vertebrate-pollinated or -dispersed clades occur in an additional 21 families for a minimum of

37 families (about 17% of tropical angiosperm families) that have evolved closely with vertebrate mutualists. The majority of these families (29) are pantropical in distribution; four each are endemic to either the Old or New World tropics. Nine of these families have evolved with avian mutualists (especially with hummingbirds in bird-pollinated families); 13 have evolved with mammalian mutualists (especially bats); and 15 have evolved with both birds and mammals. Overall, vertebrate-associated plant families are often among the most conspicuous and ecologically important families in their respective habitats. We will return to these families in our discussion of angiosperm phylogeny in chapter 6.

Speciation in Tropical Plants

As we have indicated previously, we assume that most speciation in tropical plants has involved allopatric speciation via geographic isolation, and there is substantial support for this in the recent literature, particularly in studies of Andean-centered plants in the Neotropics. As Gentry (1982, 1992) has pointed out, the Andes is an area of exceptional plant diversity, particularly in epiphytes, shrubs, and small trees. This region harbors about 45,000 plant species, of which 44% are endemic (Hughes and Eastwood 2006). According to Hughes and Eastwood (2006) and Graham (2009), four factors can explain the exceptional radiation of Andean plants: (1) its geographic scale is huge (the Andes mountains are over 4,000 km long); (2) Pleistocene climate fluctuations produced repeated fragmentation of Andean habitats and altitudinal shifts in the flora and fauna; (3) the region abounds in highly dissected topography and steep environmental gradients; and (4) high resource heterogeneity regarding habitats and environmental conditions occurs at mid- to high elevations. To this we can add a fifth factor: the prolonged period of Andean orogeny, which has occurred over a period of about 85 Ma. Uplift in the southern and central Andes began about 85 Ma; in the northern Andes it began about 55 Ma; final uplift in the northern Andes dates from the Late Miocene–Early Pliocene (5–10 Ma); and current altitudes are Pliocene-Pleistocene in age (Graham 2009; Young et al. 2002). Given this spatiotemporal heterogeneity, it is not surprising that the Andes have produced substantial radiations in plant groups such as Bromeliaceae, Cactaceae, Costaceae, Ericaceae, Heliconiaceae, Piperaceae, and Solanaceae, among many others (e.g., Kessler 2002; Knapp 2002; Luteyn 2002; Marquis 2004; Ritz et al. 2007). Allopatric speciation has been the standard mode of diversification in all of these groups, and we expect this mode of speciation to prevail in montane regions throughout the tropics (e.g., in equatorial and East Africa, New Guinea, etc.).

Both allopatric and ecological speciation (*sensu* Schluter 2000) appear to be important speciation mechanisms in many groups of lowland plants. For example, about 230 species of the bird-dispersed rain forest tree genus *Guatteria* (Annonaceae) occur in the Amazon basin where they have speciated by both mechanisms in response to its highly dissected and heterogeneous landscape (Erkens at al. 2007). Fine et al. (2005) studied soil habitat associations in a clade of 35 species of Burseraceae and found that speciation involved independent adaptation to different soil types in 26 species. Because sister species often occurred on different soil types, they concluded that ecological speciation rather than allopatric vicariance was driving speciation in this group. A similar pattern of edaphic specialization has been found in Neotropical Melastomataceae and Lauraceae as well as in trees in Bornean dipterocarp forests (reviewed in Fine et al. 2005), and it may generally be widespread in lowland tropical plants (e.g. Erkens et al. 2007). Classic allopatric speciation likely accounts for the evolution of endemic species in a number of more seasonal Neotropical forests in Central and South America (Pennington et al. 2004).

If many species of tropical plants are the products of allopatric speciation, to what extent has this process involved shifts in pollinators? To judge from the diversity of flower morphologies within and among plant clades, plant-pollinator interactions have been central to the diversification of angiosperms, tropical or otherwise (e.g., Baker 1961; Proctor et al. 1996; Stebbins 1970, 1974). Allopatric speciation involves the adaptation of geographically isolated populations to new abiotic and biotic environmental conditions, and an important component of the biotic environment for plants is the assemblage of potential pollinators they encounter (Johnson 1996). The South American genus *Schizanthus* (Solanaceae), for example, contains 12 species distributed in three different habitats in which they are associated with different pollinators: their flowers are bee pollinated in Mediterranean-type habitats; they are hummingbird-pollinated in the high Andes; and they are moth pollinated in desert habitats (Perez et al. 2006). Similarly, speciation in the genus *Cobaea* (Polemoniaceae) in Neotropical montane cloud forests has occurred in response to three kinds of pollinators: hawkmoths, bats, and hummingbirds (Prather 1999). One of the consequences of adaptation to new pollinators can be reproductive isolation. New World members of the forest understory herb *Costus* (Costaceae), for instance, are pollinated either by euglossine bees or by hummingbirds. Several species can coexist at a site, but there is virtually no overlap in pollinators between species of these two pollinator types. Prezygotic isolation is so strong in *Costus* that species lack postzygotic isolating mechanisms and can be hybridized readily when artificially crossed (Kay and Schemske 2003).

Abundant evidence from the New World suggests that pollinators are important drivers of speciation in many groups of tropical plants. Gravendeel et al. (2004) suggest that this is the case in many families of epiphytes, including Bromeliacaeae, Cactaceae, Ericaceae, Gesneriaceae, Marcgraviaceae, and Rubiaceae. Kay et al. (2005) indicate that it is the evolution of specialized pollination by hummingbirds that has driven speciation in many of these groups as well as in gingers, *Inga*, *Psychotria*, and Acanthaceae. In support of the hypothesis that hummingbirds are important drivers of plant speciation in the Neotropics, Schmidt-Lebuhn et al. (2007) indicate that hummingbird-pollinated clades contain more species than insect-pollinated sister clades in *Tillandsia* (Bromeliaceae), *Suessenguthia* (Acanthaceae), and in tribe Sinningieae of the Gesneriaceae. In their detailed analysis of factors influencing species richness and rates of speciation in the Bromeliaceae, Givnish et al. (unpubl. ms.) report that both of these variables are substantially higher in clades dominated by hummingbird pollination than in sister clades dominated by insect pollination. Comparable analyses are not yet available for Old World plants.

Knowing that pollinators are frequently involved in angiosperm speciation, we can ask what the distribution is of different pollinator types within particular plant clades. Two patterns might be expected: (1) different pollinator types are scattered throughout a clade owing to the independent evolution of the same type or multiple types several times; and (2) a single pollinator type characterizes an entire clade, reflecting the single evolution of a particular type with subsequent speciation occurring only within that type. We will call the first pattern the plastic pollinator model and the second pattern the constrained pollinator model because these patterns appear to reflect two fundamentally different ways in which plants have speciated in response to their pollinators. In the plastic pollinator model, a plant clade appears to have a generalized ability to respond to different pollinator environments, evolving new pollination types when the opportunity arises. An example of this kind of response can be seen in the South American tribe Sinningieae of the Gesneriaceae. This clade of about 81 herbaceous species

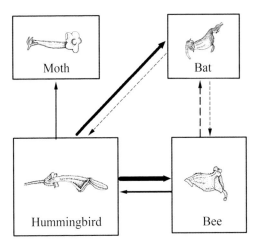

Figure 5.6.   The evolution of pollination syndromes in tribe Sinningieae, Gesneriaceae. *Solid arrows* indicate direction of changes, with arrow width being proportional to frequency of changes. *Dashed arrows* indicate transitions that have not occurred. Based on Perret et al. (2003).

is currently classified in three genera, in which most species are either bee or hummingbird pollinated but in which moth and bat pollination has also evolved (Perret et al. 2003, 2007). As shown in figure 5.6, most evolutionary transitions in this clade have occurred between hummingbirds and bees (in both directions), with bat pollination having evolved twice and moth pollination once from hummingbird pollination. Similarly, hummingbird pollination has evolved from insect pollination two or three times in Bromeliaceae; insect pollination has evolved from hummingbird pollination about seven times in subfamily Bromelioideae; and bat pollination has evolved from hummingbird pollination four times in this family (Givnish et al. unpubl. ms.). In the constrained pollinator model, in contrast, once a clade has evolved to be pollinated by one pollinator type, speciation occurs only within this type. An example of this pattern is trees and shrubs of the pantropical genus *Erythrina* (Fabaceae) in which all 112 species are bird pollinated (Bruneau 1996, 1997). South American species are basal in the genus, and 55 of the 70 New World species are hummingbird-pollinated; the remaining 15 New World species and all 42 Old World species are pollinated by passerines. Although there is considerable morphological variation in *Erythrina* flowers, especially among passerine-pollinated flowers, no other pollination mode has evolved in this genus (Bruneau 1997).

Even within a single order of plants, both the plastic pollinator and the constrained pollinator models can apply. In the mainly tropical monocot order Zingiberales (eight families, 96 genera, and more than 2,000 species; Kress 1990), a great diversity of pollinators characterize both families and genera (Kress and Specht 2005; see fig. 7.3). With regard to the constrained pollinator model, the families Heliconiaceae (Pedersen and Kress 1999) and Marantaceae are predominantly pollinated by a single class of pollinators. In the former family, the single genus *Heliconia* contains about 200 species, and all but six are pollinated exclusively by Neotropical hummingbirds. In the latter family, all 500 species are pollinated by insects, mostly bees, except one reported Neotropical hummingbird-pollinated species and a few Asian moth-pollinated species. Examples of the plastic pollinator model include the Strelitziaceae, which contains one genus (*Strelitzia*; five species) pollinated by sunbirds in South Africa, a second genus (*Ravenala*, one species) pollinated by lemurs in Madagascar, and a third genus (*Phenakospermum*, one species) pollinated by phyllostomid bats in the Amazon basin (Kress et al. 1994). In the families Costaceae (seven genera and 150 species), Zingiberaceae (53 genera and 1,000 species), and Cannaceae (one genus and 10 species), insects (bees, moths, and butterflies) and vertebrates (bats and birds) pollinate various genera and species scattered across these families in the Old and New World tropics. In the most basal family, the Musaceae (three genera and 42 species), species are pollinated either by birds or bats (in one case bees), but each pollinator type is primarily found in a single clade within the genus *Musa*. Therefore, within a single order of plants there are families demonstrating the constrained pollinator model (Heliconiaceae and Marantaceae), the plastic pollinator model (Strelitziaceae, Costaceae, Zingiberaceae, and Cannaceae), or both models (Musaceae).

Although a meta-analysis of the frequency of these two evolutionary patterns in angiosperms as a whole is beyond the scope of this book, we point out that both patterns appear to be nearly equally common. Examples of these patterns from the recent literature are summarized in table 5.5. Most of these examples come from the Neotropics, but Old World examples also exist. Based on data presented in Fleming and Muchhala (2008), however, we might expect the frequency of plastic pollinator clades to be higher in the Neotropics than the Paleotropics owing to the apparently more specialized relationships between New World plants and their vertebrate (and insect?) pollinators. Much more research in this area needs to be done. Especially

Table 5.5. Examples of Two Different Pollinator-Driven Patterns (Models) of Speciation in Angiosperms

Model	Examples
Constrained	Acanthaceae, *Aphelandra*, *A. pulcherrima* group: all 50 species are hummingbird pollinated (McDade 1992)
	Agavaceae, *Agave*, subgenus *Agave*: nearly all 82 species are bat pollinated (Gentry 1982; Good-Avila et al. 2006)
	Asteraceae, *Chuquiraga*: its basal clade of 9 species is hummingbird pollinated (Ezcurra 2002)
	Cactaceae, *Stenocereus*: nearly all of 25 species are bat pollinated (Anderson 2001)
	Campanulaceae, *Burmeistera*: 19 of 20 species in Ecuador are bat pollinated (Muchhala 2006a)
	Campanulaceae, *Centropogon*: nearly all 230 species are hummingbird pollinated (Mabberley 1997)
	Ericaceae, tribe Vaccinieae: most species are hummingbird pollinated (Luteyn 2002)
	Fabaceae, *Erythrina*: all 112 species are bird pollinated (Bruneau 1997)
	Fabaceae, *Parkia*: 29 of 31 species are bat-pollinated (Luckow and Hopkins 1995)
	Heliconiaceae: *Heliconia* with 194 of about 200 species are hummingbird pollinated (Pedersen and Kress 1999)
	Marantaceae: 495 of about 500 species are bee pollinated (Kress and Specht 2005)
Plastic	Bignonieae: multiple evolutions of hummingbird and bat pollination in 104 species and 20 genera (Alcantara and Lohmann 2010)
	Bombacoideae, *Adansonia*: either hawkmoth, bat, or lemur pollination in 8 species (Baum et al. 1998)
	Bombacoideae, *Durio*: either bird or bat pollination in 28 species (Yumoto 2000)
	Bromeliaceae: multiple evolutions of insect, hummingbird, and bat pollination in *Aechmea, Bromelia, Guzmania, Pitcairnia, Puya, Tillandsia,* and *Vriesea* (Benzing 2000; Endress 1994)
	Cactaceae, *Pachycereus*: either bat, hummingbird, or moth pollination in 12 species (Fleming 2002)
	Costaceae, *Costus,* subgenus *Costus*: multiple evolutions of hummingbird pollination from euglossine bee pollination in 38 species (Kay et al. 2005)
	Gesneriaceae, *Achimenes*: multiple evolutions of insect or hummingbird pollination in 22 species (Roalson et al. 2007
	Gesneriaceae, tribe Gesneriae: multiple evolutions of hummingbird and bat pollination in 19 species (Martén-Rodríguez et al. 2009)
	Gesneriaceae, tribe Sinningiae, *Sinningia* (*sensu lato*): bee, hummingbird, hawkmoth, or bat pollination in 18 species (Perret et al. 2003)
	Loranthaceae, *Tristeryx*: either insect or hummingbird pollination in 9 species (Amico et al. 2007)
	Polemoniaceae, *Cobaea*: hawkmoth, hummingbird, or bat pollination in 18 species (Prather 1999)
	Solanaceae, *Schizanthus*: bee, hummingbird, or moth pollination in 12 species (Perez et al. 2006)
	Cannaceae: *Canna* is bee, hummingbird, or bat pollinated (Kress and Specht 2005)
	Strelitziaceae: the three genera *Strelitzia, Ravenala,* and *Phenakospermum* are bird, lemur, or bat pollinated (Kress et al. 1994)
	Zingiberaceae: the 52 genera are either insect or bird pollinated or, in a few cases, bat pollinated (Kress and Specht 2005)

Note. See text for a description of the two models.

interesting would be determining the factors that predispose plant clades to either be "plastic" or "constrained" in their response to pollinators. These undoubtedly include genetic and developmental factors (Siol et al. 2010).

We are not aware of any studies that explicitly examine the role of verte-brate frugivores in the speciation of tropical plants (but see Fleming [2004a] for a brief discussion of this topic in the genus *Piper*). Nor do we yet know whether frugivore-driven plastic and constrained models of speciation exist in angiosperms (but see chap. 8). This is another area that deserves research attention.

Speciation and Diversification in Tropical Plant-Visiting Birds and Mammals

Unlike the situation in plants, in which plant-pollinator interactions are of-ten involved in the speciation process, there is little evidence that these or other mutualistic interactions are involved in speciation in tropical plant–visiting vertebrates, at least to judge from the recent literature. Indeed, nearly all studies of speciation in these birds and mammals are devoid of considerations of plant-animal interactions, in stark contrast to the litera-ture on speciation in plant-visiting insects (e.g., Egas et al. 2005; Funk et al. 2002; Joy and Crespi 2007; Kawecki 1998). Instead, the focus of vertebrate studies is invariably on the relative importance of vicariance versus dispersal as mechanisms that produce new species. Adaptation to new environments is obviously part of the speciation process, but explicit consideration of the role of particular groups of plants (e.g., tables 5.3 and 5.4) in this process is rare. Thus, with our current knowledge, we do not know how impor-tant plant group X (e.g., *Pipers*) has been for speciation in animal group Y (e.g., *Carollia* bats). A notable exception to this is Dumont et al.'s (2011) study of speciation rates in frugivorous phyllostomid bats, particularly sub-family Stenodermatinae. Stenodermatines are strongly frugivorous and are the most recently derived and speciose group of phyllostomids. Dumont et al.'s sophisticated statistical analysis of evolutionary trends in diversifica-tion rates, morphology, and trophic level indicates that rates of speciation in this clade, whose radiation began about 15 Ma, are about 1.9 times higher than rates in other phyllostomid clades (means of 0.250 vs. 0.135 species/Ma, respectively). Once they evolved the appropriate morphological traits (see chaps. 8 and 9), these canopy-feeding bats quickly diversified and filled a new adaptive zone.

SPECIATION MODES IN TROPICAL PLANT-VISITING BIRDS AND MAMMALS

Moritz et al. (2000, table 1) list and discuss five models of speciation in tropical rain forest animals. These include: (1) the refugia model; (2) the riverine model; (3) the vanishing refuges model; (4) the disturbance-vicariance model; and (5) the gradient model. Each of these models involves allopatric speciation, but model 5 can also operate in parapatric fashion. Evolutionary mechanisms behind models 1–3 include isolation by various means (e.g., via forest fragmentation, river formation, or other major disturbances), drift, and selection; they include competition and perhaps directional selection in model 4; and they include directional selection with or without gene flow in model 5. These models may apply to speciation in plants as well as animals.

In their review, Moritz et al. (2000) found little support for the refugia and riverine models for South American rain forest animals (also see Colinvaux et al. 2001). In the case of the widely discussed refugia model, this is because centers of endemism often do not coincide with proposed Pleistocene refuges, and the estimated ages of clades based on recent molecular data indicate that much of the speciation in Neotropical birds and mammals predates the Pleistocene. In his discussion of the evolution of African rain forest plants and animals, Plana (2004) indicated that the Pleistocene refugia model is still widely held for Africa where montane climatic refuges are especially important. Riverine refuges in a drying landscape have also been important for African primates and certain groups of plants (e.g., Rubiaceae and Caesalpiniaceae [Fabaceae]). Plana (2004) points out, however, that, compared with the Neotropics, little work has been done to test the African refugial models with molecular data (but see Koffi et al. 2011 and references therein). A similar situation obtains in Southeast Asia (Lim et al. 2010). Smith et al. (2005) and Kirschel et al. (2011) examined the efficacy of the gradient model along forest-savanna ecotones in the little greenbul (*Andropadus virens*, Pycnonotidae) in West Africa. As predicted by the model, they reported significant morphological differences in savanna versus forest populations and in montane versus forest populations despite substantial gene flow between these populations. The acoustic properties of male songs, which are an important isolating mechanism, also differed among these populations. Speciation has not yet occurred in this species, however, and it remains to be seen whether the gradient model is widely applicable to birds and mammals in Africa and elsewhere in the tropics.

Whatever the mechanism, allopatric speciation appears to be the norm in birds and mammals (S. Edwards et al. 2005; Fitzpatrick and Turelli 2006; Price 2007). Because related birds often lack postzygotic isolating mechanisms, prezygotic mechanisms involving sexual selection based on plumage, songs, and displays play important roles in bird speciation (Kirschel et al. 2011 and references therein). S. Edwards et al. (2005) suggest that cryptic female choice and sperm competition may also be important in birds. They further suggest that rapidly evolving reproductive proteins that affect mating success may be involved in speciation in mammals but not in birds. Recent studies of tropical birds that support an allopatric model of speciation include those of *Meliphaga* and other honeyeaters (Driskell and Christidis 2004; Norman et al. 2007), *Nectarinia* sunbirds (Bowie et al. 2004), currasows (Pereira and Baker 2004), *Tangara* and other tanagers (Burns and Naoki 2004; Fjeldsa and Rahbek 2006), *Pteroglossus* toucans (Eberhard and Bermingham 2005; Patel et al. 2011), *Pionus* parrots (Ribas et al. 2007), and *Metallura* and *Adelomyia* hummingbirds (Chaves et al. 2006; Garcia-Moreno et al. 1999). In mammals, allopatric speciation has occurred in *Alouatta* monkeys (Cortes-Ortiz et al. 2003), and many more groups could undoubtedly be added to this list (e.g., many genera of phyllostomid bats: *Artibeus* [Larsen et al. 2007]; *Carollia* [Hoffmann and Baker 2003]; and *Glossophaga* [Hoffmann and Baker 2001]).

SEXUAL SELECTION AND SPECIATION.—Because many tropical birds are truly "ornaments of life" owing to their spectacular plumage, it would seem obvious that sexual selection has played a very important role in their speciation just as selection for different pollinators has for plant speciation (Schluter 2000; Stanley 1979). In groups of birds such as hummingbirds, sunbirds, trogons, tanagers, and birds of paradise, among others, sexual dichromatism and spectacular male plumages are common. Price (1998) indicated that, in such cases, female mate choices can reinforce geographic differences in phenotypic and ecological characteristics of populations that can lead to reproductive isolation. Despite this, the evidence that sexual selection is an important driver of speciation in birds is mixed. Three studies using sister clade analysis (Barraclough et al. 1995; Moller and Cuervo 1998; Owens et al. 1999) reported that sexual selection, as indicated by extent of sexual dichromatism or plumage ornamentation within families, promoted greater species richness, whereas two more recent studies using phylogenetically independent contrasts (Morrow et al. 2003; Phillimore et al. 2006) did not find an association between extent of sexual selection and species richness in bird genera and families, respectively. Sexual selection is especially prevalent in birds of paradise, but rates of speciation in this family appear to be similar to those in other corvidan passerines (Irestedt et al. 2009). It is important to remember that, in addition to leading to increased rates of speciation, sexual selection can also potentially result in higher rates of extinction owing to high rates of predation on conspicuous males (Doherty et al. 2003; Morrow and Pitcher 2003). If rates of speciation are balanced by rates of extinction, then sexual selection will have no net effect on rates of diversification of animal clades (Morrow et al. 2003). Regardless of whether sexual selection has led to higher diversification rates in birds, it certainly has had an impressive effect on the evolution of plumage characteristics in many groups of plant-visiting birds.

Has sexual selection promoted higher rates of diversification in plant-visiting mammals? The results of two studies (Gage et al. 2002; Isaac et al. 2005) that use sexual size dimorphism and/or the occurrence of polyandry as measures of the intensity of sexual selection in genera or families of mammals indicate that the answer to this question is no. Strongly dimorphic or polyandrous clades in groups such as carnivores, bats, and primates contain no more species than related clades that lack one or both of these characteristics.

THE EVOLUTIONARY AGES OF SPECIES OF PLANT-VISITING BIRDS AND MAMMALS.—A recurring theme in studies of the molecular evolution and speciation of tropical vertebrates is the occurrence of old taxa (i.e., early to mid-Miocene) in lowland forest areas and younger taxa (i.e., Pliocene-Pleistocene) in montane areas. This pattern has been found in African birds and in South American birds and mammals (Moritz et al. 2000; Roy et al. 1998). In western South America, many lowland bird taxa date from 4–8 Ma, whereas most upland taxa are younger, with a sharp increase in speciation rates in the Pleistocene (Weir 2006). In this region, dispersal between lowland and upland habitats occurred in the late Miocene/early Pliocene in response to Andean uplift and again in the Pleistocene in response to the effects of glaciation on vegetation zones and habitat fragmentation (Weir 2006). A particularly well-studied group of birds in this regard is the tanagers (Thraupinae), which has about 200 species whose ancestral home is South America and whose evolution began in the early to mid-Miocene (Burns and Naoki 2004; Fjeldsa and Rahbek 2006). The subfamily as a whole as well as its largest genus (*Tangara*, with about 50 species) likely evolved in the Andean foothills of northwestern South America and from there spread into Amazonia, the central Andes, the southeast Atlantic rain forest, and finally into Central America. Lowland speciation in tanagers occurred in the upper Miocene and Pliocene, with *Tangara* splitting from its sister taxa at least 6.5 Ma ago; most speciation events in this genus occurred 3.5–5.5 Ma during Andean uplift (Burns and Naoki 2004). The importance of these estimates of the ages of various groups of vertebrates will be addressed when we consider the phylogenetic distribution of pollination and dispersal systems in plants in chapter 6.

DIVERSIFICATION OF TROPICAL PLANT-VISITING BIRDS AND MAMMALS

Just as we have asked whether clades of tropical plants that interact strongly with nectar-feeding or fruit-eating birds and mammals are more species rich than sister clades that lack these mutualistic interactions, we can ask whether clades of these birds and mammals are more species rich than their nonmutualistic sister clades (or close relatives). We will use the bird and mammal phylogenies we have generated for chapter 6 plus species counts found in Dickinson (2003) and Wilson and Reeder (2005) for birds and mammals, respectively, to answer this question. As summarized in table 5.6, the answer is equivocal, primarily because of the small sample sizes available for this analysis. Only certain clades of nectar-feeding birds are more species rich than their nonmutualistic sister clades. Thus, species of sunbirds outnumber (mutualistic) flowerpeckers by a factor of 2.9, hummingbirds outnumber species of swifts by a factor of 3.5, and honeyeaters outnumber pardalotes by a factor of 43.5. However, among frugivores, only eight out of 19 clades of birds or mammals (about 42%) contain more species than their sister clades. If we use the insectivorous bat family Rhinolophidae as the closest relative of the plant-visiting family Pteropodidae, then the ratio of mutualistic to nonmutualistic species is 2.4. Similarly, within the ecologically diverse bat family Phyllostomidae, the ratio of plant-visiting species to nonmutualistic species is 2.6. Within the ecologically diverse primate family Cercopithecidae, the ratio of species of cercopithicines (mostly frugivorous) to species of colobines (mostly herbivores) is only 1.3. Finally, at the ordinal level, primates ($N = 376$ species) are strongly frugivorous and have speciated much more extensively than their traditional close relatives, the tree shrews ($N = 19$ species) and dermopterans ($N = 2$ species). From this analysis we conclude that while there are a few exceptionally species-rich

Table 5.6. Comparison of Species Richness in Mutualistic versus Nonmutualistic Clades of Tropical Birds and Mammals

Mutualism	Mutualistic Clade	Nonmutualistic Sister Clade or Close Relative	Mutualistic > Nonmutualistic?
Birds:			
Pollination	Hummingbirds (331)	Swifts (94)	Yes
	Honeyeaters (174)	Pardalotes (4)	Yes
	White-eyes (95)	Babblers (273)	No
	Sunbirds (127)	Flowerpeckers (44)*	Yes
Frugivory	Cassowaries (3)	Kiwis (3)	No
	Turacos (23)	?	?
	Cracids (50)	Grouse and relatives (218)	No
	Oilbirds (1)	Nighthawks and potoos (96)	No
	Trogons (39)	New World barbets (14)*	Yes
	Manakins (48)	Cotingids and frugivorous tyrannids (≥100)*	No
	Cotingids (96)	Pipromorph tyrannids (4)	Yes
	Bowerbirds (18)	Australian treecreepers (7)	Yes
	Old World orioles and figbirds (29)	?	
	Birds of paradise (40)	Crows (117)	No
	Bulbuls (118)	Babblers (273)	No
	Mimids (34)	Starlings (115)*	No
	Thrushes (165)	Old World flycatchers (275)	No
	Tanagers (202)	Carduline finches (42)	Yes
Mammals:			
Pollination	Honey possum (1)	Feathertail gliders (2)	± Equal
	Pteropodid bats, part (12)	Rhinolophids (77)	No
	Phyllostomid bats, part (38)	Mormoopids (10)	Yes
Frugivory	Brushtail possums (27)	Pygmy possums (5)	Yes
	Pteropodid bats, part (154)	Rhinolophids (77)	Yes
	Phyllostomid bats, part (75)	Mormoopids (10)	Yes
	Haplorhine primates (288)	Strepsirrhine primates (88)*	Yes
	Viverrids (35)	Herpestid mongooses (33)	± Equal
	Procyonids (14)	Mustelids (59)	No
	Tragulids (8)	Other artiodactyls (204)	No

Note. Number of species is shown in parentheses.
* Also a plant-visiting group.

families or subfamilies of plant-visiting birds and mammals (e.g., hummingbirds, honeyeaters, sunbirds, and phyllostomid and pteropodid bats) and a species-rich order of mammalian frugivores (Primates), overall most clades of vertebrate plant visitors have not speciated to a greater extent than their nonmutualistic relatives. Being a nectar feeder appears to have sometimes resulted in significantly greater speciation while being a fruit eater has not necessarily led to higher rates of diversification in birds and mammals, as Snow (1971) pointed out in less rigorous fashion for birds long ago.

Finally, in reference to figure 5.1, we can ask whether the frugivores that are associated with particularly species-rich genera of plants in the French Guiana flora (and elsewhere in tropical America) are particularly rich in species themselves. These frugivores include manakins and trogons among birds and three clades of phyllostomid bats (*Carollia*, *Sturnira*, and *Artibeus* and its relatives) and spider monkeys (*Ateles*) among mammals. With 48 species, manakins are not particularly species rich compared with two related groups—the frugivorous cotingids (96 species) and insectivorous tyrannid flycatchers (400 species). Trogons are an old bird group whose closest relatives are currently unknown (Cracraft et al. 2004; Ericson et al. 2006). New World members of this family are frugivorous and include 25 species compared with 14 species in the Old World, all of which are primarily insectivorous. Based on the phylogeny of Baker et al. (2003), *Carollia* (six species) is somewhat more species rich than its sister clade (*Glyphonycteris* [three species] and *Trinycteris* [one species], both of which are insectivorous). With 14 species, *Sturnira* is more species rich than *Carollia*, but it is not more speciose than its sister clade, the fig-eating bats (*Artibeus* and its relatives, 53 species). It is, however, the second largest genus in subfamily Stenodermatinae. The genus *Artibeus* itself contains at least 18 species and is the largest genus in that subfamily. Finally, the monkey genus *Ateles* contains somewhat more species (seven) than its closest relatives (*Brachyteles*, two species; *Lagothrix*, four species; and *Oreonax*, one species). Overall, these results provide limited support for the hypothesis that frugivorous vertebrate taxa that are associated with particularly species-rich fruit taxa are usually richer in species than their close relatives. One reason for this is that, like their food taxa, these birds and mammals have large geographic ranges, often covering the entire mainland Neotropics. Large ranges in both the plant and animal taxa provide more opportunities for allopatric speciation and hence more species. If, as seems reasonable, we can equate species richness with evolutionary success, then the species-rich plant genera depicted in figure 5.1 and their associated frugivores have both been particularly successful evolutionarily.

Conclusions

Ecological interactions between tropical angiosperms and their vertebrate pollinators and seed dispersers clearly have had significant macroevolutionary consequences for both plants and animals. The role of pollinators is especially clear with both hummingbirds and phyllostomid bats being important drivers of flower diversification and plant speciation in the Neotropics. No single group of Paleotropical vertebrate pollinators appears to have had as profound an effect on plant speciation as have hummingbirds.

Although tropical vertebrate frugivores are not as well-recognized as drivers of plant speciation as vertebrate pollinators, we hypothesize that they have also influenced the speciation rates of their food plants through the process of intermediate dispersibility. By this we mean that certain groups of relatively sedentary understory frugivorous birds and bats have promoted speciation in their food plants by providing occasional long-distance seed dispersal into new habitats. The clearest examples of this process are likely to be found in the Neotropics and involve shrubs and treelets of the Melastomataceae, Piperaceae, Rubiaceae, and Solanaceae that are dispersed by manakins and phyllostomid bats.

Mutualistic interactions between tropical angiosperms and their vertebrate pollinators and seed dispersers have involved different degrees of coevolution. The most common form appears to be diffuse or multispecies coevolution between groups of plants and groups of pollinators or frugivores in the absence of any close phylogenetic congruity. But we suspect that a substantial proportion of the adaptive radiation of tropical vertebrate mutualists in specific lineages has involved a somewhat more focused degree of generalized coevolution with various lineages of their core food plants. On the animal side, this coevolution has affected morphology, ecology, and behavior, as we discuss in chapters 7 and 8, and it has likely influenced analogous features of their food plants, as we also discuss in those chapters. As a result of this process, at least 37 plant families (about 17% of tropical plant families) have evolved in close association with their vertebrate pollinators and seed dispersers. Many of these families are conspicuous and ecologically important members of their communities. Finally, a few plants and their vertebrate pollinators exhibit an exceptionally high degree of specialized coevolution, especially in the Neotropics, as exemplified by the interactions between the two sexes of the purple-throated carib hummingbird and two species of *Heliconia* in the Eastern Caribbean (Temeles and Kress 2003). All such mutualisms may show quantifiable variation in the degree of the interaction across their ranges as suggested by the geographic mosaic theory of coevolution (Thompson 2005).

Diversification of tropical angiosperms has involved allopatric and, to a lesser extent, ecological speciation. The evolution of new species of plants in response to different assemblages of pollinators may be common and has involved at least two processes. In the plastic pollinator model, plant clades have repeatedly and independently evolved species attracting different types of pollinators in different habitats. In the constrained pollinator model, once a plant clade has evolved to interact with a particular kind of pollinator, all subsequent speciation occurs within that pollinator type. Both kinds of models are likely to be common in tropical angiosperms, but we predict that the plastic pollinator model may be more common in the Neotropics than in the paleotropics because of a higher degree of plant-pollinator specialization in the former region than in the latter.

Finally, while tropical vertebrate pollinators and frugivores have both had a significant effect on the species richness of their food plants, it is not as clear that tropical plant diversity has had a similar effect on their mutualist animal partners. The majority of families or subfamilies of plant-pollinating tropical vertebrates contain more species than their sister taxa, whereas this is not true for fruit eaters and seed dispersers. Overall, the effect of these vertebrate interactions on species richness appears to be asymmetrical, with plants diversifying to a greater extent than their animal mutualists. This is not surprising given the differences in the mobility and behavioral and eco-logical flexibility of plants and animals. An obvious exception to this is the hummingbirds, which with about 330 species is one of the largest bird fami-lies. Extensive plant-animal coevolution, including both diffuse and more focused kinds, in the context of a geologically dynamic region of the tropics (the Andes of South America) has had a strong positive effect on the species richness of these birds and their food plants. Many families of plants contain much greater species richness as a result of hummingbird pollination. In addition to being true ornaments of life, these fascinating birds have been major drivers in the evolution of Neotropical plant diversity.

In the next chapter, we step back and view the evolutionary interactions between tropical angiosperms and their vertebrate mutualists from a more detailed phylogenetic and biogeographic perspective before examining how these interactions have affected the morphology, behavior, and physiology of these interactors. Whereas the material we have covered in chapter 5 has dealt with evolution on a mesoscale, geologically speaking, we will shift to a macroscale of inquiry in chapter 6 before shifting back down to a meso- or microscale in chapters 7 and 8.

6

Phylogeny and Biogeography of These Mutualisms

In this chapter we will provide an overview of the evolutionary histories and geographic distributions of tropical plants and their vertebrate mutualists. As we discussed in chapters 2–4, the cast of botanical and zoological characters often changes from one biogeographic region to another, but the functional roles of plant-visiting vertebrates as pollinators and seed dispersers are superficially much the same throughout the tropics (Corlett and Primack 2011). A closer look, however, reveals that evolutionary history and biogeography do matter in these interactions. For example, New World hummingbirds and their mammalian counterparts—glossophagine phyllostomid bats—often appear to have more specialized relationships with their flowers than do Australasian honeyeaters and pteropodid bats (Fleming and Muchhala 2008). Similarly, New World frugivorous phyllostomid bats appear to play a more important role as seed dispersers in early tropical forest regeneration than do Old World frugivorous pteropodid bats (Muscarella and Fleming 2007). Finally, it is likely that hummingbirds are stronger drivers of speciation in their food plants than their Old World counterparts (e.g., sunbirds and honeyeaters; chap. 5). What factors might account for these biogeographic differences?

The major questions that we address in this chapter are historical in nature. Thus, for both plants and animals we ask: How widespread are vertebrate pollination and seed dispersal mutualisms in the phylogenies of angiosperms, birds, and mammals? How many times have these mutualisms evolved independently in these groups? To what extent is the evolution of pollination and seed dispersal mutualisms congruent or noncongruent in these phylogenies? Where have these mutualisms evolved? Are there evolutionary "hotspots" for these mutualisms and are these hotspots congruent for both mutualisms? And what are the temporal patterns of the evolution of these mutualisms and are these patterns congruent for both mutualisms? How long have specific groups of plants and their vertebrate mutualists been interacting?

Earth History Background

The mutualisms we are dealing with have evolved over the past 60–90 million years or more—a period over which much has changed geologically and climatically on Earth. Here we provide a brief synopsis of these changes as background for understanding the evolution of tropical vertebrate-plant mutualisms in time and space. Much of the following narrative is based on Morley (2000, 2003, 2007). Graham (2010) provides a comprehensive review of the history of Neotropical plant communities.

Major changes in earth history by geological period are summarized in table 6.1. Angiosperms first appear in the fossil record in the Early Cretaceous, about 130–140 Ma (Frolich and Chase 2007; Hu et al. 2008 and included references; but see Smith et al. [2010] for an earlier time estimate for this radiation). At this time, the breakup of the northern Laurasian and southern Gondwanan subdivisions of Pangea was well advanced. By 100 Ma, West Gondwana (South America and Africa) had separated from Antarctica and Australia; Africa, Madagascar, and India had separated; South America and Africa were beginning to drift apart; and Africa was not connected to Eurasia. Warm and wet conditions prevailed as far north as midlatitudes throughout the Cretaceous, and sea levels were generally high (Miller et al. 2005; but see Bornemann et al. [2008] for evidence that Antarctica may have been ice covered during part of the Late Cretaceous). Under these conditions, the early floral elements of tropical rain forests (TRFs) were developing independently in three regions of the world: (1) an equatorial belt in which lineages including families such as Arecaceae, Fabaceae, Myrtaceae, Restoniaceae, and Sapindaceae were evolving; (2) a northern midlatitude belt in North America and Europe (the boreotropical region) where lineages including families such as Bombacaceae (now Malvaceae, *pro parte*), Icacinaceae, Menispermaceae, Rutaceae, Theaceae, and Zingiberaceae were evolving; and (3) a southern midlatitude belt in South America and East Gondwana in which lineages including families such as Aquifoliaceae,

Table 6.1. Summary of Major Continental Movements and Climate Changes during The Last 150 Ma

Geological Period	Epoch	Subdivision and Ages (Ma)	Major Continental Movements	Major Climate Changes
Cenozoic	Pleistocene	1.8–0.02	. . .	Major Ice Age
	Pliocene	5.3–1.8	Panamanian portal closed ca. 4 Ma joining S Amer & N Amer	Global temps drop sharply as modern Ice Age begins
	Miocene	Late: 11–5.3		
		Middle: 16–11	All continents close to present positions; S Amer still separate from N Amer; India part of S Asia; Asia & Australia in contact via New Guinea @ 15 Ma	Global temps drop sharply; tropical forests shrink and grasslands/ savannas spread
		Early: 24–16	Africa & Asia close again @ 17 Ma	Global temps oscillate
	Oligocene	Late: 28–24	. . .	
		Early: 34–28	. . .	
	Eocene	Late: 41–34	S Amer in contact with Australia via Antarctica until 38 Ma	Global temps drop sharply; Antarctica becomes ice-covered
		Middle: 49–41	Austral & Antarctica separate but still close; Madagascar separate from E Africa; India close to S Asia; S Amer separate from Africa and N Amer	Global temperatures still high
		Early: 55–49		Global temps high and tropical/subtropical climates and habitats occur at high latitudes (e.g., 60° N)
	Paleocene	Late: 61–55	S Amer equally distant from Africa and N Amer @ 50 Ma	Increased global warming begins
		Early: 65–61	Africa separate from Asia until Miocene	. . .
Cretaceous		Late: 99–65	Austral & Antarctica still joined; Africa & S Amer beginning to separate; India still close to E Africa; Africa separate but close to Eurasia	Warm and wet conditions prevailed from equator through mid-latitudes
		Early: 142–99	W Gondwana separated from Australasia by 100 Ma; last direct connection between Africa, Mada-gascar, & India @ 100 Ma; last direct connection between Africa & S Amer @ 100 Ma	Warm and wet conditions prevailed from equator through mid-latitudes

Sources. Geological ages—Stanley (2005); continental connections and movements—Cracraft (2001); Raven and Axelrod (1974); Morley (2000); Stanley (2005); climate changes—Morley (2000); Stanley (2005).
Note. In the Cenozoic Era, the Paleogene includes the Paleocene, Eocene, and Oligocene, and the Neogene includes the Miocene, Pliocene, and Pleistocene.

Olacaceae, and Proteaceae were evolving. Late Cretaceous tropical forests likely were not closed canopy and multistratal but were instead open and contained a single stratum. Average seed sizes of angiosperms then were small (Eriksson 2008; Tiffney 1986, 2004). This scenario differs somewhat from that of Erikkson et al. (2000), who proposed that angiosperm seed and fruit sizes began to increase in the Late Cretaceous rather than in the Paleocene, suggesting that closed-canopy forests were present in Late Cretaceous.

After recovering from the floral and faunal extinctions caused by the bolide collision at the K-Pg boundary, tropical forests from the Early Paleocene on were closed canopy and multistratal, and angiosperms began to produce fruits with larger seeds dispersed by birds and mammals (Erikkson 2008; Jacobs 2004; Tiffney 2004). By Late Paleocene, South America was equally distant from Africa and North America; Australia and Antarctica were still joined; and India was drifting NNE toward Asia. Throughout the Paleocene, most long-distance plant dispersal occurred within rather than between the three TRF belts, but occasional dispersal between formerly connected continents (e.g., South America and Africa, Africa and Madagascar/India) was still occurring. With the onset of significant thermal warming in the Late Paleocene and Early Eocene, TRFs extended farther north and south of midlatitudes (e.g., to 60° N in North America), and long-distance dispersal between the three tropical belts became possible.

By the Middle Eocene, Australia and Antarctica had separated, and India was close to South Asia. When India finally collided with South Asia (by mid-Miocene), many of its plants entered Asia to eventually create a homogeneous flora throughout South Asia as far east as Sundaland. Elsewhere, equatorial forests contained many floral elements of modern aspect. Although they shared many plant families, the floras of tropical South America and Africa were evolving independently.

Global temperatures and sea levels dropped sharply in Late Eocene as Antarctica became ice covered (fig. 6.1). This caused the extinction of many tropical plants (and their animal associates?) in the northern and southern TRF belts. The absence of dispersal corridors between North and South America and between Africa and Europe prevented the northward movement of tropical elements from equatorial TRFs. Plant extinctions also occurred in the equatorial belt but were not as extensive as elsewhere. The terminal Eocene cooling also affected Indo-Asian forests with monsoonal forests replacing evergreen forests throughout the region. Not until the Early Miocene (20 Ma) did evergreen forests again become widespread in lowland

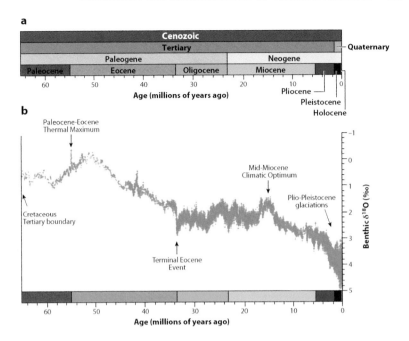

Figure 6.1.   The geologic time scale (a) and estimates of global temperatures occurring during the Cenozoic Era (b). Benthic $\delta^{18}O$ (‰) is a proxy for global ocean-water temperatures, with lower values corresponding to warmer temperatures. From Blois and Hadly (2009) with permission.

Asia. By this time, the Dipterocarpaceae, which likely evolved in the Late Cretaceous either in Africa or South America and which had become established in monsoonal forests of Southeast Asia during the Late Oligocene and Early Miocene (or even earlier; Rust et al. 2010), was the dominant plant family in lowland Asian forests. The ancestral flowering and fruiting pattern in this family was annual and seasonal. Its current pattern of multiannual and aseasonal flowering and fruiting in lowland Asian forests is a derived condition.

By Middle Miocene, all continents were close to their present positions, but South America was still isolated from North and Central America until the closing of the Panamanian portal in the Pliocene (fig. 6.2). Tropical biotas were essentially modern in composition at this time. Asia and Australia were in potential contact through New Guinea by 15 Ma; as a result, Asia gained elements of the Australian flora, especially in its mountains. Climate warming in mid-Miocene allowed equatorial TRFs to expand north and south again for a short period, but climate deterioration after mid-Miocene

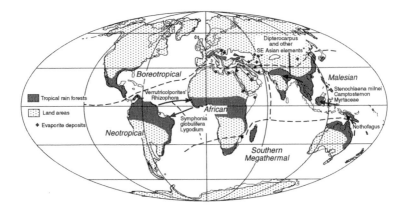

Figure 6.2. The distribution of continents during the Middle Miocene. From Morley (2000) with permission.

resulted in shrinkage of the TRF zone and substantial expansion of grasslands, savannas, and deserts. In addition to cooling and drying, geological uplift in East Africa caused many plant and animal extinctions in Africa in Late Miocene and later.

Finally, closure of the Panamanian portal about 3.8 Ma allowed extensive floral and faunal interchange to occur between North and South America. Movements of plants and animals between North and South America, probably via island hopping, had occurred prior to this, but the rate of migration increased markedly in the late Pliocene. A sharp drop in global temperatures in the Pliocene initiated the modern ice age and a series of glacial advances and retreats at high northern latitudes and at high elevations worldwide during the Pleistocene. Plio-Pleistocene climatic fluctuations had a profound effect on the extent and distribution of tropical forests and their faunas worldwide (chap. 5).

In summary, closed-canopy tropical forests have prevailed in warm parts of the globe throughout the Cenozoic. Many families of tropical plants had evolved by 90 Ma, and nearly all modern plant families and many modern plant genera had evolved by 45 Ma. By the Late Oligocene or Early Miocene, most families of modern birds and mammals had evolved. The geographic extent of TRFs changed substantially during the Cenozoic as a result of changes in mean global temperatures and the locations of continents. Their latitudinal distributions were maximal during the thermal maximum in Early Eocene, and their extent over the past 3 Ma has fluctuated strongly as a result of glacial advances and retreats.

Phylogeny and Biogeography of Tropical Angiosperms

THE PHYLOGENETIC DISTRIBUTION OF PLANT-VERTEBRATE MUTUALISMS

As a first step in our attempt to answer the questions listed at the beginning of this chapter about the evolutionary patterns of plant-vertebrate mutualisms, we assembled lists of plant families that are important as flower or fruit resources for plant-visiting vertebrates. Whenever possible, we classified plant families as major or minor food sources from the point of view of animal consumers. It is important to note that the designations major and minor do not necessarily mean that vertebrates are the only or most important pollinators or seed dispersers for these families. For example, Myrtaceae is an important flower and fruit source for many tropical birds and mammals (chaps. 3 and 5), but insect pollination and wind dispersal are probably more widespread than vertebrate pollination and dispersal in this family. Major food sources thus are families that are frequently used as sources of nectar, pollen, or fruit by particular groups of birds or mammals. These include the core resource families listed in table 3.9 plus other widely used families (table 1.3). Minor food sources include families that only occasionally provide resources for nectar- or fruit-eating birds and mammals. Although this classification is somewhat subjective, we find it useful to distinguish between major and minor plant families for both practical and theoretical reasons. On the practical side, the plant lists supplying food resources for well-studied vertebrates tend to include species in many families. For example, the lists of families containing flowers known to be visited by New World hummingbirds and Old World sunbirds both contain about 94 families (Fleming and Muchhala 2008). Similarly, the family lists for frugivorous New World phyllostomid bats and Old World pteropodid bats contain at least 40–50 families (Lobova et al. 2009; Mickleburgh et al. 1992), and family lists for New and Old World primates are even longer (e.g., Chapman et al. 2002; Russo et al. 2005; Tutin et al. 1997). These long lists indicate that opportunistic feeding often occurs in tropical plant–visiting vertebrates (chap. 3). But these animals are not totally catholic in their diets. As discussed in chapter 3, many (most?) species clearly have preferred food species and families, and detailed studies indicate that they tend to concentrate their feeding on a subset of their total plant list. We recognize this preferred subset as major plant families. From a theoretical viewpoint, this group of plants is the one with which vertebrate plant visitors are most likely to undergo more specialized coevolution (chap. 5). And it is this group of plants that is of critical importance in the conservation of tropical plant–visiting birds and mammals (chap. 10). Removal of species of Fabaceae or Moraceae from tropical forests, for example, would have severe negative impacts on many vertebrate nectar feeders and fruit eaters, respectively.

In similar fashion, we classified families of tropical birds and mammals as major or minor consumers of nectar or fruit. Most of these families are listed in table 1.1. Minor families are those containing a few species that occasionally eat nectar or fruit or those in which nectarivory or frugivory is generally uncommon (e.g., New World orioles and warblers among nectar-feeding birds and didelphid opossums and tapirs among fruit-eating mammals).

After classifying the characteristics of the plant and animal mutualists, we then mapped these features onto recent molecular phylogenies using the parsimony module in Mesquite version 2.0 (Maddison and Maddison 2007). For flowering plants, we used the APG III phylogeny (APG III 2009); for birds, the phylogenies from Mayr (2011, fig. 5) and Cracraft and Barker (2009, fig. 2); and for mammals, the phylogeny from Bininda-Emonds et al. (2007, fig. 1). We mapped characters of the plant taxa at the ordinal and family levels. We could have also mapped pollination and dispersal characters at generic levels and lower, but we chose orders and families as a reasonable approximation of major changes in lineages. In our taxon-character matrix, we initially scored all orders or families for two characters, vertebrate pollination and frugivory, each of which had three states: pollinators/frugivores not present, pollinators/frugivores in minor role, and pollinators/frugivores in major role. We also classified plant families in a number of additional ways, including scoring them for presence or absence of bird pollination, bat pollination, bird frugivory, bat frugivory, and/or primate frugivory as well as bird and bat pollination, and all combinations of bird, bat, and primate frugivory. For pollination, we also scored families at finer taxonomic levels for birds and bats (i.e., hummingbirds, sunbirds, and honeyeaters; phyllostomid and pteropodid bats). To analyze the biogeographic distributions of the major plant families that interact with tropical vertebrate mutualists, we scored all tropical (and some subtropical) families as occurring in each of four regions: the Neotropics, Africa and Madagascar, Asia, and Australasia, based on distribution maps found in Heywood et al. (2007).

We followed the same basic procedures for birds and mammals. The occurrence of nectarivory and frugivory in each animal family was mapped onto their respective phylogenies as nectarivory/frugivory not present, nectarivory/frugivory in minor role, or nectarivory/frugivory in major role. In addition, we scored each family as present or absent in each of four tropical regions: the Neotropics, Africa and Madagascar, Asia, and Australasia. Distributions followed Dickinson (2003) and Wilson and Reeder (2005) for birds and mammals, respectively.

Finally, for both plants and animals, we used these mapped phylogenies to identify the number of independent evolutionary origins of each of these vertebrate pollination and frugivory mutualisms across their evolutionary histories. An independent origin of pollination or frugivory in a family was counted whenever the mutualism was not derived from an immediate ancestor having the same mutualism. Thus, when all families in a clade shared the same mutualism and a common ancestor, then only a single independent origin was recorded. We were conservative in the scoring of mutualisms in unresolved polytomies. In these cases, we counted one-half the number of occurrences of the same mutualism in a polytomy as the number of independent evolutions. Thus, if a polytomy of five families exhibited frugivory, we scored the number of independent origins of this mutualism in these families as 2.5 rather than 5. We suspect that if we had mapped these characters at taxonomic levels lower than family, the number of independent origins would increase significantly.

This exercise of mapping pollination and dispersal mutualisms onto the phylogenies of plants, birds, and mammals was conducted as a broad-brush yet illustrative attempt to understand the evolutionary histories of these interactions. Ecological features, such as those characteristic of the species in these community mutualisms, combined with phylogenetic inferences about the taxa involved create a powerful tool for gaining insights into the patterns and processes of evolution (Harvey and Pagel 1991; Pagel 1999). Here, we have only scratched the surface in applying this methodology to the evolution of plant-animal interactions. Much more sophisticated and quantitative methodologies have been developed for not only mapping these features onto phylogenetic histories and determining ancestral character states but also determining the evolutionary rates of these historical events (e.g., Ackerly 2009). The incorporation of phylogenetics in ecological investigations is rapidly increasing with applications in community ecology (Webb et al. 2002; Cavender-Bares et al. 2009; Kress et al. 2009) and patterns of geographic diversification (e.g., Wiens et al. 2011). We hope that in the future the data given in this chapter and throughout this book will provide a starting place for more explicit and quantitative analyses of patterns of diversification in plant-vertebrate mutualisms.

ANGIOSPERM PHYLOGENY AND VERTEBRATE MUTUALISMS

ORDINAL-LEVEL ASSOCIATIONS.—The APG III classification of angiosperms recognizes 62 orders divided into the following major lineages: basal angiosperms (8 orders, 28 families), monocots (12, 83), basal eudicots (12, 64), asterids (14, 105), and rosids (17, 125; fig. 6.3). Of these lineages, only monocots, asterids, and rosids represent monophyletic clades. The other two lineages represent evolutionary "grades" rather than clades. We will call these five groups "lineages" throughout this book. Of the 62 orders, 44 (71%) contain families that exhibit either vertebrate pollination or dispersal; 29 (47%) contain families that exhibit vertebrate pollination; 39 (63%) contain families that exhibit vertebrate dispersal; and 24 (39%) contain families that exhibit both vertebrate pollination and dispersal. Thus, tropical vertebrate pollination and dispersal mutualisms have evolved in a majority of angiosperm orders.

The single and joint distributions of these two mutualisms are not uniformly distributed among orders within the five major angiosperm lineages (table 6.4, comparison A). Except for basal angiosperms, in which it is absent, vertebrate pollination occurs in 42%–75% of the orders within these lineages. Frugivory occurs in 33%–82% of these orders, with highest values occurring in asterids and rosids. The occurrence of either pollination or frugivory ranges from 42% (in basal eudicots) to 88% (in rosids) within orders, and the joint occurrence of pollination and frugivory ranges from 0% (in basal angiosperms) to 50% (in asterids). In general, both mutualisms are more common either singly or in combination within orders in the two most advanced lineages of angiosperms (asterids and rosids). Monocots are notable for having the highest occurrence of vertebrate pollination (in 75% of its orders) among these lineages.

As might be expected, percentages of families within orders that exhibit pollination or frugivory mutualisms with vertebrates are substantially lower than percentages at the ordinal level (table 6.2). Maximum percentages for both pollination and frugivory are 22% (in monocots and asterids) and 29% (in basal angiosperms), respectively. The joint occurrence of vertebrate pollination and frugivory is especially low among families (0%–13%). Not

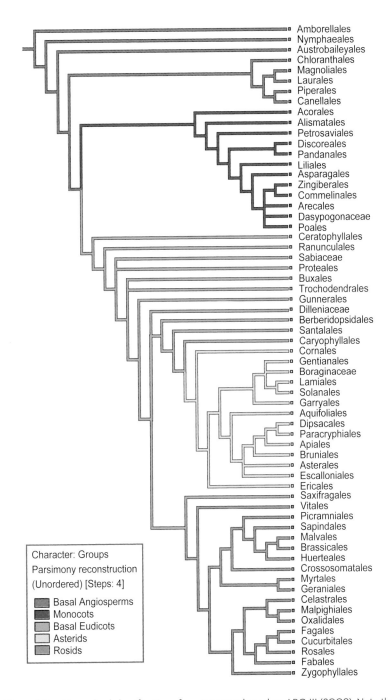

Figure 6.3. The ordinal classification of angiosperms based on APG III (2009). Note that the orders are grouped into five lineages, which are not necessarily monophyletic.

Table 6.2. Distribution of Vertebrate Pollination and Frugivory (Dispersal) Mutualisms among Orders and Families of Angiosperms

Lineage	Number	Proportion With:			
		Pollination	Frugivory	Pollination or Frugivory	Pollination and Frugivory
Orders:					
Basal angiosperms	8	0.0	0.63	0.63	0.0
Monocots	12	0.75	0.50	0.75	0.42
Basal eudicots	12*	0.42	0.33	0.42	0.33
Asterids	14	0.50	0.71	0.64	0.50
Rosids	17	0.53	0.82	0.88	0.47
Families:					
Basal angiosperms	28	0.0	0.29	0.29	0.0
Monocots	83	0.22	0.16	0.27	0.11
Basal eudicots	64	0.08	0.17	0.17	0.05
Asterids	105	0.22	0.22	0.29	0.13
Rosids	125	0.14	0.26	0.32	0.10

*Includes one family without ordinal placement (Sabiaceae; see fig. 6.3).

included in table 6.2 is information about the occurrence of "major" and "minor" families exhibiting vertebrate pollination or frugivory in these lineages. In general, minor families outnumber major families. For example, in plant families associated with birds across all lineages, minor families outnumber major families by factors of 3.3 and 5.3 in pollination and frugivory mutualisms, respectively. In families associated with bats across all lineages, minor families outnumber major families by a factor of two for both mutualisms.

We broke the ordinal and family data down further by examining the distributions of the two mutualisms by particular taxonomic groups. In this and all other analyses in this chapter, we will focus on three major groups of vertebrate mutualists—birds, bats, and primates—because they are among the most important vertebrate pollinators and/or primary seed dispersers in many tropical habitats. In general, bird pollination occurs more frequently than bat pollination at both the ordinal and family level within four of the five lineages, and the disparity is greatest (1.8-fold) among orders of rosids (table 6.3). Bird dispersal also occurs more frequently than either bat or primate dispersal at both levels in all five lineages. The disparity between the occurrence of bird and bat dispersal is greatest (fourfold) in basal eudicots; it is greatest (threefold) between birds and primates in the monocots. Bat and primate dispersal has similar frequencies in three of the five lineages. Bat dispersal is more common than primate dispersal in monocots, and primate dispersal is more common in basal eudicots. Overall, angiosperms have evolved pollination and dispersal mutualisms more frequently with birds than with bats and primates.

Table 6.3. Distribution of Vertebrate Pollination and Frugivory (Dispersal) Mutualisms among Orders and Families of Angiosperm by Animal Taxa

		Proportion With:				
		Pollination		Dispersal		
Lineage	Number	Bird	Bat	Bird	Bat	Primate
Orders:						
Basal angiosperms	8	0.0	0.0	0.63	0.25	0.25
Monocots	12	0.58	0.42	0.50	0.33	0.17
Basal eudicots	12*	0.42	0.25	0.33	0.08	0.17
Asterids	14	0.43	0.43	0.64	0.36	0.29
Rosids	17	0.53	0.29	0.76	0.41	0.47
Families						
Basal angiosperms	28	0.0	0.0	0.32	0.07	0.11
Monocots	83	0.20	0.11	0.14	0.08	0.04
Basal eudicots	64	0.08	0.03	0.16	0.02	0.05
Asterids	105	0.18	0.11	0.22	0.07	0.08
Rosids	125	0.12	0.06	0.27	0.10	0.18

* Includes one family without ordinal placement (Sabiaceae; see fig. 6.3).

At the ordinal and family levels, what kinds of associations exist either within a particular mutualism between different vertebrate taxa (e.g., how often does bird and bat pollination occur together within the same order or family?) or within a particular taxon between different mutualisms (e.g., how often does both bird pollination and frugivory occur within the same order or family?)? Answers to these questions will begin to give us some insight into the extent to which the evolution of these mutualisms has been constrained or promoted by prior evolutionary events. For example, how often is the presence of bat pollination in a particular order or family associated with the presence of bird pollination? How often is the presence of primate dispersal associated with bird dispersal? Positive associations between animal mutualists, particularly at the family level, indicate situations in which evolutionary facilitation might possibly occur (i.e., the presence of bird-pollinated flowers in a family might facilitate the evolution of bat flowers in that family), but they also might be the product of chance. The absence of such associations, particularly at the ordinal level, indicates that no facilitation has occurred.

We summarize the occurrence of these associations at both the ordinal and family levels (major and minor families combined) in table 6.4. In this analysis, we combined families across the five lineages because preliminary analyses indicated that the lineages did not differ in the frequency of the different character states (e.g., bird pollination only, bat pollination only, bird and bat pollination in comparison A; $Ps \geq 0.052$ in χ^2 tests) in the four comparisons in table 6.4. In comparison A at the ordinal level, bat pollination

Table 6.4. Associations among Vertebrate Pollination and Dispersal Mutualisms in Angiosperms

Association	Orders		Families	
	Number	Proportion	Number	Proportion
A. Bird vs. bat pollination:				
Bird pollination only	10	0.34	34	0.52
Bat pollination only	2	0.07	9	0.14
Bird and bat pollination	17	0.58	22	0.34
B. Bird vs. bat vs. primate dispersal:				
Bird dispersal only	13	0.33	41	0.46
Bat dispersal only	1	0.03	2	0.02
Primate dispersal only	0	0.0	2	0.02
Bird and bat dispersal	6	0.15	8	0.09
Bird and primate dispersal	5	0.13	17	0.19
Bat and primate dispersal	0	0.0	0	0.0
Bird, bat, and primate dispersal	14	0.36	19	0.21
C. Bird pollination vs. bird dispersal:				
Bird pollination only	5	0.12	21	0.21
Bird dispersal only	16	0.37	44	0.44
Bird pollination and dispersal	22	0.51	35	0.35
D. Bat pollination vs. bat dispersal:				
Bat pollination only	5	0.20	20	0.42
Bat dispersal only	6	0.24	16	0.33
Bat pollination and dispersal	14	0.56	12	0.26

Note. In this summary, families have been combined across all orders. Proportions are based on the number of orders or families within each comparison, not the total number of orders or families involved in these mutualisms.

was much more likely to occur with bird pollination than in its absence (0.58 vs. 0.07). A similar but smaller difference held at the family level (0.34 vs. 0.14). Even at the generic level in families such as Bromeliaceae and Campanulaceae, bat pollination is more likely to co-occur with bird pollination than it is to occur separately (Fleming and Muchhala 2008). In comparison B, bat and primate dispersal were much more likely to occur in the same orders and families as bird dispersal rather than to occur separately. Primate dispersal sometimes co-occurred alone with bird dispersal but never with bat dispersal alone at both the ordinal and family levels. Bird and bat dispersal were much less likely to co-occur in the same orders and families than bird and bat pollination. Also, a substantial number of families (19) exhibit bird, bat, and primate dispersal, particularly in the rosids. In comparison C, bird dispersal was at least twice as likely to occur as bird pollination in both orders and families, and both mutualisms co-occurred in about one-third to one-half of the appropriate orders and families. Finally, in comparison D, bat pollination and dispersal were about twice as likely to co-occur together as separately at the ordinal level but not at the family level where only one-quarter of families exhibited both bat pollination and dispersal.

Several conclusions emerge from these results. First, there was little evidence from these comparisons that the different character states differed in frequency among orders and families or among families in the different lineages. This suggests that tropical vertebrate-plant mutualisms generally have deep origins within angiosperm evolution. These mutualisms have undoubtedly become more fine-tuned within particular families through time, but their evolutionary antecedents likely have existed for much of angiosperm evolution. Second, within vertebrate pollination mutualisms, bat pollination is much less common than bird pollination, and it is more likely to evolve in orders (but not necessarily in families; see below) that also contain bird pollination than in orders that lack bird pollination. This suggests that bird pollination has sometimes facilitated the evolution of bat pollination (rather than vice versa; e.g., fig. 5.5). We discuss this conclusion in more detail in chapter 7. Third, within vertebrate dispersal mutualisms, bird dispersal is more likely to evolve independently in orders and families than either bat or primate dispersal. At the family level, bird and primate dispersal co-occur twice as often as bird and bat dispersal, which suggests that birds and primates may facilitate the evolution of each other's fruit. They are less likely to facilitate the evolution of bat fruit. Finally, in both birds and bats, pollination and dispersal mutualisms co-occur in about one-half of the orders and one-quarter to one-third of the families containing these mutualisms. Although bird-angiosperm mutualisms are at least twice as common as bat-angiosperm mutualisms, they co-occur in similar proportions at higher taxonomic levels in both groups of vertebrates.

FAMILY-LEVEL ASSOCIATIONS. — Overall, our analysis included 405 families of angiosperms. Of these, vertebrate pollination occurs in at least 63 families (major and minor families combined; 16%), and vertebrate dispersal occurs in at least 89 families (22%). As seen in figure 6.4, vertebrate pollination does not occur with equal frequencies in the five major lineages of angiosperms ($\chi^2 = 16.5$, df = 4, $P = 0.002$). It is absent in basal angiosperms and is less frequent in basal eudicots than in the other three groups. In contrast, vertebrate dispersal is relatively equally represented in the five lineages but tends to be higher in basal angiosperms (e.g., in families such as Annonaceae, Lauraceae, Myristicaceae, and Piperaceae) and in rosids (see below). In figure 6.5, we treat the two mutualisms separately and plot their frequency of occurrence by birds, bats, and primates. Figure 6.5 shows that the overall vertebrate trends are reflected by trends within birds. For both pollination and dispersal, birds are the vertebrates of choice in angiosperms, just as bees

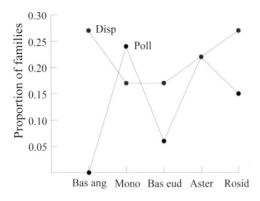

Figure 6.4. Distribution of tropical vertebrate pollination and fruit/seed dispersal among the five major angiosperm lineages. *Bas ang* = basal angiosperms; *Mono* = monocots; *Bas eud* = basal eudicots; *Aster* = asterids.

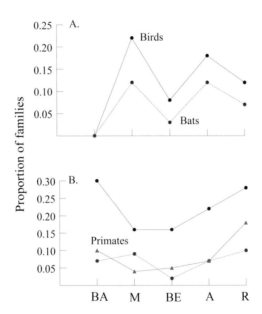

Figure 6.5. Distribution of bird and mammal pollination (A) and seed dispersal (B) among the five major angiosperm lineages. *BA* = basal angiosperms; *M* = monocots; *BE* = basal eudicots; *A* = asterids; *R* = rosids.

are the insects of choice as pollinators in most of these plants. Bird pollination is common in monocots and asterids, especially in the Zingiberales and Lamiales, respectively; bird dispersal is common in basal angiosperms, especially in the Magnoliales and Laurales. The frequency of occurrence of bat pollination mirrors that of bird pollination and is common in monocots (e.g., in the Zingiberales) and asterids. Bat dispersal is somewhat less frequent in basal eudicots than it is in the other four lineages. Finally, primate dispersal is especially common in families of rosids, especially in the Rosales and Sapindales.

Vertebrate pollination or dispersal has evolved independently rather than in phylogenetically restricted fashion in a majority of families of flowering plants (table 6.5). Our estimates of the number of independent evolutions include 46.5 out of 63 families (74%) for pollination and 61.5 out of 88 families (70%) for dispersal. The frequency of independent origins tends to be lower for both vertebrate pollination and dispersal in monocots than in the other four lineages. In monocots, bird and bat pollination and dispersal are clustered in the Zingiberales, which includes such families as Heliconiaceae (bird flowers), Musaceae (bat flowers and fruit), and Zingiberaceae (bird flowers). Phylogenetic clustering also occurs to some extent in the rosids (table 6.5), especially in the Rosales, which includes Moraceae and Urticaceae (which now includes Cecropiaceae; vertebrate fruits), and Sapindales, which includes Anacardiaceae, Burseraceae, Meliaceae, and Sapindaceae (bird and primate fruits).

Finally, we can ask, To what extent do different families of vertebrate pollinators share the same orders and families of flowering plants? We do this to gain a general understanding of the extent to which nectar-feeding birds and bats use resources in the same evolutionary lineages. To what extent have independently evolved New World and Old World vertebrate pollinators converged on the same general flower resources? To answer these questions, we focused on hummingbirds, sunbirds, and honeyeaters among birds and on phyllostomid and pteropodid bats. Results indicate that hummingbirds share about 70% of the same orders with sunbirds and phyllostomid bats whereas sunbirds share ≤33% of their orders with honeyeaters and pteropodid bats (fig. 6.6). Honeyeaters share about 33% of their orders with pteropodid bats. Finally, phyllostomid bats share fewer orders (41%) with pteropodid bats than with hummingbirds (70%). These nectar feeders clearly have not all converged on the same general set of food plants. Similarities and differences appear to have a deep biogeographic basis (see below).

Table 6.5. Summary of the Overall Occurrence of Different Vertebrate Mutualisms in Angiosperm Families by Major Lineages

Lineage/Mutualism	Number of Major and Minor Families	Number of Independent Origins	Proportion of Independent Origins
Basal angiosperms (28):			
Vertebrate pollination	0	0	0.0
Vertebrate dispersal	9	5	0.56
Bird pollination	0	0	0.0
Bat pollination	0	0	0.0
Bird dispersal	9	7	0.78
Bat dispersal	2	2	1.00
Primate dispersal	3	3	1.00
Monocots (83):			
Vertebrate pollination	18	12	0.67
Vertebrate dispersal	13	7	0.54
Bird pollination	17	11	0.65
Bat pollination	9	5	0.56
Bird dispersal	12	10	0.83
Bat dispersal	7	6	0.86
Primate dispersal	3	3	1.00
Basal eudicots (64):			
Vertebrate pollination	4	4	1.00
Vertebrate dispersal	11	9	0.82
Bird pollination	5	5	1.00
Bat pollination	2	2	1.00
Bird dispersal	10	8	0.80
Bat dispersal	1	1	1.00
Primate dispersal	3	3	1.00
Asterids (105):			
Vertebrate pollination	23	16.5	0.72
Vertebrate dispersal	23	18.5	0.80
Bird pollination	19	15	0.79
Bat pollination	12	8.5	0.71
Bird dispersal	23	19.5	0.85
Bat dispersal	7	5	0.71
Primate dispersal	8	7	0.88
Rosids (122):			
Vertebrate pollination	18	14	0.78
Vertebrate dispersal	33	22	0.67
Bird pollination	15	12	0.80
Bat pollination	8	8	1.00
Bird dispersal	34	22.5	0.66
Bat dispersal	13	11	0.85
Primate dispersal	22	15	0.68

Note. Number of families for each lineage is given in parentheses. The final column indicates the proportion of families that represent phylogenetically independent origins (see text).

How similar are the diets of these families of birds and bats at the level of plant family? In this comparison, we focus only on hummingbirds, sunbirds, and phyllostomid bats—the three families whose diets are most similar at the ordinal level. Not surprisingly, dietary similarity is lower at the family level than it is at the ordinal level (table 6.6). Thus, hummingbirds share about 33% of their plant families with both sunbirds and phyllostomid bats (compared with about 70% of their orders). In the hummingbird-sunbird

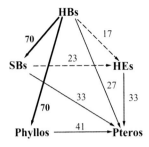

Figure 6.6. Percentage of angiosperm orders shared among three families of nectar-feeding birds and two families of bats. *HB* = hummingbirds, *HE* = honeyeaters, *Phyllo* = phyllostomid bats, *Ptero* = pteropodid bats, *SB* = sunbirds.

Table 6.6. Summary of Comparisons of the Families of Plants Providing Flowers

Lineage/Comparison with Hummingbirds	Shared Families	
	Number	Proportion
Monocots:		
Sunbirds	1/15	0.067
Phyllostomids	2/12	0.17
Basal eudicots:		
Sunbirds	1/5	0.20
Phyllostomids	1/2	0.50
Asterids:		
Sunbirds	9/15	0.60
Phyllostomids	5/15	0.33
Rosids:		
Sunbirds	4/11	0.36
Phyllostomids	5/9	0.56

comparison, percentage of shared families ranged from 7% (in monocots) to 60% (in asterids). In the hummingbird-phyllostomid comparison, percentage of shared families ranged from 17% (in monocots) to 56% (in rosids). These results indicate that, within shared orders, birds and bats usually visit flowers in different families. Dietary overlap tends to be lowest in monocots and highest in advanced dicot lineages.

FAMILIES STRONGLY ASSOCIATED WITH VERTEBRATE POLLINATORS OR SEED DISPERSERS.—In chapter 5 we identified a series of families or subfamilies and tribes whose pollination or dispersal biology is strongly associated with tropical vertebrates (table 5.4). Here, we place those families in a phylogenetic context by asking how these families are distributed among angiosperm lineages. Are they concentrated in (or absent from) particular lineages? Results indicate that, except for basal angiosperms in which there are no strongly associated families, families that rely heavily on vertebrate

Table 6.7. Distribution of Families, Subfamilies, and Tribes Strongly Associated with Tropical Vertebrate Pollinators or Seed Dispersers

	Pollination		Seed dispersal	
Lineage	No. of families	Proportion of Families	No. of families	Proportion of Families
Basal angiosperms (28)	0	0.0	4	0.14
Monocots (83)	3	0.04	1	0.01
Basal eudicots (64)	4	0.06	3	0.05
Asterids (105)	3	0.03	3	0.03
Rosids (125)	6	0.05	10	0.08

Note. Number of families for each lineage is shown in parentheses. Based on table 5.4.

pollination represent 3%–6% of families in these lineages (table 6.7). This is consistent with Hu et al. (2008), who indicated that basal angiosperms lack evidence of vertebrate and other specialized forms of pollination. In contrast, basal angiosperms have the highest percentage of families and so forth (14%) strongly associated with vertebrate seed dispersers. Rosids are also notable for having the highest number of families strongly associated with vertebrate seed dispersers (10) and pollinators (6). Thus, whereas there is a clear trend for vertebrate seed dispersal to be concentrated in the most basal and most advanced lineages of angiosperms, this trend is not apparent for vertebrate pollination.

THE CHRONOLOGY OF ANGIOSPERM EVOLUTION

The origin and diversification of angiosperms has been a long-standing evolutionary problem, and current research efforts in this area using molecular and fossil data are substantial (reviewed in Frolich and Chase 2007; Hedges and Kumar 2009; Soltis et al. 2005). Until recently, the fossil record for angiosperms from the Cretaceous was meager, and the oldest fossil angiosperm pollen has been dated at 141–132 Ma (Frolich and Chase 2007; Wikstrom et al. 2001). Crepet et al. (2004), among others, point out that increasing numbers of nonpollen fossils, including charcoalified Cretaceous flowers, have recently increased our knowledge about the early diversification of major angiosperm clades. Based on a critical review of this evidence, Crepet et al. (2004) concluded that numerous clades of angiosperms, including various magnoliids, the monocots, and eudicots such as early asterids and rosids, had evolved by 90 Ma. Frolich and Chase (2007), in contrast, suggested that many of these clades had evolved by 125 Ma. Overall, the fossil record suggests that many angiosperm clades underwent rapid diversification between about 130–80 Ma, well before the K/Pg boundary and well before the evolution of most modern birds and mammals.

In addition to the fossil record, which in many cases is likely to provide minimum estimates of the age of different clades, reconstruction of the evolution of angiosperms has relied heavily on DNA sequence data for the past two decades. As a recent example of this approach, Bell et al. (2005) used three genes and multiple fossil constraints to construct a dated phylogeny for angiosperms. Based on two analytical methods, they estimated that angiosperms first evolved 180–140 Ma, some 5–45 Ma years before the first fossil appearance of flowering plants. Their estimates of the ages of groups of modern angiosperms include basal angiosperms such as magnoliids and Chloranthaceae at 149–130 Ma; the monocots at 135 Ma; and early eudicots at 125 Ma. Most researchers feel that further refinements of angiosperm age estimates will be forthcoming from a combination of more sequence data (especially at the genomic level), better taxon sampling, and a better fossil record (e.g., Hedges and Kumar 2009; Smith et al. 2010; Soltis et al. 2005, 2011).

Because about 86% of the 29 families of basal angiosperms are insect pollinated, it is likely that insect pollination is ancestral in angiosperms (Hu et al. 2008). Whereas early angiosperms are usually thought to have had generalized pollination systems involving beetles and flies (e.g., Crepet and Friis 1987; Proctor et al. 1996), recent fossil evidence indicates that both generalized and specialized pollination systems involving insects had evolved in basal angiosperms and basal monocots by mid-Cretaceous. Specialized Cretaceous floral morphologies included enclosed floral chambers, viscin threads, nonnectar floral rewards, and sympetaly (Crepet et al. 2004). Zygomorphy, the ultimate refinement in pollination syndromes, evolved somewhat later than these features and characterizes some of the most diverse clades of modern angiosperms (e.g., asterids and papilionoid and caesalpinoid legumes). Zygomorphic flowers, however, have not been the only route to evolutionary success in angiosperms, and other modern groups with nonzygomorphic flowers such as rosids, ranunculids, and mimosoid legumes are also very diverse (Crepet et al. 2004). Davies et al. (2004) point out that the history of angiosperm evolution has featured frequent shifts in diversification rates among clades that have no simple explanation (e.g., the evolution of zygomorphic flowers as a key innovation). On page 1908 they state: "The pattern is not consistent with a simple model in which diversification is driven by a few major key innovations but rather argues for a more complex process in which propensity to diversify is highly labile: there are 'winners' and 'losers' at all levels, and shifts occur repeatedly." Vamosi and Vamosi

(2010) reached this same conclusion in their analysis of factors associated with angiosperm diversification. As we have seen (chap. 5), the evolution of pollination or dispersal mutualisms with tropical vertebrates has promoted the diversification of some groups of angiosperms but not others, which is consistent with this view.

Dated molecular phylogenies of particular families or clades of angiosperms are becoming increasingly common in the literature, and we have summarized available data for many families with vertebrate-pollinated or -dispersed species in table 1.3. Chapters in Hedges and Kumar (2009) were the source of these data. We will use these data to produce a rather broad-brush overview of the chronology of the origins of some of these families. Given that vertebrate dispersal is relatively common in families of basal angiosperms and that vertebrate pollination is most common in advanced families of monocots and dicots, we expect to see a curve of cumulative origins of vertebrate-dispersed families to reach 50% "saturation" at an older age than that of vertebrate-pollinated families. As shown in figure 6.7, however, this is not the case. The 50% saturation point occurs at about 80 Ma, well before the K/Pg boundary, in both sets of families. By definition, the appearance of crown groups in these families should be more recent than the appearance of stem groups, but detailed data are not available for all of these families. In many cases, it is likely that tens of millions of years elapsed between the appearance of stem and crown members of particular families. For example, the New World basal eudicot family Cactaceae, whose advanced subfamily Cactoideae is vertebrate pollinated and dispersed, is sometimes thought to be Cretaceous in age. Recent studies, however, suggest that its modern radiation dates from the Oligocene (about 30 Ma), when the northern Andes were undergoing uplift (Arakaki et al. 2011; E. Edwards et al. 2005; Nyffeler 2002). Similarly, although molecular data suggest that the Neotropical family Bromeliaceae originated over 100 Ma, its crown radiation likely dates from about 23–19 Ma (Anderson and Janßen 2009; Givnish et al. 2004, 2011). Three families, including one that is vertebrate pollinated and two that are vertebrate dispersed, are notable for their relatively young ages (table 1.3). These include the New World arid zone endemic Agavaceae, with many bat-pollinated species, and the pantropical Burseraceae and Myrsinaceae, with bird- and primate-dispersed fruits. Overall, however, most families that interact with vertebrate mutualists appeared relatively early in angiosperm evolution. Since bird and mammal pollination tend to be derived conditions in most families (chap. 7), it

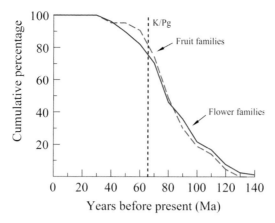

Figure 6.7.   Cumulative temporal distribution of origins of families of angiosperms that contain vertebrate-pollinated or -dispersed species. Estimates of times of origin are based on time-calibrated molecular phylogenies and come from chapters in Hedges and Kumar (2009).

is likely that these mutualisms are significantly younger (probably from the mid-Tertiary on) than are vertebrate-dispersal mutualisms, which are more likely to occur in both stem and crown members of their families. Notable exceptions to this are the four "banana families" in the Zingiberales in which vertebrate pollination is basal and dates from the Early Cretaceous (fig. 5.6 and chap. 7). It is not yet clear who their early vertebrate pollinators were.

ANGIOSPERM BIOGEOGRAPHY

In their classic review of the biogeographic history of angiosperms in the context of plate tectonics, Raven and Axelrod (1974) concluded that angiosperms first evolved in West Gondwana (South America/Africa) and that current distribution patterns are largely the product of vicariance resulting from the breakup of Gondwana. According to their analysis, most of the widely distributed families that are important sources of flowers and fruit for tropical vertebrates had a Gondwanan origin and few originated in Laurasia. Recent phylogenetic studies based on morphology, DNA sequences, and fossils, however, are changing this picture. To be sure, some old families and orders (e.g., Annonaceae, Lauraceae, Melastomataceae, Myristicaceae, Myrtaceae, Piperaceae, Sapotaceae, and Zingiberales) likely diverged as a result of Gondwanan vicariance (Doyle et al. 2004; Jaramillo and Callejas 2004; Kress et al. 2001; Morley and Dick 2003; Smedmark and Anderberg

2007; Sytsma et al. 2004). But other families or groups (e.g., Bignoniaceae, Burseraceae, some Lauraceae, Meliaceae, Menispermaceae, Moraceae, some Myrtaceae, and Rubiaceae) had a Laurasian origin and attained their current southern distributions by overland migration and/or via long-distance dispersal (e.g., Pennington and Dick 2004; Renner et al. 2001 and included references; Smedmark and Anderberg 2007; Zerega et al. 2005). Prior to the late Eocene–early Oligocene global cooling event, for example, Laurasian groups such as Malpighiaceae, Moraceae, and the Sideroxyleae clade of the Sapotaceae were members of the boreotropical florals whose species migrated extensively overland between Eurasia and North America and thence into South America (Smedmark and Anderberg 2007; Zerega et al. 2005 and included references).

In addition to vicariance and overland migration, long-distance dispersal has had an important influence on angiosperm distributions. Renner (2004), for example, listed 110 genera in 53 families of angiosperms that have South American–African amphi-Atlantic distributions involving long-distance migration. Dick et al. (2003), Pennington and Dick (2004), and Richardson et al. (2004) describe additional examples of trans-Atlantic long-distance dispersal. Renner (2004) concluded that most west-to-east dispersal likely involved wind transport, whereas most east-to-west dispersal likely involved water transport. Although she also concluded that transport of seeds across the Atlantic by seabirds or by frugivorous birds was unlikely, other studies suggest that long-distance dispersal mediated by frugivorous birds is possible. For example, the single species of *Rhipsalis* (Cactaceae) that has colonized Africa and Madagascar from South America produces sticky, mistletoe-like seeds that are unlikely to be dispersed by wind or water (Nyffeler 2002). Various species of Melastomataceae have also undergone long-distance dispersal between South America and Africa and from Southeast Asia to Madagascar and Africa and back again as a result of either bird or wind dispersal (Morley and Dick 2003; Renner et al. 2001). Long-distance seed dispersal by birds, especially by hornbills and fruit pigeons, has played an important role in the biogeography of tribe Aglaieae of Meliaceae and other large-seeded plants in the Indomalesian, Australasian, and western Pacific (Muellner et al. 2008; Viseshakul et al. 2011). Finally, crowberries (*Empetrum*) have a bipolar distribution, and Popp et al. (2011) postulated that a single long-distance dispersal by a bird best explains this distribution. Overall, there is now considerable evidence from molecular phylogenies that long-distance dispersal and subsequent radiation on different continents is

Table 6.8. Summary of the Biogeographic Distributions of Major Families of Angiosperms That Interact with Tropical Vertebrate Mutualists

Lineage and Mutualism	Number of Families			
	Neotropics	Africa and Madagascar	Asia	Australasia
Basal angiosperms	18	14	17	20
Pollination	0	0	0	0
Dispersal	4	3	4	4
Monocots	41	49	48	56
Pollination	5	4	3	3
Dispersal	6	2	3	5
Basal eudicots	36	44	44	31
Pollination	2	2	1	2
Dispersal	8	3	3	7
Asterids	63	56	62	61
Pollination	11	11	4	6
Dispersal	11	12	13	13
Rosids	78	76	78	66
Pollination	6	5	3	7
Dispersal	25	22	22	23

Source. Plant distributions come from Heywood et al. (2007).
Note. These data include the total number of families per region and the number of families that exhibit vertebrate pollination and/or vertebrate seed dispersal.

not an uncommon feature of angiosperm evolution (e.g., Dick et al. 2007; Erkens et al. 2007; Muellner et al. 2008; Namoff et al. 2010; Pennington and Dick 2004; Wen and Ickert Bond 2009). Pennington and Dick (2004), for example, suggested that about 20% of the species in an Amazonian forest plot in Ecuador belong to immigrant lineages. Transoceanic dispersal, in addition to Gondwanan vicariance, may account for the floral similarity of many tropical regions.

The current distributions of angiosperm families that interact with vertebrate mutualists in four tropical biogeographic regions are summarized in table 6.8. The general distribution of angiosperm families is quite uniform among regions. For example, the number of families of basal angiosperms per region ranges from 14 in Africa (including Madagascar) to 20 in Australasia. Likewise, the number of families of rosids per region ranges from 66 in Australasia to 78 in the Neotropics and Asia. The number of families per region varies more among major angiosperm lineages than among regions. A similar pattern occurs within the two mutualisms. Within each lineage, the number of families that are either vertebrate pollinated or dispersed is more or less constant among regions. Finally, within three of the five lineages the number of vertebrate-dispersed families is generally greater than the number of vertebrate-pollinated families in each region.

While each biogeographic region has its endemic vertebrate-pollinated or -dispersed families (e.g., Agavaceae, Bromeliaceae, and Cactaceae in the Neotropics; we discount the presence of one species each of Bromeliaceae and Cactaceae in West Africa as disqualifying these families from being classified as Neotropical endemics), many of the important families in each region have pantropical (often cosmopolitan) distributions (table 1.3). Among vertebrate-pollinated families, eight of 63 (13%) are pantropical in distribution. None of these families occurs in the basal angiosperms or monocots, and from 17% to 25% occur in the other three lineages. Among vertebrate-dispersed families, 39 of 88 families (44%) have pantropical distributions. Percentages vary significantly among lineages ($\chi^2 = 11.1$, df = 4, $P = 0.03$) and range from 15% in monocots to 64% in rosids. Pantropical families are of considerable evolutionary interest because they interact with taxonomically and behaviorally very different pollinators or seed dispersers in different biogeographic regions. In the New World, for example, their flowers interact primarily with small, hovering birds and bats, whereas they interact with larger, nonhovering birds and bats in the Old World (Fleming and Muchhala 2008). Similarly, bat-dispersed fruits interact with small, hovering phyllostomids in the New World, whereas they interact with larger, nonhovering pteropodids in the Old World (Muscarella and Fleming 2007). Seeds dispersed by monkeys in the Neotropics are usually swallowed and defecated, whereas they are often spit out by paleotropical monkeys. We will explore the botanical consequences of these differences in chapters 7 and 8.

At least five of the 42 pantropical families (12%) are both vertebrate pollinated and dispersed. These families include Loranthaceae in the basal eudicots, Verbenaceae and Rubiaceae in the asterids, and Fabaceae and Myrtaceae in the rosids. Not included in this list are other families (e.g., Cactaceae and Solanceae in the Neotropics, Musaceae in the Paleotropics) that are regionally important as sources of both flowers and fruits for vertebrates. These families and the handful of pantropical families represent the exception rather than the rule in the evolution of mutualistic interactions between tropical vertebrates and their food plants. Most families of tropical angiosperms that interact with vertebrates provide either flowers or fruits but not both as food for their mutualists.

The converse of the pantropical families are families that are only regionally important food sources for vertebrates. These families are either endemic (i.e., restricted to a single biogeographic region) or not, and it is of interest to know how common these two classes of families are among regions and major lineages. Only three of 45 flower families (7%) and five of 78 fruit families (6%) are endemic. All of these families except for Balanopaceae, which is Australasian, are Neotropical (pollination: Agavaceae, Bromeliaceae, Cactaceae, and Marcgraviaceae; dispersal: Bromeliaceae, Cactaceae, Lacistemaceae, and Marcgraviaceae). Among flower families, 35 of 42 pantropical families (83%) are important in only one or two regions; among fruit families, 37 of 73 pantropical families (51%) are important in only one or two regions. These two proportions differ significantly ($\chi^2 = 10.8$, df = 1, $P = 0.001$), which indicates that widely distributed flower families are more likely to be important food sources for nectar-feeding birds and mammals in a regionally restricted fashion than are fruit families. Put another way, flower families interact with vertebrate mutualists in a more geographically restricted (patchier) fashion than do fruit families. This is a pattern we would expect to see if flower-pollinator interactions are more specialized, on average, than fruit-frugivore interactions.

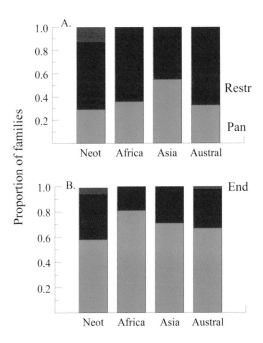

Figure 6.8.   Frequency of geographic distribution patterns of plant families among four biogeographic regions by pollination (A) and frugivory (B) mutualisms. *End* = endemic, *Pan* = pantropical, *Restr* = restricted.

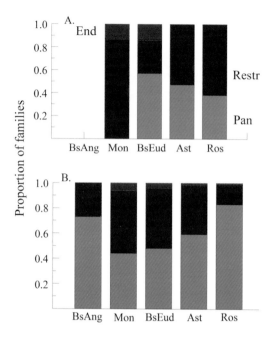

Figure 6.9.   Frequency of geographic distribution patterns among the five major angiosperm lineages by pollination (*A*) and frugivory (*B*) mutualisms. *End* = endemic, *Pan* = pantropical, *Restr* = restricted. *BsAng* = basal angiosperms, *Mon* = monocots, *BsEud* = basal eudicots, *Ast* = asterids, *Ros* = rosids.

We break down the major flower and fruit families by region and lineage in figures 6.8 and 6.9. As expected from the above results, pantropical families are more common among fruit families than among flower families, and regional differences in these proportions are relatively small; proportions of regionally restricted families also do not differ conspicuously among regions (fig. 6.8). Regarding distributions among the five lineages, major flower families are absent in basal angiosperms, and pantropical flower families are absent in monocots (fig. 6.9). The few endemic flower families occur only in monocots and basal eudicots. Pantropical fruit families are in the majority in three of the five lineages, especially in the rosids. In sum, the biogeography of tropical vertebrate mutualisms is complex and reflects the opportunistic nature of the evolution of these mutualisms. Historical and geographic context are clearly important for understanding this evolution. The importance of historical contingency is best illustrated by pantropical angiosperm families that have evolved pollination or dispersal interactions with birds and mammals in only parts of their geographic ranges.

Phylogeny and Biogeography of Vertebrate Mutualists

PHYLOGENY AND BIOGEOGRAPHY OF BIRDS

PHYLOGENY.—The phylogenetic structure of birds is rather complex and is incompletely resolved currently but can be broken down into three major clades or radiations (Cracraft et al. 2003, 2004). These include: Paleognathae (tinamous, cassowaries, and other flightless birds; six orders, six families); Galloanserae (chickens and their relatives, ducks and geese; two orders, nine families); and Neoaves (all other birds; 23 orders, about 181 families [Cracraft et al. 2003]; fig. 6.10A). Passeriformes, which includes about 97 families and nearly two-thirds of all species of modern birds, can be divided into suboscine and oscine (true songbird) clades. Cracraft et al. (2003) divide the oscine passerines into three clades: basal passerines, corvidans, and passeridans, and we will follow this classification here (fig. 6.10B). As an alternate view of avian phylogeny, Fain and Houde (2004) and Ericson et al. (2006) divided Neoaves into two monophyletic clades: Metaves and Coronoaves. Metaves ("near-birds") is basal to Coronaves ("crown birds") and contains a small group of nonpasserine orders and families; Coronaves contains many orders and families of birds, including both nonpasserines and passerines. The reality of this classification has been strongly questioned (reviewed in Mayr [2011]).

Nectar feeding is not a common adaptation in birds. In table 1.1 we listed 13 families of pollinators, eight of which (two nonpasserines and six passerines) are strongly nectarivorous. The nonpasserines include hummingbirds and lorikeet parrots. Nectar feeding occurs only in Philepittidae among suboscine passerines; it is found in only one family of basal oscines (honeyeaters); and it is concentrated in passeridan songbirds (fig. 6.10). Ten families of passeridans (five major and five minor) contain nectar eaters. The major families include sugarbirds, flowerpeckers, sunbirds, white-eyes, and Hawaiian honeycreepers (which are included in Fringillidae [or sometimes the Cardinalidae] in fig. 6.10B); the minor families include New World icterids (orioles) and parulid warblers and two additional small families or subfamilies (table 1.1). Seven of the 13 occurrences (54%) of nectarivory in birds represent independent origins of this feeding mode. The other six families occur in two phylogenetic clusters: one containing sugarbirds, sunbirds, flowerpeckers, and fairy bluebirds in the Old World and another containing icterids and warblers in the New World; nectarivory is widespread only in the Old World cluster.

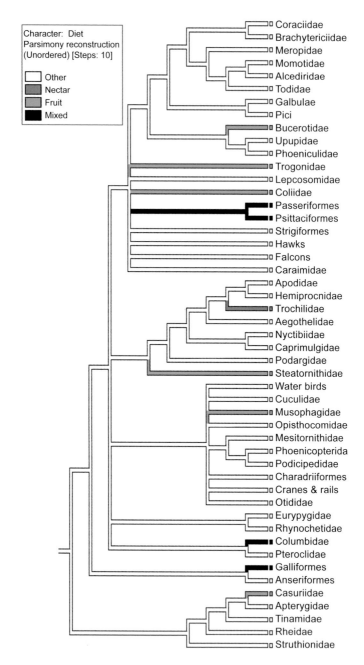

Character: Diet
Parsimony reconstruction
(Unordered) [Steps: 10]

- ☐ Other
- ▨ Nectar
- ▨ Fruit
- ■ Mixed

Coraciidae
Brachytericiidae
Meropidae
Momotidae
Alcediridae
Todidae
Galbulae
Pici
Bucerotidae
Upupidae
Phoeniculidae
Trogonidae
Lepcosomidae
Coliidae
Passeriformes
Psittaciformes
Strigiformes
Hawks
Falcons
Caraimidae
Apodidae
Hemiprocnidae
Trochilidae
Aegothelidae
Nyctibiidae
Caprimulgidae
Podargidae
Steatornithidae
Water birds
Cuculidae
Musophagidae
Opisthocomidae
Mesitornithidae
Phoenicopterida
Podicipedidae
Charadriiformes
Cranes & rails
Otididae
Eurypygidae
Rhynochetidae
Columbidae
Pteroclidae
Galliformes
Anseriformes
Casuriidae
Apterygidae
Tinamidae
Rheidae
Struthionidae

Figure 6.10. Cladograms showing (A) the major clades of birds based on Mayr (2011, fig. 5) and (B) the Passeriformes based on Cracraft and Barker (2009, fig. 2). The major families of nectar- and fruit-eating birds are indicated. "Mixed" groups include nectar and fruit as well as other dietary habits.

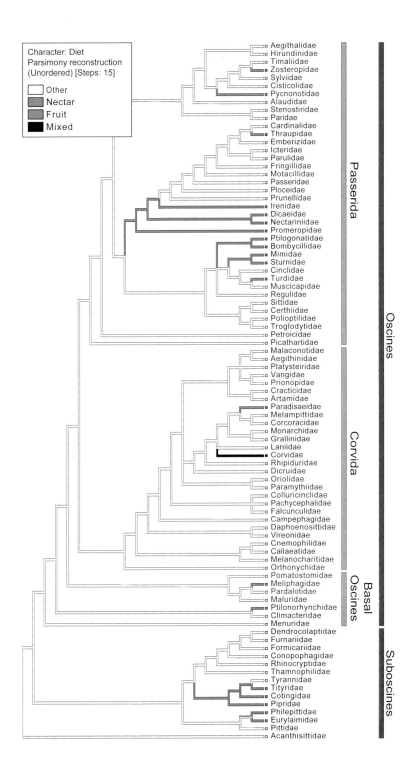

Character: Diet
Parsimony reconstruction
(Unordered) [Steps: 15]

☐ Other
■ Nectar
■ Fruit
■ Mixed

Aegithalidae
Hirundinidae
Timaliidae
Zosteropidae
Sylviidae
Cisticolidae
Pycnonotidae
Alaudidae
Stenostiridae
Paridae
Cardinalidae
Thraupidae
Emberizidae
Icteridae
Parulidae
Fringillidae
Motacillidae
Passeridae
Ploceidae
Prunellidae
Irenidae
Dicaeidae
Nectariniidae
Promeropidae
Ptilogonatidae
Bombycillidae
Mimidae
Sturnidae
Cinclidae
Turdidae
Muscicapidae
Regulidae
Sittidae
Certhiidae
Polioptilidae
Troglodytidae
Petroicidae
Picathartidae
Malaconotidae
Aegithinidae
Platysteiridae
Vangidae
Prionopidae
Cracticidae
Artamidae
Paradisaeidae
Melampittidae
Corcoracidae
Monarchidae
Grallinidae
Laniidae
Corvidae
Rhipiduridae
Dicruridae
Oriolidae
Paramythiidae
Colluricinclidae
Pachycephalidae
Falcunculidae
Campephagidae
Daphoenositthidae
Vireonidae
Cnemophilidae
Callaeatidae
Melanocharitidae
Orthonychidae
Pomatostomidae
Meliphagidae
Pardalotidae
Maluridae
Ptilonorhynchidae
Climacteridae
Menuridae
Dendrocolaptidae
Furnariidae
Formicariidae
Conopophagidae
Rhinocryptidae
Thamnophilidae
Tyrannidae
Tityridae
Cotingidae
Pipridae
Philepittidae
Eurylaimidae
Pittidae
Acanthisittidae

Passerida

Corvida

Basal
Oscines

Oscines

Suboscines

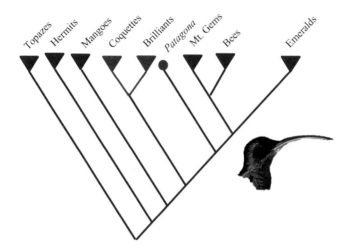

Figure 6.11. Basic phylogeny of hummingbirds showing the major clades. Red clades are strongly associated with the Andean uplands. Based on McGuire et al. (2007). The inset shows a rufous-breasted hermit (*Glaucis hirsutus*) from Schulenberg et al. (2007) with permission.

In terms of evolutionary ages, most of the major lineages of avian nectar feeders originated in the early Cenozoic (i.e., in the Eocene and Oligocene), but their major radiations date from the Miocene on (Barker et al. 2004). For example, although the oldest fossil hummingbirds date from the Oligocene of Europe (Bochenski and Bochenski 2008; Louchart et al. 2008; Mayr 2004, 2005a, 2007), their major radiation occurred in South America, beginning in the Miocene (Bleiweiss 1998a, 1998b; McGuire et al. 2007). Much of this radiation has occurred in association with Andean orogeny, and McGuire et al. (2007) estimated that this radiation involved at least 10 colonizations of the Andes from the South American lowlands (fig. 6.11). Basal members of this family are lowland forest dwellers, but the trochilines have radiated widely into many different lowland and highland habitats throughout the Neotropics, including the Caribbean, and seasonally in North America (Stiles 2004).

Two other families of Paleogene nectar-feeding birds—lorikeets and honeyeaters—occur mostly in Australasia. Recent molecular evidence indicates that lorikeets are advanced members of an Australasian clade of parrots (de Kloet and de Kloet 2005; Schweizer et al. 2010). Fossil parrots date from the Lower Eocene, and the family can be divided into three monophyletic clades: (1) a basal New Zealand clade, (2) an Australasian clade in which the Loriinae (lorikeets) is advanced, and (3) a geographically widespread clade containing Australasian palm cockatoos, all South American parrots, and some African/Madagascan parrots. Basal members of the Meliphagidae (honeyeaters) are Australian, and Driskell and Christidis (2004) suggest the family evolved in the heathlands of Australia in the mid-Tertiary (Late Eocene or Early Oligocene?). The honeyeaters of New Guinea represent several independent dispersal events from Australia, probably in the Pliocene (Norman et al. 2007).

According to Barker et al. (2004) and Beresford et al. (2005), the earliest passerines probably evolved in Gondwana in the late Cretaceous, some 30 Ma before their earliest fossil evidence in the Southern Hemisphere and 60 Ma before their presence in the fossil record in the Northern Hemisphere. New World suboscines began to diversify in South America near the K/Pg boundary, and basal oscines (including honeyeaters) and ancestral corvidans diversified in Australasia or in Australo-Papua shortly thereafter (Jonsson et al. 2011). The passeridan radiation began outside of Australasia 47–45 Ma ago. Several passerine lineages occurred in Africa in the Eocene, which suggests that this region played an important role in passerine evolution (Fjeldsa and Bowie 2008; Jonsson and Fjeldsa 2006). One of these clades includes the sugarbirds (Promeropidae), which is basal to the sister group of sunbirds and flowerpeckers; this clade dates from about 39 Ma. Basal flowerpeckers occur in the Indo-Malayan region with Australo-Malayan species evolving rapidly and more recently (Nyári et al. 2009).

As discussed in chapter 1, frugivory is more common in birds than nectarivory. At least 29 families or subfamilies contain frugivores, of which 13 are nonpasserines and 16 are passerines (table 1.1; fig. 6.10). Nonpasserine avian frugivores include cassowaries, cracids, oilbirds, turacos, fruit pigeons, Old World barbets, toucans, New World barbets, trogons, and hornbills. Unlike the nectar feeders, frugivores are well-represented by four families of New World and one family of Old World suboscine passerines (fig. 6.10B). Frugivorous corvidan passerines include birds of paradise. About 28% of the frugivorous families or subfamilies (eight of 29), including turdids, sturnids, bulbuls, and tanagers, are passeridan (advanced) passerines. Twenty of 29 families or subfamilies (69%) represent independent origins of frugivory. Phylogenetic clustering of frugivory is greatest in New World suboscines in which it has evolved in four closely related families (manakins, cotingids, tityrids, and certain tyrannid flycatchers). Phylogenetic clustering in passeridan songbirds has occurred in mimids and starlings and in flowerpeckers and sunbirds. Three or four passerine families (honeyeaters, sunbirds, flowerpeckers, and bananaquits—which are sometimes treated as a subfamily of Emberizidae) contain both nectar and fruit eaters.

Because it is a less-specialized dietary mode (chaps. 1 and 8), frugivory undoubtedly evolved earlier than nectarivory in birds. For example, it occurs in cassowaries among the ratites and in cracids in the Galloanserae, that is, in the two most basal groups of extant birds (fig. 6.10A). As it is in angiosperms, dating the evolutionary origins of modern birds (and mammals, see below) is currently an area of very active research and of some controversy. Based on the fossil record, some researchers (e.g., Feduccia 1996, 2003) claim that all modern birds originated after the K/Pg boundary. Using both molecular and fossil evidence, Ericson et al. (2006) reached a similar conclusion, saying that although some diversification of Neoaves occurred before K/Pg, most of the higher-level splits in this group occurred in the early Tertiary. Brown et al. (2007) disputed this assessment and, using similar evidence, concluded that at least 24 lineages of Neoaves, including such frugivorous groups as pigeons, oilbirds, turacos, parrots, and trogons, originated before K/Pg. Ericson et al. (2007) disagreed with this conclusion.

Recent molecular studies provide us with a detailed look at the phylogeny and biogeography of a number of families of fruit-eating birds. A molecular phylogeny of pigeons, for instance, indicates that this cosmopolitan family consists of three monophyletic clades: (1) a basal clade of New World pigeons; (2) a clade of New World ground doves; and (3) a geographically widespread clade containing species found in Africa, Asia, Australasia, and New Zealand. The Australasian fruit pigeons are advanced members of the latter clade (Pereira et al. 2007). Stem members of this family likely evolved 110–87 Ma in West Gondwana (South America), with modern genera of clade 1 evolving 55–37 Ma and genera of clade 3 evolving 57–41 Ma (Pereira et al. 2007). Trogons are another old nonpasserine family whose modern forms began to evolve in the Oligocene. Based on a molecular analysis, de los Monteros (2000) concluded that the basal members of this pantropical family evolved in Africa. Using a more extensive database, however, Moyle (2005) proposed that New World quetzals, which are generally considered to be specialized frugivores (e.g., Snow 1981; Wheelwright 1983), are basal to the insectivorous African and Asian forms. Phylogenetic analyses by Moyle (2005) and Hosner et al. (2010) indicate that New World trogons are not monophyletic but instead include three clades (quetzals, the Greater Antillean genus *Priotelus*, and the genus *Trogon*). *Trogon*, in turn, is sister to the Old World trogons in which African species are basal to Asian species. Fossil trogons are known from the Eocene to Miocene of Europe and appear to be unrelated to modern forms, which suggests that the biogeographic history of this family is likely to be complex (Mayr 2005b; Moyle 2005).

Moyle (2004) examined phylogenetic relationships between Neotropical, African, and Asian barbets, a group of birds whose closest relatives are woodpeckers (Picidae) and honeyguides (Indicatoridae). A maximum likelihood phylogenetic tree based on mitochondrial and nuclear DNA data indicated that each continental radiation is monophyletic (and are treated as separate families in table 1.1) and that Asian barbets are basal to the other clades. At least four clades of barbets exist in Africa, suggesting that the evolution of this group is complex. Depending on the phylogenetic position of *Semnornis* (toucan barbets), New World barbets and toucans may or may not be sister groups. The phylogenetic structure of these birds (Moyle 2004, fig. 4a) suggests that their modern forms first evolved in the early Cenozoic.

Finally, Paleotropical hornbills are represented by two families: the ground-feeding and carnivorous Bucorvidae and the canopy-feeding and frugivorous Bucerotidae. According to the phylogenetic hypothesis of Viseshakul et al. (2011), the stem groups of these birds evolved about 76 Ma and their crown groups date from 52–47 Ma, with most extant lineages radiating from the late Oligocene on. Perhaps Laurasian in origin, ground hornbills evolved mostly in Africa, whereas the more diverse frugivorous hornbills are mostly Asian in distribution. Viseshakul et al. (2011) propose that the strong-flying frugivorous hornbills, along with fruit pigeons, were important long-distance dispersers of large-seeded Indian plants that colonized mainland tropical Asia as well as the islands of Sundaland.

Among passerines, cotingids and their relatives, bulbuls, and starlings/mimids have been the subjects of recent molecular phylogenetic studies. Cotingids, along with their sister family Pipridae (manakins), are frugivorous New World suboscines that are closely related to tyrannid flycatchers and tityrids (formerly a subfamily within Tyrannidae). Ohlson (2007) could not resolve the relationships among these four groups but did find a strongly supported monophyletic clade containing four subclades within Cotingidae. Because of his inability to resolve relationships among these groups, Ohlsen (2007) suggested that they underwent rapid radiation but did not suggest a time for this radiation. Barker et al. (2004) estimated that the split between

New World and Old World suboscines occurred about 69 Ma, just before the K/Pg boundary. If this is correct, then it is likely that cotingids and their relatives radiated in the early Cenozoic.

Among corvidan passerines, the Australasian birds of paradise (Paradisaeidae) are conspicuous frugivores. Recent molecular studies indicate that this family, which has traditionally been thought to contain two subfamilies (the basal Cnemophilinae and advanced Paradisaeinae; Frith and Beehler 1998), is paraphyletic with the cnemophilines being basal to the "core Corvidans" in which paradisaeines are advanced members (Barker et al. 2004; Irestedt et al. 2009). Sibley and Ahlquist (1990) estimated that, within Paradisaeidae (*sensu stricto*), early forms (e.g., manucodes) originated in the Miocene (20–18 Ma); this estimate is congruent with the scenario presented in Barker et al. (2004, figs. 1 and 2). More recently, Irestedt et al. (2009) estimated that this family originated about 24 Ma and that the core of the polygynous genera originated about 15 Ma.

Radiations within major families of frugivores in the Passerida (e.g., bulbuls, turdids, sturnids, mimids, and thraupids) likely took place later in the Cenozoic than they did with other frugivores. Barker et al. (2004) suggested that mimids and emberizine sparrows, among other families, dispersed into the New World from Eurasia 28–20 Ma and that emberizine lineages quickly diversified there 18–16 Ma. The ancestors of tanagers entered South America relatively recently (in the Miocene?) and diversified there during Andean uplift (chap. 5). Although the phylogeny of the sister clade of Old World starlings and New World mimids is well-resolved (Lovette and Rubenstein 2007, fig. 5), the age and biogeographic history of this group is still poorly known. Two sister clades occur within Sturnidae: one clade containing South Asian and Pacific starlings plus *Rhabdornis* in the Philippines and another clade of African, Madagascan, and Eurasian starlings. Relationships within the latter clade are not well-resolved. In the New World, the Mimidae contains two clades: a "catbirds and Caribbean thrasher" clade and a "mockingbird and continental thrasher" clade. Barker et al. (2004) placed the origin of the sturnid-mimid clade as Miocene in age. Finally, another Old World passeridan family of frugivores—bulbuls (Pycnonotidae)—contains two clades, with the African clade being basal to the Asian clade (Moyle and Marks 2006). Judging from data in Barker et al. (2004), this family probably radiated in the Miocene.

We summarize the chronology of the first appearances of families of avian pollination and frugivory mutualists based on data in table 1.1 in figure

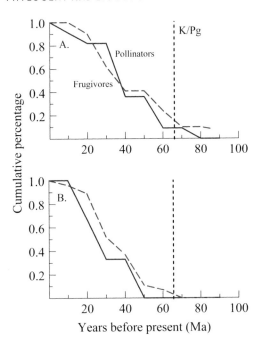

Figure 6.12. Cumulative temporal distribution of origins of families of plant-visiting birds (*A*) and mammals (*B*). Estimates of times of origin are based on time-calibrated molecular phylogenies and come from chapters in Hedges and Kumar (2009).

6.12*A*. Chapters in Hedges and Kumar (2009) were sources of these data. As expected, the midpoints of the family accumulation curves in both mutualisms occur well after the K/Pg boundary. These midpoints are virtually identical for both curves and occur at about 37 Ma (i.e., in the Late Eocene).

BIOGEOGRAPHY.—Families of avian nectar feeders are nearly equally represented among major tropical regions with two to four families per region (Fleming and Muchhala 2008). In the Neotropics, hummingbirds clearly dominate avian pollinator niches, with New World warblers and icterids representing minor pollinators. Sunbirds are the major avian pollinators in Africa; sunbirds and flowerpeckers in Asia; and honeyeaters and lorikeets in Australasia. The biogeography of hummingbirds is interesting because recent studies indicate that both stem and crown forms were present in Europe in the early Oligocene (ca. 32 Ma; Mayr 2004, 2005a, 2007). Mid- to Late Miocene climatic deterioration may have led to their extinction in the Old World. By this time, hummingbirds were well-established in South

America and were radiating extensively in the Andes (Bleiweiss 1998a, 1998b; McGuire et al. 2007). Parrots are also interesting because, except for lorikeets, which are confined nearly exclusively to Australasia, they are seed predators rather than nectarivores or frugivores throughout the tropics. The early Cenozoic radiation of Australasian Myrtaceae with its accessible and nectar-rich flowers probably provided an attractive food resource for these mobile, gregarious parrots.

All of the New World nectar-feeding birds and most of those in Australasia are currently endemic to those regions, but this is not the case for African and Asian nectarivores. Irwin (1999) suggested that basal sunbirds are Asian in origin, but their major radiation has occurred in Africa. Like sunbirds, flowerpeckers probably first evolved in Asia and then dispersed into Australasia; their greatest diversity currently occurs in New Guinea. Distributions of these two relatively recently evolved Old World families reflect the relative ease of dispersal among different Paleotropical regions in the mid-to-late Cenozoic (Beresford et al. 2005). Finally, basal honeyeaters occur in Australia where their major radiation has taken place, probably in Late Eocene or Early Oligocene. They dispersed independently to New Guinea several times in the Pliocene and are widely distributed on islands in the Pacific as far east as Hawaii (Driskell and Christidis 2004; Norman et al. 2007).

Turning to avian frugivores, with 11 families (in nine independent clades) the Neotropics contains the greatest taxonomic (and species) diversity and Asia (with five families in five independent clades) the least. Africa and Australasia contain six and seven independently evolved frugivorous families, respectively. The only cosmopolitan family of frugivorous birds is Turdidae (thrushes). Trogons are nearly pantropical in distribution but are strongly frugivorous only in the New World. Similarly, pigeons are cosmopolitan in distribution but are only truly frugivorous (rather than being seed predators) in Australasia (including the South Pacific), where a clade of fruit pigeons containing 10 genera and about 124 species has evolved (Goodwin 1983; Steadman 1997). The evolution of mutualistic relationships with Australasian plants in two groups of birds that are normally plant antagonists (seed predators)—parrots and pigeons—likely reflects the biogeographic isolation of that region through much of the Cenozoic. In the absence of other frugivores such as primates and hornbills, pigeons were able to feed on nutrient-rich, large-seeded fruits such as those of the Lauraceae and Myristicaceae as well as the Moraceae without experiencing intense competition from other diurnal frugivores. As mentioned above, the extensive radiation of one particular plant family (Myrtaceae) in Australia likely provided an important food resource for relatively small parrots such as lorikeets as well as basal passerines such as honeyeaters.

The Moraceae and Myristicaceae are also important food plants for another conspicuous radiation of Australasian birds, the insectivorous/frugivorous birds of paradise. Unlike honeyeaters, which evolved in Australia in the early Cenozoic and later colonized New Guinea, the birds of paradise, which are members of the corvidan group of passerines, evolved in forested habitats in New Guinea, likely in Early Miocene, and only colonized Australia very recently, when sea levels were low enough so that Australia and New Guinea were connected by continuous forest (Frith and Beehler 1998).

A major feature of the evolution of avian frugivores in tropical forests worldwide has been the radiation within and between families of canopy and understory feeders (chap. 2). From a biogeographic perspective, parallel radiations have occurred in the New and Old World tropics. Thus, in the New World two sister families (toucans and barbets) have evolved as large and small canopy frugivores, respectively. Their Old World ecological counterparts are canopy-feeding hornbills and barbets (which are not related to each other). In the Neotropics, two other frugivorous sister families—cotingids and manakins—have basically radiated in the canopy and understory, respectively. A similar radiation has occurred within the ecologically diverse Neotropical tanagers in which many genera (e.g., *Chlorophonia, Euphonia,* and *Tangara*) are canopy feeders and others (e.g., *Conothraupis* and *Tachyphonus*) feed primarily in the understory (Isler and Isler 1999). We are not aware of similar canopy/understory sister family radiations in the Old World, where a majority of frugivorous birds are canopy feeders. Starlings (Sturnidae), for example, have radiated into a wide range of habitats in Africa and elsewhere in the Old World, but, with only one exception, all African forest species feed in the canopy (Feare and Craig 1999). Finally, as in the case of avian pollinators, avian frugivores exhibit a higher degree of family endemism in the Neotropics than in the Paleotropics. Except for trogons and turdids, families of Neotropical frugivores are endemic to that region. In contrast, Africa and Asia share a number of families (e.g., hornbills, barbets, starlings, and bulbuls), and Asia and Australasia share others (e.g., orioles and flowerpeckers). Again, the occurrence of several families of

Paleotropical avian frugivores in two or more biogeographic regions reflects the existence of dispersal routes among Old World continents in the latter half of the Cenozoic.

PHYLOGENY AND BIOGEOGRAPHY OF MAMMALS

PHYLOGENY.—With six major clades or superorders, the basic phylogenetic structure of mammals is seemingly much less complex than that of birds (fig. 6.13). These clades include: Monotremata (platypus and echidna; one order, two families); Marsupialia (marsupials; seven orders, 21 families); Xenarthra (armadillos and anteaters; two orders, five families); Afrotheria (African placentals; six orders, eight families); and Boreotheria, which

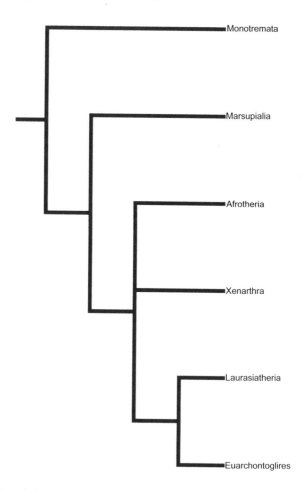

Figure 6.13. Cladogram showing major clades (superorders) of mammals.

includes two subclades (often treated as superorders)—Laurasiatheria (northern shrews, bats, carnivores, ungulates, etc.; eight orders, 63 families) and Euarchontoglires (tree shrews, primates, rodents, lagomorphs; five orders, 54 families; Wilson and Reeder 2005). Families of mutualistic plant-visiting mammals occur in four of these clades; they are absent in monotremes and xenarthrans (although extinct ground sloths probably ate fruit; Guimaraes et al. 2008).

As in birds, the number of mammalian families containing specialized nectar feeders is low and includes two families of bats and one marsupial (table 1.1). Flower visiting has been reported in an additional four families (Mystacinidae in bats, Lemuridae in primates, Muridae in rodents, and Procyonidae in carnivores). Of these families, two families of bats—the New World Phyllostomidae and Old World Pteropodidae—are by far the most important mammalian pollinators (Fleming and Muchhala 2008; Fleming et al. 2009). Based on the molecular phylogeny of Baker et al. (2003), nectar-feeding phyllostomids occur in two closely related subfamilies—Glossophaginae and Lonchophyllinae—with the former subfamily containing more genera and species than the latter (Simmons 2005). We call these subfamilies "glossophagines" collectively throughout this book. Whereas specialized nectar-feeding pteropodids have traditionally been classified in subfamily Macroglossinae (e.g., Koopman 1993), recent morphological and molecular research (Giannini and Simmons 2005; Kirsch et al. 1995) indicates that nectarivory has evolved independently at least three times in this family (twice in Asia/Australasia and once in Africa) and that the Macroglossinae is paraphyletic. The African nectarivorous pteropodid (*Megaloglossus*) is a more derived member of this family than the Australasian and Asian genera (Giannini and Simmons 2005). Both of these families of bats also contain fruit eaters as well as nectar feeders. The other family of specialized mammalian nectarivores is the marsupial Tarsipedidae (honey possums) of southwestern (nontropical) Australia. Containing a single species, this family is closely associated with flowers in the Myrtaceae and Proteaceae.

At least 29 families of mammals in four of the six major clades contain frugivorous species (table 1.1). Eighteen of these families (62%) are independently evolved. Families of frugivorous mammals are strongly concentrated in Euarchontoglires, which contains over twice as many families of frugivores (19) as Laurasiatheria (eight). Nonindependent evolution of frugivory occurs in Primates, which contains several strongly frugivorous families and in which frugivory may or may not be the ancestral feeding mode (e.g.,

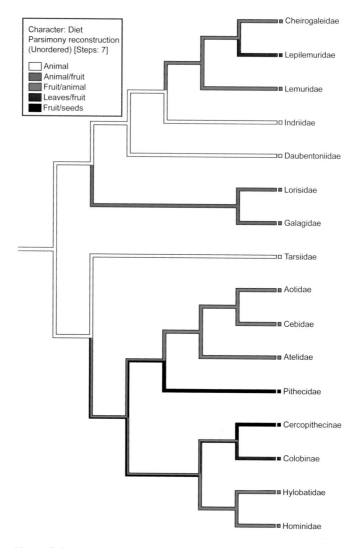

Figure 6.14. Cladogram of Primates based on Fabre et al. (2009) and their general diets.

Martin et al. 2007; Soligo and Martin 2006; Sussman 1991). Based on diet information in Rowe (1996), we estimate that 10 or 11 of 15 extant families of primates are strongly frugivorous, but our phylogenetic hypothesis suggests that insectivory, rather than frugivory, may have been the ancestral feeding mode (fig. 6.14). Fruit eating in these primates is strongly concentrated in forest canopies and subcanopies, but both canopy/subcanopy and understory frugivores have evolved in phyllostomid and pteropodid bats.

Muscarella and Fleming (2007) point out that frugivorous phyllostomids have evolved a higher degree of stratum-based feeding specialization than pteropodids in which both large and small species often feed in the canopy, especially on figs (Hodgkison et al. 2004b; Shanahan et al. 2001). In addition to evolving in primates and bats, frugivory has evolved in mammalian carnivores and in certain ungulates. In carnivores, it has evolved independently in New World procyonids and in Old World palm civets, both of which feed on canopy fruits. A higher diversity of large, terrestrial frugivore/browsers occurs in the understory of African and Asian forests than in Neotropical forests (Cristoffer and Peres 2003). These mammals include elephants, tragulid deer, rhinocerosids (mainly in Asia), and tapirs (Asia only) in the Old World and tapirs in the New World.

The evolutionary ages of different mammalian clades and their orders and families is currently undergoing strong debate based on different lines of evidence (e.g., Bininda-Emonds et al. 2007; Meredith et al. 2011; O'Leary et al. 2013; Wible et al. 2007; see Donoghue and Benton [2007] for a review of methods being used to calibrate phylogenies and the Tree of Life). Three models have been proposed to describe evolution within mammalian orders: the short-fuse model, the long-fuse model, and the explosive model. These models, of course, are not restricted to mammals but are likely to occur in all forms of life. The short-fuse model proposes that the radiation of crown (extant) families occurred shortly after the evolution of the order in the Late Cretaceous. The long-fuse model proposes that many mammalian orders first evolved in the Late Cretaceous but did not radiate until after the K/Pg boundary. The explosive model posits that most orders of placental mammals and their families did not evolve and radiate until after the K/Pg boundary.

Researchers working primarily with molecular data favor the Cretaceous origin of many mammalian orders. Bininda-Emonds et al. (2007), for example, proposed that the four major placental clades evolved around 100 Ma and that nearly all extant orders of placentals were present by 75 Ma, well before the K/Pg boundary. Their data suggest that extant mammals can be placed in each of the three models. Placental orders containing at least 29 species conform to the short-fuse model; long-fuse model groups include monotremes (but see Rowe et al. [2008] for evidence that this model does not apply to monotremes), xenarthrans, and low-diversity placental groups such as proboscideans and perissodactyls; and explosive model groups include marsupials, which radiated at the K/Pg boundary. According to Bininda-Emonds et al. (2007), the time of origin of Primates and bats, two short-fuse groups, was about 91–89 Ma. In the case of bats and primates, at least, this is far earlier than the fossil record indicates; the earliest bats and modern primates date from the Eocene (Martin et al. 2007; Simmons and Geisler 1998; Simmons et al. 2008).

In contrast, researchers working with fossils and morphology dispute the Cretaceous age and radiation of many groups of placental mammals. Working with an extensive morphological data set of extinct and extant mammals, Wible et al. (2007) concluded that placental mammals did not arise and radiate until after the K/Pg boundary and that the explosive model best describes their evolution. The extensive morphological and genetic analysis of O'Leary et al. (2013) supports this. In addition to differing about which models best describe the evolution of different groups of mammals, researchers differ as to the location of early mammalian evolution. Based on fossil evidence, Wible et al. (2007) proposed that placental mammals first evolved in Laurasia rather than Gondwana. Working with genomic-level DNA data, however, Wildman et al. (2007) concluded the opposite. Their results suggested that a Gondwanan clade of Afrotheria + Xenarthra is basal to a Laurasian clade of Euarchontoglires + Laurasiatheria, which puts the evolution of placentals as occurring first in Gondwana, a conclusion that Cracraft (2001) also reached for birds. Nishihara et al. (2009) suggested on the basis of a retroposon analysis (i.e., rare genomic insertions and indels that occur after chromosomes have been reverse transcribed from an RNA molecule) that the six major clades of mammals can be grouped into three lineages—Afrotheria, Xenarthra, and Boreotheria originating in Africa, South America, and Laurasia, respectively—that arose nearly simultaneously about 120 Ma with the breakup of Gondwana and Laurasia. More recently, Meredith et al. (2011) used an extensive DNA and amino acid database plus many fossil calibrations to conclude that the evolution of mammals is consistent with a long-fuse model in which interordinal diversification occurred in the Cretaceous and intraordinal diversification occurred mostly in the Cenozoic.

Despite the ongoing debate about the evolutionary origins of different superorders and orders of mammals, it seems clear that a substantial number of orders of modern mammals were present in the early Paleogene and that mammals have had mutualistic relationships with tropical and subtropical plants for much of the Cenozoic. Estimates of the origins of the two families of plant-visiting bats—Pteropodidae and Phyllostomidae—based on molecular data are 58 and 36 Ma, respectively; the crown groups in both families date from 26–24 Ma (Teeling et al. 2005). Davalos (2009) estimated that glossophagine phyllostomids first evolved about 20 Ma, and strongly frugivorous phyllostomids (e.g., subfamilies Carolliininae and Stenodermatinae) are likely somewhat younger than this based on the phylogenetic hypothesis of Baker et al. (2003; also see Datzmann et al. [2010]). We currently lack estimates of the ages of nectarivorous bats within the Pteropodidae. Based on the phylogenetic hypothesis of Giannini and Simmons (2005), however, frugivory is basal and nectarivory is derived in this family, which first evolved in Asia (Teeling et al. 2005). Frugivorous pteropodids have therefore existed for much of the Cenozoic. Within Chiroptera, two families—Phyllostomidae and Molossidae—have had especially high rates of diversification (Jones et al. 2005). As we know, many phyllostomids are strongly associated with flowering plants, and this association is likely responsible for much of this diversification, especially in fig-eating bats of the genus *Artibeus* and its relatives (Datzmann et al. 2010; Dumont et al. 2011). In contrast, Molossidae is a cosmopolitan family of fast-flying insectivores; factors behind its high rate of diversification are currently unknown.

One hypothesis of evolutionary relationships among primates is that of Fabre et al. (2009), which is an update of Purvis (1995; fig. 6.14). Two suborders of Primates have traditionally been recognized: Strepsirrhini, which includes lemurs, galagos, and lorises, and Haplorhini, which includes tarsiers, New World platyrrhines, and Old World catarrhines. Evolutionary diversification within Primates according to this and other recent phylogenies has taken place within the Cenozoic. The split between strepsirrhines and haplorhines, for example, was set at 57.5 Ma by Masters et al. (2006). More recent estimates of divergence times within this order, however, are somewhat different (Spoor et al. 2007). These include: separation of Primates from (Scandentia + Dermoptera)—86 Ma; separation between strepsirrhines and haplorhines—77 Ma; base of the lemuriforms—62.7 Ma; base of lorisiforms—55 Ma; base of the haplorhines—55 Ma; separation of platyrrhines and catarrhines—43.6 Ma; base of the platyrrhine radiation—25 Ma; separation between cercopithecoids and hominoids—34.7 Ma; and separation between hylobatids and hominids—15 Ma. Within Primates, the Cercopithecidae with its two subfamilies, which have different feeding modes (mostly frugivory in Cercopithecinae and mostly herbivory in Colobinae), has had an especially high rate of diversification (Heard and Cox 2007; Purvis et al. 1995). Finally, the supermatrix analysis of primate relationships conducted by Fabre et al. (2009) is mostly consistent with recent molecular analyses and indicates that strepsirrhines first appeared and radiated in the Eocene/Oligocene whereas families of platyrrhines, hominids, and cercopithecids did so in the Miocene/Pliocene. New World primates have sometimes been classified in four families (e.g., table 1.1), but a recent phylogenetic analysis at the generic level suggests that night monkeys (*Aotus*) are basal members of Cebidae (Wildman et al. 2007).

We summarize current estimates of the chronology of first appearances of mammalian nectar feeders and frugivores in figure 6.12*B* based on data in table 1.1. Chapters in Hedges and Kumar (2009) were the source of these data. As in birds, first appearances of mammalian frugivores predate those of nectar feeders, and the midpoints of these family accumulation curves occur long after the K/Pg boundary. The frugivore midpoint occurs at about 31 Ma, in the Early Oligocene. This was a time of major diversification in mammals (Stadler 2011).

BIOGEOGRAPHY.—Different tropical biogeographic regions currently contain one to three families of mammalian pollinators. Throughout the tropics, bats are the major mammalian pollinators. Their species richness is highest in the New World (table 1.1; chap. 2). In Old World pteropodids, parallel independent radiations of specialized nectar feeders have evolved in Asia and Australasia, which have a total of 14 species in five genera, and more recently in Africa, which has one species. The Asian genera are likely to be older than the African genus (Giannini and Simmons 2005). In addition to "macroglossines," many other pteropodids (e.g., species of *Pteropus* in Asia and Australasia) visit flowers, often in opportunistic fashion (Fleming and Muchhala 2008). While we consider rodents, especially those in the large family Muridae, to be minor pollinators, it should be noted that a few species in the New and Old Worlds, especially in subtropical South Africa, are regular flower visitors and can be effective pollinators (e.g., Ackermann and Weigend 2006; Carthew and Goldingay 1997; Cocucci and Sersic 1998; Fleming and Nicholson 2002; Johnson et al. 2001).

Different biogeographic regions contain three to 13 families of strongly frugivorous mammals (table 1.1). The greatest number of frugivorous families occurs in Africa and Asia, each of which contains 13 families, and the lowest number occurs in Australia. With the exception of tapirs, which have a disjunct Asian/Neotropical distribution, families of New World frugivorous mammals are endemic to that region, as are two of three families in Australasia. As in the case of other vertebrate mutualists, Africa and Asia share a number of families of frugivorous mammals (e.g., pteropodid bats, cercopithecid and hominid primates, viverrids, tragulids, and elephants), again reflecting the relative ease of dispersal between these two regions during much of the Cenozoic (Kappelman et al. 2003).

Primate biogeography is of special interest because of the importance of monkeys and apes as major seed dispersers. Taxonomic and ecological diversity of primates is clearly greater in the Old World tropics than the New World tropics (chap. 2). Old World primates exhibit a much greater range of body sizes, locomotory abilities, and diets than do Neotropical forms. This is especially true of the Malagasy primates, which underwent an impressive adaptive radiation in isolation from other Old World primates, beginning in the Eocene (Fleagle and Reed 1999). The biogeography of extant primates includes two long-distance dispersal events: one from East Africa to Madagascar about 40 Ma and one from West Africa to South America about 35 Ma (Masters et al. 2006; Purvis 1995). Compared with the Malagasy radiation, the South American morphological and ecological radiation of platyrrhine monkeys was much more modest despite its occurrence on a much larger isolated continent. Nonetheless, a majority of New World primates are strongly frugivorous and are important seed dispersers in many Neotropical forests (chap. 4).

CHRONOLOGY OF THE RADIATION OF VERTEBRATE MUTUALISTS AND THEIR FOOD PLANTS

With estimates of the chronologies of families of vertebrate mutualists and their major food plants in hand (figs. 6.7 and 6.12), we can now ask, to what degree are these chronologies congruent? That is, how closely did the first appearances of vertebrate nectarivores and frugivores match the first appearances of their major food families? To address this question, we focused on those animal taxa and their core plant families found in table 3.9, reasoning that if temporal congruence is going to occur, it will most likely occur between taxa that have formed strong mutualistic associations. We can envision two possible scenarios: (1) the chronologies of plants and their animal mutualists correspond closely to each other or (2) the chronologies of animals lag behind those of their families of food plants. In scenario 1, plants and their mutualists interact for most of their histories; in scenario 2, mutualists don't begin to interact with their food plants until well after the plant's first appearance, as discussed in chapter 5. Given that plant-pollinator interactions tend to be more specialized than plant-frugivore interactions

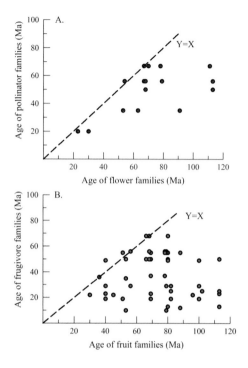

Figure 6.15. Correlations between the estimated ages of families of nectar-feeding (*A*) and fruit-eating (*B*) birds and mammals and their core plant families. See text for further explanation.

(chaps. 1 and 7), we might expect scenario 1 to apply to the former interaction and scenario 2 to apply to the latter.

Data addressing this question are shown in figure 6.15. These data strongly support scenario 2. In both mutualisms, most animal families are younger than their core plant families. This is especially true in frugivores in which a number of prominent contemporary families (e.g., tanagers, birds of paradise, various haplorhine primates, and phyllostomid bats) have evolved in the last 15 Ma—much more recently than their major fruit families (Fleming and Kress 2011). Nonetheless, a number of points fall on or close to the Y = X line in both mutualisms, and it is of interest to identify these plant-animal pairs. Among pollinators, these pairs include Acanthaceae and Costaceae and hummingbirds, Bombacoideae and pteropodid bats, and Cactaceae and Agavaceae and glossophagine bats. Among frugivores, these pairs include Elaeocarpaceae and Myrtaceae and cassowaries, Loranthaceae and honeyeaters, Burseraceae and cotingids, Meliacae and hornbills, Anacardiaceae and pteropodid bats, and Cactaceae and phyllostomid bats. Despite the temporal congruence in the first appearances of these groups, it remains to be seen whether this represents true "causal" coradiation or merely temporal correlation. We suspect that this congruence is correlational rather than causal in most cases because we lack strong evidence that most of these taxa have closely coevolved with each other. Nonetheless, these correlations are intriguing and certainly merit closer examination. Our overall conclusion here is that degree of temporal congruence between the evolution of vertebrate mutualists and their food plants has been rather low. To be sure, certain vertebrates and their food plants have established close, coevolved relationships (chap. 5). But at the family level, temporally close origins appear to be the exception rather than the rule. We will revisit this topic and its evolutionary implications in chapter 9.

General Summary and Conclusions

From our review of the phylogeny and biogeography of tropical angiosperms and their vertebrate mutualists, it is clear that current plant-animal interactions have had a long and complex evolutionary history. For plants, this history began at least 90 Ma, and by Middle Eocene (45 Ma) most of the families and genera that currently interact with tropical vertebrate pollinators or seed dispersers had evolved. Initially occurring in a warm, "greenhouse" world that was characterized by large, interconnected land masses, angiosperm evolution subsequently occurred in an increasingly fragmented world whose climate has slowly deteriorated from mid-Eocene on. Angiosperm evolution occurred against a geological background of land subsidence and uplift, mountain building, formation and disappearance of land bridges and islands, the rise and fall of sea levels, and the waxing and waning of polar ice sheets. Processes such as vicariance, overland migration, island hopping, and occasional long-distance dispersal have all played a role in creating the patterns of plant distribution that characterize the world today. As a result of these processes, contemporary tropical plant communities contain a mixture of plant lineages that have evolved either (1) in situ or close by (i.e., syntopically within the same continent or biogeographic region) or (2) some distance away (i.e., allopatrically in a different hemisphere or biogeographic region). Tropical plant communities in different locations thus differ strikingly in their degree of provinciality and taxonomic diversity. Provinciality is greatest in continents such as South America and Australia, which have had long histories of geographic isolation, and least in Africa and Asia, which have been connected by migration for large parts of the Cenozoic. Despite its long isolation (but see Pennington and Dick [2004] for a discussion of how isolated it really was), South America is notable for its outstanding biotic diversity. We will examine this diversity in more detail below.

Overall, the plant families that presently provide flower and fruit resources for tropical vertebrates represent a mixture of evolutionary lineages and geographic distribution patterns. Most of the families providing flowers for birds and mammals come from advanced angiosperm lineages—advanced monocots, asterids, and rosids; basal angiosperms provide no floral resources and basal eudicots provide relatively few (fig. 6.5A). In contrast, families providing fruits for birds and mammals come from all five of the major angiosperm lineages and include basal angiosperms as well as the two advanced eudicot clades (fig. 6.5B). These differences suggest that vertebrate frugivores have interacted with angiosperms for a longer period of time than have vertebrate nectar feeders, although the family accumulation curves in figure 6.7 don't necessarily support this. Many of the families providing flowers for vertebrates have broad geographic distributions but interact with vertebrates in only a part of their ranges (fig. 6.16A). Relatively few vertebrate-pollinated families are endemic or have pantropical distributions in which they interact with vertebrate pollinators throughout their range. In contrast, families providing fruits for tropical vertebrates are about equally divided among pantropical families and widespread families that interact with them in only part of their range (fig. 6.16B). In both a phylogenetic and geographic sense, then, vertebrate pollinators interact with their flower families in patchier fashion than do vertebrate frugivores. Only relatively advanced families of angiosperms interact with nectar-feeding vertebrates and they do so in a geographically restricted fashion. This is not the case in the interaction between angiosperms and vertebrate seed dispersers. Fruit-eating vertebrates often interact with basal lineages of angiosperms, and many pantropical families provide fruit for birds and mammals throughout their ranges. These patterns emphasize the historical importance of biotic dispersal for angiosperms, beginning with the evolution of closed-canopy tropical forests and large seeds in the Late Cretaceous and early Cenozoic.

Like their food plants, tropical nectar- and fruit-eating birds and mammals have had a complex evolutionary history. Stem lineages of many modern birds and mammals originated around the K/Pg boundary or somewhat

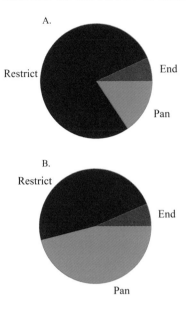

Figure 6.16.   Frequency distribution, by pattern of geographic distribution, of plant families providing flowers (A) or fruits (B) for tropical birds and mammals. *End* = endemic, *Pan* = pantropical, *Restrict* = restricted.

earlier (by 75 Ma), and the crown groups of most families had evolved by Late Oligocene/Early Miocene (28–16 Ma). Reflecting the long-standing importance of biotic dispersal, families of vertebrate frugivores likely are older, on average, than are families of nectar feeders. This is especially true in birds in which 13 of 28 strongly frugivorous families (46%) are nonpasserines compared with only two of 13 families (15%) of nectarivores. While most families of nectarivores and frugivores have evolved independently, there are two conspicuous examples of phylogenetic clustering (nonindependence) among vertebrate frugivores: one in four families of Neotropical suboscine passerine birds and one in the mammalian order Primates, in which most families are frugivorous. In terms of numbers or proportions of angiosperm families, bird pollination and dispersal is more common than mammal pollination and dispersal. Nearly 60% of all vertebrate-pollinated families are exclusively bird-pollinated and another 27% are pollinated by both birds and bats (fig. 6.17A). Similarly, nearly 50% of all vertebrate-dispersed families are bird-dispersed, and another 40% are dispersed by some combination of birds and primates or birds, bats, and primates (fig. 6.17B). As we've mentioned previously, angiosperms have targeted birds to a greater extent than

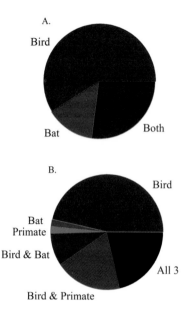

Figure 6.17. Frequency distribution of plant families that are pollinated (*A*) or dispersed (*B*) by different groups of birds and mammals.

mammals as their vertebrate pollinators and seed dispersers of choice. We will explore possible reasons for this in chapters 7–9.

Unlike their food plant families, families of vertebrate nectar and fruit eaters generally have restricted geographic distributions. There are no pantropical families of avian or mammalian nectarivores and only one family of pantropical avian frugivores (Turdidae). A few pantropical families of birds contain either nectar or fruit eaters in part of their geographic ranges (e.g., Australasian lorikeets, Neotropical trogons, and Australasian fruit pigeons), but these families are the exception rather than the rule. The rule is that most families of vertebrate nectarivores and frugivores are restricted to a single biogeographic region, or at most two. Endemism and provincialism, rather than cosmopolitanism, characterizes communities of tropical vertebrate mutualists. At first glance this pattern may seem paradoxical given the obvious differences in mobility between plants and animals. There are at least two possible explanations for this pattern. First, it is likely that most tropical plant families are older, on average, than their vertebrate mutualists and evolved before Gondwana and Laurasia had become strongly fragmented. This would have allowed plant lineages to have broad geographic distributions early in their evolution. Many families of modern birds and mammals, in contrast, evolved after the breakup of Gondwana and Laurasia and hence originated in a more fragmented world. Alternatively, or in addition, plant propagules might be more vagile or more capable of surviving long-distance dispersal events than most species of birds and mammals. Recent molecular evidence clearly indicates that long-distance dispersal is not uncommon in the evolutionary histories of many plant lineages (Thorne 2004). Intercontinental migration, often via wind or water dispersal, appears to be much more common in tropical plants than in their vertebrate pollinators or seed dispersers. If this is true, it would explain the lower degree of provincialism that we see in tropical plants compared with their vertebrate mutualists, even in a post-Gondwanan/Laurasian, geologically fragmented world.

In a recent paper, Richard Corlett (2007b) asked, What's so special about Asian tropical forests? In answering this question, he described tropical Asia's unique geology (it's a hodgepodge of plate fragments that make this region geologically younger than other tropical regions), flora (its lowland forests are dominated both in the understory and canopy by a single family, Dipterocarpaceae), and fauna (its forests lack specialized understory frugivores but are rich in squirrels, carnivorous mammals, and gliding vertebrates). To this list, Corlett could have added that some Asian forests (e.g., Lambir in Borneo) are extremely rich in species of figs, whose life forms include geophytes, shrubby pioneers, lianas, hemiepiphytes, epiphytes, and free-standing giants (Harrison 2005). Clearly, Asia's unique geological history has had a significant impact on the extent to which its plants interact with nectar-feeding and fruit-eating birds and mammals, and the same is true for other tropical regions.

Rather than belabor this point for all other regions, we will end this chapter by asking: What's so special about one additional region—the Neotropics? As we discussed in chapter 2, the Neotropics is notable for its extremely high plant and animal diversity. This diversity is well-illustrated by comparing the species richness of pantropical families of fleshy-fruited understory plants in different biogeographic regions. Among basal angiosperms, for example, the Piperaceae contains about 1,150 species in the Neotropics, 15 species in Africa, 300 species in Asia, and 40 species in Australasia despite first evolving in the Paleotropics (Jaramillo and Callejas 2004). Among asterids, Solanceae is represented in the Neotropics by 50 endemic genera containing over 400 species; much lower diversity occurs in Africa and Asia (D'Arcy 1991). Among rosids, Melastomataceae contains about 3,000 species in the Neotropics, 250 in Africa, and 1,000 in Asia despite possibly being Laurasian in origin (Renner et al. 2001). One important plant genus that runs counter to these examples is *Ficus* (Moraceae), which originated in the Neotropics and is represented by 132, 112, and over 500 species in the Neotropics, Africa, and Indo-Pacific, respectively (Harrison 2005; Jousselin et al. 2003). Similarly, only one of the 50 or so genera in the advanced monocot family Zingiberaceae occurs in the Neotropics. Despite these exceptions, it is striking to note the strong disparity in species richness between the Neotropics and other tropical regions in these and many other families of plants, including Arecaceae, Bignoniaceae, Bombacoideae (= Malvaceae, part), Lauraceae, and Verbenaceae (Mabberly 1997).

In addition to high species richness in many families, the Neotropical flora is notable for its tremendous diversity of life forms that provide important resources for birds and mammals (chap. 2). Along with woody plants such as trees, shrubs, and lianas, these include large understory herbaceous monocots (Costaceae, Heliconiaceae), succulent xerophytes (Agavaceae, Cactaceae), and a variety of epiphytes (Araceae, Bromeliaceae, Gesneriaceae, Marcgraviaceae). It is interesting to note that of four Gondwanan families of Caryophyllales that have adapted to arid habitats in different biogeographic regions (i.e., pantropical Aizoaceae, Neotropical Cactaceae, African and Madagascan Didiereaceae, and cosmopolitan Portulacaceae), only Cactaceae has evolved extensive mutualistic relationships with vertebrate pollinators and seed dispersers. Why have these Neotropical plants evolved extensive mutualisms with birds and mammals?

In a classic paper, Gentry (1982) reviewed the geological/geographic history of the Neotropical flora and provided a general explanation for its exceptional richness (also see Antonelli and Sanmartin 2011 and Pennington et al. 2010). He first pointed out that this flora contains two major elements: a Laurasian montane element that is relatively young and a Gondwanan lowland element that is older. This latter flora can be further divided into two historic segments: a series of Amazonian-centered families of canopy trees and lianas and a series of Northern Andean–centered families of understory shrubs, epiphytes, and palmettos. Among their many differences, Laurasian and Gondwanan plants differ significantly in their modes of pollination, with about half of the Laurasian families being wind pollinated, while nearly all of the Gondwanan families are animal pollinated. Many of the Andean-centered families—including Acanthaceae, Bromeliaceae, Campanulaceae (subfamily Lobelioideae), Costaceae, Ericaceae (a Laurasian family), Gesneriaceae, Heliconiaceae, Loranthaceae (subfamily Loranthoideae), Marcgraviaceae, and Tropaeolaceae—are largely pollinated by hummingbirds. Other Andean-centered plants, including understory shrubs such as Melastomataceae, Piperaceae, Rubiaceae, and Solanaceae, are dispersed by birds and bats. The Amazonian-centered plants, in contrast, are largely pollinated by bees and other insects, and hummingbirds and bats are often only minor pollinators. Except for wind-dispersed lianas, their fruits are often dispersed by birds, bats, and primates.

As a result of their different pollination modes, the various floral elements differ considerably in modes and rates of speciation (Gentry 1982). Wind-pollinated Laurasian taxa have not speciated extensively in South America, even in the Andes. As discussed in chapter 5, speciation in woody Amazonian trees and lianas has involved classic allopatric as well as ecological speciation. In contrast, "explosive" speciation has often occurred in Andean-centered epiphytes, shrubs, and palmettos in which over 100 genera contain more than 100 species apiece. According to Gentry (1982, 587–88), "Microgeographic, perhaps even more or less sympatric, speciation is probably the rule rather than the exception" in these plants. He further states (588): "I propose that high diversity in epiphytes and other Northern Andean–centered groups results mostly from recent very dynamic speciation" often involving coevolution between plants and their pollinators. He concludes his review by saying (589): "Thus, the historical accident of the Andean uplift, with the concomitant opportunity for explosive speciation among certain taxa of Gondwanan plants having the evolutionary potential for exploiting epiphytic, palmetto, and understory shrub strategies, may largely explain the 'excess' plant species diversity of the Neotropics" compared with the Paleotropics.

The Neotropical bird and mammal fauna is also exceptional in its taxonomic and ecological richness. In addition to hummingbirds, which we consider to be the flagship animals in the Neotropical zoological ark, high species richness occurs in other plant-visiting birds, including toucans and barbets, the cotingid/manakin/tyrannid complex, and tanagers. These birds and their mammalian counterparts (phyllostomid bats and primates) have radiated largely in response to a diverse Neotropical flora. Nowhere else in the tropics have nectar- and fruit-eating birds and mammals collectively evolved as many specialized feeding niches as they have in the Neotropics. As one example of this exceptional specialization, midelevation habitats in the northern Andes house a hummingbird (*Ensifera ensifera*) with the world's longest bill relative to its size as well as a glossophagine bat (*Anoura fistulata*) with the world's longest tongue relative to its size among mammals (Muchhala 2006b; Snow and Snow 1980). Both of these species are the sole pollinators of extremely long-corolla flowers (*Passiflora mixta* for *Ensifera* and *Centropogon nigricans* for *Anoura*). We are unaware of comparable examples of such extreme ecological and evolutionary specialization in Paleotropical vertebrates and their food plants.

In conclusion, there clearly is something special about how plants and their avian and mammalian pollinators and seed dispersers have evolved in the Neotropics. We believe that Al Gentry (1982) was correct when he postulated that Neogene Andean orogeny has had an exceptionally important effect on the evolution and diversification of Neotropical plants. It is equally clear that coevolution between plants and their vertebrate mutualists has occurred to a much higher degree in the Neotropics than elsewhere in the tropics. During this coevolutionary radiation, both plants and animals have applied strong selective pressures on each other and have influenced each other's rates of diversification. Gentry's explosive speciation in Andean-centered plants could not have occurred without the assistance of vertebrate pollinators and seed dispersers. We discussed the taxonomic consequences of this coevolution in chapter 5. We will discuss the adaptive consequences of this coevolution for plants and animals in the next two chapters.

7   The Pollination Mutualism

In chapter 3 we discussed the nutritional basis of pollination and seed dispersal mutualisms between tropical nectar- and fruit-eating birds and mammals and their food plants. In it we summarized data on the nutritional rewards of vertebrate-pollinated or -dispersed flowers and fruit, described patterns of resource production, and indicated that resource levels often appear to control the biomass of plant-visiting birds and mammals. In this chapter and the next we return to flowers and fruit but from an evolutionary rather than a purely ecological perspective. Our major goal in these chapters is to discuss the evolutionary and biological consequences of these mutualisms for both plants and animals—that is, the nuts and bolts of these interactions. We describe the morphological, physiological/biochemical, and behavioral characteristics associated with these mutualisms in plants and animals and ask how these characteristics evolved. What character transformations are involved in the evolution of bird- or bat-pollinated flowers and vertebrate-dispersed fruits, and what selective pressures were involved in these transformations? To what extent does the evolution of these mutualisms involve geographic or phylogenetic convergence and plant-animal complementarity?

We focus on the plant-pollinator mutualism in this chapter. As we indicated in chapter 1, this is the more "precise" or specialized of the two mutualisms examined in this book because the payoffs to plants and animals are unambiguous. Plants use their flowers as conspicuous visual signals to acquire conspecific pollen grains residing on the feathers and fur of birds and mammals on their stigmas while providing a nutritional reward to their pollinators. As a result of these interactions, well-known morphological (*sensu lato*) syndromes associated with this mutualism have evolved in both plants and animals. (Though well known, these morphological syndromes are controversial—see Ollerton et al. [2009] and Armbruster et al. [2011] and included references) Botanical traits associated with bird and bat pollination are summarized in table 1.4. Bird-pollinated flowers (or more simply bird flowers) usually differ from insect or bat flowers in one or more of the following characteristics: time of opening, size and shape of the flower and the color of its corolla, floral scent (or lack thereof), and quantity and quality of the nutritional reward, among other traits. Not reflected in this table are important geographic differences in this syndrome that are associated with biogeographic and historical differences in the kinds of birds and mammals that visit flowers. As discussed by Fleming and Muchhala (2008), for instance, the characteristics of flowers visited by Neotropical hummingbirds and Australasian honeyeaters often differ substantially as do the flowers visited by New World glossophagine and Old World pteropodid bats. In reality, there is not a universal set of characteristics that describe all bird-or bat-pollinated flowers. Thus, flower characteristics are contingent on which particular birds and mammals interact with particular plants (Fleming et al. 2009). Therefore, in addition to elucidating the general trends associated with the evolution of vertebrate-pollinated flowers, we will discuss geographic and historical (phylogenetic) variation in these trends.

After describing in detail the flower syndromes associated with vertebrate pollination, we describe complementary and convergent adaptations in nectar-feeding birds and mammals. Thus, most nectar-feeding birds and mammals have elongated bills or rostrums and long, brush-tipped tongues, and many other aspects of their biology differ from those of their nonnectarivorous relatives. Again we seek to determine the general trends associated with the evolution of these feeding adaptations in tropical birds and mammals and to describe geographic and phylogenetic variations in these trends. Ideally, we would like to be able to use phylogenetic analyses to ascertain the mechanisms behind these trends for both plants and animals. In reality, however, such an approach for studying these mutualisms is in its infancy, especially on the animal side, and we will have to be content to present a few exemplary cases here. Much more research needs to be done before we have a complete understanding of the evolution of these mutualisms.

The Evolution of Vertebrate-Pollinated Flowers

As reviewed in chapter 1, pollination by vertebrates occurs in at least 29 orders and 70 families of flowering plants. A detailed analysis of bat pollination indicates that bat flowers occur in 28 orders, 67 families, and about 530 species of angiosperms (Fleming et al. 2009). These families are distributed throughout angiosperm phylogeny except for the earliest evolving lineages in the basal families (chap. 6) in which no genera have flowers pollinated by birds or bats. Most angiosperms are pollinated by insects, although wind pollination dominates in a few families, such as those in the Fagales and Poales. In most lineages of animal-pollinated tropical angiosperms, insect pollination is the ancestral condition and vertebrate pollination is a derived condition. The question then becomes, what evolutionary changes are involved in the transformation of an insect-pollinated flower into a bird- or bat-pollinated flower? Are these changes easy or difficult to make either genetically or developmentally and do they limit the extent to which plants interact with nectar-feeding vertebrates? To what extent are these changes constrained by phylogeny (see chap. 5)? Do the characters involved in these changes evolve in coordinated or patchwork fashion? For example, does flower morphology change before nectar quality and quantity or vice versa? Using table 1.4 as our guide, we will discuss eight of these characters: time of flower anthesis, seasonal phenology, flower size, flower shape and method of presentation, flower color, flower scent, nectar, and pollen. In addition, we will discuss the role that tropical vertebrates have played in the evolution of plant sexual and mating systems. To set the stage for this discussion, we begin by reviewing the evolution of pollination systems in three groups of plants at different taxonomic levels: in the order Zingiberales (an advanced lineage of tropical monocots), in the family Gesneriaceae (a common herbaceous and shrubby tropical group), and in the genus *Ruellia* (the most speciose tropical genus in the Acanthaceae with exceptional floral diversity). Each of these case studies provides an example of the types of character transformations that take place in adapting to vertebrate pollinators. We use these examples because evolutionary transitions within them are especially well-studied and represent model systems for understanding the evolution of tropical vertebrate pollination systems.

Pollinator Diversification and Flower Evolution in the Tropics: Case Studies in the Zingiberales, the Gesneriaceae, and *Ruellia*

THE ORDER ZINGIBERALES

The wide variation in the basic floral ground plan among the eight families of the Zingiberales (fig. 7.1) suggests that a significant diversity exists in the floral biology and pollination systems of these plants. The presence of conspicuous petals and sepals arranged around the five fertile, pollen-producing stamens in the banana families (Musaceae, Strelitziaceae, Lowiace, and Heliconiaceae) contrasts strongly with the relatively inconspicuous petals and sepals and reduction to a single fertile stamen in flowers of the four ginger families (Zingiberaceae, Costaceae, Canaceae, and Marantaceae). In these families the four remaining fertile stamens are modified into diverse sterile, petal-like organs. These contrasting floral structures provide the morphological basis for evolutionary transformations among and within the plant families of the various types of pollination mechanisms, including adaptation to visitation by both vertebrates and insects (fig. 7.2).

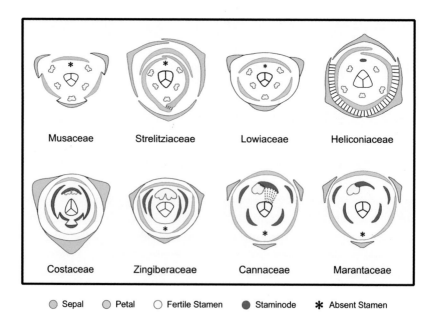

Musaceae   Strelitziaceae   Lowiaceae   Heliconiaceae

Costaceae   Zingiberaceae   Cannaceae   Marantaceae

○ Sepal   ○ Petal   ○ Fertile Stamen   ● Staminode   ✱ Absent Stamen

Figure 7.1.   Floral diagrams representing the eight families of the Zingiberales showing perianth whorls, fertile stamens, staminodia, and carpels. (Not drawn to scale; from Kress 1990).

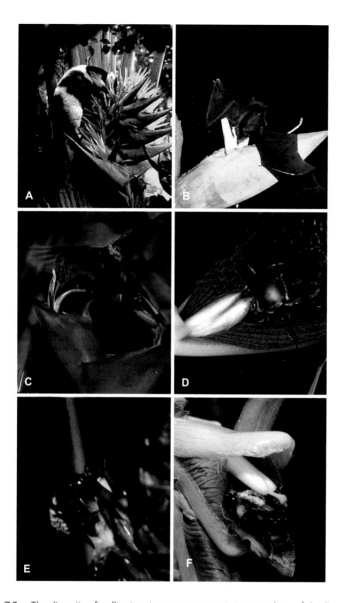

Figure 7.2. The diversity of pollinators in some representative members of the Zingibera-les. *A,* The black-and-white ruffed lemur (*Varecia variegata*) visiting the flowers of *Ravenala madagascariensis* (Strelitziaceae). *B, Phyllostomus hastatus* pollinating the flowers of *Phena-kospermum guyannense* (Strelitziaceae). *C,* The long-tailed hermit hummingbird (*Phaethornis superciliosus*) taking nectar from the flowers of *Heliconia trichocarpa* (Heliconiaceae). *D,* A dung beetle entering the flower of *Orchidantha inouei* (Lowiaceae). *E,* A euglossine bee hov-ering at an inflorescence of *Calathea latifolia* (Marantaceae). *F,* A carpenter bee visiting the flowers of *Alpinia* (Zingiberaceae). Photo in *D* provided by S. Sakai; all other photos by W. J. Kress.

VERTEBRATES.—All of the major groups of known vertebrate pollinators have been shown to be important floral visitors of various taxa in the Zingiberales (Kress and Specht 2005). Both pteropodid and phyllostomid nectar-feeding bats pollinate flowers of species in several families of the order, especially in the banana families. In Asia some of the earliest accounts of pollination by bats documented the relationship between nectar feeding fruit bats and the genus *Musa* and possibly *Ensete* (Itino et al. 1991; Nur 1976; van der Pijl 1936). Pollination of South Pacific species of the genus *Heliconia* by pteropodids has also been shown (Kress 1985). In the Neotropics, the Amazonian monotypic *Phenakospermum* of the Strelitziaceae is dependent on phyllostomid bats for successful pollination (Kress and Stone 1993). Although field observations are still lacking, specialized floral characteristics such as large flower size and cuplike shape, sweet floral odor, and nocturnal flowering, also suggest that some members of the Zingiberaceae (e.g., species of *Alpinia* in the Pacific islands) and Cannaceae (e.g., *Canna liliiflora* of Bolivia and Peru) may also be primarily bat-pollinated (J. Kress, unpubl.).

Pollination by nonflying mammals has also been demonstrated for some members of the Musaceae and the Strelitziaceae. In Madagascar, the endemic *Ravenala madagascariensis* is pollinated by lemurs, especially the black-and-white ruffed lemur (*Varecia variegata*). These animals extract large amounts of nectar from the robust flowers, which are held under tension until touched by the animals; the five anthers open explosively, transferring pollen onto the fur (Kress et al. 1994). In addition, tree shrews (species of *Tupaia*) often visit flowers of wild species of *Ensete* and *Musa* in Asia (Itino et al. 1991; Nur 1976), although their effectiveness as pollinators has not been adequately documented. Floral characteristics and preliminary observations suggest that tree shrews also visit flowers of at least one species of the genus *Orchidantha* (*O. fimbriata*; Lowiaceae) in the lowland wet forests of Malaysia (J. Kress, unpubl.), but as in the Musaceae, data on actual pollination are absent. Interestingly, in contrast to all other members of the order, these flowers lack nectaries. Instead, they attract floral visitors with a specialized medial petal called the labellum, which is modified into a fruitlike structure that produces a sweet, sugary substance and probably serves as the attractant and reward for these scansorial mammals (see below for observations of insect pollination in the Lowiaceae).

Birds are the most diverse group of vertebrate pollinators for members of at least seven of the eight families of the Zingiberales. Sunbirds (Nectariniidae) pollinate the blue, white, and orange flowers of species of *Strelitzia* in South Africa (Frost and Frost 1981) and many species of the genus *Musa* in Asia that have erect, brightly colored pink and red inflorescences (Itino et al. 1991; Kato 1996; Liu et al. 2001). In Borneo, sunbird and spiderhunter pollination has been reported for a number of wet forest understory genera of the Zingiberaceae with red and yellow floral bracts and flowers, including *Amomum, Etlingera, Hornstedtia,* and *Plagiostachys* (Classen 1987; Kato 1996; Kato et al. 1993; Sakai et al. 1999). The wattled honeyeater of Samoa and Fiji (Meliphagidae: *Fulihao carunculata*) is the primary pollinator of two native species of *Heliconia* (Pedersen and Kress 1999). Honeyeaters have also been documented as pollinators of the genus *Hornstedtia* in tropical Australia (Ippolito and Armstrong 1993) and may play a role in the pollination of the poorly known genus *Tapeinochilos* in the Costaceae (Gideon 1996).

Hummingbirds (Trochilidae) have evolved the most intricate relationships as pollinators in several families of the Zingiberales, primarily the Heliconiaceae, Costaceae, Zingiberaceae, and Cannaceae. In the genus *Heliconia*, the ecology and evolution of the features of both the hummingbirds and the plants that link them in a highly coevolved relationship are well-documented (e.g., Feinsinger 1983; Kress 1985; Kress and Beach 1994; Linhart 1973; Stiles 1975; Temeles and Kress 2003; Temeles et al. 2000). The length and curvature of the floral tubes, position of anthers, daily phenological patterns, and nectar volume and composition are all closely associated with hummingbird bill size and shape as well as the pollinator's energetic requirements.

INSECTS.—Even though pollination by vertebrates is common in the Zingiberales, only two families, the Strelitziaceae and Heliconiaceae, are exclusively vertebrate pollinated. Many different groups of insects play important roles as pollen vectors in the remaining six families. Only relatively recently have beetles been conclusively demonstrated to pollinate the flowers of species of *Orchidantha* in the Lowiaceae. These flowers have no nectaries and deceive their pollinators by producing a foul, dunglike odor, thereby attracting dung beetles to their flowers (Sakai and Inoue 1999). It has been suggested that in other species of the genus the white labellum mimics a fungus that also deceives specialized beetles, which normally lay their eggs in fungi, to visit the flowers (L. Pedersen, unpubl.). Another species, *O. chinensis*, also produces foul odors, but these flowers attract small dung flies that mate in the flowers and appear to facilitate pollination (J. Kress, unpubl.). Fly pollination has not been reported in other Zingiberales nor has floral visitation by lepidopteran insects (butterflies and moths) been adequately documented in these plants. In some genera in the Zingiberaceae, such as *Hedychium*, *Hitchenia*, and *Curcuma*, flowers possess long floral tubes, open in the evening, and may produce a strong, sweet fragrance—all characteristics suggesting pollination by long-tongued hawkmoths (Mood and Larsen 2001; J. Kress, unpubl.).

Bees are by far the most common insect pollinators at the species level in the Zingiberales and probably are responsible for pollination in the majority of genera in the order. Bee pollination is present in only one genus of the banana families, *Musella* in the Musaceae where it is clearly secondarily derived from vertebrate-pollinated ancestors (Liu et al. 2002). Pollination by a number of families of bees is found in all four of the ginger families and probably occurs pantropically, although it is surprisingly poorly documented, especially in Africa. In Southeast Asia, pollination by small halictid and medium-sized anthophorid (*Amegilla*) bees has been demonstrated in a number of genera of the Zingiberaceae (e.g., *Alpinia*, *Amomum*, *Boesenbergia*, *Elettaria*, *Elettariopsis*, *Globba*, *Plagiostachys*, *Zingiber*), Costaceae (*Costus*), and Marantaceae (*Phacelophrynium*, *Stachyphrynium*; Kato 1996; Kato et al. 1993; Sakai et al. 1999). In the Neotropics, pollination by bees is especially common in species of *Costus* (Costaceae; Schemske 1981; Sytsma and Pippen 1985), *Renealmia* (Zingiberaceae; Kress and Beach 1994; Maas 1977), and at least a few species of *Canna* (Cannaceae; Kress and Beach 1994). Members of the Marantaceae possess very specialized and highly complex flowers with explosive secondary pollen presentation and are pollinated almost exclusively by euglossine bees in the American tropics and *Amegilla* and halictid bees in the Old World tropics (Kennedy 1978, 2000). Two staminodia, which are petaloid in flowers of the other three ginger families, have evolved into a triggerlike mechanism in flowers of the Marantaceae. These structures allow only a single effective pollinator visit to a flower and may promote outcrossing in these plants (Kennedy 1978).

GENERAL PATTERNS OF FLORAL EVOLUTION AND DIVERSIFICATION OF POLLINATION SYSTEMS.—By mapping the types of pollinators onto a phylogenetic tree of the Zingiberales, a broad picture of the evolutionary origins and transitions between the different pollinator types emerges (fig. 7.3). Vertebrate pollination is clearly linked to large flowers with highly conspicuous sepals and/or petals, copious nectar, five or six fertile stamens, and abundant pollen. These flowers are in every case enclosed in tough floral bracts that

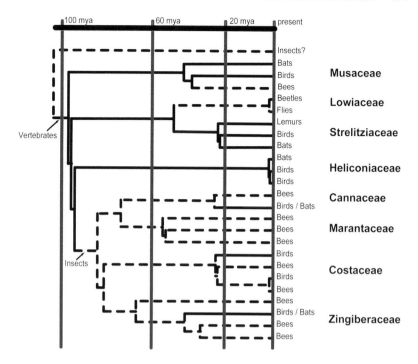

Figure 7.3.   Chronogram of the Zingiberales, with major pollinators indicated for each clade. Terminal taxa have been replaced with pollinator types known to be found in each family. Incidental or unsubstantiated reports of pollinators are not included here but are discussed in the text. *Solid lines* indicate vertebrate pollination; *broken lines* indicate insect pollination. Modified from Kress and Specht (2005).

protect them from potentially rough pollinators. Taxa with these characteristics, including heliconias, bananas, and birds of paradise, are pollinated by both birds and bats. In contrast, the transformation to smaller flowers in the ginger families, in which the sepals and petals are inconspicuous and most of the fertile stamens are modified into showy petal-like structures, is associated with insect pollination. These latter flowers produce much less nectar and pollen. Although more field work and observations are needed, the most complex floral mechanisms, such as the trip flowers of the Marantaceae in which the stamens have been transformed into nonfertile floral structures (Kennedy 1978) and flexistyly in the gingers (Li et al. 2001), have evolved in insect-pollinated lineages.

In contrast to this broad-scale documentation of pollinator and flower phylogenetic transformation across the order Zingiberales, Sakai and colleagues (1999) have investigated the ecological partitioning of floral biology in a sympatric group of gingers in the lowland wet forests of Sarawak. Forty-four species of the Zingiberaceae and the closely related Costaceae occur at this study site. Of those species, Sakai et al. were able to make observations on the pollinators of 29 taxa in eight genera. Each species was pollinated either by nectar-feeding spiderhunters (Nectariniidae), medium-sized bees in the genus *Amegilla* (Anthophoridae), or small halictid bees (Halictidae). Using canonical discriminant analyses, they demonstrated that at least four floral characters of the plants are related to the three specific pollinator types: width of the petaloid labellum, width of the stigma, length of the single anther, and length of the floral tube (see fig. 7.12A). Other characters that were linked to specific pollinators included significantly greater amounts of nectar secreted by flowers pollinated by spiderhunters, the red color of the spiderhunter flowers versus orange, yellow, or white flowers of the bee-pollinated taxa, and the number of flowers per inflorescence. This study demonstrates that pollinators, including vertebrates and invertebrates, may also partition floral types across ecological space similarly to the phylogenetic partitioning that has occurred at the ordinal level in the Zingiberales.

With regard to the timing of pollinator diversification in the Zingiberales, based on the prevalence of vertebrate pollinators (bats, birds, and nonflying mammals) in the three basal banana-family lineages (Musaceae, Strelitziaceae + Lowiaceae, and Heliconiaceae), parsimony suggests that the late Jurassic and early Cretaceous common ancestor of the order (some 150 Ma ago) was also pollinated by an early vertebrate taxon. However, it is difficult to envision which vertebrates may have been pollinators at that time because most vertebrate groups that serve as extant pollinators evolved much later in the Tertiary (Bleiweiss 1998a; Nowak 1991; Sibley and Ahlquist 1990). Possible candidates include early mammalian multituberculates or even small dinosaurs. Although this conclusion is not unreasonable, little evidence exists to support it.

If vertebrate pollination represents the plesiomorphic state in the Zingiberales, then pollination by insects is a derived condition in the specialized ginger families as well as in the derived genera *Musella* (Musaceae) and *Orchidantha* (Lowiaceae), and some derived species of *Canna* (fig. 7.3). This evolutionary pattern, in which ancestral taxa are pollinated by vertebrates and derived taxa by insects, contradicts the generally accepted notion that insect pollination systems have usually given rise to more specialized bat- and bird-pollinated taxa (e.g., Faegri and van der Pijl 1979). In some families and genera in the order, this latter pattern is indeed the case. For example, in three of the families where most taxa are pollinated by insects, bird-pollinated taxa (Zingiberaceae: *Etlingera, Hornstedtia, Amomum*; Costaceae: *Costus* sect. *Ornithophilus, Tapeinochilos*; and Marantaceae: *Calathea timothei*) and bat-pollinated taxa (e.g., Zingiberaceae: a few species of *Alpinia*) appear to have evolved independently from insect-pollinated taxa. Thus, vertebrate pollination is both a basal and a derived condition in the Zingiberales.

THE FAMILY GESNERIACEAE

Our second case study highlights the transformation of pollination systems between genera and species within a single family of tropical plants. Members of the Gesneriaceae possess a diversity of floral types differing in color, shape, size, and timing of anthesis. In a study centered in the Caribbean islands, Martén-Rodríguez et al. (2009) demonstrated that species in the tribe Gesnerieae possess suites of flower features that correspond to pollination by either bats or hummingbirds and in a few cases, pollination by both types of vertebrates. In a multidimensional scaling analysis of 19 species in three Caribbean genera (plus four species from Costa Rica), they found that four species were specialized for bat pollination and had flowers with wide yellow or green corollas, low nectar concentrations, and nocturnal anthesis (see fig. 7.12*B*). In contrast, nine of 10 species that were exclusively hummingbird-pollinated had flowers with red or orange tubular corollas, more concentrated nectar, and diurnal anthesis. A few species with flowers that resembled bat-pollinated species but were distinguished from them by a distinctive constriction in the corolla were visited by both nocturnal and diurnal animals.

When additional species from Costa Rica classified in other tribes of the Gesneriaceae were analyzed in the same fashion, Martén-Rodríguez et al. (2009) were able to classify both bat- and hummingbird-pollinated species into the same floral categories as defined for the Caribbean taxa. A phylogenetic study based on DNA and morphological evidence indicated that hummingbird pollination and its associated flower characteristics are ancestral in the Gesnerieae and that bat-pollinated flowers or flowers pollinated by hummingbirds, bats, and bees have each evolved independently at least two times in this lineage (Martén-Rodríguez et al. 2010). Their observation that the same floral features can evolve numerous times within the Gesneriaceae supports the concept that convergent evolution has resulted in "pollination syndromes" in which the same suite of floral features has originated independently in many separate lineages of angiosperms as a result of natural selection and adaptation to particular types of pollinators.

THE GENUS *RUELLIA*

Perhaps the taxonomic level at which investigations can be most illuminating about natural selection and adaptation of floral characters associated with the transitions between pollination systems is among species within a single genus. *Ruellia*, the second largest genus in the primarily tropical family Acanthaceae, is exceptional in the diversity of its floral characters and respective pollinators. In their investigation of the spectrum of pollinators in *Ruellia*, Tripp and Manos (2008) demonstrated that four main floral types are associated with specific pollen vectors. Species visited by insects, primarily bees, possess diurnal flowers with short, open floral tubes and large purple lobes; their sexual organs are located inside the open tube. Hummingbird-pollinated species possess diurnal flowers with long red floral tubes with reflexed lobes; their stamens and stigma are located outside the floral tube. The few species that are pollinated by hawkmoths produce white flowers with very long corolla tubes with broadly spreading lobes and stamens positioned at the entrance to the flower. The fourth type is bat-pollinated species with nocturnal, yellow-to-green flowers possessing broadly open flowers with wide throats and long-exserted sexual organs.

Tripp and Manos (2008) analyzed the level of differentiation as well as the directionality of transformations among these four floral types using principal components analyses and phylogenetic hypotheses for 115 species of *Ruellia* (and one outgroup species). Their results for this relatively recent and rapidly evolving lineage of tropical plants demonstrated that floral evolution has been highly labile with numerous transitions between floral and pollinator types (fig. 7.4). Most interesting was that although transformations between some floral types, such as from hummingbird to bee pollination, bee to moth pollination, and hummingbird to moth pollination, were relatively frequent, other transitions were either infrequent, such as bee to bat pollination, hummingbird to bat pollination, and bee to hummingbird, or nonexistent, such as moth to bat, hummingbird, or bee pollination and bat to hummingbird, bee, or moth pollination. They interpreted these infrequent or nonexistent transformations as indications that bat and moth pollination were evolutionary dead ends as pathways to develop other floral types and pollination systems. As we will see, however, this conclusion is not universally true among vertebrate-pollinated species.

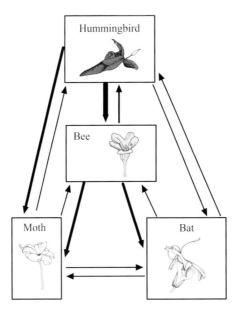

Figure 7.4. Evolutionary transitions among pollination systems in *Ruellia* (Acanthaceae). Width of the lines indicates the relative probability of each transition. Based on Tripp and Manos (2008).

Another of their conclusions was that each of these four pollination types has evolved multiple times and that the number of transitions to a particular type was dependent on ancestral states. That is, species with red or purple flowers and pollinated by hummingbirds or bees, respectively, were ancestral to other floral systems much more commonly than either moth- or bat-pollinated taxa, which were ancestral to no other floral types. Tripp and Manos (2008) point out that their results indicate that some pollination systems, such as hummingbird pollination, which were formerly considered to be specialized mutualisms that were only derived from insect pollination (e.g., Stebbins 1989), are in fact quite labile and can give rise to many other types of pollinator interactions. This observation is similar to the conclusion of Kress and Specht (2005) for the order Zingiberales that insect pollination is in general a derived state originating from vertebrate-pollinated ancestors, although this is not universally true in this tropical plant group (see above).

These three case studies demonstrate that, by combining phylogenetic and statistical analyses with morphological data and ecological investigations, robust and dynamic models of evolutionary diversification and character transformation in both flowers and their pollinators can be developed. Additional studies of plant-pollinator systems that employ such methodology are described throughout this chapter.

The Evolution of Flower Characteristics

FLOWER ANTHESIS AND SEASONAL PHENOLOGY

Birds and mammals differ in their daily activity cycles and hence we expect to see differences in the timing of anthesis in the flowers they pollinate. Like many insects (e.g., bees and butterflies), birds are diurnal and so are their flowers. Anthesis generally occurs at or after sunrise and most pollination of these flowers occurs during daylight hours. In contrast, most moths and mammals are nocturnal, and their flowers do not begin to open until or just after sunset. Not surprisingly, flowers that rely on both birds and bats for pollination (e.g., the columnar cacti *Carnegiea gigantea* and *Weberbauerocereus weberbaueri*) are open during the day as well as at night (Fleming et al. 1996; Sahley 1996).

As is the case for most tropical flowers, most bird- or bat-pollinated flowers usually last only a single day (or night—sometimes for as little as six hours in the bat-pollinated *Oroxylum indica*; Gould 1978). When bat flowers remain open for more than one day, they are usually protandrous (e.g., species of *Agave* and *Burmeistera* and the palm *Calyptrogyne gheisbreghtiana*; Cunningham 1995; Howell 1979; Muchhala 2006a), which is not generally the case with bird flowers. Although individual flowers of tropical plants tend to be short-lived, the inflorescences of species whose flowers are pollinated by traplining animals (e.g., certain bees, hummingbirds, and bats) tend to be long-lived (Endress 1994).

As discussed in chapter 3, the duration of flowering seasons in vertebrate-pollinated plants generally reflects plant growth habits and/or position with respect to forest canopy. Canopy trees, as well as columnar cacti, tend to have relatively short "cornucopia" flowering seasons (*sensu* Gentry 1974) whereas understory shrubs and epiphytes tend to have prolonged "steady state" flowering seasons. The beginning and end of flowering seasons is known to have a genetic basis in a variety of plants, and Sandring and Ågren (2009) have demonstrated experimentally that pollinators of the perennial temperate herb *Arabidopsis lyrata* can provide positive phenotypic selection for timing of the end of the flowering season as well as number of flowers per plant; pollinators apparently do not provide selection on flower petal size in this species. In general, we expect selection on plant reproductive traits provided by pollinators to increase as pollinator limitation increases, and pollinator limitation is widespread in plants (chap. 4). Therefore, we expect that experimental studies with tropical plants will indicate that pollinators play an important role in the evolution of when they flower as well as many other floral traits. Such research is yet to be done.

In addition to pollinators and seed dispersers, seed predators can help determine the flowering and fruiting schedules of plants. The importance of predators can perhaps be seen most clearly in the masting flowering and fruiting behavior of Southeast Asian dipterocarps. As discussed in chapter 3, these plants and many others in their communities flower and fruit every 2–10 years. In their analysis of the demography of the West Malaysian dipterocarp *Shorea leprosula*, Visser et al. (2011) used stochastic matrix population models to show that under conditions of intense seed predation by insects and vertebrates (which is the case in these forests), mast fruiting results in higher individual reproductive success and higher population growth rates than annual flowering and fruiting. Dipterocarps do not use vertebrates to disperse their seeds, so mast fruiting in these plants does not lower the carrying capacity of habitats for vertebrate frugivores. In contrast, in plants that rely on animals to disperse their seeds, selection is less likely to favor mast fruiting unless annual levels of seed predation are extremely high. While these results are intuitively pleasing, they still raise the question of why so many animal-dispersed trees undergo mast fruiting in synchrony with dipterocarps in lowland forests of Southeast Asia. Clearly, there is a need to look at the evolution of masting phenological behavior at the community level in this area.

FLOWER SIZE AND SHAPE

Vertebrate-pollinated flowers come in a wide variety of sizes and shapes, reflecting their broad distribution across angiosperm phylogeny (Endress 2011; chap. 6). Various authors have tried to organize this variation into discrete morphological types, and here we will focus on the system used by Dobat and Peikert-Holle (1985) in their detailed discussion of bat flowers. Dobat and Peikert-Holle (1985) recognized 12 different flower types that ranged from "shaving brushes," characterized by many long stamens and very accessible nectar, to tubular or bell-shaped flowers with shorter, less numerous stamens and more restricted access to nectar (by legitimate pollinators). For simplicity, Fleming et al. (2009) broke this classification down into three basic types for bat flowers: (1) "shaving brush" or "stamen ball" flowers with many projecting stamens (e.g., many Mimosoideae and Myrtaceae); (2) "bell-shaped" flowers whose corollas form a narrow tube (e.g., *Bauhinia, Musa, Vriesea*), and (3) "cup-shaped" flowers with open corollas (e.g., many columnar cacti, *Caryocar, Ipomoea*, and various Bombacoideae). Nectar is accessible to a wider array of potential pollinators in flower classes 1 and 3 than in class 2. Many of the flowers visited by morphologically specialized vertebrate pollinators such as hummingbirds, sunbirds, and glossophagine bats are tubular in shape (class 2) whereas those visited by more generalized honeyeaters, lorikeets, pteropodid bats, and opportunistic phyllostomid bats are more likely to belong to classes 1 and 3 (Fleming and Muchhala 2008; Fleming et al. 2009; Helversen 1993). In their study of several species of African bat-pollinated flowers, Pettersson et al. (2004) remarked that, compared with the many tubular flowers pollinated by glossophagine bats, African flowers pollinated by pteropodid bats are less specialized morphologically and are accessible to a wider variety of pollinators. They suggested that echolocating glossophagines have provided stronger selective pressure for unique floral morphologies and scents than have visually orienting pteropodids. Alternatively, many tubular flowers pollinated by glossophagine bats are derived from tubular hummingbird flowers (e.g., in Acanthaceae, Bromeliaceae, Campanulaceae, Gesneriaceae, etc.) so that tubular phenotypes have not had to evolve de novo in many bat-pollinated flowers (reviewed in Fleming et al. 2009).

The sizes of vertebrate-pollinated flowers as measured by corolla length or width generally correlate with the size of their pollinators and tend to be larger than related insect-pollinated flowers. Representative size ranges of related bird and bat flowers are summarized in table 7.1. Because bats are larger than most avian pollinators, their flowers also tend to be larger in overall size and are broader and shallower to accommodate the snouts of the bats. Conversely, bird flowers, especially species pollinated by hummingbirds with exceptionally long bills and tongues, are more tubular, with narrower and longer perianths, which often restrict other (legitimate) floral visitors from accessing nectar at the base of the tube. Of course this is not always the case, as exemplified by the exceptionally long-tongued *Anoura fistula* that pollinates exceptionally long tubular flowers in the genus *Centropogon* (Muchhala 2006b; table 7.1). The vertically oriented flowers of sunbird-pollinated flowers of African Marantaceae have a deterrent effect on bees specialized for pollinating horizontally oriented flowers (Ley and

Table 7.1. Mophological Characteristics Of Bird- or Bat-Pollinated Flowers in Selected Genera and Families of New World Plants

Genus and/or Family/ Pollinator	No. of Species	Corolla Length (mm)	Corolla Width (mm)	Shape	Color	Source
Vriesia, Bromeliaceae:						
Bats	4	27.4–43.4	...	Tubular	Orange or yellowish white	Sazima et al. (1999)
Hummingbirds	3	31.0–50.0	...	Tubular	...	Buzato et al. (2000)
Pachycereus, Cactaceae:						
Bats	5	70–100	...	Funnel-shaped or tubular	White	Anderson (2001)
Hummingbirds	1	40	...	Funnel-shaped	Red	
Stenocereus, Cactaceae:						
Bats	4	60–120	...	Funnel-shaped or tubular	White to pale rose	Anderson (2001)
Hummingbirds	1	100	30	Tubular	Red	
Burmeistera, Campanulaceae:						
Bats	9	7.7–14.1	5.6–10.4	Tubular	Various, dull	Muchhala (2006a)
Hummingbirds	1	7.2	10.2	Tubular	Red	
Gesneriaceae, 6 genera:						
Bats	8	18.1–39.7	7.8–16.0	Campanulate/ subcampanulate	Green or yellow	Martén-Rodríguez et al. (2009)
Hummingbirds	15	15.7–34.8	2.5–12.6	Tubular/ subcampanulate	Red, orange, yellow	
Abutilon, Malvaceae:						
Bats	1	34.2	...	Bell	Pinkish-purple	Sazima et al. (1999)
Hummingbirds	3	16.0–32.7	...	Dish	...	Buzato et al. (2000)
Marcgravia, Marcgraviaceae:						
Bats	2	12.2–20.5	...	Brush	White	Sazima et al. (1999)
Norantea, Marcgraviaceae:						
Hummingbirds	1	23.8	...	Dish		Buzato et al. (2000)
Hillia, Rubiaceae:						
Bats	1	63.6	...	Tubular	...	Sazima et al. (1999)
Rubiaceae, 2 genera:						
Hummingbirds	4	11.9–42.9	...	Tubular	...	Buzato et al. (2000)

Classen-Bockhoff 2009). Flower size, shape, and orientation thus can serve as "pollinator filters" in the evolution of pollinator syndromes.

Comparisons between vertebrate-pollinated flowers and close relatives that are not vertebrate pollinated provide us with insights into the evolutionary transitions involved. Species-level phylogenies are critical for this. For example, in the study cited earlier by Tripp and Manos (2008) on pollinator transformations within the genus *Ruellia*, the transition from bee to bat or hummingbird pollination always entails the enlargement of the flowers whereas the transition from hummingbird to bee pollination always entails reduction in corolla length. Similar are transformations of flower color in which some changes are likely and labile, such as between red hummingbird flowers and purple bee flowers, and others never occur, such as the evolution of red or purple flowers from bat-pollinated white or green flowers. In *Burmeistera* (Campanulaceae), the transformation from bat to hummingbird pollination involves a shift in corolla length and width, degree of stigma exsertion, and flower color and odor (Muchhala 2006a). As more species-level investigations using phylogenetic hypotheses are completed, the evolutionary rules governing the directionality of these changes in floral size, shape, and color will become better defined and tested.

MODES OF FLOWER PRESENTATION

Flowers are presented to their pollinators in many different ways, including whether flowers occur solitarily or are grouped in inflorescences, whether they are located close to or away from foliage (e.g., via flagelliflory), and whether they include strong perches (reviewed in Endress 1994; Faegri and van der Pijl 1979; Westerkamp 1990). All of these features have evolved independently many times within angiosperms and likely reflect the size and foraging behavior of different kinds of pollinators. In plants pollinated by hovering birds and bats, for example, solitary flowers prevail; when their flowers occur in inflorescences, only one flower usually opens per day or night. Bat flowers tend to be positioned farther away from foliage—sometimes dramatically so—than bird flowers (e.g., in *Marcgravia*; Tschapka et al. 2006; see additional examples in Fleming et al. 2009). In contrast, flowers pollinated by perching birds usually occur in groups in which many open at once and are accessible from a single perch. These flowers are usually less tubular than those pollinated by hovering vertebrates. Perches used by birds are positioned outside inflorescences on the ground, on a nearby structure, or within the inflorescence itself, either at its base or distal end. A dramatic example of a ground-level perch adapted for perching-bird pollination occurs in the "rat's tail" plant, *Babinia ringens* (Iridaceae), in South Africa. Anderson et al. (2005) demonstrated experimentally that sunbirds pollinate this plant's flower more effectively from the plant's tail-like sterile inflorescence than from the ground. In addition to perches on or near inflorescences, large flowers (e.g., those of *Strelitzia*) sometimes provide perches, as can buds, spent flowers, or pedicels.

To our knowledge, phylogenetically based studies of inflorescence structure in relation to pollinator type have rarely been conducted. One example is the century plant *Agave*, which contains two subgenera, *Littaea* and *Agave*. *Littaea*, whose bee-pollinated flowers occur in a single tall racemose spike, is considered to be ancestral to *Agave*, whose bat-pollinated flowers occur in branched (paniculate) umbels (Rocha et al. 2006). Likewise, in pantropical *Erythrina* (Fabaceae), inflorescence structure is correlated with kinds of avian pollinators. According to Bruneau's (1997) phylogenetic analysis, pollination by passerine birds is basal, and hummingbird pollination has evolved independently at least four times. Inflorescences with flowers pollinated by perching birds are oriented horizontally, and their flowers are robust and produce sticky pollen. Those with flowers pollinated by hummingbirds are oriented vertically on a long peduncle; their floral standards (enlarged outer petals) form a pseudotube that encloses the anthers and stigma; and their pollen is dusty, not sticky (Bruneau 1997; Westerkamp 1990). These examples as well as the existence of multiple inflorescence types in many angiosperm families indicate that, like many other floral features, inflorescence structure is evolutionarily labile and is responsive to selection pressures exerted by pollinators.

FLOWER COLOR

Flowers come in many different colors, reflecting, in part, the diversity of their animal pollinators. Major flower pigments include anthocyanins, favonols, and carotenoids, and there is a multitude of ways by which plants produce different flower colors using these pigments (Lee 2007). Most work on the evolution of flower colors has focused on biosynthetic pathways associated with anthocyanin-based pigments, and little is currently known about the evolution of carotenoid-based floral pigments (Rausher 2008).

As discussed in detail below, birds and mammals have different kinds of color vision, and birds can see a wider range of colors than most mammals. Thus, we would expect the color palette of their flowers to be much more diverse than that of bats. Red is the flower color most often associated with bird pollination, perhaps because long visible wave lengths fall outside the visual spectrum of most insects (Chittka and Thomson 2001; Chittka et al. 1994; Cronk and Ojeda 2008). Birds are thus less likely to compete with bees for nectar when their flowers are red. Bird flowers are not exclusively red, of course, but actually come in a wide variety of colors, including shades of yellow and orange as well as white and green. However, even if a flower is not red in color, often an associated plant part is red, such as a floral bract, inflorescence axis or even a red spot on a leaf positioned near the flower as is found in many gesneriads.

Research on the evolution of flower color in tropical bird-pollinated plants is in its infancy. It is much more advanced for North American temperate plants. In many of these plants, bee pollination of blue or purple flowers is ancestral and red hummingbird flowers are derived. Rausher (2008) and Streisfeld and Rausher (2009) review studies of the genetic basis for this transition. Macroevolutionary trends involved in this transition include the transition from blue to red flowers is asymmetric (i.e., it is biased from blue to red rather than the reverse); a similar bias occurs in the transition from pigmented to white flowers (e.g., in the evolution of bat flowers). This transition also involves inactivation of branches of the anthocyanin biosynthetic pathway via loss-of-function mutations. Although blue-to-red and pigmented-to-white transitions are biased, reversals are also known (Kay et al. 2005; Martén-Rodríguez et al. 2010; Muchhala 2006a; Perret et al. 2003; Smith et al. 2008). In plants of the genus *Ipomoea* (Convolulaceae), for example, there are three possible biosynthetic pathways in the transition from cyanidin-type (blue/purple) to pelargonidin-type (red) anthocyanin pigments. Using phylogenetic reconstruction and character mapping, Streisfeld and Rausher (2009) showed that red flowers have evolved independently four times in clade Astripomoeinae of the tribe Ipomoeeae and that the same biochemical pathway produced this change in at least three of these transitions—an example of evolutionary parallelism at the biochemical level.

Although it is generally believed that pollinators select for the color of their flowers, Rausher (2008) warns us that this is not necessarily true. Other potential factors involved in flower color evolution include pleiotropy of flower color genes (i.e., selection acts directly on other aspects of color genes that affect fitness and only indirectly on flower color itself) and selection from floral herbivores. Rigorous proof that pollinators are the major agents behind flower color evolution is currently lacking, but the ubiquity of flower color associations with particular kinds of pollinators makes it hard to believe that pollinators have not had an important selective effect on the colors of their flowers.

Most nocturnal mammals, including bats, have no or poorly developed color vision so we expect the colors of their flowers to differ significantly from those of bird flowers. Although white or light-colored flowers are common in bat plants, these colors are by no means universal. Other colors associated with bat flowers include brown, green, pink, fuchsia, and yellow (Fleming et al. 2009). This variation likely reflects the varied ancestry of bat flowers. For example, as discussed above, flowers of various species of *Ruellia* (Acanthaceae) are pollinated by bees, hawkmoths, hummingbirds, and bats. Phylogenetic analysis and character mapping has revealed that purple bee flowers or red hummingbird flowers are ancestral in two clades and that white or yellow bat flowers are derived in these clades (Tripp and Manos 2008). Whereas red hummingbird flowers can give rise to purple, white, and yellow flowers and purple flowers can give rise to white and yellow flowers, reversals from bat flowers to pigmented hummingbird or bee flowers apparently have not occurred. In these plants, specialization on bats through the loss of flower pigments is an evolutionary dead end (Tripp and Manos 2008). A similar situation occurs in tribe Sinningieae of Gesneriaceae in which bat pollination is also nonreversible (Perret et al. 2003). In contrast, *Burmeistera rubrosepala*, a red-flowered hummingbird flower, is derived from dull-colored bat flowers, which indicates that bat pollination is reversible in at least some plants (Muchhala 2006a).

FLORAL SCENTS

As described in detail by Knudsen et al. (2006), floral scents can be defined as a mosaic product of biosynthetic pathway dynamics, phylogenetic constraints, and balancing selection due to pollinator and florivore attraction. Scents include at least seven major classes of compounds: aliphatics, benzenoids and phenyl propanoids, C5-branched compounds, terpenoids, nitrogen-containing compounds, sulfur-containing compounds, and miscellaneous cyclic compounds. Having molecular weights of <300, most scent compounds are small and lipophilic. Several genes and enzymes localized in floral tissues are known to be involved in their biosynthesis. Compounds containing nitrogen or sulfur are the products of amino acid metabolism.

In his review of the ecology and evolution of floral scents, Raguso (2008) points out that floral scent is an ancient channel of communication between flowering plants and their pollinators and that scents should be recognized as being as important as visual cues (e.g., flower color and shape) in plant-pollinator interactions. He also reminds us that in addition to having a pollinator attraction function, floral scents can have an antiherbivore repellant function (e.g., via calyx odors). Knudsen et al. (2006) indicate that scent compounds can also be toxic to microbes and can provide protection against physiological stresses such as extreme temperatures. They suggest that the defensive and physiological functions of plant scents predate the origin of angiosperms because volatile compounds are also present in Gnetales, Cycadales, and Pinales. These compounds may have been used as mating-site signals for potential pollinators such as beetles. Finally, "floral scents have converged in chemical composition . . . across plant orders in species sharing a suite of morphological and phenological characters adapting them to pollination by one group of pollinators (pollination syndromes), for example, by moths or bats . . . or production has ceased, as in hummingbird-pollinated species" (Knudsen et al. 2006, 12). As we will see, the production of floral scents is phylogenetically labile, even at the level of species within a genus, which provides strong evidence that these complex chemical traits are adaptive and are responsive to selection pressures provided by specific kinds of pollinators.

The olfactory senses of nectar-feeding birds and mammals differ significantly and the scents (or lack of scent) of their flowers reflect this. Compared with relatively strong scents of bee- or moth-pollinated flowers, bird flowers typically produce no scent, a rare condition in flowers generally. Knudsen et al. (2004), for example, analyzed the floral scents of 17 species (in 14 families) of hummingbird-pollinated plants from a variety of habitats in Ecuador. They found no evidence of scent in nine species and only trace amounts of scent in the other eight species. Similarly, in their study of four species of Brazilian *Passiflora*, Varassin et al. (2001) reported that bee- or bat-pollinated species had two to four times more volatiles and a much higher diversity of scent compounds (16–78 compounds) than the single hummingbird-pollinated species (one compound) they studied. In contrast, Kessler et al. (2008) found that North American temperate zone hummingbirds were attracted to flowers of native tobacco (*Nicotiana attenuata*) containing benzyl acetone but were repelled by those containing nicotine, suggesting that at least some birds are responsive, both positively and negatively, to floral scents. In general, however, New World hummingbird-pollinated flowers lack scents.

The situation in Old World bird-pollinated flowers apparently has not yet been studied in detail. Azuma et al. (2002) analyzed the floral scents of eight species of Asian mangroves and found five floral scents in bird-pollinated *Bruguiera gymnorrhiza* (Rhizophoraceae). Compared with the flowers of bat-pollinated *Sonneratia alba* (Lythraceaae) and other mangroves, however, floral scents of *B. gymnorrhiza* were weak. Flowers of the mangrove *Kandelia candel* (Rhizophoraceae) are pollinated by bees and butterflies and produce a chemical (methyl anthranilate) that is apparently repellent to birds. Thus, as in the New World, there is evidence that some Old World flowers produce scents that possibly attract or repel nectar-feeding birds.

In contrast to birds, mammals live in a rich olfactory world, and their flowers often reflect this by having strong scents. The scent of bat flowers is often described as being unpleasant ("musky" or "skunky") to the human nose. An interesting biogeographic dichotomy exists in the scents of bat flowers. Flowers produced by New World bat plants often have a scent that is dominated by sulfur compounds, for example, by dimethyl disulfide or 2,4-dithiapentane. Sulfur-scented flowers are generally uncommon in nature but have evolved independently in many kinds of New World bat plants (Helversen et al. 2000; Knudsen et al. 2006). In contrast, most Old World bat flowers do not produce sulfur-rich scents. Exceptions to this include the Asian mangrove *Sonneratia alba* whose floral scent contains four major compounds, including the volatile sulfur compound 2,4-dithiapentane, and the African baobab tree, *Adansonia digitata*, whose sharp scent contains dimethyl disulphide and S-methyl esters (Azuma et al. 2002; Pettersson et al. 2004). Several other African bat flowers (e.g., *Maranthes aubrevillei*, *Pentadesma butyracea*, and *Parkia bicolor*) also contain trace amounts of S-containing compounds, but other compounds (e.g., ethyl acetate) are dominant in their scents (Pettersson et al. 2004). The New World–Old World dichotomy in the scents of bat flowers can even occur within a single genus, reflecting evolutionary plasticity in scent production. Flowers of some New World species of *Parkia*, for example, contain strong-smelling sulfur compounds whereas those of Old World species do not (Helversen et al. 2000; Pettersson and Knudsen 2001). An exception to this is Neotropical *P. pendula*, which is pollinated by a nonglossophagine bat (*Phyllostomus discolor*) and does not produce sulfur-rich floral scents (Piechowski et al.

2010). Experimental flower choice experiments using the glossophagine bat *Glossophaga soricina* suggest that a preference for sulfur-rich scents may be hardwired in these bats (Helversen et al. 2000). In contrast, when given a choice of seven different odor compounds found in its food items, the ptero-podid bat *Cynopterus sphinx* showed the strongest positive response to ethyl acetate and weakest positive response to dimethyl disulfide (Elangovan et al. 2006). While these results suggest that New and Old World plant-visiting bats differ fundamentally in their responses to the odors of flowers or fruits, tests of additional species of glossophagines and pteropodids will be needed before we can reach a firm conclusion about this.

FLORAL ACOUSTIC CHARACTERISTICS

In addition to vision and olfaction, nectar-feeding glossophagine bats have another sensory modality—echolocation—that they can use to locate flowers. Two studies coming from the laboratory of the late Otto von Helversen have revealed how plants can use auditory cues to attract pollinating bats. The bat-pollinated vine *Mucuna holtonii* (Fabaceae) has a typical papiliona-ceous flower with a keel, two lateral petals, and an upward-pointing trian-gular "standard" or vexillum, which is raised when the flower first opens. Bats have to land on a flower to obtain nectar and are hit with pollen from an exploding staminal column during the flower's first visit. Newly opened flowers release about 100 μL of nectar on first visit but only 10–20 μL on subsequent visits. Helversen and Helversen (1999) showed experimentally that the vexillum acts as a small concave mirror that reflects the energy from echolocation calls back toward the direction of incidence. Flowers whose vexilla had been removed or modified with a small cotton ball received about one-quarter the number of visits as intact flowers. These researchers concluded that these flowers produce acoustically conspicuous echoes that increase their attractiveness to bats. Interestingly, *Mucuna* flowers pollinated by African pteropodid (nonecholocating) bats lack a raised and concave vexillum.

The second example of an acoustically conspicuous flower involves an-other vine, *Marcgravia evenia* (Marcgraviaceae), from Cuba. Unlike the bat-pollinated Central American *Marcgravia* studied by Tschapka et al. (2006), this plant has a dish-shaped leaf located above the ring of flowers and nectar cups on each inflorescence. Simon et al. (2011) show that this leaf has unusual acoustic properties: its echoes are strong, multidirectional, and spatially invariant and differ significantly from echoes from general foliage.

Captive (non-Cuban) *Glossophaga* bats were able to find feeders equipped
with these "acoustic beacons" twice as quickly as those that lacked a "bea-
con." From this Simon et al. (2011) concluded that the dish-shaped leaves are
"signaling bracts" whose function is analogous to the conspicuous colors of
diurnally pollinated flowers. They predicted that more examples of this kind
of floral adaptation are likely to be found in the Neotropics but should be
absent in Old World bat-pollinated flowers.

NECTAR AND POLLEN CHARACTERISTICS

Nectar and pollen are the two floral resources harvested by vertebrate nectarivores. Because of their relatively large size, substantial energy budgets, and
large pollen-carrying capacities compared with insect pollinators, we might
expect to see evidence for strong selective pressures by birds and mammals
on the nectar and pollen characteristics of their food plants. Available data
strongly support this hypothesis for nectar but not for pollen.

NECTAR.—We learned in chapter 3 that vertebrate-pollinated flowers tend
to produce more nectar than related insect-pollinated flowers, reflecting
the general rule that nectar volume is positively correlated with flower size
(fig. 3.1). Like many other plant traits, nectar characteristics such as volume,
percent sugar, and sugar composition are evolutionarily plastic and can evolve
in response to selective pressures provided by pollinators. For example, bat-
pollinated flowers generally produce substantially larger volumes of nectar
than do bird-pollinated flowers (e.g., table 3.4), and vertebrate-pollinated
flowers tend to produce more dilute nectar than do related insect-pollinated
flowers (e.g., Baker and Baker 1983; Feinsinger 1987; Kim et al. 2011). Within-
family or within-genera comparisons indicate that bat flowers generally produce greater volumes of nectar than bird flowers even when flower sizes are
similar (chap. 3).

In addition to being influenced by pollinator type, nectar production
rates, defined as the volume of nectar produced per flower per unit time,
have a strong phylogenetic signal. Ornelas et al. (2007) collected data on
daily nectar production, sugar concentration, and corolla length for 289 species of New World plants from 56 families; each species was classified by
pollinator type, which included hermit and trochiline hummingbirds separately, passerine birds, bats, and insects. Their results indicate that certain
families or subfamilies in which bat pollination is common (e.g., Agavaceae,
Bombacoideae, and Cactaceae) are "super producers" with mean daily nec-

tar production rates of 112–156 μL. In contrast, families in the Myrtales, including Lythraceae and Onagraceae, and in the Lamiales, including Acanthaceae, Bignoniaceae, Gesneriaceae, Lamiaceae, and Scrophulariaceae, that contain many hummingbird-pollinated species are relatively poor producers with mean daily nectar production rates of 7.8–20.8 μL. Controlling for phylogeny, daily nectar production rates and daily sugar production were positively correlated with corolla length. Although this correlation might simply reflect a nonadaptive allometric relationship, Ornelas et al. (2007) suggested that this allometry is reinforced by selection from pollinators with New World bats and passerines selecting for larger, more productive flowers than hummingbirds.

We can further explore changes in nectar characteristics among pollinator types by focusing on particular plant genera or families. Within the genus *Passiflora*, for instance, the nectar of two bat-pollinated species differs from that of a bee-pollinated species in volume per flower and in percentage of sugar (26% vs. 45%) and in having much higher levels of cholesterol and triglycerides and a higher ratio of P:Na; nectar in the bee-pollinated flower contained three times more P than that of bat- or hummingbird-pollinated flowers (Varassin et al. 2001). Similar differences in the nectar characteristics associated with different kinds of pollinators occur within Bromeliaceae, Sinningieae (Gesneriaceae), and five families of Gentianales (Krömer et al. 2008; San Martin and Sazima 2005; Wolff 2006). Results of these studies support the hypothesis that selection by different kinds of pollinators results in qualitatively and quantitatively different nectars. The genetic basis behind this selection remains to be determined.

Johnson and Nicolson (2008) proposed that a major dichotomy exists in the nectar characteristics of plants pollinated by specialized versus generalized avian pollinators in both Africa, where sunbirds are specialists and birds such as weavers, bulbuls, and Old World orioles are generalists, and the New World, where hummingbirds are specialists and New World orioles and warblers are generalists. Results of their study indicate that the nectar of flowers pollinated by sunbirds and hummingbirds is similar in volume (10–30 μL per flower), sugar concentration (15%–25% sugar weight/weight), and sucrose content (40%–60%) and differs significantly from flowers visited by generalized avian pollinators whose nectars are characterized by large volume (40–100 μL), low sugar concentration (8%–12% weight/weight), and low sucrose content (0%–5%). A similar dichotomy, regarding nectar volume, at least, occurs in New World bats in which glossophagines visit smaller flowers, on average, than those visited by opportunistic phyllosto-mids (Helversen 1993). Such a dichotomy is not as likely to occur in Old World pteropodid specialists and opportunists because of their generally less-specialized relationships with flowers (Fleming and Muchhala 2008; Fleming et al. 2009; Petterson et al 2004). In the case of birds, nectar dif-ferences between specialist and generalist flowers reflect differences in sizes (e.g., generalist flower-visiting birds in Africa are more than three times heavier than sunbirds) and the absence of sucrase in the intestines of many generalists; a similar situation occurs in the New World (Johnson and Nicol-son 2008). These results strongly support the hypothesis that there has been convergence in many of the nectar (and other) characteristics of flowers pol-linated by specialized vertebrate nectar feeders and that more opportunistic flower visitors select for a different set of nectar (and flower) characteristics.

Finally, as reviewed in chapter 3, there has been much discussion of the sugar composition of flower nectars and fruit. Flowers pollinated by hum-mingbirds, sunbirds, and pteropodid bats are rich in sucrose whereas those pollinated by New World bats and generalist birds are hexose rich (Baker et al. 1998). These differences can occur even within a genus. In *Erythrina*, for example, in which pollination by passerines and hexose-rich nectars are ancestral, hummingbird-pollinated species produce sucrose-rich nectar (Lotz and Schondube 2006). Similarly, when certain plants in the Canary Islands switched from insect- to passerine bird–pollination, nectar compo-sition changed from sucrose rich to hexose rich (Lotz and Schondube 2006). Again, as in many other flower characteristics, the sugar composition of nectar is evolutionarily plastic and appears to be responsive to selection for particular kinds of pollinators, although evidence for this can be equivocal (see below).

POLLEN.—Far less attention has been paid to the characteristics of vertebrate-dispersed pollen than to nectar. These characteristics include the size and external morphology of pollen grains as well as their chemical contents. Compared with insect- or wind-dispersed pollen, we might expect vertebrate-dispersed pollen to be larger and rougher in external morphology and to be especially rich in amino acids in species whose pollen is regularly digested by specialized flower-visiting birds or mammals. In his comprehen-sive review, Stroo (2000) examined the morphological characteristics of the pollen of 75 species of bat-pollinated plants from 23 families and compared them to the characteristics of related, non-bat-dispersed pollen. He reported that bat pollen was somewhat larger than that of related pollen (means were 72 μm and 64 μm, respectively; size range of bat pollen was 17–170 μm). The greater size of bat flower pollen was attributed to the generally longer length of the styles of these flowers and not to a specific association with the pollen vector itself. When the allometric relationship between pollen grain size and flower style length was taken into account, bat pollen was no larger than nonbat pollen. A good example of this is the especially long styles of bat-pollinated members of the Bombacoideae, which are associated with giant pollen grains in this subfamily. Stroo (2000) also found that bat pollen has no special ornamentation and hence is no rougher than nonbat pollen. He concluded that pollen architecture is strongly constrained phylogenetically and that it does not closely reflect pollinator syndromes.

Except for studies of bee-pollinated plants for which there is a large literature (e.g., Hanley et al. 2008 and included references), systematic studies of the chemical composition of pollen grains in relation to pollinator modes are also scarce. As mentioned in chapter 3, the contents of pollen grains tend to be relatively rich in proteins (16%–30% or more protein), but the pollen of zoophilous plants is no richer in protein than that of anemophilous plants when plant phylogeny is controlled for (Roulston et al. 2000). In their extensive literature review, Roulston et al. (2000) also reported that the protein content of pollen of bird and bat flowers covers the same range as that of bee flowers. They concluded that the need for growing pollen tubes is more important than nourishing pollinators in determining the protein content of pollen. As originally proposed by Darwin (1884), especially important is the need for pollen grains to penetrate the stigma using autotrophic resources before traveling through the style using stylar (heterotrophic) resources (Cruden 2009). Cruden (2009) suggests that the correlation between pollen grain size and style length actually results from the correlation between these two variables and stigma depth. In any case, it appears that selection has not favored the evolution of pollen grains that are especially large or rich in proteins in flowers pollinated by vertebrates.

SEXUAL AND MATING SYSTEMS

About three-quarters of all angiosperms have hermaphroditic sexual systems, and physiological self-incompatibility—which prevents self-fertilization—is common in tropical plants, especially trees. As discussed in chapter 4, birds and bats are often excellent at moving pollen between plants and can sometimes carry pollen substantial distances between conspecifics in both canopy and understory plants. Bats appear to transport pollen longer distances, on average (up to about 18 km; Ward et al. 2005), than birds. Given this situation, we would predict that dioecy, in which male and female flowers occur on separate plants and which is a sexual system that forces plants to outcross, should be much less common in vertebrate-pollinated plants than in insect-pollinated plants. Data presented in Renner and Ricklefs (1995) support this prediction. They reported that about 6%–7% of all angiosperm genera and species are dioecious and that this sexual system is most strongly associated with monoecy (as an ancestral condition), wind pollination, and a climbing habit. In animal-pollinated plants, dioecy is associated with small, insect-pollinated flowers. Only four of 786 dioecious genera (two in Liliaceae and one each in Balanophoraceae and Cactaceae) are pollinated by birds or bats. Whereas dioecious taxa are usually associated with wind or insect pollination, they are often associated with biotic seed dispersal via fleshy fruits (Chen and Li 2008; Vamosi et al. 2003, among others). Vamosi et al. (2003) suggest that a relatively low investment in flowers and inflorescences may allow dioecious plants to invest more heavily in fleshy, vertebrate-dispersed fruits.

As discussed in chapter 4, plant mating systems have traditionally been classified as primarily inbreeding, primarily outcrossing, or mixed. Michalski and Durka (2009), however, suggest that in an era in which the degree of outcrossing is being routinely estimated using genetic markers, this classification can be more finely quantified. According to these authors, a better way of describing a plant's mating system would be to quantitatively express its degree of outcrossing on a scale of 0–1. Self-compatible and inbred plants would fall at the low end of this scale; self-incompatible outcrossers would fall at the high end; and species with a mixed mating system would fall broadly in the middle. Based on our current knowledge, canopy and subcanopy trees in the tropics occur higher on the outcrossing scale than understory shrubs (chap. 4).

Although nectar-feeding birds and bats are often good pollen dispersers, their food plants are still likely to experience reduced fruit set as a result of pollen limitation (chap. 4). What effect do these two factors—pollen dispersal and pollen limitation—have on the evolution of plant mating systems? Cheptou and Massoi (2009) used a metapopulation modeling approach to address this question. In their model, dispersal was broadly defined to include the rate at which both pollen and seeds moved among subpopulations. Results of their model indicated that the optimal (i.e., equilibrium) mating strategy was a high dispersal rate and a low selfing rate. When pollination limitation was strong, however, selection favored a high selfing rate and low dispersal rate, and when the abundance of pollinators varied stochastically (resulting in a variable degree of pollen limitation), a mixed mating system was favored. Results of this study suggest that mixed mating systems (i.e., those with intermediate outcrossing rates) should be common in animal-pollinated tropical plants. The implication here is that food-limited populations of nectar-feeding birds and mammals (chap. 3) can have an important evolutionary effect on the mating systems of their food plants. They should select for mixed mating systems.

Pollen-ovule (P: O) ratios are also correlated with plant mating systems and vary substantially among plants. Among self-incompatible, animal-pollinated plants, this ratio averages about 1,200:1 to 8,000:1 (range: 500:1 to 200,000:1; Cruden 2000). Two general patterns are well-established: (1) self-compatible plants have lower P: O ratios than self-incompatible plants (Cruden 1977) and (2) wind-pollinated plants have higher P: O ratios than animal-pollinated plants (Michalski and Durka 2009). Woody perennials (trees and shrubs), which are often self-incompatible, tend to have higher P: O ratios than understory herbs, which are often self-compatible. After reviewing data on P: O ratios and a variety of other reproductive factors for 107 species of angiosperms, however, Michalski and Durka (2009) concluded that a plant's pollination mode has a stronger influence on its P: O ratio than on its mating system.

What effect, if any, do vertebrate pollinators have on the P: O ratios of their food plants that is not directly related to the mating system? We can envision two reasons why these ratios might be higher in vertebrate-pollinated plants than in related insect-pollinated plants: (1) certain birds and mammals ingest pollen as a source of nutrients, thereby reducing the number of pollen grains available for fertilization; greater pollen production per flower would then be selected for to compensate for this loss; and (2) these animals have larger pollen-carrying capacities than most insects; hence, they can deliver more pollen grains to plant stigmas per visit which could result in greater reproductive assurance (Muchhala and Thomson 2010). According to Cruden (2000), however, neither of these predictions is supported by available data. In most plant families that have been examined, P: O ratios of insect-pollinated species are similar to those of bird- or bat-pollinated species. An exception to this appears to be tropical monocots in which species pollinated by beetles or flies (which feed on pollen grains) tend to have higher (not lower) P: O ratios than those pollinated by bees, birds, and bats. Overall, however, relatively few plant families have been studied, and variation in P: O ratios has seldom been studied using species-level phylogenies. Thus, we cannot yet reach firm conclusions about whether vertebrate pollinators have had a selective effect on the amount of pollen produced by the flowers they visit.

In summary, vertebrate pollination has had a substantial impact on the evolution of flowers in the tropics. Although many plant species in temperate zone habitats may be generalists with regards to pollinators, in the tropics flowers pollinated by vertebrates as well as some insect-pollinated taxa are more likely to be specialists for particular pollen vectors and to have a relatively narrow suite of features that match the characteristics of their floral visitors. This relationship is true with regard to flower color, size, shape, and presentation as well as floral scents, nectar qualities, and phenological patterns. This relationship appears to be less specific with regard to pollen features and breeding systems. Overall, there is abundant evidence indicating that convergence in floral features has evolved independently many times during angiosperm evolution in response to selection by particular kinds of pollinators. We will now turn to the other partners in this mutualism to explore the consequences of a diet of nectar and pollen.

The Evolution of Nectar-Feeding Vertebrates

Birds and bats are the primary vertebrate pollinators of tropical flowers, although a variety of other mammals, including honey possums (*Tarsipes*) in Australia, certain Neotropical opossums (e.g., *Caluromys*), New and Old World primates (e.g., tamarins and lemurs), rodents, and omnivores such as kinkajous and palm civets, occasionally pollinate flowers (Carthew and Goldingay 1997; Corlett 2004). At least six families or subfamilies of tropical or subtropical birds are strongly adapted for nectar feeding whereas only two families of tropical bats contain flower visitors; morphologically specialized nectar feeders are in the minority in both bat families (table 1.1; Fleming and Muchhala 2008). Overall, at least 500 genera of plants contain bird-pollinated species, whereas about 250 genera of plants contain bat-pollinated species (Sekercioglu 2006; Fleming et al. 2009). Here we focus on the morphological, physiological, and behavioral consequences of nectarivory for birds and mammals, placing these adaptations in a phylogenetic context whenever possible. Our major motivating question here is: To what extent have different nectar-feeding vertebrates converged in the ways in which they acquire and process nectar and pollen?

MORPHOLOGICAL ADAPTATIONS IN NECTAR-FEEDING BIRDS

From many perspectives, hummingbirds (Trochilidae) are the most specialized nectar-feeding birds in the world. Their small size, iridescent plumage, long, thin bills, rapid flight, and amazing hovering ability make them unmistakable members of many New World communities where they have had a profound influence on the evolution of Neotropical flowers (chap. 6). We will focus most of our attention on them in this section and will compare them with their closest Old World ecological analogues, sunbirds and honeyeaters.

GENERAL MORPHOLOGY.—Hummingbirds differ strongly from their closest relatives, aerial insectivores of the swift family (Apodidae), and indeed the majority of other birds in most aspects of their biology. Having diverged in the early Cenozoic, swifts and hummingbirds have had long independent evolutionary histories (Bleiweiss 1998a). Swifts have a cosmopolitan distribution but hummingbirds are currently restricted to the New World, although fossil hummingbirds from stem as well as crown lineages are known from the Oligocene of Europe (Mayr 2004, 2005a, 2007). Both families have reduced tarsi and feet and long, pointed wings, but aside from this, their differences are substantial. Swifts, for example, are much larger, on average, than hummingbirds; range of masses in swifts is 9–150 g compared with 2–20 g in hummingbirds (Perrins and Middleton 1985).

Sunbirds are usually considered to be the second most specialized family of nectar-feeding birds. They are relatively advanced passerines (in the Passerida clade) whose current diversity is concentrated in Africa. Their ancestors were likely thin-billed nectarivores or insectivores (Johansson et al. 2008). Honeyeaters arose early in the radiation of passerines and are Australasian in distribution. Collins and Paton (1989) suggested that both sunbirds and honeyeaters evolved from foliage-gleaning insectivores, an evolutionary route that is hard to envision for weak-legged hummingbirds. Unlike hummingbirds, neither sunbirds nor honeyeaters are adept at hovering and usually perch on or near flowers to feed.

Major morphological features associated with nectar feeding in these three bird families are summarized in table 7.2. Hummingbirds are the

Table 7.2. Comparison of the Major Morphological Features of Three Families of Nectar-Feeding Birds

Feature	Hummingbirds (Trochilidae)	Sunbirds (Nectariniidae)	Honeyeaters (Meliphagidae)
Mass (g)	5.2 (2–20.2)	11.3 (4.7–38.4)	33.4 (7.4–152.0)
Bills and tongues:			
Length (mm)	7–105	12–39	11–47
Description	Tip is bifid and tubular; edges are fringed. Bill shape variable.	Tip is bifid, often tubular; edges are serrated. Bills decurved.	Tip divided into 2, 4, or 8 parts with trough near end; tips are brushy. Bills decurved.
Tarsus and feet	Tarsus reduced; feet small but toes large for perching; most cannot walk.	Tarsus and feet not reduced; strong perchers and walkers.	Tarsus and feet not reduced; strong perchers and walkers.
Flight morphology	Wings highly specialized: short humerus and forearm bones, elongated hand bones; entire wing rotates without flexion of elbow and wrist for forward and backward flight and hovering.	Wings nonspecialized. Cannot hover easily.	Wings nonspecialized. Cannot hover easily.
Plumage	Males of many species highly iridescent. Iridescence covers entire visible spectrum and is produced by light refraction by melanin granules in specialized feathers.	Males of most species highly iridescent. Iridescence produced by internal structural relationships of keratin and melanin granules.	Males are noniridescent.
Vision	Can see in near UV	Can see in near UV	Can see in near UV

Sources. Cheke and Mann (2001) Johnsgard (1997); Paton and Collins (1989).
Note. Mass shown is mean, with range given in parentheses.

smallest in size and honeyeaters the largest of these families. Maximum size of honeyeaters is about seven times greater than that of hummingbirds. Pyke (1980) noted, however, that the most nectarivorous honeyeaters tend to be small and that large species are either frugivorous or omnivorous. Sunbirds are closer in size to hummingbirds than to honeyeaters. Nonetheless, their maximum size is about twice that of hummingbirds. Like honeyeaters but unlike hummingbirds, many sunbirds also include fruit as well as insects in their diets (Cheke and Mann 2001). Their small size, delicate bills, and ability to hover allow hummingbirds to feed at smaller, more delicate flowers (and eat smaller insects?) than larger, nonhovering sunbirds and honeyeaters.

Members of all three families have long, thin bills and long extensible tongues whose distal ends are troughlike and whose tips are split and fringed or brushy (fig. 7.5). Median bill lengths are about 20 mm in all three families, but hummingbirds have a greater range of bill lengths and shapes than the other two families (Fleming and Muchhala 2008; Paton and Collins 1989). The two-dimensional morphospace occupied by these families in terms of their bill sizes and shapes is shown in figure 7.6. Sunbirds and honeyeaters

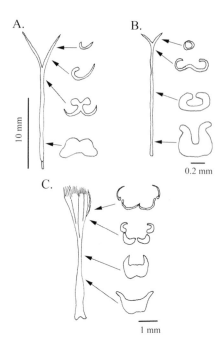

Figure 7.5. The tongues of a hummingbird (*A*), a sunbird (*B*), and a honeyeater (*C*). Cross sections are shown at the right of each tongue. Based on Paton and Collins (1989).

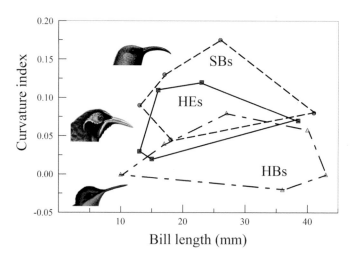

Figure 7.6. Plot of the morphospace occupied by three families of nectar-feeding birds by bill length and curvature. The curvature index is defined as the maximum perpendicular height of the bottom edge of the culmen above the chord from the gape to the tip of the bill divided by the length of this chord. Abbreviations include honeyeaters (*HEs*), hummingbirds (*HBs*), and sunbirds (*SBs*). Based on Paton and Collins (1989).

differ from hummingbirds in consistently having decurved bills. The bills of many hummingbirds are straight, and at least one species (the avocetbill) has an upturned bill. Not included in figure 7.6 is the swordbill hummingbird (*Ensifera ensifera*) whose maximum bill length of about 105 mm far exceeds that of any other nectar-feeding bird.

Unlike sunbirds and honeyeaters, which have strong legs and feet, hummingbirds have greatly reduced legs, and most (but not all) species have lost the ability to walk. Instead, they have evolved a sophisticated flight mechanism featuring highly specialized wings whose feathers are supported primarily by elongated hand bones and flight muscles that allow them to rapidly fly forward and backward and to hover. During forward flight, their wingbeats describe a vertical oval as in other birds. During hovering, their primary feathers are reversed by rotary movements of the shoulder, wrist, and the hand bones supporting them, and their wings move horizontally in a figure eight. Variation in wing size and shape is substantial among hummingbirds and depends on foraging strategy (e.g., territorial vs. traplining) and altitudinal distribution (Feinsinger and Colwell 1978; Stiles 2004). Territorial species tend to have shorter, more heavily loaded wings for maneuverable flight; trapliners tend to have longer, less heavily loaded wings for fast but less maneuverable flight; and high-elevation populations and species have longer, less heavily loaded wings than their lowland counterparts (Feinsinger et al. 1979; Stiles 1981). A detailed examination of the wing disc–loading concept (which can be defined as weight/wing cord length) by Altshuler et al. (2004), however, questions the former two generalizations. Their analysis found that wing disc loading was not a good predictor of competitive ability in territorial species or foraging strategy (territorial vs. traplining) in several hummingbird communities.

As discussed above, the biogeographic dichotomy between hovering (New World) and nonhovering (Old World) in nectar-feeding birds apparently results more from differences in plant morphology than from intrinsic differences in these birds. When a perch is available, for example, hummingbirds, like sunbirds and honeyeaters, will often use it rather than hover (Miller 1985). Hummingbirds often perch to feed at hummingbird feeders. Conversely, malachite sunbirds (*Nectarinia famosa*) hover about 80% of the time when feeding at flowers of the invasive New World hummingbird plant *Nicotiana glauca.* Two smaller sunbirds do not hover at this plant but instead pierce corollas to obtain nectar while perched (Geerts and Pauw 2009).

Although hummingbirds usually hover while feeding, whereas sunbirds and honeyeaters perch on or close to flowers, these birds extract nectar from flowers in similar amounts per visit (Collins 2008; Paton and Collins 1989). Birds extend their tongues into nectar rapidly at rates of 10–15 licks/sec. Because hummingbirds have longer bills and tongues than the other two families, they can extract nectar from longer corollas. Supported by long hyoid bones, the tongues of hummingbirds can be extended 20–30 mm or more beyond the tips of their bills compared with an extension of only 5–20 mm in honeyeaters (Paton and Collins 1989). Tubular tongues and extensive surface areas at the tips allow these birds to acquire nectar rapidly via capillarity or dynamic nectar trapping (Rico-Guevara and Rubega 2011). Paton and Collins (1989) suggest that the different tongue morphology seen in honeyeaters (fig. 7.5) is related to (1) a larger ancestral size and wider tongues (tongue width is an allometric function of body mass) that likely function better with a brushy, moplike tip and (2) the prevalence in their diets of brush-type flowers (e.g., eucalypts and banksias) that present nectar in a thin film rather than in a tube.

A conspicuous morphological feature of two of the three families is the occurrence of iridescent males. Indeed, male hummingbirds and sunbirds are among the most brilliantly colored terrestrial animals on earth and clearly merit the designation "ornaments of life." Iridescence in hummingbirds is restricted to the head and back in most species; in sexually dichromatic species, heads of males range in color from intense violet to fiery red. Iridescence is mostly restricted to trochiline hummingbirds in which males usually undergo sexual displays in well-lit habitats; it is uncommon in the understory-foraging and -displaying hermits. This iridescence is structurally based and results from the refraction of light by the keratin surfaces of specialized feathers (barbules) after it interacts with stacks of melanin granules within these feathers. Like trochiline hummingbirds, many male sunbirds have iridescent blue, purple, or green heads and backs; bright reds and yellows are common dorsal and ventral colors. Like hermit hummingbirds, spiderhunters (*Arachnothera*) are less gaudily plumaged, although they are not necessarily restricted to foraging in forest understories. Iridescence in sunbirds is also structural and results from the interaction of light with feather keratin and melanin granules. In contrast to hummingbirds and sunbirds, both sexes of most honeyeaters are dull green, gray, or brown in color, often with black, white, or yellow markings. A few species have bright red, yellow, or blue bare facial regions.

Vision and the ability to see colors are obviously important for successful foraging in nectar feeders and fruit eaters. As discussed in detail in Gold-smith (2006), Bennet and Thery (2007), and Hart and Hunt (2007), among others, birds differ from most mammals in having tetrachromatic, rather than dichromatic, vision. Their retinal cones contain four different kinds of opsins (visual pigments) whose peak spectral sensitivities differ; these peaks occur over a spectral range of 300–700 nM and include peaks in the near UV as well as in short (violet), medium (blue-green), and long (orange) visible wavelengths. Ultraviolet discrimination has been detected in many different kinds of birds and is presumed to be present in all terrestrial, diur-nal birds. It is known to occur in hummingbirds, sunbirds, and honeyeat-ers (as well as other plant-visiting birds). Interestingly, hummingbirds and honeyeaters, along with other non-Passerida birds, have slightly different UV sensitivity compared with Passerida (advanced passerine) birds (Ödeen and Håsted 2010). The occurrence of four visual pigments, along with the presence of oil droplets containing three types of color pigments (red and yellow carotenoids plus colorless) in cone cells, exposes birds to a much more diverse color palette than most mammals. Ultraviolet sensitivity in plant-visiting birds allows them to detect conspicuous UV color patterns in flowers (i.e., nectar guides) and UV-reflecting fruits against green foliage. It also likely enables females to discriminate among subtle differences in the colors of bright (e.g., iridescent) plumages of males during mate choice (Altshuler 2001; Goldsmith 2006).

Brain size, not included among the morphological features listed in table 7.2, might be expected to differ among different dietary classes within birds. According to Bennett and Harvey (1985), however, ecology explains very little of the variation in brain size among groups of birds. Nealen and Ricklefs's (2001) extensive analysis of brain-mass to body-mass relationships in birds indicated that the slope (b) of the allometric equation relating these two variables ($Y = aX^b$, where Y is brain mass and X is body mass, both in g) is 0.593 (i.e., negative allometry) and varies significantly among orders. Relative brain size varies from 2% to 9% in birds and is about 4% in hum-mingbirds (Gill 1990; Johnsgard 1997). Comparative data for sunbirds and honeyeaters apparently are not available. As we will see below and in chap-ter 8, brain size in mammals is significantly influenced by feeding habits.

THE EVOLUTION OF BODY AND BILL SIZE IN HUMMINGBIRDS.—As in-dicated above, hummingbirds differ from other specialized nectar-feeding birds in their small size and in the diversity of size and shape of their bills.

Since body size and bill size play very important roles in flower choice and nectar extraction efficiency in these birds (e.g., Temeles et al. 2009; Wolf et al. 1972), it becomes important to understand the phylogenetic basis of this variation. To what extent does this variation reflect the phylogenetic history of the family? To answer this question, we determined mean values of wing cord length (a surrogate for body size) and bill length for 64 genera of hummingbirds included in the molecular phylogeny of Trochilidae by McGuire et al. (2007). Morphological data came from the database in Fleming et al. (2005). As in other phylogenetic analyses (chap. 6), we used Mesquite (Maddison and Maddison 2007) to construct a phylogenetic hypothesis for eight of the nine major clades of hummingbirds (McGuire et al. 2007, fig. 3); we combined the Mountain Gem and Bee clades—two small and closely related clades—in our hypothesis. Using the parsimony reconstruction module, we then mapped categorical values of wing cord length and bill length separately onto the phylogeny. Character states for wing cord length were: (0) small—maximum value for any genus in a clade was ≤70 mm; (1) medium—maximum value 71–100 mm; and (2) large—maximum value >100 mm. Character states for bill length were: (0) small—maximum value for any genus in a clade was ≤26 mm; (1) medium—maximum value was 27–100 mm; and (2) large—maximum value was >100 mm.

   Results of this analysis are presented in figure 7.7. With respect to body size (wing cord length), the two basal clades are medium sized, with wing cord lengths averaging 74 mm and 60 mm for Topazes and Hermits, respectively (fig. 7.7A). Four of the six clades of trochilines, including the basal Mangoes, are small-sized. Exceptionally large size has evolved independently twice: once in *Ensifera* (in the Brilliant clade) and again in *Patagona*; both of these taxa occur at high elevations in the Andes. Coquettes, another clade strongly associated with the Andes (McGuire et al. 2007), are relatively small in size (fig. 7.7A). Regarding bill size, medium-sized bills predominate throughout the family (fig. 7.7B). Small bills have evolved twice (in the Coquettes and Emeralds), and *Ensifera* has an extraordinarily long bill. It might be argued that bill size simply reflects body size, possibly because these two variables are highly correlated in these birds. Bill length, however, is only modestly correlated with wing cord length ($r^2 = 0.15$, $P = 0.002$ in all 64 genera; $r^2 = 0.28$, $P < 0.001$ after removal of *Ensifera*, a strong outlier) and accounts for only a small proportion of variation. The modest correlation and the lack of concordance in the two character-state mappings indicate that bill size is an evolutionarily plastic feature of hummingbirds and likely reflects adaptation to local flower communities in different habitats.

MORPHOLOGICAL ADAPTATIONS IN NECTAR-FEEDING MAMMALS

Bats are the principal nectar-feeding mammals, and extensive flower visiting has evolved independently in two families—in the New World Phyllostomidae and the Old World Pteropodidae (table 1.1). In the phyllostomids, a single clade that includes four subfamilies (Glossophaginae, Lonchophyllinae, Brachyphyllinae, and Phyllonycterinae collectively called "glossophagines") contains the most specialized flower visitors (Baker et al. 2003; Wetterer et al. 2000). The first two subfamilies are mostly (Glossophaginae) or entirely (Lonchophyllinae) restricted to the Neotropical mainland of Mexico and Central and South America. The other two subfamilies are West Indian endemics. Specialized pteropodid nectar feeders do not belong to a single clade but have evolved independently at least three times: once relatively recently in Africa and twice in Asia/Australasia (Gianinni and Simmons 2005). In addition to the specialized nectar feeders, both families of bats contain a variety of other taxa that are opportunistic flower visitors (Fleming et al. 2009). Indeed, many frugivorous phyllostomids (e.g., bats in the genera *Carollia* and *Artibeus*) and pteropodids (e.g., *Cynopterus*, *Epomophorus*, and *Pteropus*) often include nectar in their diets. Opportunistic flower visitors in both families tend to be larger than their more specialized relatives (up to 90 g in phyllostomids and ≥1,000 g in pteropodids), and they almost always land on or near flowers to feed. Consequently, they tend to visit larger and more robust flowers than their more specialized relatives (Fleming et al. 2009; Helversen 1993).

Molecular phylogenies are available for both of these families, and we used them to infer patterns of dietary evolution in phyllostomids using Mesquite (fig. 7.8). Insectivory is clearly the ancestral feeding mode in phyllostomids, whereas frugivory is likely to be ancestral in pteropodids (Giannini and Simmons 2005; Rojas et al. 2011; Wetterer et al. 2000). Interestingly, Baker et al.'s (2003) phyllostomid phylogeny (fig. 7.8) suggests that frugivory is derived from nectarivory in the New World, whereas the converse is likely to be true in Old World pteropodids. Rojas et al's (2011) recent analysis, however, suggests that both frugivory and nectarivory in Phyllostomidae are derived from insectivory.

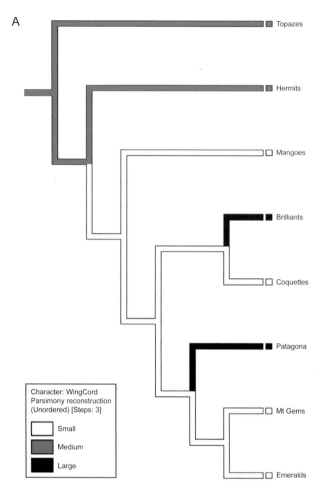

Figure 7.7. Cladograms of the major lineages of hummingbirds showing the distributions of (*A*) wing chord length (i.e., body size) and (*B*) bill length, based on the phylogeny of McGuire et al. (2007).

GENERAL MORPHOLOGY.—Like their avian counterparts, specialized nectar-feeding bats are relatively small compared with their relatives and have elongated rostrums and long, brush-tipped tongues; their teeth are reduced in size and number. The range in body mass in phyllostomid nectar bats is 7.5–30 g; it is 13.2–82.2 g in pteropodid nectar bats. In figure 7.9 we provide a general overview of the distribution of nectar-feeding and fruit-eating phyllostomids and pteropodids in multivariate morphospace. Data included in this analysis are generic means of four morphological

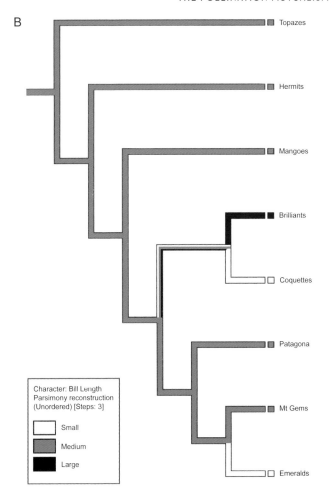

variables (forearm length [overall size], length of the maxillary tooth row [upper jaw length], breadth across the upper molars [upper jaw breadth], and the ratio of jaw length to jaw breadth [relative jaw length]) based on data in Swanepoel and Genoways (1979) and Andersen (1912). We subjected these data to PCA. The first two PCA axes accounted for 99.1% of the variation in these data; axis 1 represented general size, and relative jaw length loaded heavily on axis 2.

Each of the four dietary groups occupies a different region of this morphospace (fig. 7.9). As expected, phyllostomids are generally smaller than pteropodids. Within phyllostomids, nectar feeders have substantially longer jaws than fruit eaters, which are generally short jawed. Within pteropodids,

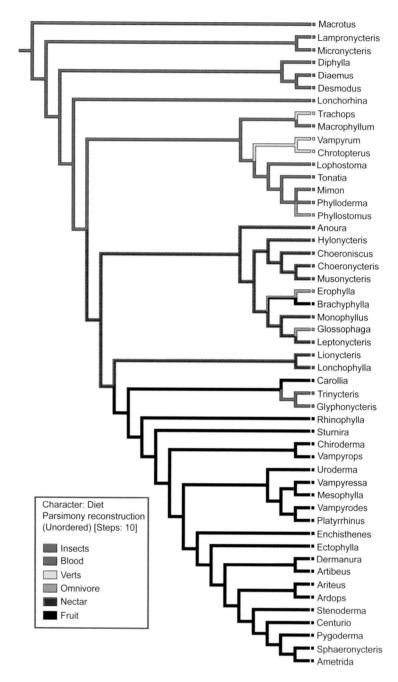

Figure 7.8. Cladogram of phyllostomid bats by general diet class based on the phylogeny of Baker et al. (2003).

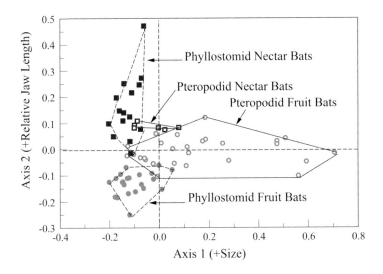

Figure 7.9. Ordination of the skulls of four groups of plant-visiting pteropodid and phyllostomid bats by principal components analysis. PCA axis 1 is positively correlated with skull length (i.e., body size); axis 2 is positively correlated with the ratio of jaw length to jaw width. Points represent generic means. See text for further explanation.

nectar feeders occupy the small end of the size spectrum but do not have especially long jaws compared with phyllostomid nectar feeders, which occupy a much larger area of morphospace than their pteropodid counterparts (see further discussion of this in Fleming and Muchhala [2008]).

In addition to elongated rostrums and narrow palates, nectar-feeding bats have reduced tooth size and number and long tongues. The teeth of both pteropodids and plant-visiting phyllostomids are much smaller in area than bats with other feeding habits (Dumont 1997; Freeman 1995, 1998). Pteropodid nectar bats generally have the same number of teeth as frugivores, but their lower incisors tend to be smaller and are sometimes reduced in number (Andersen 1912). Compared with their insectivorous or carnivorous relatives, phyllostomid nectar bats generally have fewer and much smaller teeth (Freeman 1995). Minimum number of teeth in phyllostomid nectar bats is 24 whereas it is 30 in insectivores and carnivores (Phillips et al. 1977). Missing teeth in glossophagines include all lower incisors (in advanced forms with especially long tongues) and certain molars. Although most teeth are reduced in size, the canines of glossophagines are particularly long for supporting their long tongue when it is extended. When the jaws are closed, the upper and lower canines interlock and form a rigid arch through which the tongue is extended. The elongated and anteriorly fused lower jaw of nectar bats in both families has changed the lower jaw from a tooth-bearing structure to a tongue-supporting structure (Freeman 1995).

As in nectar-feeding birds, nectar bats have long, brush-tipped tongues for quickly extracting nectar from flowers in visits that generally last <2 sec in both families. Compared with their nonnectarivorous relatives, the tongues of nectar bats of both families are tipped with dense patches of filiform papillae that serve as a nectar mop via capillary action or as a scoop via the upturned tip (Birt et al. 1997; Winter and Helversen 2003). Tongue extension involves both muscular and vasohydralic processes involving the migration of the extrinsic tongue musculature to the tongue's base and hyoid apparatus and an enlarged lingual artery and vein (Griffiths 1978, 1982). Tongue lengths and extension can be extreme in glossophagines. Operational tongue length (which includes the extended tongue plus length of the snout inserted into test tubes 9 or 15 mm in diameter) in eight species ranged from 55 mm to 77 mm and was correlated with each bat's palatal length (Winter and Helversen 2003). The recently described glossophagine *Anoura fistulata* can extend its tongue 85 mm compared with 37–39 mm for two congeners (Muchhala 2006b). Its tongue is so long that it has to be stored in a special glossal tube inside its thoracic cavity when not extended. Opportunistic phyllostomid flower visitors such as *Carollia perspicillata* and *Artibeus jamaicensis* can only extend their tongues 5–6 mm beyond their mouth tip (Nicolay and Winter 2006; Winter and Helversen 2003).

Bat wing design basically reflects the general foraging strategies of species or lineages. Bats that forage in open air away from clutter, for example, tend to have long, narrow wings for fast, agile flight whereas those that forage in and around vegetation tend to have short, broad wings for high maneuverability (Altringham 1996). The former bats are said to have high aspect ratio and heavily loaded wings whereas the latter have low aspect ratio and lightly loaded wings. Aspect ratio is measured as wingspan^2/wing area; wing loading is measured as mass/wing area. Compared with all other bats, pteropodids have below-average aspect ratio and average-to-high wing loading, with nectar-feeding species having especially low aspect ratios (Norberg and Rayner 1987). The nectar bat *Eonycteris spelaea* is an exception and has high aspect-ratio wings for long-distance, fast commuting flights. In glossophagines, aspect ratio is also low and wing loading high compared with other phyllostomid subfamilies. They also have long wing tips and large wing tip

areas for hovering. Again, because it is a fast-flying, long-distance com-
muter, species of the glossophagine genus *Leptonycteris* have high aspect
ratio and heavily loaded wings (Sahley et al. 1993).

Pteropodid nectar bats differ fundamentally from phyllostomids in the
way they visit flowers. Like their Old World avian counterparts, they land on
or near flowers before beginning to feed whereas glossophagine nectar feeders, like hummingbirds, usually hover at flowers. Opportunistic flower visitors in both families land on flowers and do not hover. Just as hummingbirds
have unique hovering abilities among birds, the ability to hover in glossophagines is unique among bats. This involves an upward turn (supination)
of the hand wing during the backstroke with the wing tip forming a distal
triangle that provides lift. During wing tip reversal, the wing tip forms a bent
figure eight compared with a horizontal figure eight in hummingbirds (Helversen and Winter 2003). Although it might seem to be an energy-expensive
way to forage, hovering in both bats and hummingbirds is not much more
costly than slow forward flight; the weight-specific cost of hovering in glossophagine bats is somewhat lower than that of hummingbirds and hawkmoths (Dudley and Winter 2002; Helversen and Winter 2003).

Whereas birds have tetrachromatic color vision, most mammals have
only two cone types that are sensitive to either short or long wavelengths. We
will discuss the evolution of trichromatic color vision in primates in chapter 8. Since bats are nocturnal, their retinas are dominated by rods; where
they exist, cones represent ≤0.6% of all photoreceptors (Muller et al. 2007).
The eyes of pteropodids are unique in having a highly papillate choroid that
is thought to provide nourishment to the avascular retina (Suthers 1970).
Muller et al. (2007) used immunocytochemistry to determine the presence
of cones in the eyes of four genera of African pteropodids. They found two
cone types (short wavelength and long wavelength) in *Pteropus* and only
long wavelength cones in the other three genera. *Pteropus* may have dichromatic color vision, whereas members of the other genera are color-blind.
Ultraviolet vision is common in birds but is uncommon in bats. Winter et al.
(2003) demonstrated experimentally that the glossophagine *Glossophaga
soricina* has photoreceptors that are UV sensitive but is color-blind. Unlike
some mammals, this species does not have a separate UV receptor but has
a receptor that is sensitive at two wavelength peaks, 365 (UV) and 510 nM.
Some New World bat-pollinated plants are known to reflect in the UV, and
UV vision may enhance their perception by bats (Helversen and Winter
2003; Winter and Helversen 2001).

Brain size does not vary substantially with feeding habits in birds, but it does in bats. In an early study, Eisenberg and Wilson (1978) showed that plant-visiting bats have significantly larger brains than insectivorous bats, correcting for body size. Relative brain sizes in pteropodid and nectarivorous and frugivorous phyllostomids were similar. Their neocortex, olfactory and optic region are larger than those of other bats. These authors suggested that foraging strategies involving the location of rich but patchy food resources require relatively large brains. More recently, Hutcheon et al. (2002) used conventional and phylogenetically based analyses to examine the effect of foraging ecology, corrected for body size, on the volume of three brain regions—the hippocampus (a correlate of spatial memory?), olfactory nucleus, and auditory nucleus—in a variety of bats. Compared with insectivores, the brains of plant-visiting pteropodids and phyllostomids were larger in overall mass and in volumes of the hippocampus and olfactory bulbs (also see Ratcliffe 2009). Not surprisingly, because they lack the ability to echolocate, pteropodids have smaller auditory nuclei than phyllostomids; otherwise the brains of members of the two families are very similar. Pteropodids clearly use olfactory and visual information to locate food; in addition, phyllostomids also use auditory information (e.g., Helversen and Helversen 1999).

THE EVOLUTION OF BODY SIZE AND JAW LENGTH IN NECTAR BATS.—
As we did for hummingbirds, we used recent phylogenies to infer evolutionary trends in body size and jaw length, an indication of the degree of specialization for flower visiting in these bats (Freeman 1995; Koopman 1981). Since nectar feeding has evolved independently at least three times in the Pteropodidae, our pteropodid analysis is based on all 34 genera, including six genera of nectar feeders and 28 genera of frugivore/opportunistic flower visitors, found in the Giannini and Simmons (2005) phylogeny. As in the PCA above, we used forearm length as a general indication of body size and the ratio of upper jaw length divided by upper jaw width as a measure of relative elongation of the rostrum. We used Mesquite to trace the history of body size and relative jaw length on the pteropodid phylogeny. Character states for forearm length were: (0) small, 43–60 mm; (1) medium, 61–100 mm; and (2) large, >100 mm. Character states for the jaw length ratio were: (0) small, ≤1.09; (1) medium, 1.10–1.32; and (2) large, >1.32.

Results of these analyses are shown in figure 7.10. In terms of body size, the ancestral and most common state in pteropodid bats is medium sized.

Relatively small bats have evolved independently at least five times—three times in Asia and Australasia and twice in Africa (fig. 7.10A). A monophyletic group of large bats evolved mostly in Australasia and Asia, including large islands; except for *Eidolon* and *Hypsignathus* (not included in the phylogeny), the radiation of African pteropodids involved mostly medium-sized bats. Three of the six genera of nectar bats are small and three are medium sized.

A somewhat different pattern is seen in the distribution of relative jaw sizes in pteropodids (fig. 7.10B). Again the ancestral condition appears to have been medium-sized jaws. A clade of small-jawed bats evolved early in Asia and once late in the family's history in Africa. Relatively long-jawed bats—including all six genera of nectar bats plus *Pteropus* (which is widespread in the Paleotropics) and *Myonycteris* (in Africa)—evolved independently at least four times throughout the geographic range of this family. In contrast to the situation in hummingbirds, in which there is a modest correlation between body size and bill length, there was no correlation between forearm length (body size) and relative jaw length in pteropodid bats ($r^2 = 0.030$, $P = 0.32$). We interpret this to indicate that relative jaw length is an evolutionarily plastic feature of these bats and probably reflects an adaptation to primary diet type (i.e., either flowers or fruit).

We used the phylogeny of Simmons and Wetterer (2002), which has a more complete coverage of glossophagine genera than that of Baker et al. (2003), for the analysis of body size and relative jaw size in nectar-feeding phyllostomid bats. Unlike the pteropodids, these bats form a monophyletic clade. Using Mesquite we mapped the character states of forearm length and relative jaw length for 16 genera from our morphological database onto the Simmons-Wetterer glossophagine phylogeny. Character states for forearm length were: (0) small, 31–40 mm; (1) medium, 41–50 mm; and (2) large, >50 mm. Character states for relative jaw length were: (0) small, ≤1.50; (1) medium, 1.51–2.00; and (2) large, >2.00.

Results of these analyses are presented in figure 7.11. Medium or large body sizes in nectar bats tend to be strongly associated with occurrence on islands (*Brachyphylla*, *Erophylla*, *Phyllonycteris*, and *Monophyllus*) or in arid habitats (*Platalina*, *Leptonycteris*, *Musonycteris*, and *Choeronycteris*). Otherwise, small body size prevails. More generally, large body size predominates among basal members of the Phyllostomidae (i.e., in genera whose diets include insects, blood, or vertebrates), and the small sizes found in many nectarivores and frugivores are derived. A different pattern emerges for relative

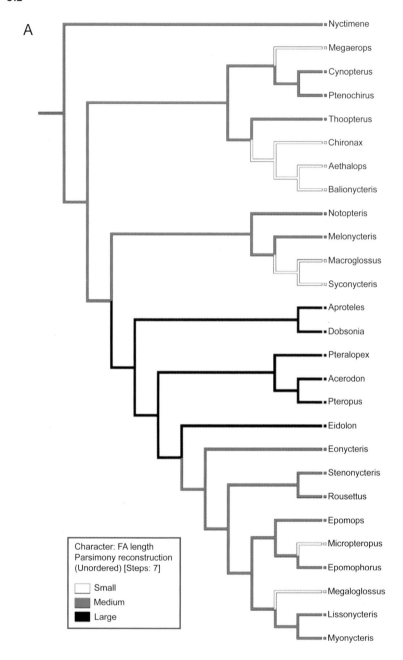

A

Nyctimene
Megaerops
Cynopterus
Ptenochirus
Thoopterus
Chironax
Aethalops
Balionycteris
Notopteris
Melonycteris
Macroglossus
Syconycteris
Aproteles
Dobsonia
Pteralopex
Acerodon
Pteropus
Eidolon
Eonycteris
Stenonycteris
Rousettus
Epomops
Micropteropus
Epomophorus
Megaloglossus
Lissonycteris
Myonycteris

Character: FA length
Parsimony reconstruction
(Unordered) [Steps: 7]

☐ Small
▨ Medium
■ Large

Figure 7.10. Cladogram of pteropodid bats showing the distributions of (*A*) forearm length (i.e., body size) and (*B*) relative jaw length, based on the phylogeny of Giannini and Simmons (2005).

B

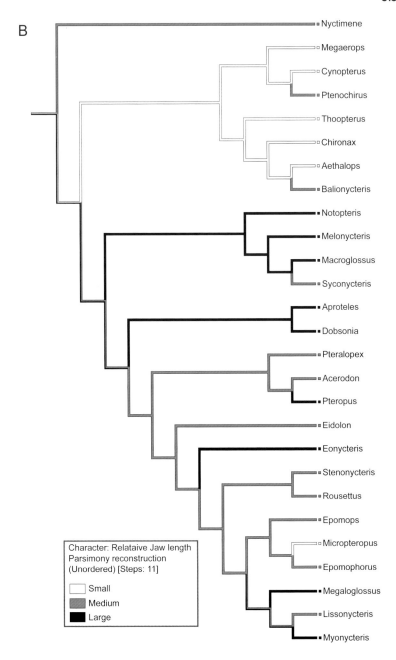

Character: Relataive Jaw length
Parsimony reconstruction
(Unordered) [Steps: 11]

☐ Small
▨ Medium
■ Large

Nyctimene
Megaerops
Cynopterus
Ptenochirus
Thoopterus
Chironax
Aethalops
Balionycteris
Notopteris
Melonycteris
Macroglossus
Syconycteris
Aproteles
Dobsonia
Pteralopex
Acerodon
Pteropus
Eidolon
Eonycteris
Stenonycteris
Rousettus
Epomops
Micropteropus
Epomophorus
Megaloglossus
Lissonycteris
Myonycteris

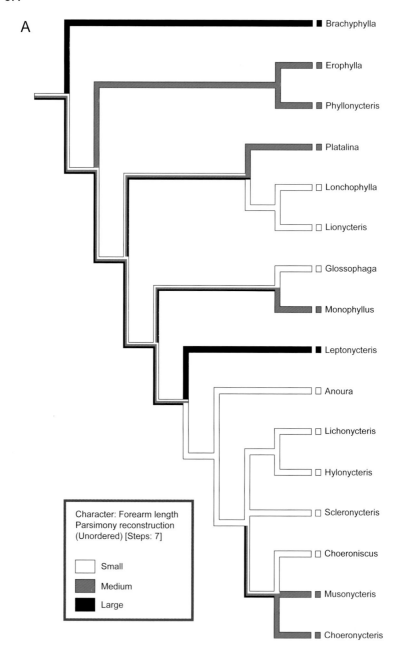

Figure 7.11. Cladogram of glossophagine bats showing the distributions of (A) forearm length (i.e., body size) and (B) relative jaw length, based on the phylogeny of Simmons and Wetterer (2002).

B

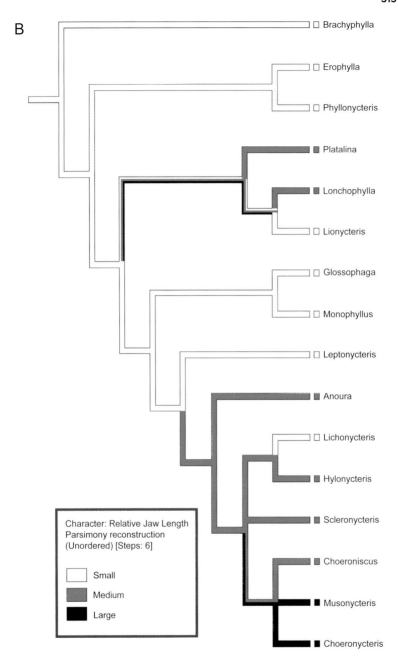

Brachyphylla

Erophylla

Phyllonycteris

Platalina

Lonchophylla

Lionycteris

Glossophaga

Monophyllus

Leptonycteris

Anoura

Lichonycteris

Hylonycteris

Scleronycteris

Choeroniscus

Musonycteris

Choeronycteris

Character: Relative Jaw Length
Parsimony reconstruction
(Unordered) [Steps: 6]

☐ Small

▨ Medium

■ Large

jaw length in glossophagines. Except in the medium-jawed *Platalina*, basal members of subfamilies are short jawed. Advanced members of the glossophagine clade (*sensu stricto*) are medium sized, and extremely long jaws have evolved once in two closely related Mexican genera (*Choeronycteris* and *Musonycteris*). Relatively long jaws are found only in the nectar-feeding clade of phyllostomids. All other clades, including insectivores, sanguinivores, carnivores, and frugivores, are relatively short jawed (e.g., fig. 7.9). As in pteropodids, there was no correlation between forearm length (body size) and relative jaw length ($r^2 = 0.23$, $P = 0.57$) in glossophagines. Rostral length is thus an evolutionarily plastic feature in these bats and probably reflects average flower size in different habitats. Fleming et al. (2005), for instance, reported that rostrum length in glossophagines is significantly larger in xeric habitats than in wet habitats. Flowers of columnar cacti are substantially larger than most flowers visited by glossophagines in wet forests (e.g., Fleming 2002; Sazima et al. 1999; Tschapka 2004).

DIGESTIVE AND METABOLIC ADAPTATIONS OF NECTAR-FEEDING BIRDS

Compared with other potential food types (e.g., seeds, insects, vertebrates), nectar is easy to procure and digest. Yet it is a physiologically challenging food characterized by relatively low energy density and very low densities of nutrients (electrolytes) in relatively high volumes of water. Morphological adaptations for dealing with this diet in nectar-feeding birds include a shortened and simplified alimentary tract and a nonmuscular, distendable stomach. Physiological adaptations include biochemical mechanisms that allow sugar molecules to be absorbed rapidly through the gut. As a result, digestion of nectar is rapid (in <30 mins) and digestive efficiency in nectarivores is very high (>95%; Lotz and Schondube 2006). Furthermore, because of its rapid assimilation, recently ingested nectar provides up to 80% of the fuel used by hummingbirds while hovering (Welch et al. 2006, 2007).

Nectarivorous vertebrates must deal with at least two major physiological challenges posed by a nectar diet—low nitrogen availability and a surfeit of water. As discussed in chapter 3, nectar contains very low amounts of proteins and amino acids. As a result, most nectar-feeding birds, including hummingbirds, obtain their protein by eating insects. Ornelas (1994) reported that 28 genera and 69 species of hummingbirds have serrations on their bills. Four genera have hooked bill tips with serrations extending up to 50% of the bill's length; bill serrations in the other 24 genera are not as extensive. Ornelas suggested that these serrations help hummingbirds capture insect prey or to pierce flower corollas in nectar robbers. Some species of sunbirds also have fine serrations on the tip of their bills, presumably for holding insects. Pollen is another potential protein source, but hummingbirds, at least, have very little ability to extract and digest the chemical contents of pollen grains (Brice et al. 1989). Sunbirds, honeyeaters, and lorikeets are better able to extract protein from pollen. Experimental studies indicate that Costa's hummingbird (*Calypte annae*) extracts and presumably digests the contents of only 5% of the pollen grains it ingests compared with 19% for the lesser double-collared sunbird (*Nectarinia chalybea*) and 5%–42% for various species of lorikeets (Brice et al. 1989; Roulston and Cane 2000; van Tets and Nicolson 2000; Wooller and Richardson 1988).

Nectar-feeding birds also deal with the low protein content of their primary diet physiologically. Compared with other birds, hummingbirds, sunbirds, and honeyeaters have low nitrogen requirements. The nitrogen requirement of a 4 g Costa's hummingbird, for example, is only 4.5 mg/day; it is 6.8 mg/day for the 6 g lesser double-collared sunbird and 2.8 mg/day in the 6.9 g tufted sunbird (*Nectarinia osea*; Brice and Grau 1991; Roxburgh and Pinshow 2000; van Tets and Nicolson 2000). These values are less than half of the allometrically predicted values for other birds. Low rates of fecal and urinary nitrogen excretion associated with a liquid diet apparently account for this exceptional nitrogen economy.

A second major physiological challenge faced by nectarivorous birds is water regulation and electrolyte conservation. As discussed above, nectar produced by bird-pollinated flowers is relatively dilute (typically 20%–25% sugar by weight) and is mostly water that has to be dealt with physiologically. In general, nectar-feeding birds have kidneys with a large cortex and small medulla for processing large amounts of water but with low urine concentrating ability. This reflects the fact that nectar feeders ingest at least five times their body mass in water per day when feeding on relatively dilute nectars. Hummingbirds apparently cannot simply excrete most of this water as cloacal fluid without first absorbing it through the gut and passing it through their kidneys. Although their kidneys have low urine concentrating ability, hummingbirds and other nectar feeders can sequester important electrolytes in the cortex of their kidneys (Fleming and Nicolson 2003; Martínez del Rio et al. 2001). Unlike hummingbirds, sunbirds and honeyeaters can adjust the amount of water their intestines assimilate to reduce water stress on their kidneys.

The sugar composition of nectar varies among bird-pollinated flowers, and as a result, the sugar preferences of nectarivorous birds have been studied extensively. Reviews of available data (Fleming et al. 2008; Lotz and Schondube 2006) indicate that sugar preferences of hummingbirds and other nectar feeders change with sugar concentration. At relatively high sugar concentrations ($\geq 20\%$ weight/volume), these birds prefer sucrose over glucose and fructose or are indifferent to different sugars; at low concentrations, they prefer hexose-rich sugars. Lotz and Schondube suggest that avoidance of osmotic stress may produce this pattern. Hexose-rich nectars have twice the osmotic potential as calorically equivalent sucrose nectars and will produce more osmotic stress at higher concentrations. In contrast to hummingbirds, sunbirds, honeyeaters and lorikeets, which can easily digest sucrose because they possess the enzyme sucrase, opportunistic nectar feeders in the mimid-sturnid-turdid clade of passerines have a strong aversion to sucrose in nectar and fruit pulp because they lack sucrase; ingestion of sucrose gives them osmotic diarrhea.

A variety of experimental studies have shown that, except at very low sugar concentrations and/or low air temperatures, nectar-feeding birds generally maintain a constant energy intake and energy balance by adjusting their feeding rates when feeding on nectars of different energy densities (Fleming et al. 2004; Martínez del Rio et al. 2001; Nicolson and Fleming 2003). Feeding rates in these birds vary inversely with the energy density of nectar in their flowers. Nectar feeders, however, cannot maintain their energy balance when confronted with low air temperatures and nectars with low sugar concentrations. Under these conditions, these birds generally lose body mass. Unlike most sunbirds and all honeyeaters, however, hummingbirds undergo torpor to reduce their metabolic needs when energy densities in their environment and/or air temperatures are low.

In birds and mammals, metabolic rates and rates of daily energy expenditure are negative allometric functions of body mass with birds generally having higher daily energy expenditures than similar-sized mammals (Nagy et al. 1999). Hummingbirds are the world's smallest birds and most other nectar feeders are also small, which means that although they have low absolute daily energy requirements, their mass-specific energy requirements are high. Compared with passerine nectar feeders, hummingbirds and their relatives (swifts, caprimulgids, and nightjars) have low rates of basal metabolism (McNab 2002). However, because of their energy-intensive mode of foraging, hummingbirds have substantially higher daily energy expenditures than sunbirds and honeyeaters. Weathers et al. (1996), for example, reported that the daily energy expenditures of three species of honeyeaters weighing 8–21 g were only about 43% of the values predicted for hummingbirds of a similar size. They also reported that energy expenditures of five species of hummingbirds ranged from 29.1 to 81.7 kJ/day compared with values of 52.9–77.6 kJ/day in three species of honeyeaters and 66.2 kJ/day in one species of sunbird.

Unlike most birds, hummingbirds have the ability to undergo daily torpor to help reduce their overall energetic costs. During daily torpor, small birds and mammals weighing <1 kg allow their body temperatures to drop close to ambient levels and reduce their metabolic rates substantially. This metabolic strategy is more common in small mammals than in small birds. It occurs only in a few caprimulgids, swifts, mousebirds, swallows, a few high-elevation sunbirds, and a few titmice and manakins when they are faced with low ambient temperatures (McNab 2002). The ability to undergo daily torpor has allowed hummingbirds to radiate extensively at mid- to high elevations in the Andes in the face of low (nighttime) temperatures and flowers that produce dilute nectar (McGuire et al. 2007). In addition to undergoing daily torpor, certain Andean hummingbirds, such as species of *Oreotrochilus*, manage to thrive at elevations as high as 4,500 m by perching, rather than hovering, to feed and by roosting and nesting in sheltered microhabitats such as caves and tunnels (McNab 2002).

DIGESTIVE AND METABOLIC ADAPTATIONS OF NECTAR-FEEDING MAMMALS

As in their feeding morphology, specialized nectar-feeding bats share many physiological features with their avian counterparts. To store their liquid diet, glossophagines have larger stomachs with an enlarged fundic caecum than their insectivorous ancestors; as in other bats, their intestines are short and lack a caecum (Forman et al. 1979). They quickly and efficiently digest nectar and pollen. Digestive efficiency of nectar in glossophagines can be as high as 99%, and they digest the contents of >75% of the pollen grains they ingest (compared with <75% in occasional nectar-feeding phyllostomids; Herrera 1999; Herrera and del Rio 1998; Kelm et al. 2008). Law (1992b) reported similar digestive efficiencies in the specialized nectar-feeding pteropodid *Syconycteris australis*. Like avian nectarivores, glossophagine bats also have low nitrogen requirements and are highly conservative of ingested ni-togen. Metabolic fecal nitrogen and exogenous urinary nitrogen, two major routes of nitrogen loss, are much lower in *Glossophaga soricina* than in other eutherian mammals (Herrera and Mancina 2008).

Hovering is a unique foraging method in hummingbirds and glossoph-agine bats, and the biochemistry and physiology involved in hovering is remarkably similar in both groups. Like hummingbirds, these bats use re-cently ingested and quickly assimilated carbohydrates to fuel up to 80% of the cost of hovering (Suarez et al. 2009). Because of their longer and less heavily loaded wings, however, the cost of hovering in glossophagines is less than that of hummingbirds on a mass-specific basis and is comparable to the cost of slow forward flight (Suarez et al. 2009; Voigt and Winter 1999). The pectoralis muscles of hovering birds and bats are also richer in mitochon-dria and have much higher rates of oxidation of glucose and fats than their nonhovering relatives (Suarez et al. 2009).

Unlike hummingbird flowers that typically produce sucrose-rich nectar, flowers attracting nectar-feeding bats are usually hexose-dominated (Baker et al. 1998). Nonetheless, like hummingbirds and other nectar-feeding birds, phyllostomid bats such as *Anoura geoffroyi* (a glossophagine), *Artibeus ja-maicensis*, and *Sturnira lilium* (both frugivorous stenodermatines) do not necessarily prefer equicaloric solutions of fructose and glucose over those of sucrose. Instead, like most mammals, these bats prefer sucrose solutions (Herrera 1999). Rodríguez-Peña et al. (2007) reported that two other glos-sophagines (*Glossophaga soricina* and *Leptonycteris curasoae*) did not prefer sucrose-dominated nectars over hexose-dominated nectars. Instead, they always preferred nectars with high sugar concentrations regardless of sugar type. Law (1992a) reported a similar result for the pteropodid *Syconycte-ris australis*. These results suggest that, contrary to common belief, factors other than sugar composition may determine the food preferences of nectar bats and that these bats don't necessarily provide strong selection pressure on the nectar characteristics of their food plants. Finally, as in nectar-feeding birds, the "intake response" (i.e., the amount of nectar ingested per unit time) of glossophagine bats such as *G. soricina* and *L. curasoae* is sensitive to the sugar concentration in nectars: intake rate is inversely related to sugar concentration and is independent of the sugar composition of experimental nectar solutions. Ayala-Berdon et al. (2008) concluded from this that neither sucrose hydrolysis nor rate of hexose assimilation by the intestine influences the intake responses of nectar-feeding birds and bats. Instead, dealing with large amounts of water at low sugar concentrations likely has a stronger effect on intake response than digestive factors.

As noted above, water regulation and electrolyte conservation are major physiological challenges for nectar feeders. As in nectar-feeding birds, kidney structure of nectar bats (and frugivores) differs from that of their insectivorous relatives. Specifically, these bats have smaller renal medullas, where urine concentration and electrolyte reabsorption occur, relative to overall kidney size (Schondube et al. 2001). As a result, nectar bats produce much more dilute urine than insect-eating bats. Detailed analysis of renal function in *G. soricina* indicates that it cannot regulate the amount of water that is absorbed by the gastrointestinal tract; its water load increases with an increase in water intake (Bakken and Sabat 2007). Instead, under a heavy water load this bat reduces the fraction of water reabsorbed by the kidney, which results in an increase in the production of dilute urine. Unlike nectar-feeding birds and certain other glossophagines, this species also lacks an exceptional electrolyte-conserving capacity and apparently must eat insects to obtain its electrolytes.

Finally, diet plays an important role in determining basal rates of metabolism in bats. In an early review, McNab (1982) reported that, relative to the "Kleiber curve" for eutherian mammals (i.e., the curve relating mass-specific basal rates of metabolism [Y] to body mass [X]), glossophagines have higher than expected mass-specific basal metabolic rates whereas frugivorous phyllostomids and pteropodids have "expected" basal rates of metabolism (also see McNab 2003). Unlike the situation in birds, where hummingbirds have lower than expected basal rates of metabolism, low rates in bats are found mostly in insectivorous families. More recent data indicate that two endemic West Indian nectar bats (*Monophyllus redmani* and *Erophylla sezekorni*) have substantially lower basal rates of metabolism than other phyllostomid nectar feeders (Rodríguez-Duran 1995). With few exceptions (e.g., the New Guinean blossom bat *Syconycteris australis*; Bonaccorso and McNab 1997), nectar-feeding bats also differ from hummingbirds in that they do not undergo daily bouts of torpor. Nor do they hibernate like most temperate zone bats. Their inability to use torpor to conserve energy has undoubtedly constrained glossophagines from radiating at mid- to high elevations in the Andes, in strong contrast to the situation in hummingbirds. Finally, the DEEs of glossophagine bats are comparable to those of similar-sized nectar birds and range from 40 to 56 kJ/day, which they obtain in 4–5 hr of flight per night (Horner et al. 1998; Kelm et al. 2008).

FORAGING AND SOCIAL BEHAVIOR
IN NECTAR-FEEDING BIRDS

Compared with their insectivorous ancestors or relatives, nectarivores are faced with a food supply that 'wants to be found.' Consequently, we expect these birds to spend less time per day searching for food than their insectivorous relatives. Because of their small size, we also expect hummingbirds to spend less time foraging each day than the larger sunbirds and honeyeaters. Available data support these predictions. Hummingbirds spend far less time foraging each day than swifts, their closest relatives, which spend much of their lives on the wing. They usually spend ≤40% of their day foraging for nectar and insects compared with 60–75% in sunbirds and honeyeaters (Collins and Paton 1989).

Being highly conspicuous and usually clumped in space, flowers are easily defended resources. As a result, territorial behavior involving defense of flowers is common in hummingbirds, sunbirds, and honeyeaters. Territorial behavior is widespread among male hummingbirds, and females are usually subordinate to males, even in species in which females are larger than males (Altshuler et al. 2004). In tropical forests, territorialists tend to be of medium size (4–5 g) and have straight bills. They tend to defend clumps of flowers that offer moderate energy rewards per flower. In both sunbirds and honeyeaters, interspecific territorial defense is based on size with large species being dominant over small species. Large species tend to feed at the high end of flower resource density gradients and small species at the low end (Feinsinger and Colwell 1978; Ford and Paton 1977).

In addition to territorial behavior, Feinsinger and Cowell (1978) recognized four other foraging modes in tropical hummingbirds: (1) high-reward trapliners of relatively large size with curved bills that match particular flowers with which they are coevolved; this group includes many hermit hummingbirds; (2) low-reward trapliners of medium size with smaller, straighter bills that are more generalized in the flowers they visit (e.g., *Chlorostilbon*); (3) territory parasites that are either large, aggressive species that can temporarily displace territorial species (e.g., *Anthracothorax*, *Florisuga*) or small furtive ones (e.g., *Philodice*, *Lophornis*); and (4) generalists of medium size with small bills. It should be noted that these foraging modes are not necessarily fixed characteristics of species but can be context dependent. Different species change their foraging behavior in response to changes in hummingbird community composition and/or resource levels and dispersion patterns

(Feinsinger and Colwell 1978). A similar degree of foraging specialization does not appear to exist in sunbirds or honeyeaters.

Flowers tend to be patchily distributed in both time and space, and as a result, high mobility, expressed as seasonal habitat shifts, altitudinal and latitudinal migrations, or nomadic behavior, is common in nectar-feeding birds (reviewed in Fleming 1992). Tropical understory insectivorous birds, in contrast, tend to be much more sedentary than nectarivores (Levey and Stiles 1992). As a result of this mobility, communities of tropical humming-birds undergo substantial seasonal changes in composition as different spe-cies enter and leave in response to changes in flower density (e.g., Feinsinger 1976, 1980; Stiles 1980; chap. 3). Australian honeyeaters tend to be highly nomadic because they live in a world of especially volatile flower resources (Ford 1985; Keast 1968).

Most birds are socially monogamous, but hummingbirds are a conspicu-ous exception. Although monogamy is known in a few species, most hum-mingbirds are polygynous with males providing only sperm and no parental help to females. It has been suggested (e.g., Johnsgard 1997; Stiles and Wolf 1979) that a major reason for this is the nature of hummingbirds' food sup-ply. Reduced foraging time frees males to devote more time to courting fe-males. And a limited food supply as well as small body size restricts females to laying small clutches of one or two eggs that they can care for alone. One interesting variation on the theme of male polygyny is lek mating, in which adult males congregate in traditional display sites to attract mates using vo-cal displays (in hummingbirds). Lek mating is particularly common in her-mit hummingbirds, which usually forage via traplining in the understories of tropical forests. These males spend most of the day "singing" or sitting quietly on their leks, and females visit leks only to mate. Unlike the more brightly adorned and sexually dichromatic trochilines, hermits are usually duller in color and sexually monomorphic.

Finally, lorikeets are Australasian nectar-feeding parrots. They differ from other specialized nectar-feeding birds in being larger (20–240 g) and more gregarious. They share a brush-tipped tongue and gaudy plumage with smaller avian nectar feeders. Compared with their seed-eating rela-tives, lorikeets have longer, thinner bills, and reduced gizzards. Like most parrots, lorikeets are sexually monochromatic and monogamous. In much of their geographic range, which includes Indonesia, Papua New Guinea, Australia, and South Pacific islands, these strong-flying birds tend to be highly nomadic in response to substantial spatiotemporal variation in their food supplies. Despite being gregarious, individuals can be highly aggressive among themselves when feeding in trees and shrubs which makes them noisy and highly conspicuous members of tropical and subtropical Australasian communities.

FORAGING AND SOCIAL BEHAVIOR IN NECTAR-FEEDING MAMMALS

As discussed in chapter 4, both territorial and nonterritorial foraging behaviors occur in nectar-feeding bats. Territorial feeding appears to be more common in pteropodids than in phyllostomids and occurs in both solitary- and gregarious-roosting species. Based on our current knowledge, solitary pteropodids such as *Syconycteris australis*, *Macroglossus minimus*, and *Melonycteris melanops* appear to defend patches of understory flowers whereas gregarious pteropodids such as various species of *Pteropus* defend small patches of flowers (and fruits) within the canopies of flowering trees (Bonaccorso et al. 2005; Richards 1995; Law 1996; McConkey and Drake 2006; Winkelmann et al. 2003). In contrast, nonterritorial traplining behavior is thought to be common in glossophagine bats, although this has not yet been confirmed with radio-tracking studies. Most of these bats forage solitarily. An exception to this is the glossophagine *Leptonycteris yerbabuenae*, which sometimes forages at flowers of columnar cacti and paniculate agaves in small groups (Horner et al. 1998; Rocha et al. 2006).

Like their avian counterparts, nectar-feeding bats can be highly mobile annually and undergo altitudinal or latitudinal migrations. In western Mexico, for instance, many females of *L. yerbabuenae* spend the fall and winter living in tropical dry forest before migrating 1,000 km north to the Sonoran Desert in the spring prior to giving birth; few adult males of this species migrate away from tropical dry forest (Fleming and Nassar 2002). Seasonal migrations by plant-visiting bats are also known to occur at low latitudes in Costa Rica, Central Africa, Malaysia, and Australia (Hodgkison et al. 2003; Richards 1995; Richter and Cumming 2006; Thomas 1983; Tschapka 2004). Like honeyeaters and lorikeets, some Australian pteropodids (e.g., *Pteropus scapulatus*) exhibit nomadic behavior in response to spatiotemporal variation in the availability of eucalyptus bloom (Vardon et al. 2001).

Polygyny is by far the most common mating system in mammals, and bats are no exception to this (Clutton-Brock 1989). Recent reviews of the mating systems of tropical bats indicate that harem or haremlike structures, in which single males defend a group of females either year round or seasonally are common in many families (McCracken and Wilkinson 2000; Wilkinson and McCracken 2003). This mating system is likely to be common in gregariously roosting pteropodid and phyllostomid bats. Most data on mating systems in these two families come from frugivorous species, and the mating systems of nectar feeders are poorly known. In phyllostomids, leklike mating has been reported in the Greater Antillean nectarivore/frugivore *Erophylla sezekorni* (Murray and Fleming 2008), and the highly gregarious *Leptonycteris curasoae* appears to have a polygynous (probably promiscuous) mating system (Nassar et al. 2008). Lek mating is known or strongly suspected to occur in a clade of African fruit- or fruit/flower-feeding pteropodids (i.e., *Hypsignathus, Epomophorus, Epomops*). Overall, however, roosting ecology rather than feeding ecology appears to be the major determinant of mating systems in bats.

Few other species of mammals feed regularly at flowers and are effective pollinators. Lemurs visit the large flowers of *Ravenala madagascariensis* in Madagascar, and kinkajous and opossums visit large flowers of species of *Ochroma* in the New World tropics (Janson et al. 1981; Kress et al. 1994). Corlett (2004) reported that, in addition to pteropodid bats, a variety of other mammals, including tree squirrels, flying squirrels, macaques, and civets, are known to visit flowers in the Indomalayan region. Except for bats, we view most of these species as opportunistic flower visitors that take advantage of the relatively large nectar rewards offered by some bat-pollinated flowers. Exceptions to this include certain murid rodents in South Africa (e.g., *Acomys subspinosus, Aethomys namaquensis*) and the Australian honey possum (*Tarsipes rostratus*), which appear to be dedicated flower visitors and are likely to be effective pollinators (Fleming and Nicolson 2002; Johnson et al. 2001; Kleizen et al. 2008; Wiens et al. 1979; Wooller et al. 1993).

General Discussion and Conclusions

Our major goal in this chapter has been to provide a detailed account of the pollination mutualism between tropical and subtropical plants and their avian and mammalian pollinators and to discuss its evolution. This mutualism encompasses at least 750 genera of plants from over 60 families and a substantial number of species of birds and bats (table 1.1). It provides many compelling examples of evolutionary convergence and complementarity. Below we discuss these concepts in detail and conclude that substantial evidence exists to support the hypothesis that this mutualism has had a profound effect on the evolution of flowering plants and their vertebrate pollinators.

THE EVOLUTION OF VERTEBRATE-POLLINATED FLOWERS

Flowers pollinated by birds and bats differ in many traits, including their size and shape and the quantity and quality of nectar, from those pollinated by insects. As seen in figure 7.12, vertebrate-pollinated flowers clearly occupy different regions of multivariate floral "niche space" from those pollinated by other animals (also see Ollerton et al. 2009). For example, the flowers of Asian gingers pollinated by spiderhunters (Nectariniidae) have longer and narrower floral tubes and more robust pistils and stamens than those pollinated by halictid and *Amegilla* bees (fig. 7.12*A*). Likewise, flowers of Caribbean Gesneriaceae pollinated by hummingbirds differ from those pollinated by bats in shape (tubular vs. campanulate), nectar concentration (8.3%–25.3% vs. 8.9%–18.2%), and color (red vs. yellow or green), among other characteristics (fig. 7.12*B*). From a macroevolutionary perspective, birds and bats (and other mammals) represent distinctly different "adaptive zones" or, in the terminology of Thomson and Wilson (2008), they are different "evolutionary attractors" for flowering plants. In evolving to attract nectar-feeding vertebrates as their main pollinators, flowers of many families (over 60 families for both birds and bats; Fleming and Muchhala 2008; Fleming et al. 2009) have converged in many of their features to form relatively distinct pollination syndromes. Here we discuss the extent of this convergence and the extent to which it can be attributed to selection by vertebrate pollinators.

Convergence has occurred in many aspects of floral biology, including timing of anthesis, flower size, shape, method of presentation, color, scent, and nectar; it has occurred at many taxonomic levels—within genera, subfamilies, and families; and it reflects the opportunistic and labile nature (within certain phylogenetic limits) of floral evolution. There can be no doubt that most of this evolution has been driven by coevolution between plants and their pollinators. The strongest evidence for this comes from the independent evolution of different flower syndromes multiple times within particular plant genera and subfamilies. Compelling examples of this occur within *Salvia* (Lamiaceae) and *Erythrina* (Fabaceae), in which somewhat different floral morphologies have evolved each time a pollinator switch has

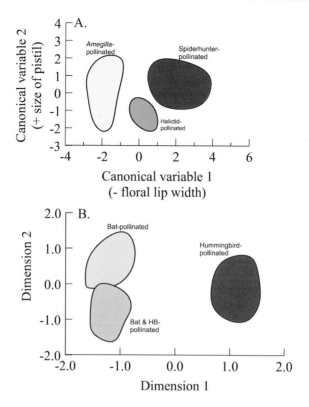

Figure 7.12. Examples of multivariate plots of flowers pollinated by different groups of animals. *A*, 44 species of gingers pollinated by two kinds of bees and spiderhunters (sunbirds) based on canonical correspondence analysis. *B*, 23 species of Gesneriaceae pollinated by bats, hummingbirds, or both based on a multidimensional scaling analysis. Based on (*A*) Sakai et al. (1999) and (*B*) Martén-Rodríguez et al. (2009).

occurred between bees and hummingbirds or perching birds and hummingbirds, respectively (Bruneau 1997; Wester and Classen-Bockhoff 2007).

A pervasive theme in the evolution of flowers is that vertebrate pollination has evolved from insect pollination more often than the reverse. The general trends associated with this evolutionary transition include: (1) an increase in flower size, including sexual parts; (2) an increase in volume of nectar; and (3) a decrease in the relative sugar content of nectar. Trends 1 and 2 reflect the larger size and larger energy budgets of birds and mammals compared with most insects; trend 3 reflects differences between insects and birds and bats in how nectar is extracted from flowers. Many insects extract nectar slowly from flowers using a hollow proboscis whereas birds and bats use a brush-tipped tongue and capillary action to quickly remove nectar from flowers. Within these general trends, there are many variations that reflect plant phylogeny (e.g., basic flower and inflorescence ground plans; Endress 1994, 2011) and biogeography (e.g., New World vs. Old World nectar-feeding birds and bats; Fleming and Muchhala 2008).

There are two additional themes associated with the evolution of vertebrate-pollinated flowers. One is that the evolution of these flowers often involves modification of existing phenotypes rather than evolving new phenotypes from scratch. Thus, in several New World families, tubular bat flowers have evolved numerous times from tubular hummingbird flowers (Fleming et al. 2009) and the colors of bird- or bat-pollinated flowers involve modifications of existing color phenotypes, sometimes as a result of simple allelic substitutions (e.g., Bradshaw and Schemske 2003). This theme emphasizes the importance of evolutionary facilitation, which we define as one group of organisms setting the stage for and promoting the evolution of another group of organisms. We will discuss this concept in greater detail below and in chapter 9. Another theme is the existence of evolutionary biases in which some transitions or changes are more likely than others (e.g., from insect flowers to hummingbird flowers rather than from bat to hummingbird flowers or the evolution of unpigmented flowers from pigmented flowers rather than the reverse). These biases probably reflect genetic and developmental constraints within particular plant lineages as exemplified by biochemical changes involved in anthocyanin-based pigment changes (Streisfeld and Rausher 2009).

Finally, biological differences within and between birds and mammals have had an obvious and strong influence on floral evolution. Birds and mammals differ in time of daily activity, color vision, and the importance of olfaction for food detection, among other traits. Within birds, nectar feeders differ significantly in size, hovering ability, and digestive physiology. Within bats, phyllostomids and pteropodids differ in the ability to echolocate, visual acuity (and the ability to see colors), and hovering ability. All of these differences are reflected in the size, form, and color of flowers as well as in their nutritional rewards. They result in taxon-specific differences (usually at the vertebrate family or subfamily level) in the distributions of flowers in multivariate ecomorphological space. Although many authors have discussed differences between bird- and bat-pollination syndromes, in reality there are multiple syndromes within both birds and bats. This multiplicity of flower forms (*sensu lato*) is the result of historical (biogeographic) differences in

the interactions between flowering plants and their vertebrate pollinators
(Fleming et al. 2009).

THE EVOLUTION OF NECTAR-FEEDING VERTEBRATES

In some respects, nectar (and pollen) is an odd food for vertebrates to be
eating. Compared with alternative foods such as insects, foliage, and seeds,
nectar is often more patchily distributed in space and time, is nutritionally
incomplete, and poses physiological problems associated with the overinges-
tion of water. In its favor, nectar is conspicuously advertised, usually non-
toxic, and very digestible. Clearly, the advantages of evolving a nectar-based
diet outweigh its disadvantages for many groups of animals, including vertebrates, but specializing on this diet has involved the evolution of rather
specialized morphology and physiology. Here we discuss the extent to which
birds and bats have converged on similar evolutionary solutions to the challenges posed by a diet of nectar and pollen. We organize this discussion
around three aspects of feeding: finding, acquiring, and processing food.

Finding food involves recognizing and responding to appropriate flower
cues and associated resources. Birds and bats differ fundamentally in the
way they do this. Birds rely solely on vision and visual cues to find flowers,
whereas bats rely on olfaction and, in the case of phyllostomids, echolocation, in addition to vision. Birds and bats also differ in their ability to see
colors, and hence bird flowers routinely contain color pigments, whereas bat
flowers are often unpigmented. It is likely that both birds and bats have excellent spatial memories so that they can relocate widely spaced flowers via
traplining behavior. Finally, highly concentrated floral resources often elicit
territorial resource defense behavior in nectar-feeding birds and pteropodid
bats; phyllostomid nectar bats, in contrast, seldom defend floral resources.
In sum, nectar-feeding birds and bats often rely on different sensory cues to
find food but often search for and defend it in similar fashion. Convergence
in these two groups is greatest in actual foraging behavior and is less strong
in sensory perception associated with food finding. Sensory differences, of
course, reflect basic biological differences between birds and mammals that
are independent of diet.

There has been extensive convergence within and between birds and bats
in the ways they acquire nectar. This convergence directly reflects the way in
which nectar is packaged and presented in flowers and hence is ultimately
botanically driven. Nearly all specialized avian and mammalian nectar feeders have elongated bills or rostrums and elongated tongues for extracting

nectar. An exception to this is lorikeets, which lack the long thin bills of other avian nectarivores. Although thinner than those of other kinds of parrots, lorikeet bills are still clearly parrotlike (a phylogenetic constraint). Without exception, all specialized nectarivores have brush-tipped tongues for rapid uptake of nectar, and the tongues of hummingbirds, sunbirds, and honeyeaters are tubular in shape; the tongues of nectar bats are not tubular, although some have lateral grooves. Tongue length varies within birds and bats and is generally correlated with bill or jaw length. It is likely correlated with the corolla lengths of a species' core set of flowers (see chap. 3). Finally, whereas nectar-feeding passerine birds, pteropodid bats, and nonspecialized nectar-feeding phyllostomid bats land on or near their flowers before feeding, hummingbirds and glossophagine bats usually hover at flowers and possess unique (but different) flight morphologies for doing this—an outstanding example of functional convergence.

Once nectar is ingested, it is quickly digested and assimilated by all vertebrate nectarivores. Digestive efficiency is very high, and hummingbirds and glossophagine bats quickly use nectar-derived energy for fueling their hovering flight behavior—an example of biochemical convergence. Unlike the highly efficient processing of nectar, however, the ability to extract proteins from pollen grains differs significantly among vertebrate nectarivores. Glossophagine and pteropodid bats frequently groom themselves and can extract the chemical contents of most of the pollen grains they ingest. In contrast, hummingbirds do not routinely ingest pollen and have little ability to digest the contents of pollen grains; instead they rely on insects for their protein. Sunbirds, honeyeaters, and lorikeets appear to be somewhat more proficient at digesting the contents of pollen grains than hummingbirds, but they are not as proficient at this as bats.

Vertebrate nectarivores differ in some of the ways in which they process the water they ingest with their nectar diet. They have not all converged on the same physiological solution to this problem. Compared with insectivores, the kidneys of nectar feeders have a reduced renal medulla and lack the ability to produce concentrated urine. However, unlike sunbirds and honeyeaters, hummingbirds and certain glossophagine bats cannot control the amount of water they assimilate from their intestines and thus pass large amounts of water through their kidneys. Many nectar-feeding birds and glossophagine bats can conserve the electrolytes they ingest in nectar in the renal cortex, but the glossophagine bat *Glossophaga soricina* lacks this ability.

The sugar composition of bird and bat nectars often differs—it is sucrose rich in most bird and pteropodid flowers but is hexose rich in glossophagine flowers—but birds and bats often show similar responses to equicaloric solutions that differ in sugar composition. At high sugar densities, most species prefer sucrose-rich solutions but prefer hexose-rich solutions at low sugar densities, perhaps for osmotic reasons. Thus, despite encountering nectars that differ substantially in their sugar compositions in nature, nectar birds and bats do not appear to be hardwired to prefer their "expected" sugar type. Instead, their sugar preferences appear to depend on sugar concentrations. This has led some researchers to question whether these animals have exerted strong selection pressure on the sugar composition of the nectar of their flowers.

Finally, there are metabolic consequences associated with a diet of nectar and pollen. The foods eaten by vertebrate nectarivores and frugivores are generally low in nitrogen, and as a result, these animals have evolved low nitrogen requirements. They conserve dietary nitrogen by having low excretion rates of urinary and fecal nitrogen. Many vertebrate nectarivores also have low basal metabolic rates compared with those of other dietary classes, although this is not true of most glossophagine bats. Low basal rates reduce daily energy costs and are perhaps an adaptation for coping with food supplies that often occur in low densities and are spatiotemporally variable.

Hummingbirds are unique among birds in that many species have the ability to undergo energy-saving daily bouts of torpor. Their small size undoubtedly predisposes them for this, but many other small birds, including most sunbirds and honeyeaters, lack this ability (McNab 2002). The use of torpor by hummingbirds is likely to be a consequence of their diet in addition to their small size (Johnsgard 1997). This physiological trait has been salutary in at least two respects. First, it allows these birds to feed on small packets of energy found in small flowers (e.g., many bee-pollinated flowers) in addition to feeding at larger flowers; that is, it has expanded their potential resource niche space to include a wider range of flowers than fed on by most other vertebrate nectarivores. Second, it has allowed them to colonize and radiate extensively in a wider range of habitats, particularly thermally challenging habitats at mid- to high elevations, than other nectar feeders. Glossophagine bats, by contrast, generally lack the ability to undergo torpor, and, as a consequence, they feed on a more restricted range of flower sizes and types and have radiated to only a modest extent at mid- to high elevations.

In summary, nectar-feeding birds and bats have converged in many aspects of their biologies, especially in morphological traits associated with food acquisition. This is not surprising given the degree to which their flowers have converged on similar suites of morphological and nectar characteristics. Degree of convergence is less striking at the physiological level, particularly in terms of pollen digestion and in the processing of water and electrolytes. Here, phylogenetic differences between birds and bats undoubtedly come into play as they also do in sensory physiology. And the use of torpor, which would benefit all small vertebrate nectar feeders by reducing their daily energy costs, is widespread only in hummingbirds. What constrains nectar-feeding passerines and phyllostomid and pteropodid bats from using torpor to save energy deserves further attention.

Although there are many similarities (convergences) in the biologies of nectar-feeding birds and bats, there still remains a strong imprint of earth history (biogeography) on these animals. Fleming and Muchhala (2008) emphasized this point by proposing that there are three avian pollinator "worlds"—a hummingbird world, a sunbird world, and a honeyeater world—but only two chiropteran worlds—a phyllostomid world and a pteropodid world. These worlds are based on degree of morphological and ecological specialization between pollinators and their food plants. New World nectar birds and bats appear to be more specialized than their Old World counterparts, with sunbirds of Africa and southern Asia being more specialized than Australasian honeyeaters. Throughout the Old World, pteropodid bats are less specialized flower visitors than phyllostomids.

New World hummingbirds and glossophagine bats have converged to an amazing degree in their morphological adaptations for harvesting nectar. These adaptations include features of their skulls (i.e., long bills/jaws and tongues) and their ability to hover at flowers. In part, this convergence is likely to be the result of evolutionary facilitation. The presence of hummingbirds and their flowers in many tropical habitats has facilitated the evolution of glossophagine bats by providing them with new floral resources to exploit. Whereas the "core" plant families for glossophagine bats are relatively large-flowered families or subfamilies such as Agavaceae, Cactaceae, Bignoniaceae, Bombacoideae, and Fabaceae, the diets of these bats also include species of small-flowered families such as Acanthaceae, Bromeliaceae, Campanulaceae, and Gesneriaceae—families that are usually associated with hummingbird pollination. In fact, it is likely that many of the bat-pollinated species in these latter families are derived from hummingbird-pollinated species (e.g., Fleming et al. 2009; Givnish et al., unpubl. ms.; Perret et al. 2003; Tripp and Manos 2008). Species in these families have relatively small tubular flowers that can be accessed most easily via hovering. Feeding at these flowers has undoubtedly selected for hovering ability in glossophagine bats. Thus, the presence of small hovering birds and their coevolved flowers in Neotropical habitats has facilitated the evolution of small hovering bats in the same habitats.

A similar close degree of convergence between nectar-feeding birds and bats has not occurred in the Old World. A major reason for this is that the "evolutionary theater" in which these animals have evolved is much larger and more complex biogeographically. Originally evolving in Asia/Australasia, pteropodid bats have only recently colonized Africa, where the bulk of sunbirds occur. Only one morphologically specialized pteropodid nectar bat currently occurs in Africa, and its diet, as well as the diets of more generalized African pteropodids, is substantially different from that of sunbirds (Fleming and Muchhala 2008; chap. 6). Throughout the Old World, pteropodid bats feed primarily on large flowers (or large inflorescences) produced by canopy trees or large lianas rather than flowers produced by shrubs, vines, or epiphytes. Sunbirds, in contrast, feed on a much broader range of flower types, including those produced by shrubs, herbs, and hemiparasites, in addition to flowers produced by trees and vines (Fleming and Muchhala 2008). The diets of Australasian pteropodids and honeyeaters are more similar taxonomically than are the diets of African nectar bats and birds, and in Australasia both birds and bats have relatively generalized relationships with their food plants. Therefore, in neither Africa nor Australasia has the presence of nectar-feeding birds facilitated the evolution of specialized nectar-feeding bats the way it has in the New World.

It is tempting to suggest that another reason for the lack of convergence and/or facilitation between Old World nectar birds and bats lies in differences in the degree of evolutionary plasticity found in phyllostomid and pteropodid bats. Feeding diversity in the Phyllostomidae, which includes species that eat insects, blood, vertebrates, flowers, and fruit, is easily the highest among all bats if not among all mammals (Baker et al. 2012). As seen in figure 7.9, the morphospace occupied by nectar- and fruit-eating phyllostomids along the axis of relative jaw length (Y) is much greater than that of plant-visiting pteropodids, which have radiated more in overall size (the X axis) than in jaw diversity. Including other phyllostomid feeding modes in figure 7.9 would greatly expand the overall morphospace occupied by this family (see fig. 6–3 in Fleming 1991a). Plasticity in skull characteristics is unusually high in phyllostomids (e.g., Freeman 1988, 1995), which has allowed them to radiate into several distinctly different feeding zones. Like most other families of bats, pteropodids lack this morphological plasticity. This situation is reminiscent of differences in the evolutionary radiation of carduline finches (Fringillidae) and thrushes (Turdidae) in the Hawaiian Islands. In Hawaii, thrushes have undergone a modest, morphologically restrained radiation that has produced five species living allopatrically on different islands, whereas in the same period of time, finches underwent an extensive morphological radiation that produced about 50 species of Hawaiian honeycreepers (Ziegler 2002). Lovette et al. (2002) showed that, like phyllostomid bats, carduline finches worldwide are much more variable morphologically than thrushes and hence are evolutionarily more plastic.

Conclusions

The pollination mutualism is a relatively specialized plant-animal interaction in which pollen must be placed rather precisely on stigmas in order for it to "work." As a consequence, many flowers have evolved relatively specialized morphology to maximize the capture of conspecific pollen grains on their stigmas. This morphology has, in turn, selected for relatively specialized morphology in nectar- and pollen-feeding vertebrates and other kinds of pollinators to extract nectar efficiently from flowers. Much of the "macroevolution" associated with this mutualism has involved a high degree of convergence and complementarity in both plants and animals. This macroevolution has important conservation implications. Being a relatively specialized interaction, the pollination mutualism is likely to be more vulnerable to disruption if either of its botanical or animal participants should go extinct than is the fruit-frugivore mutualism that we discuss next. To be sure, both pollinators and seed dispersers are needed to maximize the reproductive success of tropical plants, but without pollinators, the importance of seed dispersers becomes moot. In the end, however, both mutualisms need to be conserved in order to have sustainable tropical forests.

8   The Frugivory Mutualism

Unlike vertebrate-pollinated flowers, which are most common in advanced lineages of angiosperms, fleshy fruits eaten by vertebrates are distributed throughout angiosperm phylogeny—in basal as well as in advanced lineages. A major reason for this is that consumption of fruits and dispersal of their seeds by vertebrates does not necessarily require specialized morphology or physiology. At certain times and places, all major groups of contemporary vertebrates eat fruits and are potential seed dispersers, and vertebrates have been doing this long before the evolution of modern groups of frugivores (Tiffney 2004).

As in angiosperm flowers, the evolution of taxon-specific frugivory syndromes has resulted from coevolution within mutualistic networks as well as generalized coevolution between vertebrate frugivores and angiosperm fruits (chap. 5). Fruits dispersed by different groups of birds and mammals often differ in color, odor, size, and nutritional reward (table 1.5). Bird-dispersed fruits tend to be smaller, more colorful (or black), and lack distinctive odors compared with mammal-dispersed fruits. As discussed in detail below, many plant families have converged independently on these fruit characteristics. On the animal side of this mutualism, frugivorous birds and mammals have evolved a variety of adaptations to deal with a diet of fruit. As we did in chapter 7, here we describe the nuts and bolts of this mutualism by focusing on the concepts of convergence and complementarity as they apply to fruit-producing plants and frugivorous animals. As an introduction to this topic, we will first provide an overview of Neotropical bats and the fruits they consume before examining in detail the evolution and adaptation of fruits and their vertebrate consumers.

Interactions between Frugivorous Bats and Fruits in the Neotropics

Bats of the family Phyllostomidae are major consumers of fruit and dispersers of seeds in the New World tropics (chaps. 2, 4). Their fruit diets and the morphological characteristics of their fruit are particularly well-known in central French Guiana as a result of intensive research by scientists at the New York Botanical Garden. The following account is based on their recent monograph (Lobova et al. 2009).

Although many species of phyllostomid bats are known to occasionally include fruit in their diets, members of two subfamilies, Carolliinae and Stenodermatinae containing a total of about 76 species, have traditionally been considered to be specialized frugivores. Size range in these two subfamilies is 5–73 g, which makes them relatively small seed dispersers. In central French Guiana, these two subfamilies contain three and 20 species, respectively, and their diets are known to include at least 112 species of fruit from 50 genera and 31 families. These plants represent about 6% of the flora of the region, but if an additional 179 species that are suspected to be bat dispersed are added to this list, about 15% of the native flowering plants of this region rely on bats as their principal seed dispersers. Bats are clearly major dispersers of seeds in this region as well as throughout the New World tropics and subtropics.

Bat-dispersed fruit species represent a nonrandom segment of the Guianan flora from several different perspectives. First, they represent a nonrandom selection of plant families. In terms of number of bat-dispersed species, the top 10 families in decreasing order are Solanaceae, Piperaceae, Moraceae, Araceae, Myrtaceae, Clusiaceae, Cecropiaceae (part of Urticaceae in APG III), Arecaceae, Cactaceae, and Chrysobalanaceae. These families are broadly distributed across angiosperm phylogeny and include basal angiosperms, monocots, and advanced eudicots. Missing from this list are Fabaceae, the dominant tree family in most Neotropical forests, and Melastomataceae and Rubiaceae, two important understory shrub families that are mostly bird dispersed (chap. 2). Second, in terms of plant growth form, bat-dispersed species occur more frequently among epiphytes, shrubs, and treelets and less frequently among canopy trees and terrestrial herbs than expected. Their relatively small size prevents phyllostomid bats from eating large-seeded fruits produced by canopy trees, such as those that occur in the diets of trogons, toucans, and primates. The maximum size of a *Ficus* fruit that the 50–70 g *Artibeus lituratus* can carry is less than 35 g (Kalko et al. 1996). Third, although bats eat fruit produced by both early and late successional plants, a higher percentage of species are early successional than late successional. Fruits of *Piper*, *Solanum*, *Vismia*, and *Cecropia* are staples in the diet of *Carollia* and *Sturnira*, two common understory-feeding phyllostomids, as are the fruits of Moraceae (e.g., *Ficus*) in the diets of canopy-feeding stenodermatids (chap. 5). These bats are especially important in the regeneration of Neotropical forests (Muscarella and Fleming 2007). Fourth, bats eat a wide variety of fruit types, including arillate fruits (e.g., *Lecythis*) and those with swollen pedicels (e.g., *Anacardium*) or swollen perianths (e.g., *Cecropia*), but fleshy berries containing small seeds are the most common type. Finally, green is the predominant color of bat-dispersed fruit whereas white or black fruits, which are primarily bird dispersed, are more common in the flora of central French Guiana.

Because the diets of fruit-eating birds and primates have also been studied in central French Guiana, Lobova et al. (2009) were able to determine the extent to which 111 species of bat fruits are also eaten by birds and arboreal mammals. They reported that 35 of these species (32%) are also eaten by primates, 18 species (16%) are eaten by birds, and 10 species (9%) are eaten by opossums and kinkajous. Bats and primates shared fruit species most commonly in Araceae, Chrysobalanaceae, and Moraceae; bats and birds shared fruit species most commonly in Arecaceae, Cecropiaceae, Marcgraviaceae, and Melastomataceae. Families whose fruit are nearly exclusively eaten by bats include Clusiaceae, Cyclanthaceae, Piperaceae, and Solanaceae. Overall, 66 of the 111 species (59%) were exclusively eaten by bats whereas the other 45 species (41%) were shared by one or more other groups of vertebrates. Available data do not allow us to determine how effectively birds and arboreal mammals disperse seeds of shared fruit species compared with bats, nor do they provide any insights into whether vertebrates compete for shared species. These data do indicate, however, that a substantial number of bat fruits are consumed by other vertebrates and that the morphological or nutritional characteristics of bat (or bird or primate) fruits do not prevent other kinds of animals from eating them. Opportunistic feeding clearly is not uncommon among tropical vertebrate frugivores (also see chap. 1). As a result, the evolution of fleshy tropical fruit is likely to involve coevolution

involving animals that differ greatly in size, morphology, and foraging behavior to a much greater extent than the evolution of vertebrate-pollinated
flowers.

The Evolution of Vertebrate-Dispersed Fruits

PHYLOGENETIC CONSIDERATIONS

Many more plant families produce fleshy, vertebrate-dispersed fruits than
vertebrate-pollinated flowers, and fleshy fruits dispersed by vertebrates have
been common in angiosperms for more than 65 Ma (Eriksson et al. 2000;
Friis et al. 1987; Tiffney 2004). For example, Lobova et al. (2009) reported
that fruits eaten by phyllostomid bats occur in at least 62 families whereas
flowers pollinated by these bats occur in about 44 families (Fleming et al.
2009). As discussed by van der Pijl (1982), fleshy fruits consumed by modern
vertebrate frugivores come in a variety of forms produced by a diverse array
of developmental pathways (also see Coombe 1976, fig. 1). He classified these
fruit into four basic types, based on the plant tissue providing the nutritional
reward: seeds with fleshy sarcotestas (seed coats); arilloid fruits; pulpa (i.e.,
intralocular tissue produced by intense meristematic activity); and pericarp
fruit (table 8.1). These types and many others have evolved repeatedly during angiosperm evolution. Spujt (1994) thoroughly discusses the diversity of
angiosperm fruit types.

To begin to understand this evolution, we ask, What is the distribution
of different fruit types among the five major lineages of angiosperms? To
answer this, we used our biogeographic database of 102 major plant families
producing fleshy fruit eaten by tropical birds and mammals (chap. 6) and
data in Heywood et al. (2007) to determine the kind(s) of fruit found in each
family. We focused on the following fruit types in this analysis: berries and
berry-like fruits (e.g., baccate fruits and syconia); drupes and syncarps; arillate fruits; dry fruits such as capsules, schizocarps, and nuts; and mixed fruit
types, including both dry and fleshy fruits. For each of the five angiosperm
lineages, we tallied the number of families characterized by each of these
fruit types. Our database of plant families contained the following overall
distribution of fruit types: berries, 18% of the families; drupes, 20%; arillate fruits, 5%; capsules, 6%; and mixed fruit types, 51%. The distribution of
these fruit types was not random among the five plant lineages. Most families of basal angiosperms and monocots producing fleshy fruits are characterized by a single fruit type (i.e., berries, drupes, or arillate fruits) whereas

Table 8.1. Summary of Some Basic Angiosperm Fleshy Fruit Types

Fruit Type	Source of Flesh or Pulp	Occurrence in Angiosperms
Seeds with fleshy sarcotesta	Seed coat	Basal angiosperms (e.g., Annona-ceae, Magnoliaceae); monocots (e.g., Arecaceae, Liliaceae); eudicots (e.g., Euphorbiaceae, some Fabaceae, Melia-ceae, Sapindaceae)
Arilloid	Multiple sources, including swelling of the raphe, swelling near the micro-pyle, an encircling structure around the exostome of the micropyle (an arilloid), and a true aril near the top of the funicle and around the seed	Many families in both monocots and dicots; common in all 3 subfamilies of Fabaceae
Pulpa	Fleshy endocarp protected by a hard pericarp	Basal angiosperms (e.g., Annonaceae); monocots (e.g., Marantaceae, Musaceae); eudicots (e.g., Clusiaceae, Fabaceae, Sapindaceae)
Pericarp	Berries and drupes in which the carpel has taken over nutrition from sarcotestas and arils	Many families throughout angiosperm phylogeny

Source. Based on van der Pijl (1982).

a majority of families of basal eudicots, rosids, and asterids (≥64%) contain mixed fruit types (fig. 8.1). In three of the four lineages containing mixed families (monocots, rosids, and asterids), the most common combination was dry capsules and drupes and/or berries (in ≥65% of those families) whereas only one of six mixed families (17%) in basal eudicots contained this combination. Thus, about half of the families of advanced angiosperms whose fruits are eaten by tropical vertebrates produce a variety of different fruit types, most commonly capsular fruits as well as fleshy fruits. The evolution of fruit types and seed dispersal strategies has clearly been complex in these families.

What is the taxonomic distribution of fruit types within the mixed families? Do dry fruits such as capsules or samaras usually occur in different subfamilies or tribes than fleshy fruits (i.e., how often is there an intrafamilial phylogenetic signal?) or do dry and fleshy fruits usually co-occur within the same subfamilies or tribes? These questions are similar to the ones we asked in chapter 5 in our discussion of pollinator-related speciation patterns in tropical plants. In that chapter, we identified two basic speciation models: a "plastic" model, in which clades contain species adapted to several different kinds of pollinators and a "constrained" model, in which clades contain species attracting only one kind of pollinator. Here we ask an analogous question about the distribution of fruit types within mixed families: to what extent is fruit evolution plastic or constrained in these families? We used fruit data and the within-family classifications recognized by Heywood et al.

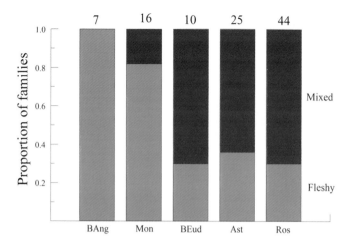

Figure 8.1. Proportions in a sample of 102 vertebrate-dispersed angiosperm families that produce only fleshy fruits (berries, drupes, or arils) or a mixture of dry (capsular) and fleshy fruit types in major angiosperm lineages. Numbers above the bars indicate number of families. *BAng* = basal angiosperms, *Mon* = monocots, *BEud* = basal eudicots, *Ast* = asterids, *Ros* = rosids. Data come from Heywood et al. (2007).

(2007) to address this question. Of the total of 46 families in this analysis, 13 (28%) appear to have phylogenetically constrained distributions of fruit types, 15 (33%) appear to be plastic in the distribution of their fruit types; we could not answer this question for 18 families (39%) because of the absence of well-recognized subfamilies or tribes. The percentage of families with constrained fruit distributions in monocots, basal eudicots, rosids, and asterids was 25%, 0%, 29%, and 38%, respectively. Assuming that the "unknown" families in these lineages are more likely to be plastic than constrained, it appears that fruit types in angiosperms are more likely to evolve without intrafamilial phylogenetic constraints than with such constraints. Our overall conclusion here, then, is that, among advanced lineages of angiosperms, the evolution of fruit types, and hence dispersal mechanisms, is evolutionarily labile, at least at the level of subfamily or tribe. We discuss this topic in more detail below.

Which families exhibit constrained evolution of fruit types, and what are the geographic distributions of their fruit types? Is there a biogeographic pattern to the evolution of fleshy fruits in mixed angiosperm families? In monocots, one of three subfamilies (Bromelioideae) of the New World Bromeliaceae produces berries; the other two subfamilies (or six according to Givnish et al. 2011) produce capsules. Among rosids, one of five subfamilies of Apocynaceae (pantropical Rauvolfioideae) produces drupes or berries (but see Simoes and Marques [2007] for a reclassification of this family). Berries are produced in four of six tribes in the Clusiaceae; these tribes occur in both the New and Old Worlds. In Meliaceae, the New World subfamily Swietenioideae produces only dry fruits, whereas the Old World subfamily Melioideae produces both fleshy and dry fruits. Similarly, in the two sub-families of Myrtaceae, most New World taxa produce fleshy fruits, whereas Australian fruits are dry. In the two subfamilies of Ulmaceae, the pantropical Celtioideae produces fleshy fruit, and the temperate Ulmoideae produces dry fruit. One of seven tribes of Zygophyllaceae (Balanitae, Old World trop-ics) produces drupes. Among asterids, one of three subfamilies of Capri-foliaceae (the pantemperate Caprifolioideae) produces fleshy fruit. One of four tribes of Loganiaceae (Strychneae, Old World tropics) is fleshy-fruited. In Oleaceae, one of three subfamilies (Oleoideae, mostly Old World tropics) contains fleshy-fruited tribes. One of two tribes of the Old World Pittospo-raceae (Billardiereae) contains fleshy fruit. Finally, Knapp (2002) recognized seven subfamilies in the Solanaceae and reported that four of these contain a single fruit type. Three subfamilies produce capsular fruits, while Schizan-thoideae (one genus of small Chilean shrubs) produces drupes; the large subfamily Solanoideae produces both fleshy and dry fruits. Overall, these data suggest that fleshy-fruited subfamilies or tribes are in the minority in the mixed families. Only 16 of 44 constrained subfamilies or tribes (37%) are exclusively fleshy-fruited or nearly so; eight out of 10 of these taxa are either pantropical (four taxa) or Old World (four taxa) in distribution. If there is a biogeographic bias in the occurrence of fleshy-fruited lineages in mixed families, it is apparently toward the Old World tropics, but additional data are needed before firm conclusions about this can be made.

EXAMPLES OF FRUIT EVOLUTION WITHIN MIXED FRUIT FAMILIES AND FACTORS INFLUENCING THIS EVOLUTION

Molecular phylogenies allow us to examine in detail the evolution of fruit types within mixed-fruit families. These phylogenies can reveal ancestral fruit conditions (e.g., dry capsular or fleshy fruits), how often fleshy fruits have evolved independently, and whether the evolution of fleshy fruits from capsular fruits is reversible. Results of available studies clearly support the hypothesis that the evolution of fruit types is labile in these families. In Sola-naceae, for example, Knapp (2002) recognized six basic fruit types (capsule, berry, berry with stone cells, drupe or pyrene, noncapsular dehiscent fruit, and mericarps [nutlets]) and mapped these fruit types onto a molecular phylogeny. Based on this phylogeny, capsular fruits are basal in the family and drupes occur in two near-basal clades. Berries are derived and have evolved independently at least three times. Only berries occur in the large genus *Solanum*, but they exhibit a diverse array of colors, sizes, and degree of woodiness. Dry capsular fruits have evolved several times from fleshy fruits so that the evolution of fruit type is reversible in this family. Pabón-Mora and Litt (2011) discuss the anatomy and development of these fruit types in detail.

Similar trends occur in Rubiaceae and Melastomataceae. Five fruit types (capsules, nuts, drupes, berries, and leathery "Gardenia" fruits with fleshy pulp) occur in Rubiaceae, and many-seeded capsules are basal (Bremer and Eriksson 1992). Many-seeded fleshy fruits have evolved at least 12 times from capsules, but evolutionary reversals are unknown, perhaps because of the complex morphology of capsules. Fossil evidence suggests that fruit morphologies in this family have undergone little evolution since their time of origin (i.e., the drupes of contemporary Psychotrieae are similar to those from the Oligocene), a situation that is likely to be common in many families of angiosperms (Herrera 1986, among others). Bremer and Eriksson (1992, 91) conclude: "The overall picture of evolution of fleshy fruits in the Rubiaceae is that basic changes in fruit type most likely resulted from shifts in the disperser fauna during the early Tertiary, but changes in the fauna since that time have had minor effect on fruit characteristics." As we discuss below, the idea that changes in fruit type are frugivore driven is controversial.

In Melastomataceae, basic fruit types include capsules, dry berries, fleshy berries, and dehiscent capsule-like fruits with fleshy placentas. Capsules are basal in this family; fleshy berries have evolved at least four times and dry berries twice; and capsular fruits are known to have evolved from fleshy fruits (Clausing and Renner 2001; Clausing et al. 2000). Because their flesh has two developmental sources (i.e., the ovary wall and hypanthium, which is formed from the fused bases of sepals, petals, and stamens), not all berries are homologous in this family.

What factors are involved in the evolution of different fruit types within angiosperm families? At least two factors—habitat type and availability of dispersers—have been discussed in the literature. To provide background for this discussion, we must digress a bit and review historical trends in seed size in angiosperms. We do this because seed size and fruit size are often

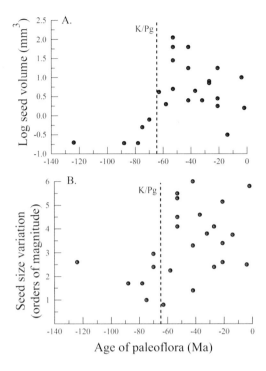

Figure 8.2.   Historical trends in (*A*) median seed size and (*B*) range in seed size (in orders of magnitude) in 25 paleofloras from the Early Cretaceous to the Pliocene. Based on Eriksson (2008).

correlated (Leishman et al. 2000; Mack 1993; Mazer and Wheelwright 1993; I. Wright et al. 2007), and seed size has been used as a surrogate for fruit size and dispersal mode in the fossil literature. In this literature, small seeds (i.e., <100 mg) are usually assumed to be abiotically dispersed whereas large seeds (i.e., >100 mg) are assumed to be biotically dispersed.

Historical trends in seed (fruit) size are shown in figure 8.2. As reviewed by Eriksson (2008), most angiosperm seeds during the Cretaceous were small (median volume = 1 mm³), but seed size and its interspecific variation increased markedly in the early Tertiary. Three nonexclusive hypotheses have been put forth to explain this pattern: (1) the "fruit-frugivore coevolution" hypothesis, in which direct coevolution between vertebrate frugivores and seeds in the early Tertiary favored large seed size; (2) the "recruitment" hypothesis, in which changes in (tropical) vegetation from open habitats in the Cretaceous to closed forest early in the Tertiary favored the evolution of large seeds and biotic dispersal; and (3) the "life-form" hypothesis, in which small shrubby Cretaceous angiosperms evolved into larger plants early in the Tertiary; trees, in turn, evolved larger seeds that required vertebrates for their dispersal.

As Eriksson (2008) indicates, there are likely to be elements of truth in each of these hypotheses. It is certainly true that (1) large plants tend to produce larger seeds and a greater range of seed sizes interspecifically than small plants; (2) on a per capita basis, large seeds are more likely to produce surviving seedlings than small seeds in shady habitats because of their greater energy reserves and tolerance of stress; and (3) large seeds are less mobile than small seeds, which would favor the evolution of biotic dispersal via the production of fleshy fruits (Leishman et al. 2000; Muller-Landau 2010). But the primary driver behind the evolution of biotic dispersal is still not clear. Was it the evolution of closed forests under the warm, wet conditions of the early Tertiary? Or the evolution of larger plants per se? Or the evolution of fruit-eating vertebrates? In the end, as Eriksson (2008) concludes, it is likely that large plant size, large seeds, and biotic dispersal evolved as coadapted traits in the early Tertiary and that the presence of fleshy-fruited plants in the Eocene set the stage for the radiation of modern groups of tropical frugivorous birds and mammals.

Returning to contemporary plants, habitat type appears to be a major factor in the evolution of dry capsular and fleshy indehiscent fruits in angiosperms. In many families (e.g., Ericaceae, Gesneriaceae, Melastomataceae, Rubiaceae, Solanaceae), dry capsular fruits containing many small seeds are produced by herbaceous or shrubby taxa in dry tropical habitats whereas fleshy berries are produced in wet forests of low or middle elevations (Bolmgren and Eriksson 2005; Clausing et al. 2000; Knapp 2002; Stiles and Roselli 1993). To test the hypothesis that the evolution of fleshy fruits is associated with habitat types and spatial predictability of disturbances, Bolmgren and Eriksson (2005) conducted a phylogenetically controlled analysis of sister groups and their outgroups from plants in 18 families and in habitats throughout the world. Results of their analyses indicate that fleshy fruits are about twice as likely to evolve from nonfleshy fruits as the reverse (36 times vs. 14 times) and that fleshy fruits are significantly associated with closed (forest) habitats and spatially unpredictable disturbances. They interpreted these results to indicate that the evolution of fleshy fruits is driven by vegetation changes in response to climate changes and not necessarily by the presence of frugivores. They concluded that fleshy fruits are associated with woody plants because of their larger energy budgets; low light (closed) habitats select for large seeds with lower dispersability than small seeds; and nutritious pulp around large seeds (or masses of small seeds) allows plants to attract biotic agents for greater seed dispersability.

Now that we've reviewed the macroevolutionary (phylogenetic) features of fleshy fruits and their seeds, we will examine the evolution of various traits associated with the syndrome of vertebrate endozoochory in more detail (table 1.5).

FRUIT AND SEED SIZE AND SHAPE

In plants, fruits and infructescences are the means for packaging seeds, the important units of reproductive success or failure for plants. As discussed above, this packaging can take a variety of different forms, depending on plant phylogeny, and sizes, depending on which animals are the primary dispersers and how they handle fruits. The maximum size of fruits that can be swallowed by birds, for example, is limited by the width of the bill gape. For many birds, this limit is a fruit diameter of about 10–11 mm, and most tropical birds are thus constrained to eat relatively small fruit. Wheelwright (1985) examined the diameter of fruit of 171 species eaten by 70 species of birds at Monteverde, Costa Rica, and compared fruit size with bird gape width (fig. 8.3). Fruit diameter ranged from 2 to 28 mm with peaks at 5–12 mm and 17 mm. Median gape width of the frugivores was 11 mm, and only large birds (e.g., cotingids, trogons, and toucans) with wide gapes ate large fruit. In this community, the mean and maximum sizes of fruits eaten by 32 species of birds was positively correlated with gape width, and

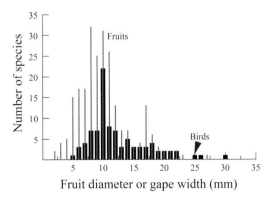

Figure 8.3. Comparison of the frequency distributions of fruit diameters and gape widths of fruit-eating birds at Monteverde, Costa Rica. Based on Wheelwright (1985).

Table 8.2. Seed Mass of Tropical Trees in a Peruvian Lowland Forest

Successional Status/Disperser	Seed Mass (g)		
	Mean ± 1 SE	n	Range
Pioneer:			
Birds	0.165 ± 0.033	55	<0.001–1.23
Mammals	2.366 ± 0.993	53	0.008–50.6
Mature forest:			
Birds	0.541 ± 0.182	24	0.001–4.0
Mammals	4.423 ± 1.775	35	0.093–54.7

Source. Based on data in Foster and Janson (1985, app. 2).

the number of bird species eating a particular fruit species was negatively correlated with fruit size. These patterns are likely to hold for many other tropical or subtropical communities (e.g., in Australian subtropical rain forest; Moran and Catterall 2010).

Many studies have shown that fruit size is positively correlated with seed size and that fruit and seed size differ between bird-dispersed and mammal-dispersed species (reviewed in Leishman et al. 2000). Bird-dispersed fruits and seeds are generally smaller than those of mammal-dispersed species. Data from a lowland Peruvian forest illustrate this point and the additional point that the seeds (and by inference, the fruits) of early successional species are smaller than those of late successional species (table 8.2). Among pioneer species in this forest, seeds in mammal fruits are 14.3 times larger than those in bird fruits; among mature forest species, this difference is 8.2-fold. Within bird fruits, seeds average 3.2 times larger in mature forest species; within mammal fruits, this difference is 1.9-fold. These data come from 203 species from many families, but similar differences in fruit size can also occur within a single genus. In *Ficus*, for instance, bird-dispersed fruits are usually <10 mm in diameter whereas mammal-dispersed fruits are generally larger than this (Lomáscolo et al. 2008; but see Sanitjan and Chen 2009).

To what extent have vertebrate dispersers had a selective effect on the size of tropical fruits? At least two hypotheses can be put forth to explain the association between fruit size (and color) and vertebrate class: (1) a nonadaptive "phylogenetic inertia" hypothesis, in which particular kinds of frugivores "fit" previously selected fruit traits by their feeding choices; this hypothesis, of course, begs the question of who or what were the original selective agents; and (2) direct selection of particular combinations of fruit traits by frugivores. There is no logical reason to think that these hypotheses are mutually exclusive. Even though contemporary frugivores may not be currently selecting for particular fruit traits, as argued by Herrera (2002), this does not eliminate the possibility that frugivores were effective selective agents during the early evolution of a plant family's fruit. Lomáscolo et al. (2008) examined these two hypotheses within a phylogenetic framework in fruits of the genus *Ficus*. They specifically sought to determine whether the size-color combinations of small/red bird-dispersed fruits and large/nonred mammal-dispersed fruits evolved together more often than expected by chance, as predicted by hypothesis 2. Their results indicated that the distribution of fruit types better fit the correlated evolution hypothesis (number 2) than the independent origins hypothesis (number 1).

Lomáscolo et al. (2010) conducted an even stronger test of the hypothesis that different kinds of frugivores have had a strong selective effect on several traits of the fruits they eat and disperse. This test involved recording the following characteristics for 42 species of figs (*Ficus*) dispersed by birds and bats in a lowland forest in Papua New Guinea: size, color and contrast against a foliage background when ripe, hardness, diversity of volatile scents, and method of presentation (within or away from foliage). They videotaped the frugivorous bird and bat visitors to 29 of these species and predicted that bird figs were likely to be smaller, softer, more visually conspicuous, nonodorous, and presented amid the foliage compared with bat figs. Results of a phylogenetically controlled PCA supported their prediction: bird and bat figs occupied different regions of fig "morphospace," with species dispersed by both groups occupying an intermediate position between the two disperser guilds. These authors concluded that their results provide strong support for the hypothesis that fruits produced by different clades of figs have converged on similar suites of characteristics as a result of selection by different groups of vertebrates. They caution, however, that since the PCA analysis explained only about one-half of the total variation in fig fruits, other factors are also likely to be important in the evolution of these fruits. Similar analyses of other kinds of plants with diverse arrays of fruit types (e.g., *Solanum*, *Eugenia*, Arecaceae) would be very informative about this issue.

Additional evidence that vertebrate frugivores can select for particular fruit traits comes from a comparison of habitat distributions of small-seeded/small fruits of Melastomataceae and their principal consumers, tanagers and related emberizine finches (Stiles and Rosselli 1993). As described in detail below, these birds are fruit mashers—that is, they feed by mandibulating and crushing fruit, squeezing out and swallowing the juice, and discarding the fruit husk and many seeds. They quickly discard seeds >2 mm long at parent plants but swallow and disperse seeds smaller than this. The other major consumers of melastome fruits are manakins (Pipridae), which are fruit gulpers—that is, they swallow fruits whole, ingesting both small and large seeds, and disperse seeds by regurgitation or defecation. Stiles and Rosselli (1993) proposed that because they disperse many more small seeds than large seeds, mashers should select for small-seeded fruits whose seeds are embedded in a soft or liquid pulp and that the distributions of small-seeded melastome fruits should overlap more with the habitat distributions of tanagers than with the habitat distributions of manakins, whose diets also include larger-seeded fruits (e.g., drupes of *Psychotria* and *Cepahalis* [Rubiaceae] and arillate fruits of *Ardisia* [Myrsinaceae] and *Trichilia* [Meliaceae]). Comparisons of the habitat distributions of melastomes and tanagers in Costa Rica, Venezuela, and Colombia support this prediction. Whereas gulpers are most diverse in wet lowland forests, mashers and small-seeded melastomes (and a variety of other small-seeded tanager fruits in the Ericaceae and Gesneriaceae) are most diverse at midelevations.

Hemispheric differences in fruit size within plant families with pantropical distributions also support the hypothesis that vertebrate frugivores have influenced the evolution of fruit size. As reviewed by Cristoffer (1987) and Fleming et al. (1987), avian and mammalian frugivores tend to be significantly larger in the Old World tropics than their New World relatives or ecological counterparts. Thus, hornbills are larger than toucans; pteropodid fruit bats are larger than phyllostomid fruit bats; and frugivorous cercopithecid monkeys average larger than cebid monkeys. Large terrestrial Old World frugivores such as cassowaries and elephants have no contemporary counterparts in the New World. These size differences lead to the prediction that fruits and their seeds should be larger in the Old World than in the New World if frugivores select for fruit size. Mack (1993) compared fruit sizes in the Old and New Worlds in eight pantropical tree families. His results (fig. 8.4) provide support for this hypothesis. Both mean and maximum fruit sizes are greater in Old World taxa than in New World taxa in these families. Mack (1993) interpreted these results as indicating that in the Old World tropics, fruit and seed size have evolved with smaller frugivore constraint than in the New World. He considered two other possible explanations for these interhemispheric differences—that is, fruit-bearing trees are larger, on average, in the Paleotropics and disturbance rates there are lower, which would favor larger seeds and fruit—but rejected these explanations. Forget et al. (2007) corroborated Mack's general results by showing that vertebrate-dispersed seeds from Central Africa are significantly larger than

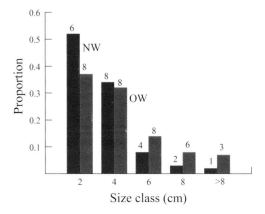

Figure 8.4. Frequency distributions of fruit lengths in genera of eight pantropical families by hemisphere. Number of genera is indicated above each bar. *NW* = New World, *OW* = Old World. Based on Mack (1993).

those from French Guiana, Thailand, and Australia and reflected differences in the average size of frugivores in those areas.

Finally, because the importance of bird dispersal tends to increase with elevation whereas that of mammal dispersal tends to decrease (chap. 2), we might expect to see a decrease in fruit and seed size with elevation if frugivorous vertebrates have had a selective effect on these plant traits. In support of this prediction, Almeida-Neto et al. (2008) reported that fruit size decreases with elevation as the frequency of bird dispersal increases in forests in southeast Brazil. In contrast, Rockwood's (1985) analysis of trends in seed size (a surrogate of fruit size?) in eight families of Costa Rican and Panamanian plant families does not support this prediction. Only in Melastomataceae were seeds smaller above 1,500 m than below (see discussion above). In five of six other families, seeds were larger above 1,500 m than below. From this and other evidence, Westoby et al. (2002) concluded that a seed size–elevation trend does not exist, at least in the tropics. Given the absence of such a trend, it appears that frugivorous vertebrates have not been a major selective factor in the evolution of seed size per se, as suggested by Mazer and Wheelwright (1993). Seed size, of course, is well-known to be affected by many biotic and abiotic factors other than dispersers (reviewed in Leishman et al. 2000; Westoby et al. 2002). Based on the Almeida-Neto et al. (2008) study, however, we cannot discount the role of frugivores in the evolution of fruit size. More elevational data are needed to test these predictions.

In addition to influencing fruit and seed size, frugivores might be expected to select for fruit shape. Because of the limited range of gape widths in fruit-eating birds, Mazer and Wheelwright (1993) proposed that there should exist a negative allometric relationship between fruit diameter and length in tropical bird-dispersed fruits. That is, fruit diameter should increase at a slower rate than fruit length so that large fruits are more elliptical in shape than small fruits. This is the pattern they found within and between several species of Lauraceae at Monteverde, Costa Rica. When they conducted a phylogenetically controlled analysis of fruits from 63 other plant families, however, they found that fruit diameter and length scaled isometrically, that is, these two dimensions increased at similar rates. They reported similar results for an analysis of seed size and shape: a negative allometric relationship between seed diameter and length in Lauraceae but an isometric relationship in a broader survey of plant families. Mazer and Wheelwright (1993) concluded that fruit shape (allometry) is an adaptation to gape-limited bird dispersers in some, but not all, plant families and that frugivorous birds likely have had a smaller selective effect on seed shape (because it is harder for frugivores to assess) than on fruit shape. Forget et al. (2007) also reported mixed results in a comparison of seed shape in four tropical forests. Seeds exhibited negative allometry in Central Africa and Thailand, positive allometry in French Guiana, and isometry in Australia.

Additional evidence that frugivorous birds can sometimes have a selective effect on fruit size and shape comes from an analysis of bird-dispersed fruit in plant genera that occur in New Zealand, Australia, and South America. Because fruit-eating birds are smaller in New Zealand than in Australia and South America, Lord (2004) predicted that fruits would be smaller and more elongate in New Zealand than elsewhere. Analysis of fruits from 20 and nine genera of plants found in New Zealand and in either Australia or South America, respectively, supported this prediction. From this, Lord (2004) concluded that the frugivore faunas of these areas have had a selective influence on the evolution of fruit size and shape in plants. A similar conclusion comes from an analysis of the size of 32 species of bird-dispersed fruit at two sites in New Zealand. Burns and Lake (2009) tested two hypotheses: a frugivore-based hypothesis, which states that fruit size matches the average size of their avian consumers, and an allometric hypothesis, in which fruit size is allometrically correlated with plant traits such as leaf size and plant height (Primack 1987). Based on a phylogenetically controlled analysis, their data supported the frugivore hypothesis and not the allometric hypothesis, and Burns and Lake (2009) concluded that frugivores could be involved in the evolution of fruit size.

SEED PACKAGING WITHIN FRUITS

In addition to influencing the evolution of fruit size and shape, we might expect vertebrate frugivores to influence the way in which seeds are "packaged" in fruits. By "packaged," we mean the number of seeds per fruit and the mass of edible flesh (the nutritional reward) relative to the mass of seeds (ballast) in a fruit. For example, it is likely that birds or mammals that feed in a masherlike fashion (e.g., tanagers, certain phyllostomid bats, and most pteropodid bats; see below) select for small seeds and hence many seeds per fruit whereas gulpers (e.g., manakins, certain phyllostomid bats, and primates) select for fewer and larger seeds per fruit. This prediction is based on the well-known tradeoff between seed size and number in many kinds of fruit (reviewed in Leishman et al. 2000; I. Wright et al. 2007, among others). In terms of the ratio of mass of flesh to seed mass, we expect frugivores to feed preferentially on the most profitable fruits, that is, those with high pulp-to–seed mass ratios. This ratio, in turn, is likely to depend strongly on fruit type (e.g., berries vs. drupes) and the nutritional content of fruit pulp. Fruits producing pulp rich in lipids and proteins (e.g., the drupes of Lauraceae and some Arecaceae and arillate fruits of Myristicaceae), for instance, are likely to have a lower pulp-to-seed ratio than fruits producing watery, sugary pulp (e.g., many berries; Grubb 1998). The evolution of an optimal pulp-to-seed ratio is an example of a conflict of interest between plants and their dispersers, with plants being selected to minimize this ratio and frugivores selecting for higher ratios. As in other such conflicts (e.g., between parents and their offspring regarding optimal parental investment per offspring; Smith and Fretwell 1974), its solution likely results in an evolutionary compromise between the antagonists such that pulp-to–seed mass ratios lie somewhere between the plant and frugivore optima.

Current evidence is mixed regarding whether frugivores have had a selective effect on the ratio of fruit pulp-to–seed mass of their food plants. Edwards (2006) hypothesized that if frugivores have had a selective effect on this ratio, then a fruit's pulp mass should increase with seed mass in positive allometric fashion; for theoretical reasons he predicted that the slope of this relationship should be 4/3 (1.3). He found support for this hypothesis in drupes of four families of tropical Australian plants (Euphorbiacae, Lauraceae, Myrtaceae, and Sapindaceae), which suggests that selection by frugivores is important. Martínez et al. (2007) found a similar pattern in their analysis of fruit pulp and seed packaging in drupes of *Crataegus monogyna* (Rosaceae) in Spain but suggested that seed predation by rodents might have a stronger overall effect on these fruits than their avian dispersers. This suggestion was strongly supported by a detailed analysis of the fate of bird-dispersed seeds of *Olea europaea* (Oleaceae) in Spain (Alcantara and Rey 2003). Previous work by Rey et al. (1997) had shown that maximum fruit size in this species is limited by avian gape width. In the present study, birds disproportionally removed and dispersed smaller-than-average fruits and seeds, but these seeds had a lower survival probability to the 2-yr-old seedling stage than larger-than-average seeds. Alcantara and Rey (2003) concluded that postdispersal events, including seed predation by rodents, probably have had a stronger selective effect on these fruits and seeds than predispersal events (e.g., fruit selection by birds). As a final example, Russo (2003) examined the effect of pulp-to-seed ratio and fruit crop size on seed dispersal in *Virola calophylla* (Myristicaceae) by spider monkeys (*Ateles paniscus*) and 17 species of birds in Peru. As discussed in chapter 4, monkeys are the major dispersers of this species. In choosing feeding trees, they responded more strongly to fruit crop size than to pulp-to-seed ratios; birds, in contrast, had the opposite response, such that fruit removal was more strongly related to pulp-to-seed ratio than to fruit crop size. These results are similar to those of Howe (1981) and Howe and Vande Kerckhove (1981) for two species of Panamanian *Virola*. Russo (2003) concluded that contrary selection by different classes of vertebrate dispersers is likely to result in weak evolutionary effects by dispersers on fruit characteristics such as pulp-to-seed ratios. Kunz and Linsenmair (2007, 2010) reached a similar conclusion from their study of fruit and seed selection by *Papio anubis* baboons in the savanna woodlands of West Africa. Their review of other primate studies further strengthens this conclusion.

In summary, the conclusion that vertebrate frugivores can have a selective effect on fruit size seems to be firmly supported by available data. Frugivores can also have an evolutionary effect on seed size in some cases (e.g., mashers vs. gulpers), but this effect does not seem to be universal. The effects of other biotic and abiotic factors can override the effect of dispersers during the evolution of seed size. Similarly, dispersers are likely to have a relatively weak effect on the evolution of pulp-to-seed ratios because of the contrary effects of different kinds of dispersers (i.e., birds vs. primates). We might expect to see a stronger effect on this ratio in fruits that are dispersed by only one kind of disperser (e.g., by either bats or birds but not both). Relatively specialized interactions (*sensu* Fenster et al. 2004) should always exhibit stronger and more directional interactions than generalized interactions. Given that fruit-frugivore interactions are more generalized than flower-pollinator interactions, it is not surprising to find that the evolutionary effect of frugivores on fruit characteristics such as pulp-to-seed ratios are relatively weak.

METHODS OF FRUIT PRESENTATION

Included in the list of plant traits associated with classical bird- and bat-fruit syndromes is method of fruit presentation. Bird fruits, which are usually harvested by birds while they are perched, tend to be located within or close to foliage, whereas bat fruits, which in the New World are usually harvested by hovering or stalling in flight, are often presented away from foliage on long peduncles or pedicels or on trunks or branches (i.e., flagellicarpy, caulicarpy). About 58% of the fruits eaten by bats in central French Guiana, for example, are presented well away from foliage (Lobova et al. 2009). Of the two species of *Cecropia* that occur in central French Guiana, the bird-dispersed species (*C. sciadophylla*) has a shorter peduncle than the bat-dispersed species (*C. obtusa*; Charles-Dominique 1986). It is likely that at least some of the phylogenetic variation in this trait can be explained by plant-pollinator coevolution rather than plant-frugivore coevolution. For example, New World *Piper* infructescences usually project above or below foliage, which increases their accessibility to small echolocating bats (Thies et al. 1998). But the spikelike *Piper* inflorescences also increase the accessibility of their tiny flowers to small halictid and megachilid bees (Fleming 1985).

Hemispheric differences in cauliflory/caulicarpy are particularly striking. Kinnaird and O'Brien (2007), for example, note that cauliflory or ramiflory (i.e., flowers and fruits produced directly on branches) occurs in a "preponderance" of trees in lowland Southeast Asia. In *Ficus*, for example, cauliflory is much more common in the Old World than in the New World (Hodgkison et al. 2007; Kalko et al. 1996; Lomáscolo et al. 2010). Kalko et al. (1996) suggest that caulicarpous fruits are easier to find by visually orienting pteropodid bats than fruits that occur among foliage. A similar situation occurs in Myrtaceae in which certain species of Australasian bat-pollinated and -dispersed *Syzigium* are cauliflorous whereas those of Neotropical *Eugenia* are not. Other examples of caulicarpy in bat-dispersed plants are described by van der Pijl (1982).

Other Fruit Characteristics

FRUIT COLOR

Fruit color is a trait that differs among the classic vertebrate fruit syndromes. As reviewed by Voigt et al. (2004) and Schaefer and Schaefer (2007), bird fruits in a variety of tropical locations are typically red, black, or white; blue, pink, purple, yellow, and orange fruits are usually less common. Primate fruits, in contrast, tend to be green, brown, yellow, or orange; red and purple are less common colors. As indicated above, the colors of phyllostomid bat fruits are predominantly green followed by yellow, white, and brown (Lobova et al. 2009). Differences in the colors of fruit dispersed by different groups of vertebrates appear to support the hypothesis that dispersers have had a significant selective effect on this aspect of fruit evolution. Before we examine evidence for or against this hypothesis, however, we will briefly review the biochemical basis for fruit color.

Fruit colors are generally the result of four classes of compounds—chlorophyll, carotenoids, anthocyanins, and betalains (Schaefer et al. 2008; Willson and Whelan 1990). Chlorophyll is responsible for the green color of many unripe fruits as well as the color of many ripe bat-dispersed fruits. The widespread occurrence of chlorophyll in unripe fruits is adaptive for at least two reasons: (1) it makes them harder to find by visually orienting frugivores and (2) it helps subsidize the growth and development of fruits and seeds (Wheelwright and Janson 1985). Carotenoids are long-chain terpenoid hydrocarbons that occur as carotenes and xanthophylls; they produce hues of light yellow to red in fruit. Anthocyanins are phenolic flavonoid compounds that produce hues such as red, blue, purple, and black in fruits. Other flavonoids produce yellows, whites, and ultraviolet reflectance. Betalins, which are uncommon in angiosperms, are alkaloids that produce hues of red to violet, blue, and yellow in fruit. From this brief survey, it is clear that the color of any particular fruit and its accessory tissues can be produced by a variety of different compounds involving a variety of different genetic pathways (Lee 2007). This biochemical diversity, in turn, has permitted the frequent convergence of unrelated fruits on similar colors. In emphasizing this convergence, Wheelwright and Janson (1985, 793) wrote: "One's first impression in a tropical forest is that bird fruits occur in an overwhelming variety of hues, but equally striking is the apparent convergence of unrelated plant species on a limited number of color patterns. Fruit displays involving both black and red probably evolved independently in the 26 different plant families in which they occur in the Neotropics."

Like the colors of flowers, fruit colors are signals that have presumably evolved as advertisements to frugivores. In addition to signaling the location of food, colors often indicate degree of fruit ripeness (Ridley 1930). They may also have an antiherbivore function, particularly in fruits from the orange-red portion of the visible spectrum, which are difficult for insects to see (Schaefer et al. 2007; Wheelwright and Janson 1985). As advertisements aimed at frugivores, however, fruit colors have evolved in a different fashion than that of flowers. Whereas fruits function both as signals and rewards, the colors of flower corollas and their accessory tissues generally provide a signal but no reward. Furthermore, effective pollination is maximized when pollinators exhibit floral constancy, and this could result in diversifying selection (i.e., the antithesis of convergence) among flower colors within the same community or region. In contrast, effective dispersal does not necessarily require fruit constancy on the part of frugivores. In fact, fruit inconstancy may increase seed mobility. If it does, then we would expect to see much greater convergence in color among different fruit species than in flower species within communities and regions (Schaefer et al. 2004).

Convergence in fruit colors among unrelated plants suggests that phylogenetic constraints in fruit color are relatively weak (Lomáscolo and Schaefer 2010; Schaefer et al. 2007, 2008 and included references). Support for this comes from speciose genera such as *Ficus* and *Solanum*, whose fruit come in a variety of different colors, ostensibly in response to selection by different vertebrate dispersers. In *Ficus*, for instance, bat-dispersed species in the New World are usually green whereas bird-dispersed species are red. In the Old World, bat-dispersed species are red, yellow, and orange in addition to green (Kalko et al. 1996). As discussed above, small red "bird figs" have evolved independently numerous times in *Ficus* (Lomáscolo et al. 2008). In *Solanum*, New World bat-dispersed species come in a variety of colors, including green (the most common color), yellow, white, black, red, and purple; bird-dispersed species are often yellow or black (Knapp 2002; Lobova et al. 2009). Other genera that produce a variety of fruit colors include *Xylosoma* (Salicaceae) and *Psychotria* (Rubiaceae; Wheelwright and Janson 1985). In contrast, fruit displays, including colors, are remarkably uniform in families such as Lauraceae and Ericaceae, indicating an apparent lack of evolutionary plasticity (or strong stabilizing selection) in these clades (Wheelwright and Janson 1985).

Cazetta et al. (2009) asked why fruits are colorful, and conducted a series of experiments in two Brazilian habitats to see which of two fruit characteristics—chromaticity (color) or achromaticity (brightness)—was more important for fruit removal by understory birds such as manakins, thrushes, and tanagers. They measured the removal rates of fruits of four different colors (red, black, white, and ultraviolet-blue) against light or dark backgrounds. In both closed forest and more open *restinga* habitats, fruits of high contrast (red, blue) were removed faster than the other two colors, and their removal rates were independent of background color. From this Cazetta et al. (2009) concluded that a fruit's color is more important for its detection than is its brightness. They also reported that understory fruits in closed forest, where light levels are highly variable, contrasted more strongly against their foliage than fruits in the more open and sunny *restinga*.

Schaefer et al. (2006) reached a somewhat different conclusion about the relative importance of color versus brightness in their study of fruit choice by captive European crows. In that study, both color and brightness influenced fruit detectability. Crows detected highly contrasting red and ultraviolet-colored blueberry fruits more readily than black fruits against leaves but detected UV blueberries and black fruits equally well against a dull background. These results suggest that both color and brightness can influence a fruit's detectability, depending on background. Under natural conditions, however, fruit color is likely to have priority over brightness in its detectability. These kinds of studies plus differences in the locations in color space of fruits eaten by different groups of frugivorous vertebrates, especially diurnal versus nocturnal species, provide strong support for the hypothesis that these animals have had a selective effect on the colors of their fruit (Lomáscolo and Schaefer 2010).

Do colors provide frugivores with information about the nutritional content of fruit? Are fruit colors correlated with the sugar composition and/or energy density of fruit? Current information provides mixed answers to these questions. Wheelwright and Janson (1985) studied the fruit characteristics of a total of 383 bird-dispersed species from Monteverde, Costa Rica, and Cocha Cashu, Peru, and reported that fruit color was not correlated with net pulp mass, seed-to-pulp ratio, sugar concentration of pulp, and energetic value per fruit. Dominy et al. (2003) reached a similar conclusion in their study of primate diets at Kibale National Park, Uganda. In a study of 45 Venezuelan bird-dispersed species, Schaefer and Schmidt (2004) found that red or black fruit provided no information about a fruit's nutrient content, but this was not true of other fruit colors. In these species, yellow and orange fruits signaled high protein content; blue fruit signaled high carbohydrate content; and the signal from white fruit was intermediate between that of other nonred or black fruit. Finally, anthocyanins and carotenoids in fruits are antioxidants that could be physiologically valuable to frugivores as oxidative stress reducers. Schaefer et al. (2008) point out that different fruit colors can provide honest signals about anthocyanin content but not about carotenoid content. Black or ultraviolet colors indicated the presence of more anthocyanins and higher caloric content than red/orange or nonultraviolet colors in a sample of 60 bird-dispersed species from Europe. Fruit detectability against natural foliage was not correlated with anthocyanin content in these species.

In summary, fruit colors provide a variety of potential "services" to frugivores. Many colors increase the detectability of ripe fruit against a foliage background. Schaefer et al. (2007) warn us, however, that the colors of ripe fruit have not necessarily been selected to maximize their detectability against their own foliage. Nonetheless, it is generally agreed that red and black fruits, which are the most common colors in many tropical floras, are highly detectable against green foliage. A second potential service is providing information about the nutritional content of fruit. Although color appears to provide honest information about anthocyanin (and hence antioxidative) content, it is less revealing about other nutritional information, particularly in the case of red and black fruits. Fruit colors are highly labile within many genera, and families and are likely to respond strongly to selection by different kinds of frugivores. Diurnal and visually orienting vertebrates (i.e., birds, primates) select for more colorful fruit than do nocturnal vertebrates (i.e., bats). Among fruits dispersed by diurnal frugivores, bird fruits tend to be "bluer" and primate fruits tend to be "greener," as predicted by models of their color vision (Lomáscolo and Schaefer 2010). As we discuss below, bats use olfaction more often than vision to locate ripe fruit, and hence they have probably had a smaller effect on the evolution of fruit colors than have diurnal species. Finally, selection by herbivores and seed predators can affect fruit color, and the reflectance properties of leaves are likely to constrain fruit colors (Burns et al. 2009). Thus, fruit color is the product of a variety of selective pressures, only one of which is the effect of visually foraging frugivores.

FRUIT SCENTS

In the classic vertebrate fruit syndromes, ripe fruits eaten by birds are less likely to have a detectable scent than those eaten by bats. Compared with the amount of research that has been done on floral scents (chap. 7), however, relatively little research has been done on this topic, and our review of fruit scents will be brief. To our knowledge, no one has yet systematically examined the array of scents produced by different clades of plants and related this variation to the attraction of different kinds of vertebrate frugivores.

As reviewed by Sanchez et al. (2006), odors associated with fruits are complex and change during the ripening process. These changes include either synthesis or breakdown of carbohydrates, lipids, and proteins as well as the production of pigments and scent-producing compounds. As a result, the flavors and scents of fruit depend on a "complex interaction of sugars, organic acids, phenols, and more specialized flavor compounds, including a wide range of volatiles. The profile of volatile compounds for a fruit is usually complex, and may include alcohols, aldehydes, esters, and other chemical groups" (Sanchez et al. 2006, 1290). Aliphatic alcohols generally indicate ripeness in fruit.

More work has been done on scents produced by fruits of *Ficus* than any other nonagricultural plants. Throughout the world, ripe *Ficus* fruits dispersed by birds usually lack a detectable odor (by humans) whereas those dispersed by bats often have species-specific, "pleasant fruity" odors that can be detected from several hundred meters away (Hodgkison et al. 2007; Kalko et al. 1996). Borges et al. (2008) compared the volatile chemical profiles of one species of Indian bat-dispersed fig (*F. hispida*) with two species of bird-dispersed figs (*F. exasparata* and *F. tsjahela*) and found that the bat fig produced a greater variety of compounds than the bird figs. One reason for this is that the bat fig is dioecious so that the scents of male and female fruits are aimed at attracting two different kinds of mutualists—pollinating female fig wasps (by male fruits) and vertebrate dispersers (by female fruits). In contrast, both monoecious and dioecious bird figs aim to attract only wasps with their scents. Their analyses revealed that the volatile chemical signatures of bat-dispersed female fruits of *F. hispida* had high levels of fatty acid derivatives, including amyl-acetates and 2-heptanone, whereas monoterpenes, sesquiterpenes, and derivatives of shikimic acid were common in male fruits of this species as well as in fruits of the bird-dispersed figs. They also reported that methyl anthranilate, a chemical known to be repellent to birds and mammals, was found only in male figs of *F. hispida*; these fruit contain pollinating wasps and thus benefit from chemical protection against potential dispersal agents. Hodgkison et al. (2007) studied the scents produced by two species of Malaysian bat-dispersed figs and reported that methyl ketones and secondary alcohols, which are common compounds in fruits of many plant taxa, were the most common compounds in ripe fruits of both species. The only unique compounds found in these species were (E)-2-penten-3-yl esters. As expected, ripe fruits produced more scent than unripe fruits. Results of these two studies suggest that systematic, phylogenetically based research on the scents of vertebrate-dispersed fruits could be very worthwhile.

FRUIT NUTRITIONAL REWARDS

We reviewed this topic in chapter 3 and will provide only a general summary here. Fruits vary widely in the chemical composition of their nutritional rewards, and a substantial part of this variation has a phylogenetic basis at the level of genus or family (Jordano 1995). Some of this variation is illustrated in figure 8.5, which shows the results of a PCA of the nutritional characteristics of 29 families of Neotropical bird-dispersed fruits based on data in Moermond and Denslow (1985, table 1). Two major points emerge from this plot. First, few families produce fruits that are rich simultaneously in protein and in lipids. Instead, the fleshy fruits of many plant families, particularly those producing berries, are rich in carbohydrates and water. Second, these families do not cluster in multivariate nutritional space by major angiosperm lineage (basal angiosperms, monocots, etc.). That is, each family's nearest neighbor in this plot is more likely to belong to a different lineage than the same lineage ($P = 0.0066$ in a binomial test). This implies that the evolution of the nutritional characteristics of fruits within clades has occurred independently of phylogeny. This result is not surprising given that there can be substantial variation in the nutritional characteristics of fruits within families (Jordano 1995). Nevertheless, the question remains, to what extent has selection by vertebrate frugivores contributed to this variation? A biologically more realistic restatement of this question is: What is the relative effect of frugivores compared with seed predators and pathogens on the chemical composition of fruit pulp?

Although we currently have limited understanding of the influence that frugivores have had on the nutritional characteristics of their fruit, available data hint that different kinds of frugivores can select for different suites of nutritional traits in fruit. In the genus *Ficus*, for example, bird figs tend to

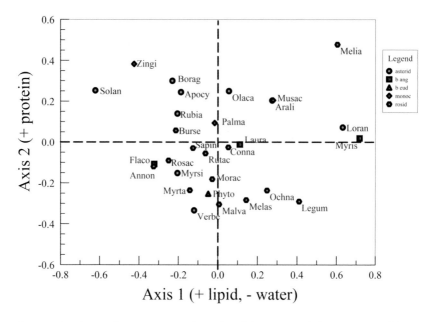

Figure 8.5. Principal components analysis plot of the nutritional characteristics of fruits from 29 families of Neotropical plants, classified by major angiosperm lineage. This analysis included six variables: percent water, percent protein, percent lipid, percent carbohydrates, Kcal/g dry weight, and the ratio of protein to carbohydrate. PCA axis 1 accounts for 46.9% of the variation and is positively correlated with lipid content and negatively correlated with water content. PCA axis 2 accounts for 23.5% of the variation and is positively correlated with protein content. Legend: b ang = basal angiosperms; b eud = basal eudicots; monoc = monocots. Plant families: Annon = Annonaceae; Arali = Araliaceae; Apocy = Apocynaceae; Borag = Boraginaceae; Burse = Burseraceae; Conna = Connaraceae; Flaco = Flacourticeae; Laura = Lauraceae; Legum = Leguminosae (Fabaceae); Loran = Loranthaceae; Malva = Malvaceae; Melas = Melastomataceae; Melia = Meliaceae; Morac = Moraceae; Musac = Musaceae; Myrsi = Myrsinaceae; Myris = Myristicaceae; Myrta = Myrtaceae; Ochna = Ochnaceae; Olac = Olacaceae; Palma = Palmae (Arecaceae); Phyto = Phytolaccaceae; Rosac = Rosaceae; Rubia = Rubiaceae; Rutac = Rutaceae; Sapin = Sapindaceae; Solan = Solanaceae; Verbe = Verbenaceae; Zingi = Zingibereaceae. Based on Moermond and Denslow (1985, table 1).

be more carbohydrate-rich and less nutritious overall than bat figs, at least in the New World (Kalko et al. 1996; Wendeln et al. 2000; chap. 3). As discussed above, Lomáscolo et al. (2008) showed that bird and mammal figs have evolved independently several times in this genus, which supports the frugivore selection hypothesis. At the plant family level, however, we know very little about the selective effect of frugivores on the nutritional composition of fruit.

Developing a deep understanding about this will not be easy. In addition to the effects of mutualists and antagonists, any theory about the evolution of fruit design and composition will have to consider the effects of plant growth form, canopy position, and successional status. That is, shrubs are likely to produce smaller and less nutritious fruit than are canopy trees, and understory species in early successional forests are likely to produce more sugar-rich fruit than primary forest understory species. Data from Schaefer et al. (2002) and Lumpkin and Boyle (2009) illustrate these points. Schaefer et al. (2002) measured the diameter and energy content of fruits from 31 species of various growth forms (shrubs, midstory trees, canopy trees) in a Venezuelan moist tropical forest. Their data indicated that while median values of size and energy of fruits in this forest did not differ greatly among growth forms, the range of values was substantially greater in midstory and canopy trees than in shrubs (fig. 8.6). That is, most fruits in each of these levels in the forest were small and energy poor, but some trees produced large, energy-rich fruits (e.g., *Cayocar glabrum*, *Kotchubaea* sp., *Pouteria* sp., and

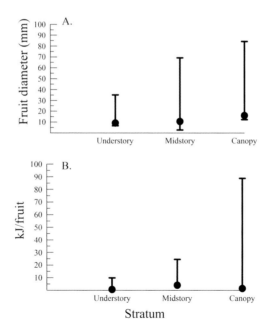

Figure 8.6.   Median values (*circles*) and range of values of fruit diameter (*A*) and energy content (*B*) in a series of species from a Venezuelan tropical moist forest by forest stratum. Based on Schaefer and Schmidt (2004).

Orthomene schomburkii). In this data set, fruit diameter and energy content were significantly correlated (Spearman's $r = 0.64$, $P < 0.01$), and this is likely to be a general pattern. In a study of the effect of successional status on the sugar content of understory plants, Lumpkin and Boyle (2009) reported that seven species (in four families) from Costa Rican wet secondary forest were somewhat richer in sugars (9.0 % vs. 6.9%) than seven primary forest species (in one family). Whether this difference is universally true is currently unknown.

In summary, the range of sizes, shapes, colors, nutritional composition, and degree of physical and chemical protection of fruits in tropical forests is enormous. Much of this variation appears to have a phylogenetic basis, especially at the family or generic level, and it is not yet clear the extent to which selection by vertebrate frugivores has contributed to this variation. Certainly some features of fruit design (e.g., size, color, scent, and nutritional content) appear to be responsive to selection by frugivores, but other features (e.g., fruit shape, ratio of pulp to seed mass, and seed size) apparently are less so. Given that many vertebrate frugivores are opportunistic feeders, this mixed bag of selective effects should not be surprising. Nonetheless, the existence of recognizable suites of fruit traits that correspond broadly to particular groups of frugivores tells us that, overall, these animals have had a significant effect on the evolution of angiosperm fruits. We next address the converse question, what effects have a diet of fruit had on the biology of vertebrate frugivores?

The Evolution of Vertebrate Frugivores

Whereas only a handful of families of birds and mammals are morphologically and physiologically adapted for feeding on nectar (and pollen), many kinds of birds and mammals (as well as other vertebrate classes) regularly consume fruits (table 1.1). Thus it is much more difficult to generalize about the adaptations associated with this food habit. Indeed, in his review of frugivory in South Asia, Corlett (1998) answered the question of which birds and mammals eat fruit in this region by saying "everyone." In order to make some sense of this taxonomic diversity, we will focus our attention on certain groups of vertebrate frugivores, including, among birds, cassowaries, trogons, toucans, fruit pigeons, hornbills, manakins, cotingids, tanagers, and birds of paradise and, among mammals, bats and primates, and will discuss the biological consequences of a fruit diet for them.

MORPHOLOGICAL ADAPTATIONS
IN FRUGIVOROUS VERTEBRATES

Whereas nectar-feeding birds and mammals have an obvious set of morphological adaptations for efficiently harvesting nectar, including small size and specialized mouthparts, the same is not true for frugivores. As indicated in chapter 1, the sizes of frugivorous birds and mammals span the entire size range of nonaquatic species in both classes. Similarly, except for the large bills of toucans and hornbills, the mouthparts of avian frugivores are not particularly specialized. The same is not necessarily true of many frugivorous bats and primates, however. Nonetheless, as we detail below, a frugivorous diet has had a relatively strong effect on many aspects of the biology and ecology of birds and mammals. In this section we will focus on morphological features associated with fruit detection, harvesting, and processing.

BIRDS

As we discussed in chapter 7, birds have tetrachromatic vision and hence can see a wide variety of colors. But whether this sophisticated color vision evolved primarily to enhance food detection or for some other reason (e.g., mate selection) is not yet clear (e.g., Hart and Hart 2007). Regardless of its origin, there is considerable experimental evidence (reviewed above) indicating that birds use color information to detect and select fruits. If the visual acuity of frugivorous birds is at least as high as that of fruit-eating primates, then they should be able to detect ripe fruits from a distance of 20 m or more, although selection of which actual fruits to ingest often occurs at distances much shorter than this (e.g., within a bird's reach).

Moermond and Denslow (1985) published an extensive review of the morphology and behavior of Neotropical fruit-eating birds and identified patterns that are likely to be applicable to the evolution of many kinds of frugivorous birds throughout the world. They first noted that birds eat fruit in three different ways: whole, piecemeal, or via mashing. While many birds swallow fruit whole and hence are limited in the size of fruits they can swallow by their gape width (e.g., Wheelwright 1985), others eat fruit piecemeal and do not experience this limitation. Birds that ingest fruits whole are often called gulpers and include species of many different sizes (from manakins to toucans and hornbills to cassowaries). An extreme example of gulping behavior is a captive cassowary that swallowed seven whole papayas at one feeding (THF, pers. observation). The ability of many small birds to eat piecemeal large fruit containing small seeds (e.g., figs and many kinds of berries) means that any correlation between the size of birds and the size of fruit that they eat is likely to be low. In fact, Moermond and Denslow (1985) suggested that many fruits or infructescences (e.g., those of *Cecropia*, *Clusia*, *Drymonia*, *Piper*, *Stemmadenia*, and *Xylopia*) appear to be designed to be eaten piecemeal because their small seeds are firmly embedded in pulp. Fruit mashing occurs when birds crush a fruit with their mandibles, swallow its juice and small seeds, and quickly discard the skin and large seeds. In birds, fruit mashing is apparently restricted to New World tanagers and their close relatives, emberizid finches. This kind of fruit handling behavior is also found in certain phyllostomid and pteropodid bats and primates.

Birds can harvest fruit either in flight or while perched on a branch or on the ground, and Moermond and Denslow (1985) point out that birds foraging in different ways often differ in the morphology of their bills, wings, and legs. In contrast to nectar feeders, there is no common pattern in bill size and shape in frugivorous birds. Aerial feeders (e.g., trogons, many cotingids, and manakins) usually have short, wide, and flat bills and swallow fruits whole. They typically harvest one fruit at a time. In contrast, perch feeders usually have longer, deeper bills and either swallow fruits whole or peck at them; they typically harvest several fruit at a time. New World toucans and their Old World counterparts, hornbills, are extreme examples of this morphotype. Not only are they excellent fruit pluckers, but toucan bills also have an important thermoregulatory function (Tattersall et al. 2009). Mashers tend to have stronger bills than other frugivores. In fruit pigeons, species of *Treron* are strong-billed and often peck at fruit whereas species of *Ducula* and *Ptilonopus* are weak-billed and swallow fruit whole. The arrangement of jaw muscles in these birds allows them to increase the intermandibular space to swallow large fruits, thus overcoming gape limitation (P. Marrero et al., unpubl. ms.). Similarly, quetzals have flexible jaws and clavicles for swallowing large fruit (Wheelwright 1983). Aerial feeders generally have lower wing loadings for more maneuverable flight than perch feeders, which usually have stronger legs than aerial feeders. Their strong legs enable perch feeders to move agilely on small branches and to reach below branches to pluck fruit.

Differences in fruit harvesting techniques and morphology usually occur at the family level and are an example of adaptive specialization in frugivores. A multivariate plot of this specialization is shown in figure 8.7, which presents the results of a PCA of morphological data found in Moermond and Denslow (1985, table 3). In this figure, families of perch feeders with large leg

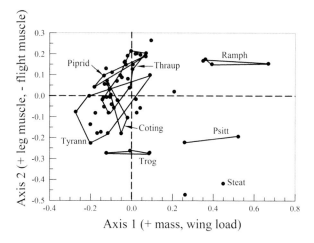

Figure 8.7. Principal components analysis of the morphological characteristics of eight families or subfamilies of Neotropical fruit-eating birds. This analysis is based on five variables: body weight, flight muscles as a percent of weight, leg muscles as a percent of weight, wing loading (g/cm²), and aspect ratio (wing span/wing chord). PCA axis 1 accounts for 42.9% of the variation and is positively correlated with mass and wing loading. PCA axis 2 accounts for 33.6% of the variation and is positively correlated with the relative size of leg muscles and negatively correlated with the relative size of flight muscles. Birds that feed from perches occur above the horizontal line; birds that are aerial feeders occur below the horizontal line. Bird families or subfamilies: *Coting* = Cotingidae; *Piprid* = Pipridae; *Psitt* = Psittacidae; *Ramph* = Ramphastidae; *Steat* = Steatornithidae; *Thraup* = Thraupinae; *Trog* = Trogonidae; *Tyrann* = Tyrannidae. Based on Moermond and Denslow (1985, table 3).

muscles and reduced flight muscles (e.g., toucans, tanagers) occur above the dashed horizontal line whereas families of aerial feeders with low wing loadings and reduced leg musculature (e.g., trogons, cotingids) occur below this line. As a result of these adaptations, different kinds of birds have differential access to a given fruit crop and harvest that crop in different ways. Thus, the fruits available to aerial feeders are not necessarily available to perch feeders and vice versa. Compared with insect-eating birds, however, fruit-eating birds have fewer ways of harvesting their food, which should ultimately limit the extent of their adaptive radiation, as pointed out long ago by Snow (1971).

Once fruit is harvested, it is processed by the gastrointestinal tract. Whereas many seed-eating birds have muscular gizzards (for grinding up food), long intestines, and slow rates of food passage, fruit-eating birds generally have thin-walled gizzards, short intestines, and fast rates of food passage (e.g., Richardson and Wooller 1986; Stanley and Lill 2002; Worthington 1989). The presence of food storage "compartments" such as an expanded esophagus or proventriculus varies among families of frugivores. Some species store fruit in their esophagi (e.g., whiteeyes, turdids), whereas others do not (e.g., manakins, quetzals). Frugivorous hornbills have large gular pouches for carrying multiple fruits; their carnivorous relatives lack these pouches (Kinnaird and O'Brien 2007). Because they can retain pulp and seeds in their guts for up to 12 hours, fruit pigeons are an exception to the trend of fast rates of food passage in frugivores. This makes them especially important long-distance seed dispersers, particularly among islands in the Pacific and Indian Oceans (P. Marrero et al., unpubl. ms.).

Like their nectar-feeding counterparts, fruit-eating birds are often spectacularly colorful. A quick glance through any tropical bird field guide will verify this (Schaefer et al. 2004). In the avifauna of Peru, for example, parrots (which are plant antagonists rather then mutualists), trogons, toucans, many cotingids, manakins, and tanagers are usually much more colorful than pigeons, flycatchers, and a host of other seed- or insect-eating families (jacamars [Galbulidae] are an exception to this; Schulenberg et al. 2007). One reason for this is that birds must ingest carotenoids to produce reds and yellows in their plumage and bare body parts, and fruits are excellent sources of these compounds. Olson and Owens (2005) analyzed the plumage colors that occur in 140 families of birds and found that the amount of red plumage is correlated with the amount of carotenoid in their diet.

Major factors influencing the evolution of plumage color in birds include intraspecific communication, which is usually associated with mate choice, and predation (Badyaev and Hill 2003; Bleiweiss 1997, among others). Along with vocalizations, conspicuous coloration is involved in mate choice in many species of birds, and examples of strong sexual dichromatism in which males are more colorful than females are common in tropical plant–visiting birds (see chap. 7 for nectar feeders). Among frugivores, strong dichromatism occurs in both nonpasserines (e.g., trogons but not in toucans) and passerines (e.g., manakins and tanagers). In New World manakins, males have colorful patches on their crowns, rumps, throats, and wings, probably as a result of sexual selection in these polygynous birds. These patches are relatively small yet are still conspicuous to nearby females but not to more distant predators. According to Doucet et al. (2007), red or orange patches are effective short-distance chromatic signals whereas blue and white patches are more effective at intermediate and long distances, respectively, in these birds. Prum (1997) analyzed variation in male display traits in manakins in a phylogenetic context and found little evidence of convergence in color patterns among different clades; that is, plumage color in these birds has evolved in a phylogenetically unconstrained fashion. A similar lack of phylogenetic constraint on male plumage evolution has been reported in birds of paradise, bowerbirds, and buntings (Irestedt et al. 2009; Kusmierski et al. 1997; Stoddard and Prum 2008). Compared with their close relatives tyrannid flycatchers, manakins are an order of magnitude more diverse in plumage and display trait characteristics despite being an order of magnitude less diverse in species richness (Prum 1997).

MAMMALS

Although many kinds of mammals occasionally eat fruit, species that are specialized morphologically and physiologically for frugivory occur primarily in two families of bats (Pteropodidae and Phyllostomidae) and in the order Primates, which is the only mammalian order in which frugivory is widespread. We will focus on these groups here.

BATS.—Nearly all fruit-eating bats are nocturnal, and, with the possible exception of certain species of *Pteropus*, all are color-blind but have a keen sense of smell (chap. 7). As a result, their fruits are generally green or dull colored but often emit a strong scent. Bats of both families clearly search for ripe fruit using olfactory cues, and they can readily distinguish between ripe and unripe fruit using only olfactory cues (Hodgkison et al. 2007; Korine and Kalko 2005; Sanchez et al. 2006; Thies et al. 1998). *Carollia* bats, common Neotropical understory frugivores, have much lower detection thresholds for a variety of monomolecular odor components such as ethyl butyrate, *n*-pentyl acetate, and linalool than insectivorous bats (Laska 1990).

In addition to vision and olfaction, phyllostomid (but not pteropodid) bats echolocate and use auditory information to locate fruit. Experimental work by Elisabeth Kalko and colleagues (Kalko and Condon 1998; Kalko et al. 1996; Korine and Kalko 2005; Thies et al. 1998) has revealed interesting differences in the use and relative importance of olfaction versus echolocation in several species of phyllostomids. *Phyllostomus hastatus* is an omnivorous bat that is not highly specialized for frugivory. When searching for the fruits of *Gurania spinulosa* (Cucurbitaceae) that hang from the ends of bare pendulous branches, it uses echolocation alone for general fruit location but both olfaction and echolocation for precise fruit location. In contrast, two species of *Carollia* use both echolocation and olfaction for general fruit location and echolocation for precise location of ripe *Piper* fruits. The fig specialist *Artibeus jamaicensis* uses both echolocation and olfaction for general orientation and fruit location but uses only odor for precise fruit location. Finally, two other fig specialists, *Artibeus watsoni* and *Vampyressa pusilla*, use echolocation and olfaction for both general orientation and precise location of ripe fruit. These results suggest that phyllostomid species that are specialized frugivores rely more heavily on olfaction for locating fruit than less specialized species. This difference is reflected in the larger olfactory lobes of their brains (Hutcheon et al. 2002).

Both pteropodid and phyllostomid bats exhibit a variety of morphological specializations for fruit eating. Frugivory is the most common diet class in pteropodid bats (chap. 7), and species across the entire size range of this family (from about 15 to 1500 g) eat fruit. As shown in figure 7.9, the skulls of these bats feature somewhat elongated and narrow rostrums compared with their phyllostomid counterparts, and their palates usually have eight transverse ridges against which fruits are crushed with the tongue (Dumont 2004). Number of teeth ranges from 24 to 34 and include small incisors, sharp canines, and low, well-spaced molars with a smooth upper surface containing a longitudinal groove for crushing fruit (Andersen 1912; Nowak 1994). Unlike most insectivorous bats, their lower jawbones are fused anteriorally, a condition also seen in phyllostomid nectar and fruit bats and in primates. In frugivores, mandibular fusion presumably aids in harvesting and processing hard, fibrous foods.

Feeding in pteropodids often occurs in masherlike fashion. That is, bats crush fruits in their mouths with their tongues and teeth, swallow the juices and a few seeds, and spit out the skin, other fibrous material, and seeds as a wedge or spat. Most wadges end up under the canopies of fruiting trees or under night roosts (chap. 4). Feeding in this fashion is generally a slow process, particularly in fig-eating species (Dumont 2003).

Compared with nectar-feeding phyllostomids with their small sizes, reduced teeth, and long rostrums and tongues, fruit-eating phyllostomids in subfamilies Caroliinae and Stenodermatinae tend to be larger and have more robust teeth, shorter rostrums, and shorter tongues. Body sizes in these bats range from about 5 to 70 g; number of teeth ranges from 28 to 32. As discussed in detail by Freeman (1988), jaw evolution in phyllostomid bats has involved a change from a V-shaped dental arcade and unfused mandibulae in animal-eating taxa to a U-shaped dental arcade and fused mandibulae in frugivores (fig. 8.8). Some species have large, daggerlike canines for plucking and holding fruit in flight, and most have flattened molars with a

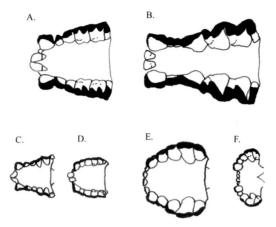

Figure 8.8. Ventral views of the palates and teeth of six species of phyllostomid bats. The stylar shelf is black. Vertebrate carnivores include *Phyllostomus hastatus* (*A*) and *Vampyrum spectrum* (*B*). Frugivores include *Carollia perspicillata* (*C*), *Sturnira lilium* (*D*), *Artibeus lituratus* (*E*), and *Centurio senex* (*F*). Note that most of the frugivores have wider palates and narrower stylar shelves than the carnivores. Based on Freeman (1988).

high labial rim that function as a mortar-and-pestle for crushing fruit. Like pteropodids, stenodermatines also crush fruit against palatal ridges with their tongue.

Fruit-eating phyllostomids can be divided into two groups regarding the way they process fruits: (1) "fast feeders" (or gulpers)—understory species such as *Carollia* and *Glossophaga* that generally consume small, nonfibrous fruits and their seeds in 1–2 min; and (2) "slow feeders" (or mashers)— members of subfamily Stenodermatinae that slowly eat more fibrous fruit and form spats (Dumont 2003). Although food passage rates are about 20– 30 mins in both types of bats, gulpers process more food per unit of feeding time than mashers because of their faster feeding rates (Bonaccorso and Gush 1987).

Fruit processing in mammals begins in the mouth, and the research of Elizabeth Dumont and colleagues has focused on how phyllostomid and pteropodid bats chew fruit (Dumont 1999, 2003, 2004, 2007; Dumont and O'Neal 2004; Santana and Dumont 2009). Although the texture of different kinds of fruit ranges from soft and mushy to very hard, it is useful to consider how bats deal with two broad classes of fruit—soft and hard—and ask if bats change their chewing behavior when dealing with these two classes. In the course of this research program, Dumont has found that bats use four different bite types. Two of these (shallow bilateral and unilateral bites) emphasize teeth at the front of the mouth, whereas the other two (deep bilateral and unilateral bites) emphasize teeth at the back of the mouth. In phyllostomids, less-derived frugivores (e.g., *Carollia*) use bilateral bites—the ancestral bite type in this family—to process both soft and hard fruits whereas more specialized taxa (e.g., *Artibeus* and its relatives) are more likely to use unilateral bites, especially on hard fruits. Members of both groups tend to use shallow bites with soft fruits and deep bites with hard fruits. Feeding flexibility appears to be somewhat lower in pteropodids than in phyllostomids. Members of two genera with specialized dentition (*Nyctimene* and *Paranyctimene*) are invariant in their use of bilateral shallow bites for both soft and hard fruits. Other species, including understory and canopy feeders, tend to use unilateral bites more often than bilateral bites, especially when eating hard fruits. This suggests that these bats have generally evolved to eat hard fruits.

Compared with animal- or nectar-eating phyllostomids, the gastrointestinal tracts of frugivorous species feature longer intestines and more complex stomachs (Forman et al. 1979). The stomach of *Carollia* is intermediate in complexity to that of glossophagines and stenodermatines; it has an enlarged cardiac vestibule and a baglike fundic caecum. Stomachs of stenodermatines are the most complex in the family and display two trends: a strongly enlarged cardiac vestibule or a long, tubular fundic caecum. The enlarged stomachs of stenodermatines are capable of storing large amounts of plant material. The stomachs of pteropodid bats also have enlarged cardiac and fundic portions (Hall and Richards 2000). The intestines of both families feature many villi (folds) covered with microvilli that greatly increase their absorptive surface areas.

PRIMATES.—In an essay entitled "The monkey and the fig," Stuart Altmann (1988, 262) concisely described the "adaptive complex" of a spider monkey (*Ateles*) as a "rapid-arboreal-locomotion, long-armed, prehensile-tailed, binocular-color-vision, terminal-branch-dispersed-feeding, seed-distributing, milk-drinking, nonterritorial, no-paternal-care, predator-fleeing complex." With a few modifications, this description could be generalized to describe many primates. Our discussion of these charismatic mammals will elaborate on several aspects of this adaptive complex. Some of these adaptations are discussed in more detail by Williams et al. (2010).

Members of order Primates range in size from 100 g (*Microcebus murinus* and *Cebuella pygmaea*) to about 200 kg (*Gorilla*); weighing up to about

90 kg, orangutans (*Pongo*) are the world's largest arboreal frugivores. Food habits are generally correlated with body size in these animals. Species that eat insects are relatively small (often <1 kg) whereas those that eat leaves are relatively large (median mass ±10 kg). Small fruit eaters supplement their diets with insects, and large fruit eaters also consume leaves (Kay 1984). Size also has important consequences for locomotory and foraging behavior. For example, terrestrial species are usually larger than arboreal species. Among arboreal New World monkeys, leaping occurs in small species (e.g., marmosets) whereas suspensory behavior occurs in large species (e.g., spider and howler monkeys). There are no terrestrial New World primates. Among Old World primates, quadrupedal walking and running appears to be independent of size (Fleagle 1988).

Locomotory behavior in primates affects many aspects of their morphology but is not closely associated with diet (Fleagle 1988). Regardless of diet, species that feed in the forest understory (e.g., certain lemurs and callitrichids) are small and often travel through this stratum by leaping between vertical supports. Canopy-feeding species, in contrast, travel via quadrupedal walking or running (e.g., many cebids and cercopithecids) or suspensory locomotion (e.g., atelids and gibbons). Fleagle and Mittermeier (1980) point out that sympatric species whose diets overlap often differ in their locomotory behavior whereas species whose diets do not overlap are often similar in locomotory behavior. Thus, the three most frugivorous monkeys in their study area in Surinam (*Ateles paniscus*, *Chiropotes satanas*, and *Pithecia pithecia*) use suspensory, quadrupedal, and leaping locomotion, respectively, when feeding whereas the two most suspensory locomotors (*A. paniscus* and *Alouatta seniculus*) are frugivorous and folivorous, respectively. Locomotory differences allow species to harvest similar resources in different ways—a pattern that we've also seen in the foraging behavior of frugivorous birds.

Primates are unique among mammals in having color vision, and the usual explanation for this involves feeding on fruit or young leaves (Dominy et al. 2001; Lucas et al. 2003; Regan et al. 2001). Trichromatic vision is universal in Old World catarrhines (monkeys and apes) and involves three visual pigments (opsins) with peak sensitivities at 430 (short, S), 530 (medium, M), and 563 (long, L) nM. An autosomal gene codes for the S pigment; the M and L pigments occur on the X chromosome and represent a duplication of an ancestral gene. Old World strepsirhines (lemurs and lorises) have only a single M/L gene on the X chromosome which is often polymorphic with alleles for two or three spectral photopigments. As a result, heterozygous females have two M/L alleles and are trichromats whereas males and homozygous females are dichromats.

Trichromatic color vision is not universal in New World monkeys (platyrrhines). Howler monkeys (*Alouatta*) have a different gene duplication on the X chromosome from that of catarrhines, which results in separate M and L pigments and trichromatic vision. Like strepsirhines, other platyrrhines have one polymorphic M/L gene on the X chromosome so that heterozygous females are trichromats and males and homozygous females are dichromats. Lucas et al. (2003) have suggested that trichromat platyrrhines have an advantage over dichromats in finding young leaves that reflect in the red-green color channel whereas both trichromats and dichromats are equally adept at finding ripe fruit in this color channel. From this they concluded that trichromatic vision evolved in primates to find young leaves rather than ripe fruit. Factors, including heterozygote advantage or frequency-dependent selection, that maintain allelic variation at the M/L locus in catarrhines and platyrrhines are not yet fully understood (Regan et al. 2001).

In addition to vision (which has a detection range of 20–30 m), primates use spatial memory, food vocalizations (with a range of up to 2 km), smell (with a range of up to 200 m), and touch to find and select fruit (Dominy et al. 2001). Ethanol is a common feature of ripe fruit, and its concentration is correlated with the sugar content of fruit. Primate sensitivity to low levels of this compound is very acute, and odor may actually be more important than color in fruit selection (Dominy 2004; Laska et al. 2006). Primates often use touch for fruit selection in species whose fruit lack an odor.

As in bats, the general morphology of the skulls of fruit-eating primates differs from that of exudate eaters and nectar-feeding mammals in several ways: they have larger teeth; the skulls are shorter and broader; and their fused lower jaws have higher coronoid processes and shorter condyle lengths (to increase gape size? [Dumont 1997]). In New World platyrrhines, skull shape is an allometric function of body size and has changed relatively little in the past 30 Ma. From an ancestral food habit of frugivory, these monkeys have entered four diet zones through the evolution of body size to a greater extent than through the evolution of skull shape. From smallest to largest body size, these zones include gum and other exudates (callitrichids), insects (cebids), seeds (pithecids), and leaves (atelids; Marroig and Cheverud 2005; fig. 8.9). Compared to primates with other food habits, the

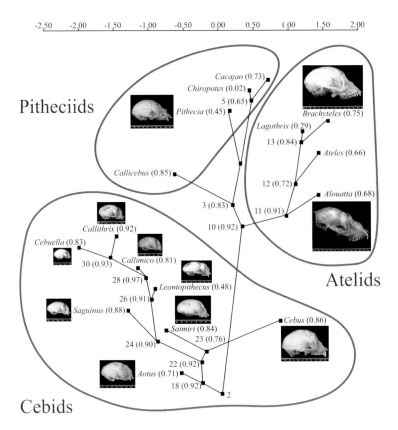

Figure 8.9. Summary of the adaptive radiation of New World primates based on multivariate skull morphology and diet. Based on Marroig and Cheverud (2005).

skulls of frugivores have larger incisors and simple molar teeth with low cusps for crushing fruit.

Primates are renowned among terrestrial mammals for having large brains relative to their body size. Brain mass as a percentage of body mass ranges from 0.4% in *Gorilla* to 3.8% in *Saimiri* (Aiello and Dean 1990). Primates share this characteristic with plant-visiting bats, which have significantly larger brains than insect-eating species (chap. 7). The implication here is that diet, specifically fruit or flower eating, has been an important factor in the evolution of large brains in these mammals. The usual explanation for this is that the spatiotemporal variability of fruit and flower resources is greater than that of other diet classes (e.g., leaves and insects; Fleming 1992) and that this variability selects for larger brains and greater cognitive ability. Several studies have shown that superior cognitive ability in terms of new, complex, or unusual behaviors (e.g., tool use) is positively correlated with size of the "executive brain" (i.e., volume of the neocortex and striatum) in a variety of large-brained birds and primates (Iwaniuk et al. 2005; Lefebvre et al. 2004; Reader and Laland 2002). The frequency of social learning is positively correlated with brain size but not with social group size in primates (Reader and Laland 2002). This suggests that a gregarious lifestyle and complex social interactions per se have not necessarily favored the evolution of large brains in these mammals. Further support for this conclusion comes from Schillaci (2008), who found a negative correlation between relative size of the neocortex and degree of male-male competition for mates in primates. He also reported that monogamous species have relatively larger brains than polygynous species.

In addition to being large, primate brains differ from those of other mammals in the relative sizes of different components. For example, Barton et al. (1995) compared the relative sizes of the olfactory and visual systems of primates, plant-visiting bats, and insectivores and found that the sizes of these systems were negatively correlated in primates but were positively correlated in bats. They also reported that time of activity and diet affect the size of brain systems in primates. Olfactory bulbs are larger in nocturnal species than in diurnal species but aren't necessarily larger in frugivores than in other primate diet classes as they are in bats. The geniculostriate visual system of diurnal frugivorous primates, however, is larger than that of other diet/activity classes.

Two additional morphological features—binocular vision arising from forward-facing eyes and manipulatable fingers—are critical for foraging success in primates. These features appear to have arisen early in the evolution of modern primates and allowed Eocene forms to forage for fruits and insects at the tips of branches of trees and on shrubs (Dominy et al. 2001; Sussman 1991). Spatulate incisors and manipulatable fingers would have allowed them to easily peel fruit with a tough skin, which characterizes many primate fruits today (Dominy et al. 2001).

As in fruit-eating birds and bats, primates process food in ways that are analogous to gulping and mashing. As we discussed in chapter 4, most frugivorous primates are gulpers that swallow pulp and seeds. They sometimes regurgitate large seeds rather than defecate them. Chimpanzees also form wadges with their lips and spit out the fibrous parts of some fruits and seeds (Gross-Camp and Kaplin 2005).

Finally, the morphology of the gastrointestinal tract varies with diet in mammals generally and in primates specifically. Chivers and Hladik (1980) compared the GI tracts of primates belonging to three diet classes: faunivores, frugivores, and folivores. The tracts of frugivores are intermediate to those of the other two classes in terms of the ratio of the stomach and large intestine to small intestine whether measured by area, volume, or mass. This ratio is low in faunivores and high in folivores. Frugivorous primates have relatively simple GI tracts characterized by a small globular stomach, a small caecum (except in some prosimians), and a short small intestine and colon. Some authors (e.g., Milton 1993) have suggested that there is a tradeoff between brain size and gut size in primates (e.g., frugivorous spider monkeys have larger brains and smaller guts than similar-sized folivorous howler monkeys), but Hladik et al. (1999) dispute this. They argue that it is difficult to compare directly the sizes of brains and guts of frugivores and folivores because of differences in their allometric relationships between gut size and body size.

PHYSIOLOGICAL ADAPTATIONS IN FRUGIVOROUS VERTEBRATES

As in nectar-feeding vertebrates, frugivorous birds and mammals consume a food that is relatively rich in water and carbohydrates but that is often deficient in proteins and lipids. In addition, a significant fraction of the mass of fruits is indigestible ballast in the form of seeds. How do these animals deal with this kind of food physically and physiologically? Because the literature on the physiology of vertebrate frugivores is much more limited than that of nectarivores, we will combine birds and bats in our discussion here and will treat primates separately.

BIRDS AND BATS.—The response of frugivorous passerines and bats to sugar-rich fruit is similar to that of their nectar-feeding relatives. These animals can easily discriminate between fruits that differ in percentage of sugar; they often prefer fruit pulp containing easily digested monosaccharides rather than disaccharides, even when they possess sucrase; they decrease their consumption of fruit as its sugar content increases to maintain a constant intake of energy; and they have high digestive efficiency (often >90%) of fruit sugars (Korine et al. 2006; Schaefer et al. 2003; Witmer and

Van Soest 1998). In addition to discriminating between fruits that differ in carbohydrate content, some frugivorous birds can discriminate among fruits based on their lipid or protein contents (Bosque and Calchi 2003; Schaefer et al. 2003). This suggests that some frugivores can select for the nutrient content of their fruits. Gut passage rates in fruit-eating birds and bats are often fast (usually <30 mins) for the nonfibrous parts of fruit. Unlike bats, which either form wadges or pass fibrous material quickly (but see Shilton et al. [1999] for a discussion of slow seed passage rates in certain pteropodid bats), birds deal with ingested seeds and plant fiber in a variety of ways, including collecting this material in the gizzard or caecum before defecating it (reviewed in Martínez del Rio and Restrepo 1993).

The lipid content of fruit pulp is generally low with some notable exceptions (e.g., fruits of Lauraceae, Meliaceae, etc.). Being more complex chemically, lipids are not as digestible as carbohydrates, and many phyllostomid bats avoid lipid-rich fruits, perhaps for this reason (Fleming 1988). Martínez del Rio and Restrepo (1993) predicted that different groups of birds are likely to deal with lipids in different ways. Birds that include many lipid-rich insects in their diets, for instance, should be better able to digest lipids in fruit than birds that eat few insects. Thus, we might expect the guts of fruit-eating tyrannid flycatchers to have higher bile and lipase activity and to digest lipids more rapidly than tanagers and euphonias. Nonetheless, the food passage rates of birds eating lipid-rich fruits are likely to be lower than those eating sugar-rich fruits (Levey and Martínez del Rio 2001).

Fruit pulp is also generally low in nitrogen-rich compounds, only a fraction of which (about 75%–80%) are proteins; the other compounds are nonprotein amino acids and secondary plant substances (Bosque and Pacheco 2000; Levey and Martínez del Rio 2001). Because of this, it is generally recognized that the low protein content of fruit pulp represents a significant dietary limitation for frugivores (Witmer and Van Soest 1998), and there has been considerable discussion of how frugivores deal with this nutritional problem. Many frugivorous birds include insects and sometimes vertebrates in their diet and often feed nonfruit items to nestlings (e.g., Morton 1973; Wheelwright 1983). Oilbirds (Steatornithidae) are an exception to this and feed lipid-rich fruit pulp to nestlings. As a result, oilbirds have smaller clutches and significantly longer nestling periods than their insectivorous relatives (Snow 1971). Frugivorous phyllostomid and pteropodid bats deal with the "protein problem" by eating insects (mostly phyllostomids) and/or leaves (both phyllostomids and pteropodids; Barclay et al. 2006; Courts

1998; Kunz and Diaz 1995; Nelson et al. 2005; Ruby et al. 2000; York and Billings 2009).

As in nectarivores, another way in which fruit-eating birds and mammals deal with the protein problem is by having low protein requirements (chap. 7). The temperate zone cedar waxwing (*Bombycilla cedrorum*, Bombycillidae), for example, is more frugivorous than two species of temperate thrushes and requires less protein as a percentage of the sugar content of fruit pulp (1.9% vs. 3.9–8.1%); its protein requirements are also less than those of similar-sized granivorous or herbivorous birds (Witmer and Van Soest 1998). Low nitrogen requirements have also been reported in a fruit-eating bulbul and Old World grackle (Tsahar et al. 2005a, 2005b). Bosque and Pacheo (2000) point out that although mass-specific daily nitrogen requirement is a negative allometric function of body mass in birds, small species have lower absolute nitrogen requirements than large birds and should be less susceptible to protein limitation from a fruit diet than large species. Their review of the protein content of temperate and tropical fruits indicates that small birds should be able to eat a higher proportion of tropical fruits (about 95%) without experiencing protein limitation than large birds (about 77%). They also suggested that, in addition to having lower daily nitrogen requirements, frugivorous birds (and probably bats) should be under selection to maximize the digestive efficiency of nitrogen and to minimize the loss of endogenous nitrogen via feces. As predicted, low fecal nitrogen losses have been reported in the above-mentioned bulbul and grackle (Tsahar et al. 2005a, 2005b).

As in small tropical birds, the protein problem may not actually be a problem for frugivorous bats. Herbst (1985) and Wendeln et al. (2000) analyzed the nutritional composition of fruits eaten by *Carollia* and fig-eating phyllostomids, respectively, and concluded that, by eating a variety of different fruits, these species can remain in positive nitrogen balance on an all-fruit diet, even when females have high energy demands during pregnancy and lactation. Delorme and Thomas (1996) also concluded from a captive study that *Carollia perspicillata* is not protein limited. Similarly, Korine et al. (1996) determined the daily nitrogen and energy requirements of captive *Rousettus aegyptiacus* pteropodids fed the fruits they normally eat in nature and found that they were significantly more energy limited than nitrogen limited, a conclusion also reached by Delorme and Thomas (1999) for this species. *Rousettus aegyptiacus* maintained a positive nitrogen balance while consuming pure diets of several fruit species. Compared with allometric expectations, the nitrogen requirements of these bats were up to 76% lower than expected. Korine et al. (1996) suggested that the low nitrogen requirements of this bat reflect low excretion rates of nitrogen (i.e., a greater degree of nitrogen economy than found in many other kinds of mammals). Results of Delorme and Thomas's (1996, 1999) studies of *C. perspicillata*, *Artibeus jamaicensis*, and *R. aegyptiacus* support this suggestion. They point out, however, that the low (ingested) fiber diets of nectar- or fruit-eating birds and mammals will naturally result in low levels of fecal nitrogen loss and that this economy does not necessarily reflect a physiological adaptation to a low-nitrogen diet.

Finally, lower than expected metabolic rates and the use of torpor are additional ways in which nectar-eating vertebrates cope with nutritionally poor food (chap. 7). However, these metabolic strategies do not appear to be common in frugivorous birds and mammals. As reviewed by McNab (2002), frugivores do not generally have lower than expected basal metabolic rates, and very few species undergo daily bouts of torpor (Audet and Thomas 1997). Thus, the energetic strategies of these animals do not appear to be exceptional.

PRIMATES.—Most arboreal and terrestrial frugivorous primates are considerably larger than their bird and bat counterparts and have broad diets that often include many nonfruit items (i.e., insects in small species and leaves in large species). Because of this, we might not expect fruit-eating primates to experience the same kinds of physiological challenges (e.g., protein limitation) faced by frugivorous birds and bats. One of the key physiological features of vertebrate frugivores is rapid food passage rates (Levey and Karasov 1992), but these rates in primates are not rapid. Lambert (1998) reviewed primate digestion and pointed out that except for very small and very large species, food passage rates are not closely correlated with body mass (fig. 8.10). These rates range from about 1.3 to 40 hrs over a size range of 1–18 kg. Lemuroids are at the low end of the range of transit times and cercopithecines and atelines are at the high end. As expected, species whose diets include large amounts of leaves (e.g., *Alouatta* and colobines) have longer transit times than frugivores. The generally short passage rates of some cebids (fig. 8.10) might preclude them from eating leaves; their "fallback" foods when fruit is scarce include nectar and insects (Terborgh 1983). In contrast, the generally long food passage rates in frugivorous cercopithecines allow them to use leaves as fallback food (Lambert 1998; Terborgh and Van Schaik 1987).

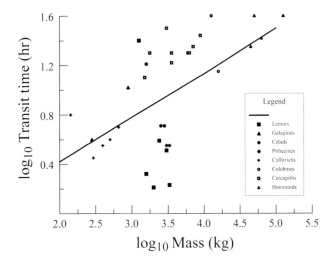

Figure 8.10.   Transit times (in hours) of ingested food as a function of body size in primates. Note the lack of a strong relationship in midsized species (i.e., those between the two vertical dashed lines. Based on Lambert (1998).

Like most other frugivores, primates have a sweet tooth and often have a preference for sugar-rich fruits. When given a choice, chimps and gorillas prefer the richer of two fructose solutions, and this preference increases with percentage of fructose (Remis 2006). Phenolic compounds also occur in fruits eaten by primates, and chimps and lowland gorillas avoid fructose-tannin solutions as percentage of tannin increases. Chimps have a lower threshold for tannin tolerance than the larger gorillas (Remis 2006).

Ethanol is a fermentation product of fruit ripening, and it is likely that frugivorous primates (and bats?) use ethanol odor plumes to locate ripe fruit crops. Dudley (2004) suggested that the association of ethanol with ripe fruit crops has given primates a sensory bias toward this compound. This bias, in turn, can lead to the overconsumption of ethanol and, in some humans, alcoholism. Because ethanol and its metabolic breakdown product, aldehyde, can be toxic, selection in frugivores should favor an increased ability to detoxify these compounds via the enzymes alcohol dehydrogenase and aldehyde dehydrogenase. Levey (2004) disputed Dudley's contention that human alcoholism has resulted from a sensory bias toward ethanol based on an ancestry of frugivory. Instead, he suggested that exposure to high levels of ethanol in humans has occurred historically from using anaerobic fermentation of sugar-rich foods to preserve them.

Compared with most other frugivores, primates are large animals and cannot usually use torpor or other means to reduce their daily energetic expenditures in the face of low food availability or unfavorable climatic conditions. Exceptions to this are certain small strepsirrhines, such as the gray mouse lemur (*Microcebus murinus*) and sportive lemur (*Lepilemur* sp.), which undergo seasonal bouts of torpor (Schmid and Speakman 2000). Orangutans are the largest arboreal mammalian frugivores, and they save energy by having one of the lowest rates of daily energy use of any eutherian mammal (Pontzer et al. 2010). Their mass-specific daily energy expenditure is even lower than that of lemurs during torpor. Pontzer et al. (2010) interpreted this as an evolutionary response to chronically severe food shortages that orangutans experience in the lowland rain forests of Southeast Asia (see chap. 4).

ECOLOGICAL AND BEHAVIORAL ADAPTATIONS IN FRUGIVOROUS VERTEBRATES

As in nectar feeders (chap. 7), many aspects of the ecology and behavior of fruit-eating birds and mammals are affected by their diet (reviewed in Corlett 2009a; Fleming 1992; Primack and Corlett 2005; Stutchbury and Morton 2001). Here we will focus on two aspects: migration and social and mating systems.

Although they are easy to locate, fruits, like flowers, can be patchily distributed in time and space (chap. 3). This spatiotemporal variation, in turn, can select for high mobility on a daily as well as a seasonal basis. Levey and Stiles (1992) noted that Neotropical avian frugivores and nectarivores are much more likely to undergo seasonal habitat shifts as well as altitudinal and intratropical migrations than insectivores. They hypothesized that these tendencies preadapted these birds to give rise to latitudinal migrants. More recently, Boyle and Conway (2007) reviewed migratory patterns within a New World clade of suboscine passerines—the Tyranni (flycatchers, cotingids, manakins, tityras, etc.)—and concluded that migration was more closely associated with habitat than diet in these birds. Frugivory does favor migration in these birds, but insectivorous species living in open habitats are also strongly migratory. In fact, open habitat insectivores are more likely to be long-distance migrants than frugivores, which tend to be short-distance and altitudinal migrants. Food-driven short-distance migrations are particularly well-known in certain cotingids (e.g., *Procnias tricarunculata*) and quetzals (*Pharomachrus moccino*), both of which specialize on fruits of Lauraceae. In Costa Rica, *P. tricarunculata* breed at midelevations (1,000–1,800 m) on the moist Atlantic slope but then spend significant amounts of time during the nonbreeding season in three other forest habitats (one montane and two lowland) separated by distances of about 200 km (Powell and Bjork 2004). Likewise, quetzals breed at midelevations in Central America but move to lower elevations during the nonbreeding season and annually use up to four montane habitats in Costa Rica and elsewhere (Powell and Bjork 1995; Solorzano et al. 2000; Wheelwright 1983).

Migrations of plant-visiting birds are also common in tropical Asia and Australia. In Asia, about 31% of landbirds are migratory, and this includes frugivores as well as insectivores among passerines (Corlett 2009a; Salewski and Bruderer 2007). Millions of partially frugivorous thrushes and chats migrate to tropical East Asia from temperate Asia annually (Corlett 2009a). In Australia, about 40% of landbird species migrate; many of these species have both resident and migrant populations. All of these species are north-south intracontinental migrants and none cross the equator. Seasonal movements of tens to hundreds of kilometers occur in at least 25% of Australian honeyeaters and 33% of parrots (Chan 2001).

In contrast to the Neotropics, Asia, and Australia, very few African frugivorous birds migrate. These include hornbills, mousebirds, parrots, and sunbirds, of which at least 85% of African species are sedentary (Hockey 2000). Hockey (2000) proposed two explanations for this: (1) African frugivores exhibit greater dietary flexibility than other frugivores owing to a long history of climatic (and food level?) fluctuations and hence don't need to migrate to find food; and (2) any migrations that do occur in these birds are likely to occur over short distances (e.g., altitudinal migrations) and hence are hard to detect.

Although they are relatively unstudied compared with birds, food-driven seasonal migrations also occur in frugivorous bats. The best-documented cases occur in Africa, where several species of pteropodids move from equatorial forests north or south several hundred kilometers to feed on seasonal pulses of fruit in savanna woodlands (Richter and Cumming 2006, 2008; Thomas 1983). Individuals of the African straw-colored bat *Eidolon helvum*, for example, annually fly at least 2,000 km round-trip from closed forests in the Democratic Republic of the Congo to open miombo woodlands of central Zambia where they feed on fruits of species of *Uapaca*, *Syzygium*, and *Parinari curatellifolia* (Richter and Cummings 2006). Seasonal migrations of hundreds of kilometers also occur in the nectar- and fruit-eating phyllostomid *Leptonycteris curasoae/yerbabuenae* in western Mexico and

Venezuela, as well as in the Australian pteropodid *Pteropus poliocephalus* and the Southeast Asian *P. vampyrus* (Eby 1991; Epstein et al. 2009; Fleming and Nassar 2002).

In addition to patterns of migration, spacing and social systems of tropical birds and mammals are also influenced by a frugivorous diet. These aspects of the behavioral ecology of Neotropical birds have been particularly well-reviewed by Stutchbury and Morton (2001), and we will base our account on their review. Many aspects of the life histories and behavioral ecology of Neotropical birds differ strongly from their temperate zone relatives. Among other things, these differences include clutch size (lower in the tropics), rates of nest predation (higher in the tropics), adult lifespan (longer in the tropics), and frequency of extrapair fertilizations (much lower in the tropics). Differences in the spacing behavior of temperate and tropical passerines are particularly striking (table 8.3). Whereas most temperate passerines occupy defended territories only during the breeding season, 43% of Panamanian species, including many frugivores, defend territories year round. Both males and females independently defend these territories against conspecifics of the same sex. The territories of frugivorous species may or may not include food resources. When their territories don't include fruit, birds leave them to feed at fruit sources that are typically not defended by other birds. These are the fruit-influenced species in table 8.3. Many canopy-feeding tanagers feed in mixed-species flocks that also include insectivores. Up to 20 species can occur in these flocks, which are particularly common at midelevations. Territories in these birds are often much larger than those of nonflocking species or insectivorous species that occur in understory flocks.

As we discussed in chapter 7, lek mating systems are particularly common in tropical plant–visiting birds. In addition to many hummingbirds,

Table 8.3. Differences in the Territorial Systems of Passerine Birds in Panama and North America

Type of Territory	Panama		North America	
	No. of species	Proportion of Species	No. of species	Proportion of Species
Breeding season only	42	0.13	224	0.94
Year long	142	0.43	15	0.06
Army ant influenced	11	0.03	0	
Mixed species flocks	65	0.20	0	
Fruit influenced	43	0.13	0	
Lek based	28	0.08	0	
Total	331		239	

Source. Based on Stutchbury and Morton (2001, table 5.1).

lekking behavior occurs in fruit-eating manakins, cotingids, birds of paradise, and bowerbirds. Leks can be considered to be tiny territories on which males display to and copulate with females. They are not usually associated with food resources. Males of lekking species provide no parental care, which allows them to potentially mate with several females during a breeding season. Snow (1971), among others, has pointed out that because fruits are often abundant and conspicuous, a frugivorous diet provides the opportunity for promiscuous mating systems such as leks to evolve in birds, whose most common mating system is social monogamy (Bennett and Owens 2002). Support for this idea comes from the ochre-bellied flycatcher (*Mionectes oleaginous*) and several congeners that are highly frugivorous and that have promiscuous or lek mating systems, unlike most other tyrannid flycatchers. But lek mating is absent in tanagers, thrushes, mimids, and other species-rich families of Neotropical frugivores. It is also absent in Old World families of frugivorous passerines (e.g., bulbuls, meliphagids, and sturnids) and nonpasserines (e.g., fruit pigeons and hornbills). Thus, a fruit diet doesn't inevitably favor the evolution of promiscuous mating systems. Whenever both parents are needed to raise nestlings successfully, promiscuous mating systems are not likely to evolve.

Research conducted by Bruce Beehler and colleagues (Beehler 1983, 1985, 1989; Beehler and Pruett-Jones 1983; reviewed in Frith and Beehler 1998) indicates that the mating systems of birds of paradise range from typical passerine monogamy to spectacular leks and are correlated with diet. Members of this family are primarily frugivorous and secondarily insectivorous. But type of fruit and degree of insectivory varies among species. Species that specialize on figs (e.g., trumpet manucode, *Manucodia keraudrenii*) are monogamous and territorial and have male parental care—the typical passerine pattern. Both sexes in these sexually monomorphic birds feed figs to their nestlings. In contrast, birds of paradise that eat the more nutritious arils of fruits of Myristicaceae and Meliaceae as well as insects are often nonterritorial lek maters. Females in these species can provide enough nutritious food to nestlings to raise them alone. These birds are strongly sexually dimorphic and include species in which males display solitarily (e.g., magnificent bird of paradise, *Cicinnurus magnificus*), in "exploded leks" in which males are spaced tens of meters apart (e.g., Lawes's parotia, *Parotia lawesii*), or in communal leks in which several males display in a single tree (e.g., Raggiana bird of paradise, *Paradisaea raggiana*). Interspecific differences in the form of leks appear to reflect the size of female home ranges. Female ranges are small and nonoverlapping in solitary-displaying species and large and overlapping in communally displaying species. Males of lek-mating species have much gaudier plumage than those of nonlekking species and have elaborate vocal and visual courtship displays. Indeed, some lek-mating males are among the most spectacular birds in the world.

Finally, although they are monogamous and form long-lasting pair bonds, frugivorous hornbills exhibit a variety of social behaviors, some of which are clearly related to diet (Kinnaird and O'Brien 2007). Large-bodied species (e.g., in the genera *Aceros, Buceros, Rhinoplax,* and *Rhyticeros*) that feed heavily on figs are nonterritorial, have very large foraging areas, and often feed in flocks. In contrast, small-bodied species (e.g., in the genera *Anthracoceros, Ocyceros,* and possibly *Penelopides*) whose diets contain more animals and lipid-rich fruits have small foraging areas that are defended by pairs or family units either seasonally or year round. Cooperative breeding in which juveniles delay dispersal and help their parents feed their offspring is an uncommon breeding system in birds but is particularly common in hornbills (Kinnaird and O'Brien 2007). At least seven species (six Asian and one African) are known to have helpers at the nest, but whether this is related to diet is currently unknown.

We discussed the mating systems of plant-visiting bats in chapter 7. Most frugivorous members of Phyllostomidae and Pteropodidae appear to have polygynous mating systems, often involving either seasonal or year-round harem polygyny. These systems are not unique to plant-visiting bats but also occur in many families of insectivorous bats (McCracken and Wilkinson 2000). An African pteropodid, the hammer-headed bat, *Hypsignathus monstrosus*, has a lek mating system in the lowland forests of Gabon (Bradbury 1977). Unlike most bats in which sexual dimorphism is modest if it exists at all, adult males of this species weigh nearly twice as much as females (424 g vs. 234 g, respectively) and have a large skull and muzzle and a greatly enlarged larynx and vocal chords for sound production and amplification. Males of this fig-eating species congregate twice a year for up to 3 mo. in groups of up to about 130 individuals. These males are spaced about 10 m apart in trees along large rivers where they produce loud, harmonically rich honking calls for several hours each night. Females visit these sites to choose a mate. Centrally located males in these linear leks tend to garner most of the matings. As in certain birds of paradise, the home ranges of females are large and overlapping, a situation that appears to favor the formation of male leks (Bradbury 1981). Although not as well-studied, males of other

African epomophorine pteropodids (e.g., species of *Epomophorus* and *Epomops*) also call loudly to attract females. Unlike *H. monstrosus*, however, these males roost solitarily while calling and may not form leks (McCracken and Wilkinson 2000). Finally, monogamy, which is an uncommon mating system in mammals, is thought to occur in *Pteropus samoensis* in the South Pacific. Other examples of monogamy in bats include species that are either carnivorous or insectivorous (McCracken and Wilkinson 2000). In summary, except for lek mating in *H. monstrosus*, the mating systems of frugivorous phyllostomid and pteropodid bats do not necessarily differ qualitatively from those of nonfrugivorous tropical bats.

As we've indicated, most primates are at least partially frugivorous, and they exhibit a diverse array of social and mating systems. Kappeler and Van Schaik (2002) provide a review of much of this literature but without much emphasis on the effects of diet per se on social evolution. Because of this, our summary of these topics will rely primarily on diet-based accounts in Fleagle (1988) and Smuts et al. (1986). Chapters in Campbell et al. (2007) also contain extensive reviews of the social systems of Primates. Ranging behavior—the size of daily and annual home ranges—is related to body size and diet in primates in the expected manner. Range size increases with body size (either individually or collectively for a social group), and for a given body size, frugivores tend to have larger ranges than folivores. These correlations reflect the energy requirements of individuals and groups as well as the density and spatiotemporal distribution patterns of ripe fruits compared with leaves. Range sizes vary over four orders of magnitude (from <0.01 km^2 to nearly 50 km^2) in various species (Fleagle 1988, fig. 8.8). And groups in many, but certainly not all, species are strongly territorial intraspecifically. For example, Cheney (1986) reported that only 50% of a sample of 45 species from most major primate lineages exhibit home range defense (i.e., territoriality). Groups that range over small areas are usually territorial whereas those with large, less readily defendable home ranges usually are not.

Less closely associated with diet and size are primate mating systems. As in bats, these systems include monogamous pairs (e.g., frugivorous gibbons, pithecids, and night monkeys; folivorous indriid prosimians; and insectivorous tarsiers), single-male groups (e.g., certain *Colobus* species, *Alouatta*, and *Gorilla*), and large multimale groups (e.g., baboons, common chimpanzees, and many cercopithecines and platyrrhines). Harvey et al. (1986) and Fleagle (1988) point out that the distribution of mating systems in primates seems to reflect phylogenetic interia because closely related species (i.e., those in the same subfamily) with different diets usually have very similar social systems and life history characteristics (also see Kappeler and Heymann 1996 and Shultz et al. 2011). Wrangham (1986) reviewed the evolution of these systems from the point of view of the costs and benefits to females and concluded that they are influenced by three ecological factors: defensibility of resources; the spatiotemporal distribution of food patches; and degree of predator pressure. Whenever resources are individually defensible, multifemale groups are not likely to form. In species in which females do form groups, food patch dynamics can determine group size and stability. Species feeding on fruit species that occur in small, scattered patches (e.g., species of *Ateles*) typically occur in smaller and less cohesive groups (i.e., fusion-fission communities) than those that feed on fruit species that occur in large patches (e.g., species of *Saimiri*). Finally, degree of predator risk, which is a function of a species' body size and habitat preference, will determine whether individuals are safe traveling alone or in groups, whether and how many males are needed to defend young, and group size that balances the costs of increased food competition with the benefits of increased predator protection. Thus, different primate lineages have responded to their abiotic and biotic environments in a variety of different ways and have not necessarily converged on similar evolutionary solutions to similar environmental challenges (Kappeler and Heymann 1996). DiFiore and Campbell (2007), however, suggest that certain New World atelines and African apes have converged in many aspects of their socioecology, including basic social organization, male-male relations, ranging patterns, and mating systems. These include species of *Ateles* and *Pan troglodytes*, which have a fission-fusion social organization in which males and females often forage separately in small groups, and species of *Brachyteles* and *P. paniscus*, which also exhibit fission-fusion but which forage in larger groups containing both sexes. Overall, however, frugivorous primates have not converged on a similar suite of adaptations regarding body size, timing of daily activity, and group size in different regions and lineages.

Shultz et al. (2011) examined the evolution of primate sociality in an explicit phylogenetic framework. Their results indicate that the ancestral social state in primates was nocturnal solitary foraging. When primates shifted to diurnal activity patterns (about 52 Ma in anthropoid primates and 32 Ma in indriid and lemurid prosimians), they formed multimale, multifemale groups; harem- and pair-based social systems are derived from these groups. The formation of large groups in diurnal primates suggests that predation pressure has been the major driver of primate social evolution. As in bats, diet has apparently played a secondary role in the evolution of sociality in primates.

General Discussion and Conclusions

As in chapter 7, we have been concerned with details of the fruit-frugivore mutualism in this chapter under the general themes of convergence and complementarity. In terms of convergence, we have asked to what extent different angiosperm families have evolved fruits that are similar morphologically and nutritionally to attract specific kinds of birds and mammals as seed dispersers. Similarly, in how many ways have frugivorous birds and mammals converged morphologically, physiologically, and behaviorally in response to characteristics of their food supplies? For both plants and animals, to what extent do their adaptations represent complementary responses to their mutualistic partners? What kinds of responses appear to be complementary?

EVOLUTION OF FRUITS

Fleshy fruits are a major, but not unique, feature of angiosperms and have undoubtedly helped promote the diversification of flowering plants. Vertebrate-dispersed fleshy fruits have evolved in many angiosperm lineages, including basal lineages, and there has been much convergence in fruit form in these lineages. Virtually all of the major kinds of fleshy fruit (berries, drupes, and arils) have evolved independently many times during angiosperm evolution (table 8.1), and these fruits have often been placed in a variety of dispersal syndromes (table 1.5). These syndromes are much fuzzier than pollinator syndromes, however, because of the opportunistic feeding behavior of many vertebrate frugivores (Heithaus 1982; Howe 1986). Nonetheless, like pollinator syndromes, dispersal syndromes tend to occupy different areas of multivariate character space (e.g., Gautier-Hion et al. 1985), supporting the hypothesis that different kinds of vertebrates have been different evolutionary attractors for fruiting plants. As we discuss in detail below, key features of birds and mammals that are reflected in these syndromes include body size, mouthparts, sensory adaptations, locomotory behavior, and time of daily activity.

Although the different types of fleshy fruit have evolved repeatedly during angiosperm diversification, the major angiosperm clades differ strikingly regarding the occurrence of families that produce single fruit types versus a mixture of fruit types (fig. 8.1). Most vertebrate-dispersed families of basal angiosperms and monocots produce only fleshy fruits whereas a majority of vertebrate-dispersed eudicot families produce both dry capsular and fleshy fruits. As we have seen, capsular fruits tend to be associated with open, disturbed habitats whereas fleshy fruits tend to be associated with closed, less-disturbed habitats. The presence of mixed fruit types in many eudicot families reflects a greater diversity of seed dispersal strategies and growth forms (e.g., trees, shrubs, vines, and herbs) than are found in most families of basal angiosperms (mostly trees and shrubs) and monocots (mostly herbs). Within mixed families, the evolution of fleshy versus non-fleshy fruit types tends to be unconstrained phylogenetically, which gives these plants considerable flexibility regarding their dispersal strategies. Dispersal flexibility, in turn, has resulted in the evolution of high species richness in these families (e.g., Ricklefs and Renner 1994).

In addition to frequent convergence in fruit types, convergence has occurred in a variety of other fruit traits including color, size, shape, and the nutritional characteristics of pulp. Thus, bird-dispersed berries around the world tend to be either red or black and scentless; they are relatively small with small seeds; and their pulp is rich in water and sugars. Larger berries or drupes in some bird-dispersed families are ellipsoidal rather than round in shape for ease of swallowing. And large-seeded drupes or arils tend to have thin, lipid-rich pulps. In contrast, bat-dispersed berries tend to be green and scented in addition to being small and having small seeds and relatively nonnutritious pulp. Bird and bat fruits tend to have these characteristics even within the same genera (e.g., *Ficus*, *Solanum*), indicating that phylogenetic constraints are often lax in these traits.

The existence of taxon-specific fruit traits implies that fruit-eating vertebrates have had important selective effects on the evolution of fruit. Traits for which there is strong evidence for this include fruit size, color, shape, scent, and method of presentation. The size of bird fruits is often limited by the width of bird gapes, and regional differences in fruit sizes reflect differences in the size of birds and other frugivores. Fruits eaten by birds and primates with color vision and diurnal activity patterns are more colorful than those eaten by nocturnal, color-blind bats. Large fruits in some (but not all) families tend to be ellipsoidal in shape rather than round as a result of gape limitation in birds. Fruits eaten by bats, which have a strong olfactory sense, are scented whereas bird fruits typically lack scents. And fruits eaten by bats tend to be presented away from foliage whereas those eaten by birds and primates are usually presented among foliage.

It could be argued that there are alternative explanations for differences in these traits other than selection by vertebrates, but taken together, these differences make such explanations implausible. Selection by herbivores and pathogens might account for some fruit traits (e.g., the presence of secondary compounds in fruit pulp), but they cannot easily explain variation in an entire suite of fruit traits that is correlated with particular kinds of frugivores (Lomáscolo et al. 2010). The most compelling evidence for selection by frugivores comes from experimental discrimination tests. When such tests are conducted based on fruit traits such as color, scent, and nutritional content, the usual result is that birds, bats, and primates can distinguish often subtle differences among fruits (e.g., in sugar or protein content). These results provide prima facie evidence that vertebrates can select for particular fruit traits.

Selection pressures exerted by vertebrate frugivores have not affected all fruit traits, including seed size and pulp-to–seed mass ratios. The evolution of seed size involves many selective factors, including life history and dispersal strategies (e.g., shade intolerant vs. shade tolerant plants), seed predators, and seedling establishment strategies (e.g., dormancy vs. immediate germination). There is some evidence that particular kinds of frugivores (e.g., fruit mashers) can select for small seeds, but overall, the selective effect of frugivores on seed size appears to be weaker than on other fruit traits. Similarly, although some kinds of frugivores (e.g., birds) appear to discriminate among plants on the basis of fruit pulp-to–seed mass ratios, other frugivores (e.g., primates) eating the same fruit species apparently do not select among plants based on this ratio. Conflicting responses by different kinds of frugivores are thought to weaken the overall selective effect of frugivores on this and other fruit and seed traits.

EVOLUTION OF FRUGIVORES

Many more kinds of birds and mammals eat fruit, at least occasionally, than nectar and pollen, and despite the existence of fruit-dispersal syndromes, there is much overlap in the diets of different kinds of vertebrate frugivores. This reflects, in part, the fact that fruit can be eaten in a variety of different ways by different animals. For example, many small birds and bats eat large fruits piecemeal so that fruit size, unlike seed size, is not necessarily a constraint in the diets of many frugivores. As a result, evolutionary interactions between fruiting plants and frugivores have been relatively diffuse. Nonetheless, the morphology, physiology, and behavior of vertebrate frugivores often differ significantly from that of their nonfrugivorous relatives. A frugivorous diet has clearly selected for specific kinds of adaptations in these animals as a result of particular features of the biology of fruits. This selection has resulted in convergence among frugivores in some of their traits but not in others. Given the substantial taxonomic diversity of vertebrate frugivores, the absence of strong convergence, especially in morphological traits, is not surprising.

Because fruits and their nutritional rewards are accessible to a greater variety of vertebrates than nectar and because fruit can be harvested and eaten in many different ways, the morphospace occupied by vertebrate frugivores is much larger than that of nectarivores. Along the dimension of body size, for instance, the size range of avian and mammalian frugivores is orders of magnitude greater than that of nectarivores (table 1.2). Despite the considerable morphological and taxonomic diversity among vertebrate frugivores, however, there are common morphological trends in their evolution. For example, birds that feed from perches tend to have longer bills, stronger legs, and higher aspect ratio wings than birds that take fruit on the wing. Frugivorous bats and primates have lower and broader molars than relatives with other feeding habits. Many frugivorous phyllostomid bats and primates, but not pteropodid bats, have short and wide faces, and stenodermatine phyllostomid and pteropodid bats have transverse palatal ridges for crushing fruit. Fruit-eating bats and primates have brains that are relatively larger overall and in specific regions (e.g., the neocortex) than nonfrugivorous relatives. The gastrointestinal tracts of frugivores tend to be simpler than those of their nonfrugivorous relatives. Frugivorous birds often have thin-walled gizzards, and frugivorous birds, bats, and primates often have short intestines with a small (or no) caecum. Finally, fruit mashing to avoid ingesting fibrous material and seeds has evolved independently in certain New World birds as well as in phyllostomid and pteropodid bats and primates.

A final example of morphological convergence in frugivores is the evolution of trichromatic color vision in primates. From a nocturnal dichromatic ancestry, color vision has evolved independently in Old and New World primates. Diurnal activity and a diet of fruit and/or colorful young leaves have probably been the major factors behind this sensory evolution.

Further examples of convergence can be found in the physiology of fru-

givorous birds and bats. These animals are similar to nectarivores in easily discriminating among fruits that differ in sugar (and lipid) content; by
quickly and efficiently digesting fruit sugars; by having low nitrogen requirements; and by minimizing fecal nitrogen losses. Frugivorous birds, bats, and
primates also supplement the nitrogen in their diets by consuming either insects (many birds, phyllostomid bats, and small primates) or leaves (certain
phyllostomid and pteropodid bats and large primates). As a result, protein
limitation is not a serious physiological problem for most frugivores.

While they are not universal in frugivorous birds, polygynous mating
systems and strong sexual dichromatism have evolved independently in
some lineages. Lek mating systems have evolved in two closely related New
World bird families, manakins and cotingids, as well as in unrelated Old
World birds of paradise and bowerbirds. In addition to characterizing some
(but not all) lek maters, colorful plumages have evolved in many other families of frugivorous birds, partly as a result of ingesting substantial amounts
of carotenoids found in fruit. Lek mating is a rare mating system in bats
(and other mammals) and has only been studied in detail in the frugivorous
African pteropodid, *Hypsignathus monstrosus.*

Finally, as in many nectarivores, many frugivorous birds and bats undergo seasonal movements among habitats and altitudes to track changes
in the distribution of their food resources. Most of these movements occur over relatively short distances but can sometimes involve migrations
of 1,000 km or more in some African pteropodids. Interestingly, frugivorous African birds tend to be far less migratory than their Neotropical or
Asian/Australasian counterparts. Rather than reflecting a more stable fruit
resource base, however, the sedentary lifestyles of African fruit-eating birds
may reflect their greater dietary flexibility. Such flexibility appears to be lacking in migratory African pteropodid bats

The many traits in which convergence has occurred in fruits and in frugivores reflect the complementary nature of this plant-animal interaction.
The adaptive zone represented by frugivory differs in many ways from that
of other feeding zones. The biology of fruits, including their morphology,
biochemistry, and phenological behavior, has selected for particular kinds of
adaptations in their consumers just as selection by frugivores has modified
the characteristics of their fruit. This evolutionary give-and-take between
fruits and frugivores has had a substantial diversifying effect on both plants
and animals (chap. 5).

Conclusions

The fruit-frugivore mutualism is pervasive in tropical and subtropical habitats around the world. This interaction is geologically old, although many modern fruit-frugivore interactions likely date from the late Oligocene and early Miocene. Despite sometimes being viewed as a "sloppy" interaction (e.g., Janzen 1983a, 1983b), there is considerable evidence on both the plant and animal side of this mutualism that it has had a profound impact on the evolution of specific traits found in fruits and frugivores. An excellent example of this is the evolution of trichromatic color vision in primates. In a vertebrate class that is notable for the absence of color vision because of its nocturnal ancestry, primates stand out by having color vision, a sensory adaptation that is common, if not universal, in other vertebrate classes. Color vision is adaptive in diurnal primates that feed on ripe fruit and/or young colorful leaves. Our own ability to see colors and to appreciate the outstanding beauty of flowers and fruits as well as the beauty of their vertebrate mutualists stems directly from the coevolution of fruits and frugivores.

9 Synthesis and Conclusions about the Ecology and Evolution of Vertebrate-Angiosperm Mutualisms

Our main thesis in this book has been that, contrary to the idea that birds, bats, and nonflying mammals are merely ornaments of life, plant-visiting vertebrates have had important ecological and evolutionary effects on their food plants in tropical and subtropical habitats. Conversely, these food plants have played important roles in the ecology and evolution of their vertebrate consumers. Based on our review of the literature and analyses, we feel confident in concluding that this thesis is supported by many lines of evidence. In terms of ecology, we know that morphologically specialized species of nectar-feeding birds and mammals, primarily bats, are important dispersers of pollen of many species of canopy trees, vines, and shrubs and that many plants rely on a wide array of birds and mammals for seed dispersal. Among vertebrates, nectar-feeding bats are especially important long-distance pollen dispersers. Similarly, many frugivorous birds and mammals, especially large species that feed on canopy fruits or large-seeded fruits that fall to the ground, can move seeds substantial distances from parent plants. These frugivores thus help to determine the density and distribution patterns of their food plants. On the plant side, we know that species that rely on vertebrates to disperse their pollen and seeds often suffer from pollen and/or seed dispersal limitation (i.e., only a portion of available ovules get fertilized and seeds do not reach all potential recruitment sites). Therefore, the abundance of vertebrate mutualists directly affects the reproductive success of many of their food plants. Conversely, resources provided by flowers and fruits can limit the population sizes of many of their mutualists. In terms of evolution, we know that plant-visiting vertebrates have had important selective effects on many aspects of flower and fruit biology and that their food plants, in turn, have had numerous effects on the evolution of their morphology, physiology, and behavior. Furthermore, vertebrate pollinators definitely, and seed dispersers likely, have influenced speciation rates in their food plants and have had a significant effect on their taxonomic diversification. Most interestingly, vertebrate pollination and dispersal have evolved across the entire phylogenetic spectrum of flowering plants and are not concentrated in or restricted to just one or a few lineages. Many different groups of plants have independently evolved reproductive interactions with vertebrates. In sum, plant-visiting birds and mammals have played significant roles in the ecological and evolutionary dynamics of tropical and subtropical plant communities for tens of millions of years.

In this chapter we will provide a summary and synthesis of previous chapters with the goal of identifying major patterns in the ecology and evolution of vertebrate pollination and seed-disperser mutualisms. Our guiding questions here are: What is the evolutionary history of the contemporary ecological interactions within communities of tropical vertebrate mutualists and their food plants? How have geological history and biogeography influenced the evolution of these interactions? What are the major evolutionary patterns of ecological traits in these communities?

We begin this synthesis by reviewing the ecological, phylogenetic, and geographic structure of contemporary communities of vertebrate pollinators and frugivores before discussing how these communities and their species and traits might have evolved. Our basic philosophy here is that the structure of contemporary assemblages should provide us with important insights into past evolutionary events, based on the assumption that evolutionary niche conservatism is widespread in lineages of these plants and animals. Considerable support for this assumption exists at both the level of individual species as well as species interactions (e.g., Crisp et al. 2009; Gomez et al. 2010; Ricklefs 2010; Wiens and Donoghue 2004; Wiens et al. 2010).

The Ecological and Phylogenetic Structure of Contemporary Communities

ECOLOGICAL STRUCTURE

As we discuss in chapters 2, 5, and 6 and below, earth history has played an important role in determining the cast of characters that interact in any particular tropical or subtropical habitat. Before examining biogeographic variations on the themes of vertebrate nectarivory and frugivory, however, we will discuss ecological communalities in the structure of contemporary communities by emphasizing their basic building blocks—functional groups of interacting populations that are repeated from one region to another—and how these groups are arranged into networks. By "functional groups," we mean both sets of plants that provide similar kinds of resources to consumers and sets of animals that provide similar kinds of mutualistic services for their food plants. As a result of convergent evolution, functional groups often contain sets of species that are distantly related and hence phylogenetically diverse. In a sense, our functional groups are similar to the guild concept of Root (1967) except that they are restricted taxonomically to include only pollinating or seed-dispersing birds and mammals. In the parlance of contemporary ecology, functional groups represent "modules" or compartments in the food web or network structure of communities. Conventional thinking often assumes that important biotic interactions such as competition or mutualism occur most frequently within these modules or functional groups. However, as we have seen, nature does not necessarily behave in such a restricted fashion. Among species of animals, at least, membership in particular functional groups is not rigid but can be fluid both within and between communities. Thus, some species (e.g., certain phyllostomid and pteropodid bats, sunbirds, and honeyeaters) function both as pollinators and seed dispersers within a habitat, and many species of plant-visiting birds and bats move among habitats on a seasonal basis. This fluidity has important ecological, evolutionary, and conservation implications.

THE STRUCTURE OF FUNCTIONAL GROUPS.—Functional groups in plants are often defined by their growth habits (e.g., trees, shrubs, vines, epiphytes, palms, etc.). This makes ecological sense because different growth forms often differ in their reproductive phenology, kinds of rewards they provide for vertebrates (e.g., flowers and/or fruit), biomass of the reward (i.e., flower or fruit crop size), and nutritional characteristics of the reward, including flower or fruit size, energy density, and other nutritional characteristics (chap. 4). Canopy trees, for example, generally produce large flowers and fruits as well as large crops of flowers and fruits but have relatively short flowering and fruiting seasons in most tropical habitats. Epiphytes and understory shrubs usually have the opposite characteristics—small flowers and fruit, small crops of flowers and fruit, and extended reproductive seasons. All tropical habitats contain most or all functional plant groups, but the relative diversity of these groups and the extent to which vertebrate mutualists interact with them tends to vary across broad geographic scales (chap. 2). The Neotropics probably has the richest array of plant functional groups providing flowers and fruits for birds and mammals, while Australian tropical forests have the least diversity. As a result, the species richness of nectar-feeding and fruit-eating birds, for example, is much higher per unit area in the Neotropics than in the Australian tropics (Kissling et al. 2009, 2011). Similarly, compared with the Neotropics, the understories of African and Asian tropical forests lack extensive floral and fruit resources, and as a result, most vertebrate mutualists are canopy feeders in these regions. For simplicity, we will recognize four major functional groups of plants based on growth habit: trees, palms, shrubs and large herbs, and epiphytes (including parasites such as Loranthaceae) and climbers. Within these groups, plants can be further classified by resource type (e.g., flowers or fruit). Thus there are eight functional groups of plants in our classification.

Functional groups of animals can be classified by at least two characteristics: trophic adaptation (i.e., feeding on nectar or fruit) and body size. Our review of the pollination and frugivory mutualisms in chapters 7 and 8 clearly shows that animals involved in these mutualisms differ strongly in size and trophic morphology, with nectarivores being smaller and having longer bills, jaws, and tongues than frugivores, which often have the opposite characteristics. Although Root's guild concept would include both birds and mammals within the same functional group, this is not ecologically realistic. Despite the existence of some overlap in their diets, nectar- or fruit-eating birds and mammals often feed on different subsets of flowers or fruits and hence merit being placed in different functional groups based on taxonomy, which becomes a third important characteristic. In addition, volant and nonvolant members of these groups often eat very different food species and could be classified in different subgroups. In sum, our animal functional groups are based on four characteristics: taxonomic classification, food resource, body size (with two groups: small [<100 g] and large[≥100 g]), and locomotory method (birds: terrestrial or volant; mammals: terrestrial, arboreal, or volant).

The structure of functional groups of contemporary vertebrate mutualists in space and time in is presented in table 9.1 and figure 9.1, respectively. Ages of the different families come from table 1.1. We realize that these age data are very broad-brush and can be potentially misleading in terms of when particular groups of vertebrates began to interact with functional groups of

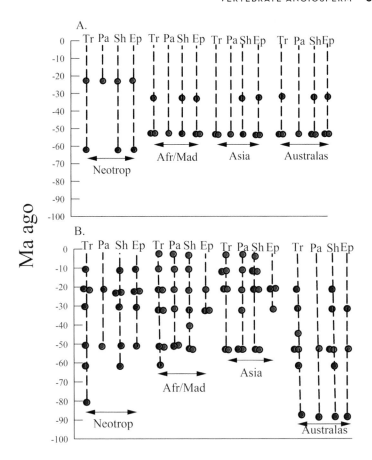

Figure 9.1. Timeline for the filling of four plant functional groups by avian (*blue*) and mammalian (*red*) nectarivores (*A*) and frugivores (*B*). Each dot represents the approximate first appearance of a family based on data in table 1.1. Abbreviations: *Tr* = trees, *Pa* = palms, *Sh* = shrubs and large herbs, and *Ep* = epiphytes.

plants (chap. 6). For example, although hummingbirds likely separated from swifts about 67 Ma ago, they probably have not been pollinating Neotropical plants since then because their earliest fossils date from the Late Oligocene in Europe (Mayr 2007 and included references). The New World radiation of hummingbirds has likely occurred within the last 30 Ma (McGuire et al. 2007). Similarly, pteropodid bats likely first evolved over 50 Ma ago, but they have not occupied their current range (all of the Paleotropics) for this entire period. They have resided in Asia and Australasia longer than they have in Africa (Giannini and Simmons 2005 and included references). As a final

Table 9.1. Functional Groups of Families of Contemporary Vertebrate Mutualists

Resource Type Plant Group, and Body Size	Birds				Mammals			
	Neotropical	Africa/ Madagascar	Asia	Australasia	Neotropical	Africa/ Madagascar	Asia	Australasia
Flowers:								
Trees:								
Small	Trochilidae	Nectariniidae, Zosteropidae	Nectariniidae, Dicaeidae, Zosteropidae	Nectariniidae, Dicaeidae, Zosteropidae	Phyllostomidae	Pteropodidae	Pteropodidae	Pteropodidae
Large	...	...	...	Loriinae	...	...	...	Pteropodidae
Palms:								
Small	...	...	...	...	...	...	...	...
Large	...	...	...	...	...	...	Pteropodidae	Pteropodidae
Shrubs and large herbs:								
Small	Trochilidae	Nectariniidae, Zosteropidae	Nectariniidae, Dicaeidae, Zosteropidae	Nectariniidae, Dicaeidae, Zosteropidae	Phyllostomidae	...	...	Pteropodidae
Large	...	...	...	...	...	...	...	...
Epiphytes and climbers:								
Small	Trochilidae	Nectariniidae, Zosteropidae	Nectariniidae, Dicaeidae, Zosteropidae	Meliphagidae, Nectariniidae, Dicaeidae, Zosteropidae	Phyllostomidae	Pteropodidae	Pteropodidae	Pteropodidae
Large	...	...	...	...	...	...	...	...
Fruits:								
Trees:								
Small	Pipridae, Thraupinae, Capitonidae	Eurylamidae, Coliidae, Dicaeidae, Lybiidae, Pycnonotidae	Megalaimidae, Eurylamidae, Dicaeidae, Pycnonotidae	Meliphagidae, Dicaeidae	Phyllostomidae	Pteropodidae	Pteropodidae	Pteropodidae

Large	Cracidae, Trogonidae, Steatornithidae, Cotingidae, Turdidae, Ramphastidae	Musophagidae, Bucerotidae, Sturnidae, Turdidae	Bucerotidae, Turdidae	Columbidae, Casuariidae, Ptilonorhynchidae, Meliphagidae, Turdidae, Paradisaeidae	Procyonidae, Cebidae*	Pteropodidae, Tragulidae, Viverridae, Lemuridae, Cercopithecidae, Hominidae, Elephantidae	Pteropodidae, Viverridae, Cercopithecidae, Hylobatidae, Elephantidae	Pteropodidae, Phalangeridae
Palms								
Small	...	Eurylamidae	Eurylamidae	...	Phyllostomidae	Pteropodidae	...	...
Large	Steatornithidae	...	...	Columbidae	Cebidae	Viverridae, Cercopithecidae, Hominidae, Elephantidae	Pteropodidae, Viverridae, Cercopithecidae, Elephantidae	Pteropodidae
Shrubs and large herbs								
Small	Pipridae, Thraupinae	Eurylamidae, Coliidae, Dicaeidae, Pycnonotidae	Eurylamidae, Dicaeidae, Pycnonotidae	Zosteropidae	Phyllostomidae	Pteropodidae	Pteropodidae	Pteropodidae
Large	Trogonidae, Cotingidae, Turdidae	Turdidae	Turdidae	Columbidae, Meliphagidae, Ptilonorhynchidae, Paradisaeidae	Cebidae	Tragulidae, Cercopithecidae, Hominidae, Elephantidae	Cercopithecidae, Elephantidae	Pteropodidae
Epiphytes and climbers								
Small	Capitonidae, Thraupinae	Sturnidae, Turdidae	Dicaeidae	Dicaeidae	Phyllostomidae	Pteropodidae	...	Pteropodidae
Large	Trogonidae, Cotingidae, Turdidae, Ramphastidae		...	Columbidae	Cebidae	Viverridae, Cercopithecidae	Pteropodidae	Pteropodidae

Source. Based on data in table 1.1.

Note. Body size classes include small (<100 g) and large (≥100 g). Within cells, families are listed from oldest to youngest. Casuariidae, Tragulidae and Elephantidae are terrestrial.

* Includes Atelidae, Aotidae, Cebidae, and Pitheciidae.

example, pigeons and doves (Columbidae) are an old group, but fruit pigeons are advanced members of this family and appear to have evolved less than 50 Ma ago (Pereira et al. 2007). These kinds of issues suggest that we must interpret the temporal component of distributions in figure 9.1 with caution because of the coarse nature of age estimates based solely on molecular data.

Overall, vertebrate mutualists occupy most, but not all, of the cells in table 9.1. As expected, large species are notably missing among most kinds of pollinators. Lorikeets and a variety of opportunistic flower-visiting pteropodid bats (e.g., species of *Pteropus* in Asia and Australasia and *Epomophorus* in Africa) are Paleotropical examples of relatively large pollinators that have no obvious ecological analogues in the Neotropics. Both large- and small-bodied species are common among vertebrate frugivores, and a substantial range of body sizes can be found within particular families (e.g., in fruit pigeons, toucans, cotingids, meliphagids, and tanagers among birds; in pteropodid and phyllostomid bats and cebid [*sensu lato*] and cercopithecine monkeys among mammals). Large terrestrial frugivores are mostly restricted to the Paleotropics currently, although a variety of large terrestrial mammals that have gone extinct in the relatively recent past in the Neotropics were fruit eaters (Guimaraes et al. 2008).

Among plants, palms have the fewest filled cells in table 9.1. Pteropodid and phyllostomid bats appear to be their major vertebrate pollinators, but they pollinate only a few genera (Fleming et al. 2009). More vertebrates, including several families of birds, bats, and nonvolant mammals, eat palm fruits and disperse their seeds than pollinate their flowers. Many palm seeds are large and/or well-protected by tough husks and hence are most accessible to species with wide gapes (in birds) or strong jaws (in mammals). Flowers of epiphytes are mainly pollinated by small species of birds and mammals, but their fruits are eaten by vertebrates of many sizes.

The temporal pattern of the filling of the functional groups of plants is shown in figure 9.1. Bird pollination likely preceded bat pollination for many plant clades in the New World; nectar-feeding phyllostomid bats are much younger than hummingbirds. This is not the case in parts of the Old World where pteropodid bats are approximately the same age as Australasian lorikeets and meliphagids. Probably not coincidentally, these three groups often feed on the same kinds of flowers (e.g., those in the Myrtaceae and Proteaceae). Avian frugivory also likely preceded mammalian frugivory in the Neotropics and in Australasia. Old avian groups in the New World include cracids, oilbirds, and trogons; old avian groups in Australasia include pigeons, bowerbirds, and cassowaries, all of which appear to be substantially older than families of African and Asian avian frugivores. Fruit-eating phyllostomids are much younger than frugivorous pteropodids with the fig-eating clade of phyllostomids (subfamily Stenodermatinae) dating from only about 10–15 Ma (Baker et al. 2012; Dumont et al. 2011). In contrast, it is likely that basal members of Pteropodidae were fig eaters. Contemporary New and Old World fruit-eating monkeys are also relatively young groups whose radiations began in the Miocene.

In terms of body size, small nectar-feeding birds and mammals likely evolved before large forms in all four biogeographic regions. Among avian frugivores, families of large birds generally appeared earlier than families of small birds in all four regions. Nonvolant frugivorous mammals are generally much larger than their chiropteran counterparts, but most of them are younger than pteropodid bats in the Old World. In the New World, frugivorous phyllostomid bats (small) as well as procyonids and cebids (both large) began radiating at approximately the same time.

Several general patterns emerge from these results: (1) Small size is selectively favored in vertebrate nectar feeders, especially those visiting morphologically specialized flowers (e.g., flowers with zygomorphic or tubular morphology). As exemplified by Australasian lorikeets, meliphagids, and opportunistic pteropodid bats, an exception to this occurs in larger species feeding at "open" flowers (i.e., those with shaving brush morphology), especially in the Paleotropics. Open flowers are accessible to a much wider size range of pollinators, including small insects as well as large vertebrates, than are most tubular flowers. (2) Body size is much less constrained in frugivorous birds and mammals than in nectar feeders. (3) Nectar-feeding or fruit-eating birds generally evolved earlier than their mammalian counterparts. (4) In birds, large species feeding on canopy fruits (or fallen fruit) generally evolved before small species. New World manakins (Pipridae), which are small understory feeders, are an exception to this. The same size trend is likely to be true in mammals except that many large, mostly terrestrial forms are now extinct (e.g., in the New World and on Madagascar). Most extant large frugivorous mammals, including elephants, evolved relatively recently (chap. 6).

THE STRUCTURE OF INTERACTIVE NETWORKS.—Networks are ubiquitous features of life (and human society) and are inevitable whenever groups of organisms (or cells, etc.) interact (Barabási 2009). Communities of pollinators, seed dispersers, and their food plants thus have a network structure (chap. 2). A variety of studies indicate that these mutualistic networks have the following basic features: (1) they are heterogeneous with most species interacting with a few species while a few interact with many species; these species are usually called feeding specialists and generalists, respectively; (2) they are nested, in the sense that specialists interact with a subset of the species that generalists interact with; (3) they often contain modules or functional groups of highly connected species; and (4) they are built on generally weak and asymmetrical links among species (reviewed by Bascompte 2009a, 2009b; Vazquéz et al. 2009). Some of these features are illustrated in figure 2.15 (also see Mello et al. 2011a, 2011b). Nestedness and modularity are generally considered to be the most important features of mutualistic networks because they impart a high degree of stability, that is, resistance to collapse as a result of the extinction of one or more species (e.g., Barabási 2009; Bastolla et al. 2009; O'Gorman and Emmerson 2009; Stouffer and Bascompte 2011; Thebault and Fontaine 2010). A predominance of generally weak interactions is also thought to impart stability to networks, although O'Gorman and Emmerson (2009) have shown experimentally that weak interactors can also play an important functional role in the stability of marine intertidal communities. Their results imply that both strong and weak interactors need to be conserved to maximize community stability.

Modularity is particularly obvious in species-rich tropical networks. Whereas the two networks described in chapter 2 (fig. 2.15) dealt only with frugivorous bats and nectar-feeding birds, in reality these networks represent modules (subsamples) of the actual frugivore and pollinators networks in their communities. At Santa Rosa, Costa Rica, for example, other modules in the frugivore network include birds and other mammals. Based on the results of Donatti et al. (2011), who described much of the network structure of frugivorous vertebrates in three habitats in the Pantanal of southern Brazil, terrestrial tropical frugivore networks are likely to contain at least four modules: two bird modules based on body size (small and large) and two mammal modules also based on body size. To these four we would add a fifth module—frugivorous bats, which were not studied by Donatti et al. (2011). Within each of the four Brazilian modules, fruit and seed size were positively correlated with animal body size, and average fruit and seed sizes differed between modules; similar size correlations are to be expected within frugivorous bat modules (e.g., Fleming 1988). Neotropical pollinator networks are likely to contain at least as many modules as frugivore networks because, in addition to two major groups of avian nectar feeders (humming-birds and passerines) and bats, they also contain one or more modules of insects (e.g., short- and long-tongued bees, butterflies, etc.).

In addition to their species connections, mutualistic networks are characterized by their phenotypic and phylogenetic structures. Phenotypic structure refers to the degree of structural complementarity or morphological matching between interacting species; species with matching phenotypes are more likely to interact than those whose phenotypes do not match. As we discussed in chapters 7 and 8, phenotypic matching tends to be greater in pollination networks than in frugivory networks owing to the more specialized nature of the former interaction. Phenotypic complementarity is usually greater than expected by chance within modules as a result of co-evolution (Guimaraes et al. 2011; Rezende et al. 2007a; Thompson 2009). This complementarity is the basis for pollination and frugivory syndromes.

Phylogenetic structure in the context of networks refers to the degree of relatedness of species within a network. Owing to niche and phenotypic conservatism, related species are more likely to resemble each other morphologically and ecologically than unrelated species, and this resemblance can effect network connections. Related species are more likely to interact with each other as potential competitors and with the same set of food species as mutualists than unrelated species. Based on a simulation model and the analysis of a Spanish fruit-frugivore network, Rezende et al. (2007a) found that phenotypic complementarity contributed more strongly to nestedness than phylogeny. They also found that phylogeny was more important in predicting nestedness in animals than in plants because co-occurring frugivores were more closely related than fruit species. These results are also likely to be true for tropical mutualistic networks, which contain much greater phylogenetic diversity among plants than among animals (e.g., Donatti et al. 2011; chap. 2). We discuss a broader view of the phylogenetic structure of communities below.

Another way to characterize mutualistic networks is to classify their species as hubs, peripherals, or connectors. Hub species are plants that often interact with many species of pollinators or frugivores. Figs are an obvious example of hub species in frugivory networks throughout the tropics, as are columnar cacti and eucalypts in Neotropical arid-zone and seasonal Australian pollination networks, respectively. Peripheral species tend to be narrow specialists (e.g., the outlying species of bats and hummingbirds in fig. 2.15) within modules, and connectors are species that interact with species in more than one module (e.g., phyllostomid bats such as *Carollia perspicillata* and *Artibeus jamaicensis*, which pollinate flowers and disperse seeds in tropical dry forest; Heithaus et al. 1975). Gilbert (1980) called connector species "mobile links" and was perhaps the first person to emphasize their conservation importance. Because of their seasonal movements, many species of tropical vertebrate pollinators and seed dispersers are connector species (chaps. 2, 7, 8). Thus, connector species are extremely important components of tropical mutualistic networks.

In his essay "The coevolving web of life," John Thompson (2009) proposed that mutualistic networks are evolutionary "vortexes" that draw other species into their interactions in a manner analogous to Thomson and Wilson's (2008) concept of "evolutionary attractors." Thompson (2009, 135) concluded that "as networks grow in size, the number of ways in which evolution can act also increases, favoring everything from extreme specialists . . . to generalists . . . that depend on the very existence of the network itself. . . . The result is a coevolutionary vortex that continues to draw more species and lifestyles into the interaction web." In this way, mutualistic networks facilitate the evolution of biodiversity, as we discuss below.

GEOGRAPHIC VARIATION IN THE STRUCTURE OF COMMUNITIES.—As indicated above, there are numerous geographic similarities and differences in the structure of communities of tropical vertebrate mutualists. Similarities at the biogeographic or hemispheric level are the product of evolutionary convergence, that is, the evolution of similar adaptations in distantly related taxa (reviewed by Losos 2011). While some similarities include relatively close morphological matches (e.g., hummingbirds and sunbirds), others involve functional rather than morphological matches (e.g., trogons and fruit pigeons that eat large-seeded fruits of Lauraceae in the Neotropics and Australasia, respectively). In general, functional convergence is more common than morphological convergence in many adaptive radiations (Losos 2011; Losos and Ricklefs 2009). Examples of interhemispheric convergence in table 9.2 and figure 9.2 provide strong support for the hypothesis that similar resources in different tropical regions provide similar selective pressures on their consumers. We discussed this topic in more detail in chapters 7 and 8.

Differences in community structure, sometimes described as "empty ecological niches," emphasize the importance of biogeography and history as major factors influencing the evolution of this structure. Obvious examples

Table 9.2. Examples of Convergence and "Empty Functional Niches" in Tropical
Bird and Mammal Mutualists

Convergence or Empty Niche	Kind of Resource or Foraging Mode	Examples
Convergence	Mistletoe flowers and fruits	Flowerpeckers in Asia and Australasia; euphonia tanagers in Neotropics
	Tubular flowers	Sunbirds in Paleotropics; hummingbirds in Neotropics
		Hummingbirds and glossophagine bats in Neotropics
	Large-seeded fruits of trees or palms	Fruit pigeons in Australasia; trogons in Neotropics
		Hornbills in Africa and Asia; toucans in Neotropics
	Canopy fruits	Palm civets (viverrids) in Africa and Asia; procyonids in Neotropics
	Small canopy fruits	Bulbuls in Africa and Asia; tanagers in Neotropics
	Figs	Old and New World barbets
		Cynopterine bats in Asia; stenodermatine bats in Neotropics
	Fallen fruits	Tragulids and antelopes in Africa; cervids and tragulids in Asia; large caviomorph rodents (agoutis, pacas) and brocket deer in Neotropics
"Empty" functional niches	Canopy tree flowers pollinated by large birds and bats	Neotropics lacks counterparts of large pteropodid bats and lorikeets found in Paleotropics (bats) or Australasia (lorikeets)
	Flowers pollinated by hovering small birds and bats	Hovering birds and bats are absent in the Old World
	Fallen fruits	Neotropics currently lacks large frugivorous terrestrial birds and mammals, including primates, compared with the Paleotropics
	Seed-spitting primates	Neotropics lacks primates with cheek pouches and seed spitting behavior

of "unfilled niches" in the New World include large (i.e., >100 g) species of
nectar-feeding birds and bats as well as large (i.e., >1,000 kg) frugivorous terrestrial birds and mammals (table 9.2). Hovering nectar-feeding birds and
bats are common in Neotropical habitats but are absent in the Paleotropics.
Seed-spitting primates and large arboreal apes are (or were) common in African and Asian forests but are absent from the Neotropics. What accounts
for the existence of these empty niches? Possible explanations include:
(1) chance, (2) the absence of appropriate evolutionary clades, and (3) significantly different selective regimes in different hemispheres. Although chance
has often played an important role in the evolution of life on Earth (e.g.,
asteroid collisions that cause mass extinction events; Schulte et al. 2010), we
suspect that it is not involved in the empty niche phenomenon we discuss

Figure 9.2. (opposite) New World–Old World examples of convergent evolution in nectar- and fruit-eating birds and mammals. *A*, New World or Madagascan nectarivores, including a hummingbird (*Eulampis jugularis* with *Heliconia caribaea*), phyllostomid bat (*Choeronycteris mexicana* with *Chelonanthus alatus*), and a Madagascan lemur (*Varecia variegata* with *Ravenala madagascariensis*). *B*, Old World nectarivores, including a sunbird (*Nectarinia* sp. with *Etlingera elatior*), pteropodid bat (*Eonycteris spelaea* with *Markhamia stipulata*), and another pteropodid bat (*Pteropus conspicillatus* with *Castanospermum australe*). *C*, New World frugivores, including a euphonia (*Euphonia violacea* with *Phoradendron serotinum*), phyllostomid bat (*Artibeus jamaicensis* with *Ficus* sp.), toucan (*Ramphastos sulphuratus* with *Virola megacarpa*), and trogon (*Trogon personatus* with *Ocotea jorge-escobarii*). *D*, Old World frugivores, including a flowerpecker (*Dicaeum cruentatum* with *Amylotheca dictyophleba*), pteropodid bat (*Cynopterus sphinx* with *Ficus* sp.), hornbill (*Bycanistes brevis* with *Myristica fragrans*), and a fruit pigeon (*Ducula badia* with *Litsea garciae*). Illustration by Alice Tangerini.

here. Instead, we favor hypotheses 2 and 3, which are not mutually exclusive, to explain this phenomenon. The absence of hovering nectar-feeding birds in the Paleotropics is particularly interesting because the fossil record indicates that both basal and advanced hummingbirds existed in (tropical) Europe in the late Oligocene (chap. 6). Why did they not increase in distribution and persist until modern times in the Paleotropics? Was their demise caused by deteriorating climatic conditions, by the absence of appropriate floral resources, or by antagonistic biotic interactions such as competition, predation, or parasitism? Without a detailed fossil record of Old World plant and pollinator communities in the late Oligocene and early Miocene, we will probably never know why Old World hummingbirds never colonized these areas or quickly became extinct. Whatever the reason for their disappearance, no other clade of Paleotropical birds has precisely filled the hummingbird morphological niche, although sunbirds and many of their flowers are relatively close functional analogues (Fleming and Muchhala 2008; Stiles 1981). Hovering in both birds and bats requires specialized shoulder joints and/or wings, and these have evolved only twice in the history of these groups (chap. 7). Similarly, as described in chapter 2, most Old World primate communities contain a more diverse array of body sizes and trophic and locomotory adaptations than those in the New World. This difference reflects the greater number of primate clades in the Paleotropics (three major clades—African/Madagascan strepsirrhines and two African/Asian anthropoid clades [cercopithecines and apes]) than in the Neotropics, where only one clade (the cebids *sensu lato*) underwent an adaptive radiation. The radiation of modern primates began earlier and over a much wider geographic area in the Old World than in the New World. Interestingly, in a situation similar to that of European hummingbirds, true primates (euprimates) were present in North American tropical habitats during the Late Eocene climatic optimum but went extinct during the global climatic deterioration that occurred 50–47 Ma ago (Woodburne et al. 2009). Primates did not occur again in the western hemisphere until South America was colonized by platyrrhines from Africa about 25 Ma ago.

Ecological opportunities and constraints have clearly played an important role in the evolution of tropical nectar- and fruit-eating birds and mammals. Similar ecological opportunities—that is, appropriate flower and fruit resources—have resulted in extensive convergence in vertebrate community structure in different geographic regions. Conversely, the absence of resources has created empty functional niches in different regions. For example, columnar cacti, paniculate agaves, and certain kinds of epiphytes (e.g., bromeliads and gesneriads) are either absent (e.g., cacti, bromeliads) or less common (gesneriads) in the Old World, and as a result, their avian and chiropteran pollinators are also missing. Similarly, bat-pollinated mangroves (e.g., *Sonneratia*) are missing from the Neotropics. As reviewed by Ricklefs and Lathum (1993), Paleotropical mangroves are older and are much more diverse phylogenetically and ecologically, including their pollination biology, than Neotropical forms. In their comparison of community structure of arboreal Australian possums and Malagasy lemurs, Smith and Ganzhorn (1996) noted that mammalian frugivores were less diverse in the former than in the latter region. They attributed the absence of arboreal mammalian frugivores in Australia principally to the predominance of bird-dispersed fruit and the seasonally restricted availability of fruit there. These results imply that niche preemption by birds and an inability to move among patches of fruiting plants in a fragmented forest landscape have contributed to Australia's modest fauna of nonvolant mammalian frugivores.

PHYLOGENETIC STRUCTURE

All ecological communities have a phylogenetic structure that reflects the co-occurrence of species from many evolutionary clades. Interest in this structure and the extent to which it reflects abiotic environmental filtering effects (i.e., interactions between the environment and organismal traits) or biotic effects (e.g., competition) has grown rapidly with the availability of detailed phylogenies (e.g., Cavender-Bares et al. 2009; Kelly et al. 2008; Pausas and Verdu 2010; Vamosi et al. 2009; Webb et al. 2002, 2008). Environmental filtering can result in phylogenetic clustering (i.e., the co-occurrence of species that are more closely related than expected by chance) whereas biotic effects can result in phylogenetic overdispersion (i.e., the co-occurrence of species that are less closely related than expected by chance, primarily as a result of interspecific competition). Ecological or evolutionary facilitation, which we discuss in more detail below, can promote either phylogenetic clustering or overdispersion (Cavender-Bares et al. 2009). We will illustrate this approach by briefly examining the phylogenetic structure of contemporary tropical communities at two taxonomic levels—families and species.

Since species-level phylogenies are still uncommon, it is easiest to examine a community's phylogenetic structure at the relatively coarse taxonomic scale of family or genus to address questions such as how many plant clades pollinators or seed dispersers interact with within their communities and how the joint use of plant resources by birds and mammals is distributed among plant clades. We will address these questions for three communities as examples of this approach.

Hummingbird and nectar-bat communities in the Atlantic forest of southeastern Brazil have been studied by Buzato et al. (2000) and Sazima et al. (1999). We took their lists of flower families and genera from one community (the coastal lowland site at Caraguatatuba) and entered it into Phylomatic (Webb and Donoghue 2005) to generate a truncated plant phylogeny based on APG III. We then imported this phylogeny into Mesquite (Maddison and Maddison 2007) and mapped the character "pollinator" with three states (birds only, bats only, or birds and bats) onto the phylogeny using the parsimony reconstruction module (see chap. 6). This community contains 15 plant families and 28 genera of vertebrate-pollinated plants and 11 species and four species of hummingbirds and nectar bats, respectively. As shown in figure 9.3, the plant community is dominated by three of the five major angiosperm lineages: monocots, rosids, and asterids. Bromeliaceae is represented by the greatest number of genera (eight) and asterids the greatest number of families (seven). Hummingbird pollination dominates this phylogeny, with bat pollination occurring mostly in rosid families. Only one bromeliad genus (*Vriesea*) contains species that are either hummingbird or bat pollinated. Overall, hummingbirds and bats are pollinating phylogenetically very different plants in this community.

We used the same approach to examine the phylogenetic structure of two communities of fruits and frugivores, one in central French Guiana studied by Lobova et al. (2009) and one in Kenya studied by Florchinger et al. (2010). Although their study was focused on bat frugivory, Lobova et al. (2009,

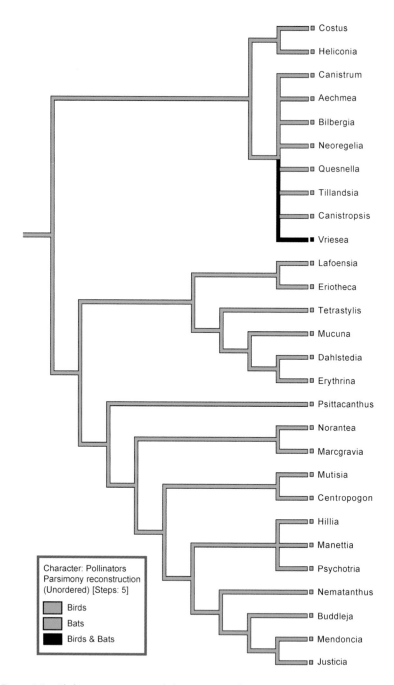

Figure 9.3. Phylogenetic structure of plant genera pollinated by hummingbirds and bats in a Brazilian lowland forest community.

table II) reported which bat fruits were also eaten by birds and primates. This community contains 27 families and 47 genera of plants distributed across all five major angiosperm lineages (fig. 9.4A). With 13 families, rosids is the best-represented linage; with four genera, Cyclanthaceae is the best-represented family. Fruits that bats share with primates are relatively evenly distributed across this phylogeny, whereas fruits that bats share with birds are concentrated in the rosids, specifically in the Melastomataceae and Meliaceae. Fruits shared by all three kinds of vertebrates are also concentrated in the rosids, particularly in the Chrysobalanaceae, Clusiaceae, Humiriaceae, Moraceae, Passifloraceae, and Urticaceae, as well as in the Marcgraviaceae in the asterids (fig. 9.4A). The picture that emerges from this study is that in this community bats share a substantial number of fruits with other kinds of vertebrates across angiosperm phylogeny. As a result, the modular structure of this frugivore network (and elsewhere in the tropics) is likely to be fuzzy rather than discrete.

A somewhat different picture emerges from the Kenya study, which included 81 species of vertebrate frugivores and 30 species of plants in 27 genera and 22 families (Florchinger et al. 2010). Vertebrate-dispersed fruits in this community were about equally concentrated in the asterids (nine families) and rosids (10 families), with Moraceae containing the greatest number of genera (four; fig. 9.4B). Bird fruits dominate this community, particularly in the asterids. Fruits shared by birds and primates are clustered in the rosids, especially in the Euphorbiaceae, Moraceae, Myrtaceae, and Salicaceae (fig. 9.4B). There are apparently no bat-exclusive fruits at this locality, and fruits shared by birds, bats, and primates occur only in the Moraceae. As in the French Guiana study, vertebrates share fruits most frequently in the rosid lineage.

Species-level phylogenetic studies that include rigorous statistical analyses are available for a number of tropical plant and animal communities. So far, plant studies have concentrated on examining the extent to which species are phylogenetically clustered or overdispersed within particular microhabitats (e.g., Kelly et al. 2008; Kress et al. 2009; Webb et al. 2008) or along successional gradients (e.g., Chazdon et al. 2003; Letcher 2010). Both patterns have been reported in a variety of tropical habitats. Sargent and Vamosi (2008) reviewed the association between light environment (i.e., forest canopy vs. forest floor) and pollinator specialization, crudely defined at the level of pollinator orders, within plant communities in a phylogenetic context. Their results revealed that (1) degree of pollinator specialization

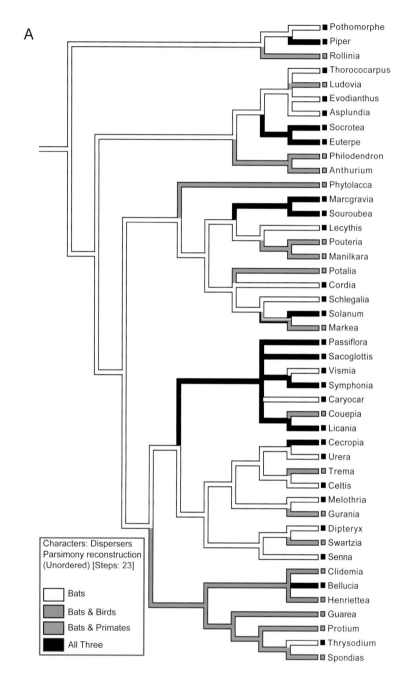

Figure 9.4. Phylogenetic structure of two communities of frugivores. *A*, Fruits eaten by bats, birds, and primates in a forest in central French Guiana. *B*, Fruits eaten by birds and primates in a Kenyan forest.

B

Piper

Dracaena
Culcasia

Cissus
Cyphostemma

Toddalia
Allophylus

Peponium
Prunus
Rhamnus
Antiaris
Trilepsium
Ficus

Harungana
Bridelia
Casaaeria

Manilkara
Maesa

Polyscias
Pittosporum

Solanum
Clerodendrom
Jasminum

Carissa
Coffea
Psychotria

Character: Dispersers
Parsimony reconstruction
(Unordered) [Steps: 9]

Birds
Primates
Birds & Primates
Birds & Bats
All 3

differed among geographic regions, with specialization being lower in India, probably because of a low frequency of bird pollination, than in other Paleotropical locations and the Neotropics; (2) pollinator specialization was greater in the forest understory than in the canopy, reflecting a lower diversity of pollinators in the understory; and (3) plants pollinated by birds tended to be more specialized than those pollinated by insects or mammals. They concluded that degree of pollinator specialization is phylogenetically labile (as we described in chap. 5) and that most plant speciation occurs without a shift in pollinators, except with shifts to bird pollination (also see Armbruster and Muchhala 2009). Such a shift can result in speciation (chap. 5).

Three recent studies examine the community structure of plant-visiting vertebrates from a phylogenetic perspective. Graham et al. (2009) and Parra et al. (2010) analyzed the phylogenetic structure of 189 hummingbird communities in Ecuador, a country notable for its high hummingbird diversity. Graham et al. (2009) reported that warm and wet lowland communities exhibited phylogenetic overdispersion (i.e., co-occurring species tended to be distantly related) whereas seasonally dry or high elevation communities exhibited phylogenetic clustering, perhaps reflecting the importance of environmental filtering in the community assembly process. Parra et al. (2010) analyzed these communities in greater spatial and environmental detail and found multiple patterns of phylogenetic structure within many communities. One hundred-forty communities contained at least one overrepresented clade compared with null expectations, and 69 communities contained at least one underrepresented clade. An analysis of the distributions of hummingbird clades (see McGuire et al. 2007; chap. 6) in multivariate environmental space indicated that different major clades occur in different bioclimatic regions (e.g., hermits occur in wet lowlands; mangoes occur in a narrow elevational band above 1,500 m; and some clades are missing from the high Andes), reflecting patterns of adaptive radiation in this species-rich family. We will return to the broader implications of this study below. Finally, Heard and Cox (2007) examined the degree of phylogenetic skewness or the degree to which related primate clades differ in species richness in South American and African communities at three geographic scales: global phylogeny, continental phylogeny, and local assemblages. They reported that skewness was relatively low for the entire primate phylogeny of 219 species. Within continents, skewness was lower in South America than in Africa, Madagascar, and Asia. Local assemblages did not differ in skewness compared with their continental phylogenies, which suggests that neither abiotic filtering nor competition have skewed their composition during community assembly. The higher skewness in the Paleotropics reflects a greater disparity in species richness in the strepsirrhine, cercopithecine, and other anthropoid clades, particularly in Africa, than in Neotropical cebids.

The study of community structure from a phylogenetic perspective is still in its infancy. Much progress can be anticipated in this area as more species-level phylogenies and new analytical tools become available (Dick and Kress 2009; Kress et al. 2009; Pausas and Verdu 2010). Based on the Parra et al. (2010) study, it is obvious that the results of community- or assemblage-level analyses must be viewed from a broad spatial and temporal perspective. Analyses of phylogenetic beta diversity (chap. 2)—that is, the extent to which species turnover along environmental gradients involves close or distant relatives—will add an important temporal element to the study of contemporary communities and should be encouraged (Graham and Fine 2008). High phylogenetic turnover reflects processes occurring over relatively long periods of time compared with low phylogenetic turnover.

The Evolutionary History of Community Structure

In considering the evolution of community structure, it is important to consider this history in a broad spatiotemporal context because different processes act at differ spatiotemporal scales (Cavender-Bares et al. 2009; Yoder et al. 2010; fig. 9.5). Density-dependent interactions such as competition, for instance, operate at small spatial scales within local communities, whereas environmental filtering and speciation operate on increasingly broader and longer spatiotemporal scales. Because it always involves close contact between interacting mutualists, trait evolution necessarily operates on a small spatial scale but will take place over relatively long periods of time (fig. 9.5). Here we will examine the evolution of community structure, primarily of plant-visiting vertebrates, at different spatiotemporal scales.

THE ASSEMBLY OF CONTEMPORARY COMMUNITIES

Contemporary communities of tropical plant–visiting vertebrates and their food plants are the products of millions of years of evolution. Their structure at any point in time is geologically ephemeral because of the limited lifespans of clades of higher vertebrates (e.g., 10–25 Ma for birds; Ricklefs 2010)

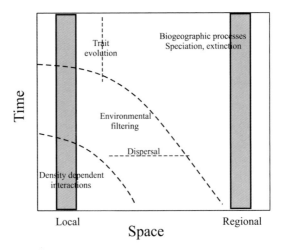

Figure 9.5. Plot of the spatiotemporal scale at which different processes that influence local community structure occur. Based on Parra et al. (2010).

and the effects of tectonic events and climatic variability. Despite species and clade turnover, however, it is likely that communities arise and are maintained with some degree of regularity or order. The existence of nonrandom network structure in extant communities supports this view. What are the rules that govern community assembly and what processes have produced the structure of contemporary communities, especially in plant-vertebrate interactions?

We suggest that one way to begin to discern these rules is to examine the structure of current communities along diversity gradients by asking, What kinds of qualitative as well as quantitative changes occur along them? Early communities of tropical nectar-feeding and fruit-eating vertebrates must have contained low taxonomic and morphological diversity just as occurs in low-diversity communities today. Does the increase in biological complexity that we see along contemporary diversity gradients (chap. 2) mirror faunal and floral buildup in the evolutionary past? We will focus on Neotropical hummingbirds and phyllostomid bats as exemplars of this approach. Morley (2007) provides some justification for this approach by indicating that patterns associated with the assembly of contemporary plant communities are similar to those in operation as far back as the Oligocene.

In both groups, islands and the limited diversity of their faunas and floras are reasonable starting places for this exercise. Among contemporary communities, islands probably best represent the conditions of early low-diversity communities of plant-visiting vertebrates and their food plants. Our account of hummingbird communities is based on Feinsinger and Colwell (1978), who classified these birds into six groups based on their foraging behavior and morphology (table 9.3). "Filchers" and "marauders" are species that either sneak into or brazenly invade territories defended by territorial species, respectively. These birds can be further classified by their seasonal residence time in particular habitats and whether they forage primarily in the understory or canopy. "Principals" are year-round residents and "secondaries" are seasonal migrants, nomads, or principals from other habitats. In general, species richness and its attendant morphological and behavioral diversity in lowland hummingbird communities increase in the following fashion: islands are less than tropical dry forest are less than tropical wet forest (Fleming et al. 2005). Only two to three species of principals live on small islands; these include trapliners and territorialists, sometimes within the same species (Temeles and Kress 2003). In more complex mainland communities, six to 10 species representing several foraging strategies occur in tropical dry forests, and 13–22 species occur in tropical wet forests; principals account for 60%–70% of these species. Mainland species form two subcommunities—understory and canopy foragers. High-reward trapliners (i.e., hermits) and generalists dominate the understory, with territorialists controlling especially rich flower clumps (e.g., clumps of *Heliconia*). Interactions between understory principals and their food plants tend to be more highly specialized (e.g., hermits and *Heliconia*s) than other groups of species. All three major foraging strategies also occur in the canopy, but trapliners are less common there than in the understory; most canopy species are short-billed opportunistic species. Open disturbed habitats are dominated by short-billed secondary species that are not highly coevolved with their

Table 9.3. Community Roles and Morphology of Neotropical Hummingbirds

Role or Foraging Strategy	Relative Size	Required Power	Bill Length	Foot Size
Low-reward trapliner	Small	Low	Short/medium	Moderate
Filcher	Small	High	Short	Large
Generalist	Medium	Moderate	Short/medium	Moderate
Territorialist	Medium	Moderate/high	Short/medium	Moderate
High-reward trapliner	Large	Moderate/high	Long	Small
Marauder	Large	High	Medium	Small

Source. Based on Feinsinger and Colwell (1978).
Note. Required power = relative cost of foraging strategy.

flowers. These same patterns are found in some island communities of hummingbirds and heliconias with restricted pollinator pools (Feinsinger 1976; Temeles et al. 2009).

Although communities of plant-visiting phyllostomid bats are less diverse in morphology and behavior than those of hummingbirds, they display some of the same general ecological trends (Fleming et al. 2005). Communities of island bats, for example, contain far fewer species than mainland communities (chap. 2), and members of the endemic West Indian subfamilies Brachyphyllinae and Phyllonycterinae are ecological generalists. These subfamilies are represented by species whose body sizes and relative jaw lengths are intermediate between those of species in the more derived (and specialized?) mainland subfamilies Glossophaginae (nectar feeders) and Stenodermatinae (fruit eaters; fig. 9.6). On the mainland, low-diversity communities, for example, in tropical dry forest, contain large, broadly distributed, and relatively generalized members of their clades (e.g., *Glossophaga soricina*, *Carollia perspicillata*, and *Artibeus jamaicensis*), whereas high-diversity communities, for example, in tropical wet forest, contain a greater array of smaller, more specialized species as well as the generalists (Fleming 1986a). As in hummingbird communities, mainland communities contain both canopy and understory species (chap. 2).

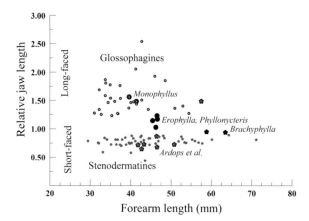

Figure 9.6. Plot of the morphospace occupied by species of West Indian brachyphylline (*black stars*) and phyllonycterine (*black circles*) phyllostomid bats in comparison to species of two mainland subfamilies, Glossophaginae (nectar feeders; *blue circles*) and Stenodermatinae (fruit eaters, *red stars*). West Indian members of these two subfamilies include *Monophyllus* (*large circles*) and the *Ardops* group (*large stars*), respectively. Each point represents a species.

The general picture that emerges from these two brief accounts is that low-diversity communities may contain both generalist and specialist species but that morphological and behavioral diversity clearly increases as the species and phylogenetic richness in communities increases. Extrapolating these trends back in time, it is likely that early hummingbird communities contained low numbers of traplining and generalist species. Extent of pollinator-flower coevolution likely was low in these communities. As the adaptive radiation of this family of birds ensued, species richness and the diversity of foraging strategies increased, in part as a result of competition for floral resources that favored ecological differentiation, including the evolution of canopy species. Eventually, these birds colonized upland habitats and radiated into a variety of montane clades that coevolved with montane flowers in the Asteraceae and Ericaceae, among other families (McGuire et al. 2007; Parra et al. 2010; Stiles 2008). Judging from the geological history of the central and northern Andes where hummingbird diversity is highest, these radiations occurred relatively recently, that is, within the past 10 Ma (Hoorn et al. 2010).

ADAPTIVE RADIATION AND THE EVOLUTION OF REGIONAL BIOTAS

Although many contemporary communities are geologically young, their antecedents often have had a long evolutionary history. For example, the assemblage of plants co-occurring at the northern edge of the Sonoran Desert, which is a northern extension of Central American tropical dry forest, is only a few thousand years old. But many biological elements of this desert date from 5–8 Ma ago, and antecedents of the Central American tropical dry forest date from the late Eocene and Oligocene (Van Devender 2002). Likewise, the high-diversity plant (and animal) communities in the lowlands of northwestern South America are about 5–7 Ma old, whereas many of their plant families have occurred in South America since the Paleocene or early Eocene (Graham 2010; Hoorn et al. 2010; Jaramillo et al. 2010; Wing et al. 2009). As a final example, the dipterocarp forests of India and Southeast Asia date from at least the Early Eocene (50 Ma), but modern forms of this forest date from about 20 Ma (Morley 2000; Rust et al. 2010).

In their review of phylogenetic aspects of community assembly, Emerson and Gillespie (2008) indicate that four major processes are responsible for producing the structure of communities at any place and point in time: immigration/dispersal, environmental filtering, biotic interactions, and speciation. The first three of these processes can occur over ecological time scales, whereas speciation generally occurs on longer (evolutionary) time scales (fig. 9.5). Here we discuss these processes, beginning with the longer evolutionary time scale.

ADAPTIVE RADIATION AND SPECIATION.—Adaptive radiations—the filling of available resource and habitat niche space by an evolutionary clade—abound throughout the history of plant-visiting vertebrate mutualisms. In the Neotropics, conspicuous examples of these radiations include hummingbirds, the manakin/cotingid/tyrannid clade, tanagers, phyllostomid bats, and platyrrhine primates. Paleotropical examples include lorikeets, hornbills, fruit pigeons, honeyeaters, sunbirds, bulbuls, birds of paradise, pteropodid bats, and strepsirrhine and haplorhine primates. Thrushes (Turdidae) are unique in that they represent a cosmopolitan radiation of insectivore/frugivores. Although these radiations differ in their details, they all likely share features that are common to many adaptive radiations (reviewed by Ricklefs 2010; fig. 9.7). These features include, first, an initial rapid filling of niche space followed by a slowing down of diversification as niche space fills up. This slowdown indicates that the rate of filling of niche space is diversity dependent. That is, there is a negative relationship between diversification rate (the difference between rates of speciation and extinction) and species richness (fig. 9.7A). This relationship implies that radiations will cease when available niche space is filled; ecological saturation is the end result of adaptive radiation. It also implies that clade size is likely to be independent of clade age, except in the case of very recent radiations. This relationship has been documented in a variety of different organisms (Rabosky 2009; Ricklefs 2010; Vamosi and Vamosi 2010). While it is logical to expect that the initial stages of adaptive radiations involve high rates of diversification, data from both the fossil record and from molecular phylogenies indicate that different clades begin to diversify at significantly different rates (e.g., the short-fuse and long-fuse models of adaptive radiation discussed in chap. 6). Given questions concerning estimated dates of origin of particular clades based on molecular or fossil data (reviewed in chap. 6), it is not yet clear whether the radiations of animals and plants discussed in this book were predominately short or long fused. We suspect that some of these radiations are likely to be long fused (e.g., in old groups such as hummingbirds, parrots, and pigeons), whereas groups that have radiated relatively recently (e.g., phyllostomid bats, anthropoid primates, tanagers, etc.) clearly are short fused.

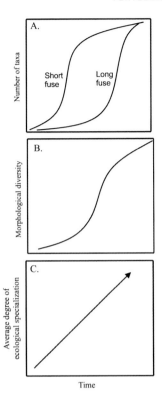

Figure 9.7. General plots of some of the major features of adaptive radiations and the filling of ecological space. *A*, accumulation of species or taxa; *B*, increase in morphological diversity; and *C*, increase in ecological specialization.

Second, as niche space fills, we expect to see an increase in the average degree of ecological specialization and the extent of species packing within clades as a result of resource competition and coevolution with food plants (fig. 9.7*B*). Both of these trends can be seen today in the structure of communities of hummingbird and phyllostomid bats described above.

Third, in addition to an increase in species richness, adaptive radiations generally involve an increase in morphological diversity, again often driven by interspecific competition for and coevolution with resources (fig. 9.7*C*). Thus, whereas generalist species usually initiate radiations, a variety of morphologically specialized species will appear during its evolutionary course. Adaptive radiations therefore involve two fundamental processes: cladogenesis (production of new species) and morphogenesis (production of new morphologies or traits). Both processes probably are diversity dependent, but their rates are likely to differ significantly. Rates of morphogenesis probably both accelerate and decelerate more slowly than rates of cladogenesis (Cavender-Bares et al. 2009; Schemske 2009; Yoder et al. 2010). That is, continued competition for resources and selection for different morphologies don't necessarily stop when speciation stops. Competition in "saturated" faunas will continue to favor morphological differences that serve to reduce intra- and interspecific competition. And plant-animal coevolution will continue to fine-tune the morphologies of closely interacting species.

IMMIGRATION/DISPERSAL.—Communities and regional faunas and floras are open systems in which species are continuously entering and leaving over time. We discussed the mounting evidence for the importance of intercontinental dispersal in the evolution of floras in chapter 6. Such dispersal movements have also been important in the evolution of vertebrate faunas. The classic example of this is the Great American Faunal Interchange, in which many families of North American birds and mammals moved into South America and South American taxa moved into Central America at or near the closure of the Panamanian Portal about 3.8 Ma ago (Bofarull et al. 2008; Dacosta and Klicka 2008; Marshall 1988; Weir et al. 2009). Other less dramatic but nonetheless historically important movements of actual or potential vertebrate mutualists among continents include three groups of birds that originated in Australia (pigeons, parrots, and passerines) as well as Madagascan strepsirrhine and Neotropical platyrrhine primates whose ancestors originated in Africa (chap. 6). The current distribution of parrots, for example, involved long-distance dispersal from their ancestral Australian region to Africa, Indo-Malaya, and South America; dispersal via Antarctica long after the breakup of Gondwana was likely involved in dispersals to Africa and South America (Schweizer et al. 2010; fig. 9.8). Movements within regions have also been important in promoting the radiation of vertebrate mutualists. Examples here include the colonization of mountains in South and Central America, East Africa, the Himalayas, and New Guinea by many clades of birds (chap. 6).

We can use the evolutionary history of hummingbirds and their communities in South America to illustrate how the processes of in situ speciation and habitat colonization have contributed to current distributional patterns and community structure. As indicated above, Parra et al (2010) found evidence for both phylogenetic clustering and overdispersion in Ecuadorian communities of hummingbirds. For example, geographically clustered

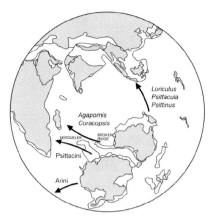

Figure 9.8. Possible dispersal routes of parrots from their ancestral home in Australia to Asia, Africa, and South America. Source: Schweizer et al. (2010) with permission.

clades include hermits, mangoes, bees, emeralds, and coquettes (fig. 9.9). They interpreted this pattern as reflecting in situ speciation in these clades. In contrast, underrepresented clades in the Andes include mountain gems, bees, and emeralds, and they interpreted this pattern as reflecting colonization from lowland habitats. Finally, certain clades are absent from some bioclimatic regions (e.g., hermits do not exist in the high Andes), reflecting a failure to colonize or radiate in these areas. In sum, it is likely that most communities of tropical plants and their mutualists contain mixtures of species that evolved either locally or somewhere else, with "somewhere else" being either from within or distant from a region. As we discussed in chapter 6, long-distance, intercontinental dispersals appear to be more common in plants than in birds and mammals. Relatively short-distance or, more rarely, long-distance dispersal movements are thus important in the community assembly process. Such movements are central to the current view that local assemblages are members of regional metacommunities connected by dispersal (e.g., Hanski 2010; Hubbell 2001; Leibold et al. 2010; McPeek 2008; Pausas and Verdu 2010).

ENVIRONMENTAL FILTERING.—Environmental filtering is also important in the community assembly process. Not all species that can potentially disperse to a habitat actually become established there, either because they lack essential morphological/physiological adaptations or because of the presence of strong competitors and/or predators and pathogens. In the case of

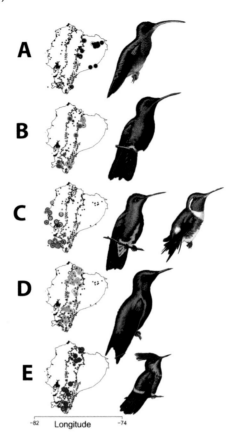

Figure 9.9. Map of Ecuador showing hummingbird communities in which particular clades are overrepresented phylogenetically (*colored circles*). Size of the circle represents increasing depth of the node on the hummingbird phylogeny of McGuire et al. (2007). Clades and a representative Ecuadorian example include: *A*, hermits (*Glaucis hirsutus*); *B*, mangoes (*Anthracothorax nigricollis*); *C*, emeralds (*left, Amazilia fimbriata*) and bees (*right, Chaetocercus mulsant*); *D*, brilliants (*Heliodoxa schreibersii*); and *E*, coquettes (*Lophornis stictolophus*). Hummingbirds are reproduced in approximate relative sizes. Maps come from Parra et al. (2010) with permission. Hummingbirds come from Schulenberg et al. (2007) with permission.

animal mutualists, the absence of specialized plant resources can also prevent species from becoming established in particular habitats (Sargent and Ackerly 2008). To date, the presence of phylogenetically clustered clades in particular habitats or microhabitats has been interpreted as the result of environmental filtering and niche conservatism (e.g., Kress et al. 2009; Webb 2000). More broadly, in situ speciation as well as a lack of ability or opportunity to disperse to habitats and local extinction can also produce phylogenetically clustered patterns, as noted by Parra et al. (2010). For many kinds of animals and plants, high montane habitats are physically rigorous and require particular morphological and physiological adaptations for successful radiation or colonization. The effects of abiotic filtering should be particularly conspicuous in such habitats (e.g., Alexander et al. 2011). In the case of hummingbirds, for example, clades that have successfully radiated at high elevations are characterized by long wings, large wing area, low wing loading, and large feet—traits that are important for foraging on short plants in low-density air (Stiles 2008). Members of other hummingbird clades that have successfully colonized high Andean habitats have converged on many of the morphological features of the endemic clades. Similarly, as noted in chapter 2, the frequency of bird dispersal increases and the frequency of mammal dispersal decreases with elevation in the tropics. As a result, higher-elevation plants in places such as the Brazilian Atlantic forest produce proportionately more fruits that conform to the bird-dispersal syndrome than plants in warmer lowland communities (Almeida-Neto et al. 2008). In this example, abiotic filters associated with elevation affect the composition of both plant and animal communities. Correlated changes in the co-occurrence of plants and their vertebrate mutualists as a result of environmental filtering are likely to be common in the tropics. This topic deserves further study.

BIOTIC INTERACTIONS.—Antagonistic biotic interactions such as competition and predation have long been considered to be important processes in the evolution of biotas and communities, and we have stressed above the importance of competition during adaptive radiation and community assembly. But other biotic interactions, including mutualism and facilitation, are obviously important in the evolutionary processes discussed in this book. From their beginnings, the radiations of nectar-feeding and fruit-eating vertebrates have occurred within communities defined by their flower and fruit resources. And from the start, processes such as coevolution and ecological fitting (i.e., the consumption of food resources by animals with which they have not coevolved) have been important in these radiations (chaps. 5, 7, and 8). The relative importance of these latter two processes during adaptive radiation and community assembly is currently being debated (e.g., Florchinger et al. 2010; Herrera 2002; Lomáscolo et al. 2008, 2010). We suspect that their importance changes systematically during the adaptive radiations of mutualists and that their importance differs significantly in pollination and frugivory mutualisms. If adaptive radiations of vertebrate pollinators and frugivores begin with generalist feeders, then they perforce feed mainly on plant resources with which they have not coevolved. That is, their initial interactions with potential food species must have been based on ecological fitting; they fed at flowers and ate fruit to which they had access and could handle but which had evolved or coevolved with other organisms (e.g., insects in the case of pollinators; the ecological antecedents of modern avian and mammalian frugivores are not as obvious but likely included archaic members of these two classes as well as fruit-eating reptiles). The process of ecological fitting likely dominated the early stages of adaptive radiations. As radiations progressed, however, the ratio of specialist-to-generalist species likely increased as did the relative importance of coevolution compared with ecological fitting. This scenario is especially likely to have occurred in radiations of pollinators and their food plants because of the more specialized nature of this interaction. In a community of specialized flowers, generalist pollinators will be less likely to compete successfully for food than coevolved specialists. Generalist frugivores, in contrast, are more likely to "fit" into communities of fruit that have coevolved to some extent with specialists than are their nectar-feeding counterparts. The greater accessibility of fruits and the multiple ways by which fruit can be eaten (e.g., whole vs. piecemeal) compared with nectar in specialized flowers should permit a greater degree of ecological fitting in communities of frugivores than in communities of pollinators (chaps. 7 and 8). As a contemporary example of ecological fitting, Corlett (2011) describes how tropical Asian frugivores readily eat Neotropical fruits such as *Cecropia pachystachya*, *Muntingia calabura*, and *Piper aduncum* in Singapore and elsewhere in tropical Asia. Similarly, it is common to see hummingbirds visiting many flowers (e.g., aloes) from the Old World.

If this scenario is correct, then we would expect to see a stronger match between the sizes (and feeding morphologies) of vertebrate pollinators and their food resources than between the sizes and morphologies of frugivores and their food resources. While a meta-analysis of available data is desirable, we will test this prediction using the nectar-eating bird and bat community from the lowlands of the Brazilian Atlantic rainforest (Buzato et al. 2000; Sazima et al. 1999) and the community of frugivorous birds and mammals from Kenya (Florchinger et al. 2010) described above. We chose these studies because the sizes of flowers and fruits and their consumers were described in both communities. In Brazil, corolla lengths of flowers pollinated by hummingbirds and bats ranged from 9.3 to 59.9 mm (a 6.4-fold difference) and 11.8 to 43.4 mm (3.7-fold), respectively. Masses of hummingbirds and bats ranged from 2 to 9 g (4.5-fold) and 11 to 24 g (2.2-fold), respectively. In Kenya, the masses of fruits eaten by birds and nonvolant mammals (squirrels and primates) ranged from 4 to 39 g (9.8-fold), whereas the masses of frugivorous birds and mammals ranged from 6.9 to 1,201 g (174-fold) and 325 to 20,000 g (61.5-fold), respectively. Vertebrate nectarivores thus exhibit a much smaller size range and more closely match the sizes of their flowers than do frugivores in these two communities. This is likely to be a general pattern in many tropical communities. Florchinger et al. (2010) concluded from their analyses that ecological fitting best describes how primates use fruit in the Kenya community. These large mammals primarily use a subset of the fruits eaten by birds, and there is no need to invoke primate-fruit coevolution in this community. Similarly, in the Brazilian nectarivore community, the largest bat species (*Platyrrhinus lineatus*) is a fruit-eating stenodermatine, not a glossophagine, and is an opportunistic flower visitor. Its use of flowers as a food source is another example of ecological fitting. While these studies support our hypothesis of greater coevolution (at least as reflected by size matching) in nectarivore communities than in frugivore communities, many more such studies and more rigorous analyses are needed to test it.

Facilitation, which we've defined as the ecological enhancement of a population of one species by the activities of another, has also played an important role in the evolution of faunas, floras, and community structure. It has both ecological and evolutionary aspects. As usually defined (Bronstein 2009), ecological facilitation generally involves close proximity between interacting species (e.g., legume trees that facilitate the establishment of columnar cacti under their canopies; Valiente-Banuet and Verdú 2007), but it doesn't have to involve close spatial proximity nor is it limited to only two interacting species. Ecological facilitation has long been recognized as being an important process in succession (e.g., Alados et al. 2010; Connell and Slayter 1977; Verdú et al. 2009; Walker et al. 2010). During primary succession, pioneer plants change their abiotic environment (e.g., by shading and changing the composition of soil and its nutrients), thereby facilitating successful colonization by other species. Ecological facilitation can also be viewed in a broader context. Thus, flowering or fruiting plants that live in the same habitat can increase (facilitate) each other's reproductive success by sharing pollinators or seed dispersers and by attracting new pollinators or seed dispersers into a habitat (e.g. Ghazoul 2006; Lazaro et al. 2009; Zeipel and Eriksson 2007). This is a more general view of ecological facilitation than one that is focused on one-on-one interactions.

Evolutionary facilitation occurs when the presence of one set of interacting taxa promotes the evolution of a new taxon (or taxa) that takes advantage of a previously established interaction. Many pollination and seed dispersal mutualisms likely evolved via evolutionary facilitation. For example, Rocha et al. (2006) proposed that the evolution of columnar cacti pollinated by birds or bats likely facilitated the evolution of vertebrate-pollinated paniculate agaves in arid regions of tropical and subtropical Mexico. Paniculate agaves likely would not have evolved in the absence of a bat-cactus interaction. Similarly, pallid bats (*Antrozous pallidus*, Vespertilionidae) can be effective pollinators of columnar cacti and paniculate agaves in arid parts of northwestern Mexico and the southwestern United States (Frick et al. 2009). Flower visiting in this insectivorous bat probably would not have occurred in the absence of bat-pollinated cacti and agaves in their habitats. The prior existence of phyllostomid bat-flower interactions has thus facilitated the incipient evolution of flower visiting in another family of bats. Similarly, hummingbird-plant interactions involving flowers of epiphytes and shrubs likely facilitated the evolution of hovering, nectar-feeding phyllostomid bats in the Neotropics (Fleming and Muchhala 2008; Fleming et al. 2009). In the absence of delicate tubular flowers pollinated by hummingbirds, hovering phyllostomids would have been less likely to evolve. More generally, it is likely that mutualistic interactions between vertebrate pollinators and seed dispersers and their food plants have facilitated or have otherwise played a major role in the evolution and ecological maintenance of many plant and animal taxa in tropical and subtropical habitats (chap. 5).

A concept that is often associated with ecological facilitation is that of foundational (plant) species (Bronstein 2009; Kikvidze and Callaway 2009). Foundational species, like keystone species (chap. 3), have a disproportionate effect on the fitness of certain other community members, including species in the same or different trophic level. Because of their ubiquity and year-round fruiting phenology (chap. 3), figs (*Ficus*, Moraceae) likely qualify as a foundational taxon in many tropical habitats. Over ecological timespans, pioneer species of *Ficus* can serve as "recruitment foci" for both pioneer and later successional plant species. During the recolonization of Krakatau, for example, bird- and bat-dispersed figs were early colonists and attracted frugivores that brought in seeds of other plant taxa (Shilton and Whittaker 2009). Thus, figs likely "jump-started" forest regeneration on these islands. Similarly, many other early successional, vertebrate-dispersed plant species (e.g., species of *Cecropia* in the New World and *Macaranga* in the Old

World) play an important role in facilitating the establishment of later successional plants (Davies and Ashton 1999; Franklin 2003; Muscarella and Fleming 2007).

In the case of figs, it is not difficult to envision them as also having a diversifying effect on both plants and their seed dispersers over evolutionary timespans. In a role analogous to "recruitment foci" in ecological succession, early-evolving plant taxa (i.e., families?) providing flowers or fruits for vertebrate mutualists likely facilitated the evolution of other taxa that took advantage of the presence of effective pollinators or seed dispersers for their reproductive success. Other foundational plant taxa for vertebrate frugivores likely include such basal angiosperm families as Annonaceae, Lauraceae, and Myristicaceae. If this scenario is correct, then foundational taxa can be viewed as being "evolutionary catalysts" in promoting the evolution and adaptive radiation of other plant and animal taxa in their habitats. During the evolution of unrelated taxa, some degree of phenotypic convergence in the morphology of flowers and fruits is likely to occur, but we do not necessarily expect them to always converge on key characteristics of the foundational taxa for at least two reasons. First, genetic and developmental differences among clades will likely constrain strong convergence in floral or fruit morphology between unrelated taxa. Superficial resemblance in floral or fruit characteristics is to be expected but not necessarily detailed resemblance. Second, as a result of competition for pollinators and/or seed dispersers, selection will favor divergent characteristics in co-occurring taxa. Thus, the presence of fleshy fruited figs (and other kinds of foundational taxa) will "encourage" other fleshy fruited taxa to evolve, but the new taxa should be expected to differ somewhat in their reproductive and morphological/nutritional characteristics from foundational species and from each other. In the end, low-diversity systems will increase in phylogenetic as well as morphological diversity through processes such as facilitation (which favors trait convergence), competition (which favors trait diversification), coevolution (which favors trait complementarity), and speciation.

THE ROLE OF EARTH HISTORY

As discussed in chapters 5 and 6, earth history, including plate tectonics, orogeny, and climate, has played a major role in the diversification and distribution of plant-visiting vertebrates and their food plants. In fact, earth history has probably played a more important role in this diversification than biotic interactions (Bofarull et al. 2008; Ricklefs 2010). Current and past distributions of landmasses have both promoted and constrained intercontinental movements of plants and animals. South America and Australia are classic examples of long-isolated landmasses on which extensive endemic radiations of plants and animals have occurred. In contrast, Africa and Asia have had an extensive land connection for at least the past 20 Ma and hence share substantial portions of their floras and faunas. Despite its long isolation, Australia is notable for having served as a source of several important radiations of birds, including pigeons, parrots, and passerines, which dispersed to other continents where they underwent further radiations (chap. 6).

Orogeny has had a profound effect on the diversification of tropical floras and faunas, particularly in South America, as we have noted repeatedly (see, especially, chap. 6). The evolutionary effects of orogeny are at least threefold: first, mountains and their rugged topography are powerful agents of geographic isolation and allopatric speciation; second, they produce a diverse array of new habitats for colonization and radiation; and third, they have a strong effect on regional climates that also results in habitat diversification. Mountains worldwide have had an especially strong effect on the diversification of birds. The Andes have generated the highest generic and species richness of birds in the world, whereas highest familial richness in birds occurs in the mountains of East Africa (Thomas et al. 2008).

Finally, climate changes have been extensive throughout the Cenozoic and have led to extensive fluctuations in the extent and distribution of habitats such as tropical forests, grasslands, and deserts. In their analysis of biome conservatism in Southern Hemisphere plants, Crisp et al. (2009) noted that major shifts in biomes (e.g., from dry to arid biomes) are rare in plant clades but that such shifts correlate with major climatic changes. Similarly, Ricklefs (2010) indicated that the clumped ages of first appearance of many clades of plants and animals correspond to significant changes in global climates (e.g., cooling and drying in the late Miocene promoted the expansion of grasslands and xeric habitats). Arakaki et al. (2011) note that the adaptive radiation of many succulent plant clades, including New World cacti and agaves, began during a period of increasing aridity and decreasing CO_2 levels in the late Miocene. Sea-level fluctuations correlated with glacial cycles have either promoted or restricted overland dispersal and degree of isolation of floras and faunas. Colonization of the West Indies from North and Central America by bats, for instance, occurred during periods of low sea levels and increased island areas (Davalos 2009). Major episodes of faunal turnover also correlate with large-scale climatic changes. The land mammal fauna of North America, for example, underwent three periods of climate-related turnover in the early Cenozoic: an increase in first appearances of taxa as a result of immigration from Eurasia during the Late Paleocene-Eocene thermal "optimum"; a further increase in diversity resulting mostly from within-continental speciation during the tropical conditions of the Early Eocene climatic optimum; and a loss of mammal diversity locally and continentally during later Eocene climatic deterioration (Woodburne et al. 2009). Finally, as Webb et al. (2008, 80) state: "The geographic distribution of most taxa will not be in equilibrium with the contemporary abiotic environment, but will represent a dynamic balance of large-scale climatic oscillations and gradients, location of species origin, rate of dispersal, and availability of dispersal routes." All of these factors support the obvious conclusion that earth history and climate change matter in the evolution of floras, faunas, and their regional and community structure. With the current dramatic and substantial changes in climatic patterns on Earth, a climate-related turnover in biotic diversity and distribution will undoubtedly occur.

The Evolution of Trait Structure

In addition to nonrandom phylogenetic structure, communities of vertebrate pollinators and frugivores and their food plants have a nonrandom phenotypic structure based on their floral and fruit traits. This structure has at least two sources: phylogeny (i.e., co-occurring plant families usually differ in at least some of their floral and fruit traits) and phenotypic complementarity and convergence caused by plant-animal coevolution. This nonrandom structure is reflected in the plant and animal traits associated with pollination and seed dispersal syndromes (chaps. 1, 7, and 8). The evolution of this structure is intimately associated with the evolution of phylogenetic diversity. Phenotypic diversity in plants is based in part on their reproductive structures, and phenotypic diversity in animals often reflects their trophic adaptations. In mutualistic systems, plant reproductive traits and animal trophic traits are linked by coevolution. A major question that arises about this process is to what extent morphological traits of plants and animals evolve in correlated fashion. We have already discussed how some traits of fig fruits (e.g., size and color) have evolved in correlated fashion (chap. 8). How general is this pattern in other flower and fruit resources eaten by birds and mammals? And to what extent do the complementary traits in these consumers evolve in correlated fashion? Genomic studies will undoubtedly help to answer these questions (Baker et al. 2012; Siol et al. 2010). For the time being, we must be content to briefly review current evidence on these issues. The plant literature is strongly biased in this regard and contains many more studies dealing with floral traits than with fruit traits. The animal literature is even more limited on this subject. Smith (2010) provides an overview of phylogenetic and statistical methods that can be used to elucidate patterns of correlated evolution in the traits of interacting plants and animals.

THE EVOLUTION OF PLANT REPRODUCTIVE TRAITS

FLOWERS.—As discussed in detail in chapter 7, the floral traits of vertebrate-pollinated plants often differ strikingly from those pollinated by other agents, and there is considerable evidence that these differences result from selection by vertebrate nectarivores. In that chapter (p. 295) we identified a series of trends that are associated with the transformation of insect- to vertebrate-pollinated flowers. Presumably, these trends have been in operation throughout much of the Cenozoic and have operated independently in many plant clades. As a result, there has been much parallelism and convergence (homoplasy) in the floral traits of vertebrate-pollinated plants.

Determining the evolutionary history of any biological trait requires knowledge of its genetic and developmental bases and how these factors respond to phenotypic selection by abiotic and biotic selective agents. Not surprisingly, there is substantial evidence that the floral traits of plants are heritable and can respond to biotic selection pressures (reviewed in Ashman and Majetic 2006). These authors note that there appears to be strong genetic covariation among corolla traits but none between corolla and inflorescence traits or floral and vegetative traits. This suggests that floral, inflorescence, and vegetative traits can evolve independently. Overall, however, our knowledge about these processes is still in its infancy, and much will be learned as we move further into the genomics era (e.g., Cronk and Ojeda 2008; Preston et al. 2011; Wake et al. 2011). Here we will describe how floral traits have evolved in a few well-studied examples by focusing on the question of to what extent floral traits evolve as integrated units as a result of selection by pollinators.

Consider the distribution of pollination systems in the cosmopolitan asterid family Solanaceae, in which pollination is mediated by bees, butterflies, moths, birds, and bats (fig. 9.10). According to Knapp's (2010) analysis, zygomorphic flowers and bee pollination are basal in the family. During this family's radiation, bird pollination (mostly by hummingbirds) has evolved at least 10 times, moth pollination eight times, and bat pollination twice; the entire subfamily Solanoideae, which contains over one-third of all species in the family, is exclusively bee pollinated. Within this family, some clades have radiated extensively in pollinator niche space (e.g., two of the clades near the base of the family), whereas others (e.g., *Solanum*) have speciated extensively without diversifying in pollinator niche space. Furthermore, sympatric species of *Iochroma*, a genus notable for its considerable morphological and color diversity (Smith et al. 2008a), often share pollinators from more than one functional group (*sensu* Fenster et al. 2004; Smith et al. 2008b). Therefore, despite their morphological diversity, flowers of *Iochroma* do not appear to be tightly coevolved with one particular group of pollinators. From this and other evidence, Knapp (2010, 457) concluded: "It is clear from this preliminary and broad-brush look at floral form and pollination in the Solanaceae that simple adaptive evolution between flowers and their pollinators does not adequately explain the great variety in floral form and broad homoplasy in pollinator types in the family." This conclusion likely applies to many plant families with diverse floral morphologies.

Detailed study of the "integrated floral unit" question has occurred in two genera of Solanaceae, *Schizanthus* and *Nicotiana*. Perez et al. (2007) studied floral integration (i.e., degree of phylogenetically independent correlation of floral parts) in eight species of *Schizanthus*, a genus of annual herbs from the Andes of southern South America. Bee pollination is basal in this genus, and other pollinators include hummingbirds and moths. They found that corolla features did evolve in an integrated fashion, with correlation patterns differing among different pollinator groups. For example, in bee-pollinated species, four corolla traits (length of wing, lateral lip, banner, and keel) were integrated, whereas only two pairs of traits (banner and lip, banner and keel) were integrated in a hummingbird-pollinated species. From this they concluded that differences in corolla shape among species are the result of pollinator-mediated selection. Bissell and Diggle (2010) used phenotypic and quantitative genetic analyses to identify morphological modules and inheritance patterns in two closely related species of *Nicotiana*, *N. alata* (hawkmoth pollinated) and *N. forgetiana* (hummingbird pollinated) and in a fourth generation hybrid. Their multivariate analyses revealed at least two morphological modules: a corolla-shaped module associated with attracting pollinators and a floral tube module associated with transferring

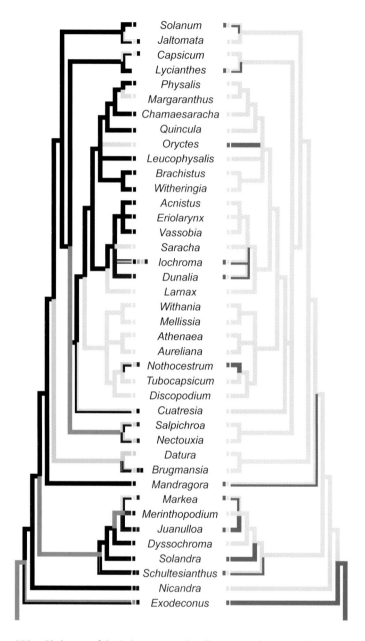

Figure 9.10. Phylogeny of the Solanaceae with pollination modes mapped onto it. The left-hand cladogram indicates pollination mode (*light gray* = general insect; *dark blue* = bee; *turquoise* = butterfly; *yellow* = moth; *red* = bird; *black* = bat; *dark gray* = equivocal). The right-hand cladogram indicates flower form (*light gray* = radial symmetry, *green* = bilateral symmetry; *dark gray* = equivocal). Source: Knapp (2010) with permission.

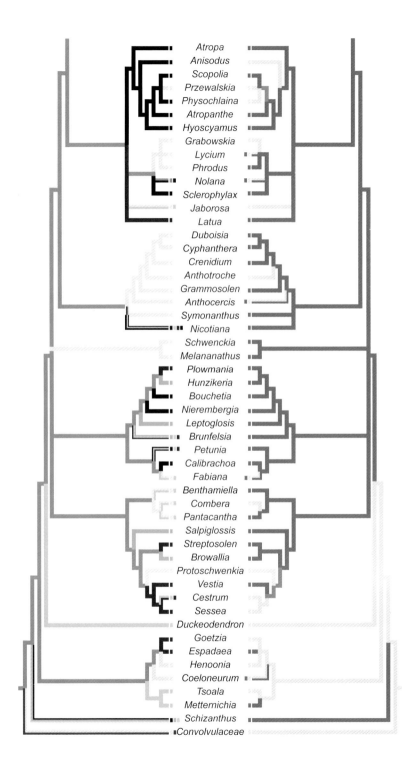

Atropa
Anisodus
Scopolia
Przewalskia
Physochlaina
Atropanthe
Hyoscyamus
Grabowskia
Lycium
Phrodus
Nolana
Sclerophylax
Jaborosa
Latua
Duboisia
Cyphanthera
Crenidium
Anthotroche
Grammosolen
Anthocercis
Symonanthus
Nicotiana
Schwenckia
Melananathus
Plowmania
Hunzikeria
Bouchetia
Nierembergia
Leptoglosis
Brunfelsia
Petunia
Calibrachoa
Fabiana
Benthamiella
Combera
Pantacantha
Salpiglossis
Streptosolen
Browallia
Protoschwenkia
Vestia
Cestrum
Sessea
Duckeodendron
Goetzia
Espadaea
Henoonia
Coeloneurum
Tsoala
Metternichia
Schizanthus
Convolvulaceae

pollen to pollinators. There was strong integration within but not between these modules, suggesting that they can respond rapidly and independently to pollinator selective pressures. Integration within these modules did not break down after four generations of hybridization, further indicating that these sets of floral traits are genetically linked and can evolve as units. Bissell and Diggle (2010) suggested that this kind of floral modularity is likely to be widespread in flowering plants. Other recent studies of floral integration in species of *Iochroma*, *Nicotiana*, *Penstemon*, *Silene*, and Rosaceae support this conclusion (Nattero et al. 2010; Ordano et al. 2008; Parachnowitsch and Kessler 2010; Reynolds et al. 2010; Rosas-Guerrero et al. 2010; Smith et al. 2008a).

Assuming that the evolution of floral modules is widespread in angiosperms, the next step is to map these modules (rather than individual traits) onto dated phylogenies to trace the chronological history of the assembly of these modules within plant clades and communities. We are not aware of any studies that have yet accomplished this task. In the absence of relevant studies, we can only speculate about how the phenotypic structure of communities evolves. For example, given that studies of the response to pollinators in contemporary plants have given us mixed results, Harder and Johnson (2009) and Parachnowitsch and Kessler (2010) have suggested that selection on floral (and fruit?) traits by animal mutualists in contemporary communities is weak and that strong selection usually occurs only during the early stages of flower (and fruit) evolution. If this is true, then the phenotypic structure of floral and fruit resources within many plant communities is likely to have been relatively stable for long periods of time and ultimately reflects phylogenetic diversity. However, colonization of new habitats and regions by plant clades and evolutionary "fine-tuning" between flowers and fruits and their mutualists, such as is common on islands, may continually introduce new plant phenotypes into communities. A beautiful example of this is the adaptive radiation of Bromeliaceae described by Givnish et al. (unpubl. ms.) in which epiphytism (and hummingbird pollination) evolved twice from basal insect-pollinated terrestrial forms; evolution of epiphytism, primarily in the Andes, increased the species richness of this family tenfold and its geographic distribution 30-fold.

FRUITS.—Like flowers, fruits consumed by different kinds of vertebrate frugivores tend to be morphologically distinct, and there has been much convergence in the traits of fruits eaten by different kinds of vertebrates (chap. 8). At the same time, there is also much overlap in the diets of some groups of frugivorous birds and mammals. From this we surmise that both coevolution at the level of guild or functional group and ecological fitting have been important processes during the history of fruit-frugivore interactions. As in flowers, fruits likely evolve in modular fashion, although there is relatively little hard evidence for this yet. As mentioned above and in chapter 8, there is evidence for correlated evolution of fruit size and color in figs eaten by birds and mammals (Lomáscolo et al. 2008, 2010). From their analysis of floral integration in four species of Rosaceae, Ordano et al. (2008) concluded that intrafloral integration could actually represent selection for fruit rather than flower characteristics. Similarly, Miller and Diggle (2007) reported that fruit size is strongly correlated with flower and ovary size and that this suite of traits covaries with degree of andromonoecy (the production of separate male and hermaphrodite flowers within inflorescences) in a group of closely related species of *Solanum.* At a coarser morphological level, Givnish et al. (2005) showed that fleshy fruits and net leaf venation have evolved in concerted (correlated) fashion in monocots. These two traits, which probably do not share a common genetic or developmental basis, are strongly associated with shady forest understories, whereas parallel venation and dry capsular fruits in monocots usually co-occur in open habitats. Both of these sets of morphological traits are subject to evolutionary reversals when plants change habitats.

Valido et al. (2011) analyzed the degree of trait integration in 111 species of Mediterranean fruit that are dispersed by birds, birds and mammals, and mammals using phylogenetically informed methods. They grouped a series of 13 fruit traits into three functional groups (morphology, nutrition, and color) and determined the degree of covariation within and between these groups. Their results indicated that: phylogeny accounted for 50%–60% of variation in fruit traits; traits within functional groups were highly correlated but were less correlated among groups (an exception to this was a positive correlation between fruit color and nutrient content); degree of phenotypic integration was higher in morphological traits than in nutrient and color traits; and degree of overall trait integration was generally low but was higher in bird-dispersed species than in species dispersed by mammals. From this, Valido et al. (2011) concluded that the morphological traits of fruits are more constrained, and hence are more highly integrated, by phylogeny than are other fruit traits; that selection by frugivores is responsible for the association between fruit color and its nutrient contents; and that frugivorous birds apply more consistent selective pressures on their fruits than frugivorous mammals, probably because of their stronger ecomorphological constraints.

In summary, there is substantial support at both the phenotypic and genetic levels for the existence of integrated trait modules in both flowers and fruits and for these modules being able to respond to selection by animal mutualists. The floral and fruit characteristics associated with these modules differ in a number of ways, including type of pollinator or seed disperser, breeding system, flower form (i.e., actinomorphic or zygomorphic), degree of autogamy, and habitat (Almeida-Neto et al. 2008; Ashman and Majetic 2006; Chazdon et al. 2003). Because fruit traits such as size often covary with flower traits, both flowers and fruits can respond to biotic selection as integrated units.

THE EVOLUTION OF ANIMAL TRAITS

In contrast to plants in which experimental studies can be done to dissect the genetic basis of particular floral and fruit traits and the response of these traits to selection, we know much less about the details of trait evolution in plant-visiting vertebrates. We have discussed at length the morphological traits associated with nectarivory and frugivory in birds and mammals in chapters 7 and 8, respectively. Nectarivores and frugivores clearly differ in many morphological, physiological, and behavioral traits from their closest non-plant-visiting relatives. It is likely that integrated groups of these traits (e.g., bill or jaw and tongue length in nectarivores, tooth size and shape in frugivores) exist in these animals, but, to our knowledge, little research has been done on this yet. Obviously much more work needs to be done in this area.

One example of this approach is Dumont et al's. (2011) analysis of rates of diversification and character change in fruit-eating phyllostomid bats, an important group of Neotropical seed dispersers (chaps. 2, 4, and 8). This family includes a basal morphological grade of fruit-eating bats with relatively long muzzles and generalized diets (subfamilies Carolliinae and Rhinophyllinae); diets of carolliinine bats include insects and nectar in addition to fruit. Subfamily Stenodermatinae represents an advanced morphological grade of fruit-eating bats with short muzzles, broad skulls, and low dentary condyles with high coronoid processes. As described in chapter 8, stenodermatines can produce higher bite forces than carolliinines and hence can eat both soft and hard fruits. The evolution of this suite of stenodermatine skull characteristics began about 15 Ma and allowed these bats to diversify quickly into a fully frugivorous adaptive zone. How this suite of skull characteristics arose was not addressed in this study, but it seems likely that it involved integrated trait evolution. Given the timing of this evolution, it seems reasonable to hypothesize that prior to about 15 Ma, Neotropical bat communities contained generalized *Carollia*-like bats that fed primarily on soft understory fruits (Freeman 2000; Muscarella and Fleming 2007). Primarily canopy-feeding species did not appear until the evolution of stenodermatines. Interestingly, basal members of the Old World Pteropodidae, the Nyctimeninae, have strong jaws with high coronoid processes. From the start these bats were apparently able to eat hard canopy fruits. The evolution of Asian understory forms in genera such as *Haplonycteris* and *Balionycteris* occurred later in pteropodid evolution (Gianinni and Simmons 2005).

Conclusions

Contemporary tropical and subtropical communities contain many clades of flowering plants and their avian and mammalian mutualists. The ecological assembly of these communities has occurred over millions of years and has been influenced by a variety of small- and large-scale ecological, geological, and evolutionary processes. Small-scale ecological processes involve such things as daily and seasonal patterns of flower and fruit production by plants and the acquisition of energy and nutrients from these resources by animals. These plant-animal interactions have, in turn, set the stage for mutualistic, facilitative, and competitive interactions within and between groups of plants and animals that have had many important evolutionary and coevolutionary consequences. Geological processes, including plate tectonics, orogeny, and large-scale climate patterns, have had both small- and large-scale effects on the distribution patterns, diversification rates, and evolutionary lifespans of groups of plants and animals. Finally, evolutionary interactions mediated by abiotic and biotic selective factors have occurred both continuously and episodically between these mutualists to produce the forms and functions that we see in tropical floras and faunas today.

As a result of these processes, tropical communities now contain an amazing array of colorful flowers and fruits along with their mobile avian and mammalian mutualists. These species are organized into modular interaction networks characterized by nonrandom phylogenetic and phenotypic structure. Theoretical studies (e.g., Bascompte 2009b; Campbell et al. 2011;

Stouffer and Bascompte 2011; Thebault and Fontaine 2010) suggest that these networks are stable and resistant to occasional extinctions. This prediction would ordinarily be difficult or impossible to test in the real world for logistic and ethical reasons. Unfortunately, owing to the burgeoning human population in the world's tropics, the experiment needed to test this hypothesis is currently underway. How many of these ornaments of life will be able to survive in a world of diminishing wildlands, increasing human pressure, and climate change? Have the grand evolutionary processes that have produced these ornaments run their course in the history of the earth? We discuss issues concerned with the conservation of these plant-animal interactions in chapter 10.

10

The Future of Vertebrate-Angiosperm Mutualisms

As described in the first nine chapters of this book, a great diversity of mammals and birds are dependent on plants for nutrition in the form of nectar and/or fruit. Likewise, many species of plants are dependent on vertebrates as vectors for pollen transfer or dispersal of propagules. In both vertebrates and plants, these mutualisms are spread across a broad phylogenetic spectrum and have evolved repeatedly in different tropical and temperate habitats. Pollination and seed dispersal, often described as "ecosystem services," are critical species interactions for the long-term functioning of ecological networks and communities. Although perhaps not as dramatic as the ecological consequences associated with the loss of apex predators in many marine and terrestrial ecosystems (Estes et al. 2011), the breakdown of these interactions will undoubtedly have a cascading effect that begins with the mutualists themselves and may eventually encompass a large segment of the entire community (e.g., Jordan 2009; Kearns et al. 1998; Kiers et al. 2010; Pauw and Hawkins 2011; Traill et al. 2010).

Many if not all of the habitats where these mutualisms are found are in the midst of profound alterations as a result of human activities. Whether these activities result from the direct degradation of environments locally or at landscape scales or from indirect modifications due to climate change, the future of these mutualisms, which have evolved over millions of years, is uncertain. The threats to plant-vertebrate mutualisms include habitat fragmentation, invasive species, diseases, and bushmeat hunting. Each of these threats varies in time and space depending on geographic location and habitat type, but taken together these human-based ecological pressures have the potential to profoundly affect the mutualistic interdependencies among tropical vertebrates and their botanical partners. As a result of these threats, many plant-visiting birds and mammals are species of considerable conservation concern today.

Our aim in this final chapter is to briefly review the conservation status of plant-visiting vertebrates and the ecosystem services they provide before we describe their major threats and their ecological consequences and provide an overview of how these threats can be mitigated. The literature on the conservation of tropical habitats and their inhabitants is enormous, and we will take a very broad-brush approach to this topic here. Entries into this vast literature can be found in Corlett (2009a), Corlett and Primack (2011), Ghazoul and Shell (2010), Laurance and Peres (2006), and Sodhi et al. (2007), among others. As an introduction to this topic, we will describe three recent studies that illustrate the ecological consequences of disrupted mutualistic interactions between vertebrates and their food plants. These disruptions, of course, also have evolutionary consequences, but ecological consequences generally occur over a much shorter time-span and will be our focus here.

Although it is well-known that populations of various pollinating and seed-dispersing vertebrates have been significantly affected and in some cases have seriously declined as a result of a variety of different threats (see discussion below), documented examples of the direct impact on their dependent plant species are few. It is relatively easy to determine if the absence of one partner in a very specialized mutualism has an immediate effect on the other member with regard to a specific service, for example, pollination or seed dispersal, during a single season. However, documentation of the long-term effects over multiple seasons or multiple generations is a much more difficult and time-intensive task, especially if the mutualisms involve multiple partners in a more generalized interaction system involving long-lived plants, as is the case of many seed dispersal mutualisms. These kinds of studies involving vertebrate mutualists are still uncommon.

One of the best documented cases of the impact of the decline of vertebrate pollinators on their host flowering plant species involves a mutualism that evolved on the North Island of New Zealand (Anderson et al. 2011; also see chap. 4). *Rhabdothamnus solandri* (Gesneriaceae) is an endemic bird-pollinated member of the forest understory. Its tubular, yellow and pink flowers depend on three species of endemic nectar-feeding birds for pollination: the bellbird (*Anthornis melanura*) and the tui (*Prosthemadera novaeseelandiae*—both Meliphagidae—and the stitchbird (*Notiomystis cincta*;

Notiomystidae). Native silvereyes (*Zosterops lateralis*; Zosteropidae) also visit the flowers but usually rob nectar rather than pollinate the flowers. Shortly after the introduction of nonnative mammal predators to the North Island in the late 1870s, two of the three species of pollinators (bellbirds and stitchbirds) were extirpated. Fortunately, all three birds are still present on several small adjacent islands where the plants also occur. By comparing the reproductive success of this shrub on the North Island versus the smaller islands, Anderson et al. were able to demonstrate that fruit set and seed number were reduced by 84% due to the absence of two of the native pollinators. Even more important, despite the persistence of adult shrubs in mainland forests, seedling recruitment was also reduced there by >50%. They concluded that even though one native species of pollinating bird remained, its services were not sufficient to compensate for the extirpation of the other two species. Furthermore, they observed that the tui often preferred to visit nonnative species of plants with more nectar-rich flowers than *Rhabdothamus*, thereby perhaps increasing the negative impact on reproductive success of this shrub.

This study has at least three important messages. First, partners in a pollinator-plant mutualism can be eliminated due to the introduction of a nonnative predator into the environment. Island mutualisms are particularly vulnerable to this threat (Traveset and Richardson 2006). Second, the extirpation of two of the three native pollinators had a dramatic effect on the pollination success of *Rhabdothamus* shrubs during a single season as well as a cascading effect on the abundance of seedlings from past reproductive generations. Finally, it showed how an introduced nonnative plant competitor with nectar-rich flowers may have further reduced the service of the sole remaining native pollinating bird in the habitat. The main conclusion of this study was that the decline and eventual extirpation of native bird pollinators in a local population had cascading effects on the reproductive success of their mutualistic plant species. We suspect that this will be a general result from the disruption of many vertebrate-plant mutualisms.

In a comparable, though more restricted study, Traveset and Riera (2005) demonstrated a similar disruption to the reproductive success of a perennial shrub on the Mediterranean island of Menorca as a result of the decline of its vertebrate seed disperser. On this island the extirpation of the frugivorous lizard *Podarcis lilfordi* (Lacertidae) over 2,000 years ago by introduced carnivorous mammals has had a significant effect on seed dispersal and seedling recruitment of *Daphne rodriguezii* (Thymelaeaceae) in coastal shrublands, where it is now considered at risk of extinction. As in the case of the extirpation of the pollinators of the New Zealand shrub, the researchers were able to show that on isolated islands near Menorca where the native lizards still persist, seedling recruitment of this plant was significantly greater.

Finally, as described in chapter 4, McConkey and Drake (2006) showed that reduced densities of *Pteropus* fruit bats on islands of the Tongan archipelago in the South Pacific have resulted in significantly reduced levels of seed dispersal in several species of large-seeded trees. Unfortunately, unlike the above two studies, the consequences of this for seedling recruitment and future plant population growth have not yet been studied but will likely be significant.

These studies are among the few detailed examples of the decline of a mutualism in a natural habitat as the result of the extirpation (or near-extirpation) of one or more vertebrate partners. However, it is not difficult to foresee how such a process could be extrapolated to similar cascades in many other environments with other species of plants, pollinators, and dispersers (see below). Most ecological models of parasites and mutualists suggest that such coextinctions may be the most common type of biodiversity decline and may result in the loss of thousands of species (Dunn et al. 2009; Koh et al. 2004). These particular cases also suggest that specialization in a mutualism may result in a greater risk of extinction of the dependent species (Sekercioglu 2011). More studies are needed, however, before any concrete generalizations can be made about such coextinctions.

The Conservation Status of Plant-Visiting Birds and Mammals

Before we discuss factors that threaten the integrity of mutualisms between tropical vertebrates and their food plants, we need to review the conservation status of families of vertebrate mutualists and the ecosystem services they provide. In this section we address the question of how endangered these animals are. The International Union for Conservation of Nature monitors the population status of the world's plants and animals, and we have used their 2004 summary (Baillie et al. 2004) to examine the status of the families of nectar-feeding and fruit-eating birds and mammals listed in table 1.1. In that table we indicate the percentage of "threatened" species in most families, as reported in appendixes 3d and 3f of Baillie et al. (2004). We summarize those data in figure 10.1. By "threatened," the International

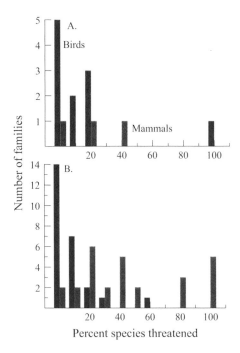

Figure 10.1.   Frequency distribution of families of nectar-feeding (*A*) or fruit-eating (*B*) birds and mammals in terms of number of "threatened" species according to the International Union for Conservation of Nature Red Book (Baillie et al. 2004).

Union for Conservation of Nature means "threatened with extinction." Of course, all species on Earth have finite life expectancies and thus face extinction sooner or later. But as a result of the direct and indirect activities of our species, extinction rates of many taxa are currently orders of magnitude greater than background levels. It is this elevated level of threat that we are concerned with here.

As seen in figure 10.1, levels of threat vary both ecologically and taxonomically. Fewer families of nectarivores contain high percentages (i.e., >50%) of threatened species than frugivores, and mammals contain more highly threatened families than birds. Among nectarivores, only Hawaiian honeycreepers (Fringillidae, Drepanidinae) are highly threatened and have suffered the loss of numerous species in historic times (Ziegler 2002). This family or subfamily is an extreme example of the precarious nature of island life, particularly once islands have been colonized by *Homo sapiens*. One reason why most vertebrate nectarivores do not currently face imminent extinction is that they are small and are not routinely hunted for food by humans. Large generalist pteropodid bats that eat both nectar and fruit are an exception to this. They are avidly hunted and eaten in many parts of Southeast Asia and the South Pacific (Mickleburgh et al. 2009). As a result, about 40% of their species are currently classified as "threatened."

Most mammalian frugivores are larger than their avian counterparts and are being severely hunted as bushmeat throughout the tropics (Fa and Brown 2009). As a result, a substantial number of their families contain many threatened species. These include families of medium-to-large primates weighing >2 kg as well as truly large mammals such as elephants and tapirs. Among frugivorous birds, only cassowaries and guans are currently highly endangered and are also important sources of bushmeat. Most species of small frugivorous birds and mammals are not currently being threatened by bushmeat hunting.

How does the threatened status of families of plant-visiting birds and mammals compare with birds and mammals in general? Is nectarivory or frugivory a riskier feeding adaptation than other feeding modes? While this question deserves a rigorous, phylogenetically controlled analysis (e.g., Davies et al. 2008), we will provide a first-order answer by simply comparing the cumulative threat-status distributions of all families of nonmarine birds and mammals with those of families of nectarivores and frugivores. Results are shown in figure 10.2. Families of avian nectarivores and frugivores appear to be no more threatened than other bird families. The cumulative percentage curves of all three groups are nearly identical (fig. 10.2A). There are too few families of mammalian nectarivores ($n = 3$) for a meaningful comparison, but it is likely that their cumulative curve is similar to the curve for all nonmarine mammals (fig. 10.2B). In contrast, the cumulative curve for frugivores lags behind that for all mammals (i.e., more families of frugivores are threatened than all families of mammals), but the difference is not significant in a Kolmogorov-Smirnov two-sample test (D_{max} [0.215] $< D_{0.05}$ [0.313]). From these results, we cannot conclude that being frugivorous is riskier in mammals than having another diet. Nonetheless, because frugivorous mammals tend to be large, they often represent a large portion of bushmeat in Central Africa and the Amazon (Fa and Brown 2009). Therefore, in an ecological sense, being a large frugivorous bird or mammal is a risky life style.

Our overall conclusion here is that being nectarivorous or frugivorous does not necessarily predispose birds and mammals to higher risks of extinction than seen in other kinds of nonmarine birds and mammals. But this conclusion is no cause for complacency. As described by Hoffmann et al.

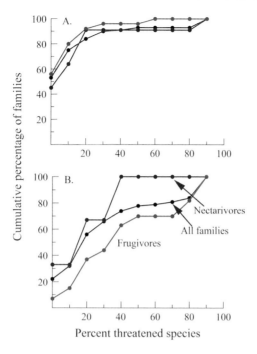

Figure 10.2.   Cumulative frequency distributions of families of birds (A) and mammals (B) in terms of number of threatened species by diet.

(2010), substantial numbers of birds and mammals are currently classified as threatened, especially in the tropics, and these numbers are steadily increasing. Increased conservation efforts are clearly needed to halt the slow but steady march of many species toward extinction. And, as we have emphasized repeatedly in this book, the loss of vertebrate mutualists will have profound ecological and evolutionary consequences for their ecosystems.

The Effects of Threats on the Mutualistic Partnership

One critical question in the conservation of mutualistic relationships is how widespread, taxonomically, the threats are to one of the partners as a result of the extinction of the other partner. We coupled the APG III phylogenetic tree (fig. 6.3) with the International Union for Conservation of Nature Red List of Threatened Animals in an attempt to understand the phylogenetic distribution of plant taxa across the angiosperms that may suffer decline and even possibly extinction as a result of the current threats to their mutualist fruit and seed dispersers. The families of animals that are particularly specialized for fruit dispersal (those in bold in table 1.1) were analyzed with respect to the most important plant orders and families associated with their dispersal services based on table 3.9. Taxonomic families of animal fruit dispersers were sorted into those that have 25%–50% of their species threatened with extinction and those that have >50% (up to 100%) threatened with extinction according to the Red List (table 1.1). Although some of the minor animal disperser families (those not in bold in table 1.1) may also have high extinction risks, their relatively small effect on plant dispersal makes them less important for this analysis. The most important families of animal dispersers, their levels of threat of extinction, and their associated plant orders and families are listed in table 10.1.

We recognize the fact that only some of the species in these orders and families are dispersed by animals threatened with extinction and that the majority is not. Nonetheless, this analysis is intended to emphasize primarily the phylogenetic distribution of plant groups that may be affected as a result of the threat to the seed dispersers of some of their species. The plant orders associated with these major dispersal groups can be categorized according to the level of threat to their dispersers. We therefore assigned each of these plant orders to one of three character states (0 = <25% dispersers threatened, 1 = 25%–50% dispersers threatened, 2 = >50% dispersers threatened) and then mapped these states onto the APG III phylogeny using Mesquite. The results suggest that four of the five major groupings of flowering plants (basal angiosperms, monocots, asterids and rosids) have at least one or more lineages that contain species that may be affected by the extinction of their dispersal mutualists (fig. 10.3). Only the basal eudicots are not affected, while the rosids (especially Anacardiaceae, Burseraceae, Clusiaceae, Combretaceae, Elaeocarpaceae, Fabaceae, Meliaceae, Moraceae, Myrtaceae, Ulmaceae, Salicaceae, and Sapindaceae) and asterids (especially Cactaceae, Ebenaceae, Rubiaceae, Sapotaceae, and Solanaceae) contain the most families with members who have fruits dispersed by animals under a significant threat of extinction. Several families in the basal angiosperms (Annonaceae, Myristicaceae, and Piperaceae) and monocots (Musaceae) also contain species that are dependent on animal dispersers that are increasingly under threat of extinction.

These results emphasize our earlier observation, discussed in chapter 8, that the dispersal of fruits by specific animal mutualists has evolved repeatedly across the flowering plants. Our results also suggest that as one side of

Table 10.1. Proportion of Threatened Species among Major Fruit and Seed Dispersal Agents and Primary Associated Plant Groups

Animal Dispersal Group	Percentage of Species Threatened	Mutualistic Orders (Families) of Plants
25–50% threatened species:		
Old World fruit bats (Pteropodidae)	40.4	Ericales (Ebenaceae), Myrtales (Combretaceae, Myrtaceae), Rosales (Moraceae), Sapindales (Anacardiaceae), Zingiberales (Musaceae)
American leaf-nosed bats (Phyllostomidae)	20.1	Caryophyllales (Cactaceae), Malpighiales (Clusiaceae), Piperales (Piperaceae), Rosales (Moraceae), Solanales (Solanaceae)
Marmosets/capuchins (Cebidae)	33.3	A subset of plants eaten by spider monkeys and howlers
Night monkeys (Aotidae)	28.6	A subset of plants eaten by spider monkeys and howlers
Sakis/titis (Pitheciidae)	25.6	Ericales (Lecythidaceae), Fabales (Fabaceae), Myrtales (Myrtaceae), Sapindales (Sapindaceae)
Spider monkeys/howlers (Atelidae)	40	Ericales (Sapotaceae), Fabales (Fabaceae), Magnoliales (Annonaceae, Myristicaceae), Malpighiales (Salicaceae), Rosales (Moraceae)
Old World monkeys (Cercopithecidae)	45.8	Ericales (Ebenaceae, Sapotaceae), Fabales (Fabaceae), Magnoliales (Annonaceae), Rosales (Moraceae, Ulmaceae), Sapindales (Meliaceae, Sapindaceae)
Raccoons (Procyonidae)	42.1	Rosales (Moraceae)
Palm civets (Viverridae)	26.5	Magnoliales (Annonaceae)
>50% threatened species:		
Cassowaries (Casuariidae)	66.7	Laurales (Lauraceae), Myrtales (Myrtaceae), Oxalidales (Elaeocarpaceae), Rosales (Moraceae)
Large lemurs (Lemuridae)	80	Gentianales (Rubiaceae), Malpighiales (Clusiaceae, Euphorbiaceae), Rosales (Moraceae)
Gibbons (Hylobatidae)	58.3	Ericales (Ebenaceae, Sapotaceae), Magnoliales (Annonaceae), Malpighiales (Euphorbiaceae), Rosales (Moraceae), Sapindales (Meliaceae)
Great apes (Hominidae)	100	Ericales (Ebenaceae, Sapotaceae), Fabales (Fabaceae), Magnoliales (Annonaceae), Malpighiales (Euphorbiaceae), Rosales (Moraceae), Sapindales (Burseraceae, Meliaceae)

Note. The animal dispersal groups are categorized by the percentage of threatened species within each family according to the IUCN Red List.

the mutualistic partnership is threatened with extinction by habitat degradation and other processes discussed in this chapter, then a similar fate may await their host plant partners distributed across the angiosperms as well.

Ecosystem Services Provided by Plant-visiting Vertebrates, Redux

In recent years it has become fashionable to discuss the conservation of nature in terms of the ecosystem services provided to our species. Daily (1997, 3) defines these services as "the conditions and processes through

Figure 10.3. Phylogenetic distribution across the flowering plants of orders containing species that will most likely be affected by the extinction of their mutualist fruit dispersers.

which natural ecosystems, and the species that make them up, sustain and fulfill human life. They maintain biodiversity and the production of *ecosystem goods* [Daily's italics], such as seafood, forage, timber, biomass fuels, natural fiber, and many pharmaceuticals, industrial products, and their precursors. . . . In addition to the production of goods, ecosystem services are the actual life-supporting functions, such as cleansing, recycling, and renewal, and they confer many intangible aesthetic and cultural benefits as well." Often these human benefits are given an economic (dollar) value, presumably to indicate why such services are essential to humans. For example, to highlight the ecological importance of bats to mankind, Boyles et al. (2011) estimated that insectivorous bats in North America annually prevent agricultural losses of more than US$3.7 billion as a result of insect suppression. But whether this knowledge will substantially increase efforts to conserve North American bats remains to be seen.

Although we have not stressed their economic benefits in this book, it is clear that the pollination and seed dispersal services provided by plant-visiting birds and mammals have played a very important role in the ecology and evolution of tropical (and subtropical) habitats. We eschew attempting to place a monetary value on these services, however, simply because in our estimation their value is inestimable. How does one really put a dollar value on biodiversity and the ecological services it provides? What is the economic value of an Amazonian hermit hummingbird and the plants it pollinates or a Sumatran orangutan and the plants it disperses? In our opinion, these animals and their habitats must be conserved regardless of their economic (or medical, etc.) contributions to *Homo sapiens*. For our species to do otherwise would be unconscionable and unethical.

Despite our view of the folly of trying to put a monetary value on wildlife and its ecological services, two recent reviews have attempted to do so for birds and mammals (Kunz et al. 2011; Whelan et al. 2008), and we will summarize some of their results here. No such reviews currently exist for primates (C. Chapman, pers. comm.; but see Astaras et al. [2010] for a review of seed dispersal differences between terrestrial and arboreal African primates and their conservation implications). Whelan et al. (2008) indicated that birds provide four essential services recognized by the United Nations Millennium Ecosystem Assessment: provisioning (i.e., production of fiber, clean water, or food), regulating (i.e., affecting climate, water, and human disease), cultural (e.g., recreation or aesthetics), and supporting (i.e., all other ecosystem services). Pollination and seed dispersal represent supporting services and, as this book emphasizes, birds are important pollinators and seed dispersers of plants worldwide, though Whelan et al. indicate that birds are important pollinators of only a few human food crops (i.e., about 5.4% worldwide based on estimates by Nabhan and Buchmann [1997]). In their table 2 they also indicate that birds are important seed dispersers worldwide but focus only on temperate North American plant genera that are bird dispersed. Economically important plants dispersed by birds in North America are mainly trees (e.g., *Pinus, Juniperus, Magnolia, Fagus, Quercus, Prunus,* etc.). A much greater diversity of plants are dispersed by birds in the tropics (see, e.g., chaps. 4 and 5). These include many trees that are likely to have significant economic value, at both local and international levels (e.g., species of Lauraceae, Meliaceae, Myristicaceae, etc.; Corlett 2009a).

Kunz et al. (2011) review the ecosystem services of bats, focusing on tropical as well as temperate systems. In table 10.2 we indicate the major groups of plants that have economic or especially important ecological value that are either pollinated or dispersed by bats, and this impressive array of plants has considerable significance as sources of food and timber locally or globally. Kunz et al. (2011, table 4) provide details about how these plants are used. However, putting a dollar value on these plants, particularly in terms of the relative contribution of bats compared with other pollinators or dispersers, is extremely difficult, if not impossible, for at least two reasons. First, as discussed in chapter 4, we lack detailed knowledge of the relative contributions of bats to pollination and/or seed dispersal for most of these plants. Second, many of these species are now in cultivation and do not need the services of animal pollinators or seed dispersers for their propagation. Commercially grown species of *Agave* and *Musa* are two good examples. The wild progenitors of commercial crops still need animal services for their reproduction and genetic diversity, but their "domestic" relatives do not.

It is even more difficult to assign a dollar value to the pollination and seed dispersal services of bats and other vertebrates for plants that play especially important roles in the regeneration and maintenance of plant community structure (e.g., table 10.2B). In our view, these services are priceless, and conservation of the species responsible for ensuring the reproduction of ecologically important plants in habitats throughout the world is critical. That being said, the fate of many of these species of birds and mammals hangs in the balance today. We now turn our attention to the threats faced by these animals and what steps need to be taken locally and globally to ensure their survival.

Table 10.2. Examples of Economically and Ecologically Important Tropical and Subtropical Plants Serviced by Bats

Plant Family and Subfamily	Taxon	Service
A. Economically important plants:		
Anacardiaceae	*Anacardium occidentale, Mangifera indica, Spondias* spp.	Dispersed
Annonaceae	*Annona* spp.	Dispersed
Araceae	*Anthurium* and *Philodendron* spp.	Dispersed
Arecaceae	*Acrocomia, Astrocaryum, Bactris, Euterpe, Phoenix,Prestoea, Roystonea, Sabal,* and *Socratea* spp.	Dispersed
Agavaceae	*Agave* spp., subgenus *Agave*	Pollinated
Boraginaceae	*Cordia dodecandra*	Dispersed
Cactaceae	Many genera in subfamily Cactoideae, tribe Pachycereeae	Pollinated, Dispersed
Caricaceae	*Carica papaya*	Dispersed
Caryocaraceae	*Caryocar*	Pollinated, Dispersed
Cecropiaceae	*Cecropia peltata*	Dispersed
Chrysobalanaceae	*Chrysobalanus icaco*	Dispersed
Clusiaceae	*Clusia, Symphonia,* and *Vismia* spp.	Dispersed
Combretaceae	*Terminalia catappa*	Dispersed
Cyclanthaceae	*Carludovica palmata*	Dispersed
Ebenaceae	*Diospyros digyna, D. kaki*	Dispersed
Fabaceae, Faboideae	*Andira inermis, Dipteryx odorata*	Dispersed
Fabaceae, Mimosoideae	*Inga vera, Parkia speciosa*	Pollinated, Dispersed
Lecythidaceae	*Lecythis pisonis*	Dispersed
Malpighiaceae	*Malpighia glabra*	Dispersed
Malvaceae, Bombacoideae	*Ceiba* spp.	Pollinated
Malvaceae, Helicteroideae	*Durio* and *Ochroma* spp.	Pollinated
Malvaceae, Sterculioideae	*Guazuma ulmifolia*	Dispersed
Moraceae	*Brosimum alicastrum; Artocarpus* and *Ficus* spp.	Dispersed
Muntingiaceae	*Muntingia calabura*	Dispersed
Musaceae	*Musa* spp.	Pollinated, Dispersed
Myrtaceae	*Anomomis umbellulifera* and *Psidium guajava; Syzygium* spp.	Dispersed
Passifloraceae	*Passiflora* spp.	Dispersed
Piperaceae	*Piper aduncum*	Dispersed
Polygonaceae	*Coccoloba uvifera*	Dispersed
Rosaceae	*Eriobotrya japonica*	Dispersed
Rubiaceae	*Coffea arabica*	Dispersed
Rutaceae	*Casimiroa edulis*	Dispersed
Salicaceae	*Flacourtia indica*	Dispersed
Sapindaceae	*Meliococcus bijugatus, Sapindus saponaria*	Dispersed
Sapotaceae	*Chrysophyllum cainito, Mimusops elengi; Manilkara* and *Pouteria* spp.	Dispersed
Ulmaceae	*Trema micrantha*	Dispersed
Vitaceae	*Vitus vinifera*	Dispersed
B. Ecologically important plants:		
Agavaceae	*Agave* spp.	Pollinated
Arecaceae	Many New and Old World genera	Dispersed
Cactaceae, Cactoideae	Many columnar cacti in several tribes of this subfamily	Pollinated, Dispersed
Cecropiaceae	*Cecropia* spp.	Dispersed
Clusiaceae	*Vismia* spp.	Dispersed
Malvaceae, Bombacoideae	*Adansonia, Bombax, Ceiba, Pachira, Pseudobombax,* etc. spp.	Pollinated
Moraceae	*Ficus* spp.	Dispersed
Piperaceae	*Piper* spp.	Dispersed
Solanaceae	*Solanum* spp.	Dispersed
Ulmaceae	*Trema micrantha*	Dispersed

Source. Kunz et al. (2011), which should be consulted for further details.

Threats to Vertebrate Pollination and Seed Dispersal Mutualisms

Many different types of changes in a particular environment may have del-
eterious effects on one or the other partner in a plant-vertebrate mutualism. These environmental alterations may be natural, such as severe geologic upheavals, including volcanic eruptions and extraterrestrial impacts,
or localized immediate catastrophes, including weather-related storms
(e.g., Rathcke 2000). However, the most significant immediate threats to the
majority of species are caused by human activities. These threats include
habitat modifications and fragmentation, introduced and invasive species,
pathogens and diseases, bushmeat hunting, and commercial wildlife trade.
Climate change, primarily in response to increased carbon dioxide in the
atmosphere, is a separate category, which many conservationists now believe may be the biggest threat to biodiversity and hence plant-vertebrate
mutualisms (Dawson et al. 2011).

ENVIRONMENTAL DEGRADATION

The primary threat to tropical plant-vertebrate mutualisms is the degradation and conversion of natural habitats to human-dominated landscapes.
Whether this conversion is due to small- or large-scale agroforestry, agriculture, logging, mining, wildfires, or simply human population expansion,
the result is often the same: disruption of the interaction between and/or
population decline and even extinction of one or more mutualistic partners
(Keitt 2009). For example, in a survey of a wet forest site in Thailand, about
one-third of the plant species were assessed to be vulnerable to extinction
because they are dispersed mostly by large frugivores, which were deemed
intolerant of any human impact on the environment (Kitamura et al. 2005).
These authors suggested that when plant extinctions take place, these forests may become dominated by plant species dispersed by abiotic vectors
or species with small-seeded fruits (also see chap. 4). A study in the Neotropics showed that plant species with bird- and monkey-dispersed fruits
were more common in intact forest habitats, whereas species with passive
fur-dispersed seeds were more common in deforested habitats (Mayfield
et al. 2006). In their review of recent studies of avian responses to forest conversion, Tscharntke et al. (2008) found two trends. First, species richness of
large frugivores and insectivores (especially terrestrial and understory species) declines in agroforests. And second, nectarivores, small-medium in-

sectivores (especially migrants and canopy species), omnivores, and sometimes granivores and small frugivores do better or thrive in agroforestry. In general, agroforest bird communities are richer in species of nectar- and fruit-eating birds than intact forests or nontree agricultural habitats, and agricultural habitats have higher proportions of granivores and lower proportions of insectivores, frugivores, and nectarivores than agroforests or intact forests. Similar results have been reported for bird communities in four regions of Costa Rica (Karp et al. 2011). Our overall conclusion here is that different trophic groups of vertebrate mutualists (chap. 9) respond differently to human-modified, formerly forested habitats. Small species of birds and mammals appear to be more ecologically flexible than large species.

The transition from primary to disturbed habitats may not always be abrupt or immediate, and hence the transition in the status of a mutualism may take place in stepwise fashion. With respect to one specific activity that is an increasing threat to tropical wet forests, namely, wildfires caused by human activities, it has been shown that specific families of trees that produce fleshy fruits become less abundant than expected in once- and twice-burned habitats in central Amazonia, suggesting that tree mortality can be nonrandom in terms of fruit dispersal type (Barlow and Peres 2006). Moreover, populations of large frugivores declined significantly in response to single fires, and most primary forest dispersal specialists were extirpated from twice-burned forest habitats. Such studies suggest that habitat effects on the disruption of plant-vertebrate mutualisms will certainly vary with the specific partners involved and the degree of specialization of the interactions.

An as yet unanswered question is how rapidly mutualisms present in tropical forests will recover after these disturbances and degradation. Although sufficient field and modeling data are lacking to make accurate predictions, one study on vertebrate dispersal agents in the Atlantic coastal forests of Brazil estimated that once a habitat is significantly altered, it may require 100–300 years to regain the percentage of animal-dispersed species (80% of the total plant species) found in mature predisturbance forests (Liebsch et al. 2008).

The conversion of primary habitats to agricultural lands is one of the principle sources of environmental degradation in both the temperate zone and the tropics. Most studies on the effects of widespread agriculture on pollinators have been focused on insects, especially bees, and have demonstrated both positive and negative responses (Burkle and Alarcón 2011; Potts et al. 2010). This mixed response is because bees are pollinators of both native plant species as well as many agricultural species. In some cases, pollinator decline is due to the absence of native species, while in others, pollinators increase in abundance because of an increase in total floral resources. Although data are limited on vertebrate pollinators and dispersers in this context, it has been shown that bird diversity and abundance decline in tropical agricultural environments, such as coffee plantations, unless these habitats are supplemented with native fruiting plants as shade trees, living fences, and windbreaks or unless forest remnants are preserved close to the plantations (Luck et al. 2003).

This same pattern has also been shown for the abundance of plant-visiting bats in the agricultural landscape. In a comparison of secondary forests, riparian forests, forest fallows, live fences, pastures with high tree cover, and pastures with low tree cover in Nicaragua, Medina et al. (2007) showed that riparian forests had the highest mean bat species density and abundance whereas the lowest values were found in pastures with low tree cover. Results of this study suggest that agricultural landscapes must retain a heterogeneous assemblage of tree species in order to maintain a diverse bat assemblage. This study contrasts with a similar investigation in lowland Amazonia where frugivores and nectarivores were abundant in areas that had been converted to agriculture (Willig et al. 2007). However, these authors cautioned that their results may have been scale dependent and that if habitat conversion to agricultural use continues to fragment the landscape, then source populations of bats may decline severely.

With respect to the effects of logging and secondary growth on volant vertebrates, several studies in the Amazon region have shown that the abundance of nectarivorous and frugivorous phyllostomid bats (Glossophaginae, Lonchophyllinae, Carolliinae, and Stenodermatinae) increased in logged sites where the canopy was more open and the understory was denser and in less-disturbed areas of *Cecropia*-dominated regrowth (e.g., Peters et al. 2006). In some cases many phyllostomids disappeared from habitats that experienced constant disturbance, whereas frugivorous stenodermatids favored heavily altered areas (Bobrowiec and Gribel 2010.). In tropical forests of southeastern Mexico, Castro-Luna et al. (2007) found that common species of frugivorous bats were very abundant in young successional stages, but rare species inhabited only primary forest and seldom foraged in secondary growth, even when it was close to primary vegetation. These results suggest that a mosaic of habitats modified by human activities, in combination with original vegetation, may be necessary to conserve the full spectrum of frugivorous and nectarivorous phyllostomid bats in an area (Castro-Luna et al. 2007).

Similarly, the abundance and diversity of nectarivorous and frugivorous birds in tree plantations established after major habitat degradation is in part dependent on the floristic composition of the regenerating vegetation. Understory frugivorous birds in Colombia appear to function at a larger spatial scale than the patchiness created by plantations, whereas nectarivorous birds respond to small-scale patchiness within secondary growth (Durán and Kattan 2005). Furthermore, patch size as well as tree species composition influence the number and duration of bird visits in forest restoration plots in southern Costa Rica (Fink et al. 2009). Investigations of the tolerance of frugivorous birds to habitat disturbance in a Costa Rican cloud forest indicate that large-bodied species are moderately tolerant of intermediate habitat disturbance but are intolerant of severe disturbance, whereas tolerance in medium- and small-bodied species is often greater in these same habitats (Gomes et al. 2008). As also seen in bats, the conservation of frugivorous birds will thus require a variety of forest types representing different levels of disturbance.

HABITAT FRAGMENTATION

The fragmentation of undisturbed habitats into a mosaic of pristine and degraded areas is an increasingly common precursor to widespread environmental destruction. Over the last several decades the role of fragmentation in biodiversity loss, including effects on mutualisms, has been widely investigated across the tropics (e.g., Laurance and Bieregaard 1997). With respect to pollination and dispersal mutualisms, it is overwhelmingly clear that plant reproduction is highly susceptible to habitat fragmentation and that fragmentation reduces plant reproductive success (Aguilar et al. 2006). Variation in both the life history traits of plants as well as habitat type has varying impacts on the effects of fragmentation. For example, trees with different-sized seeds usually respond differently; small-seeded plants are more resilient to forest fragmentation than large-seeded species (Cramer et al. 2007). In a study of fragmented habitats in Spain, the number of species with the ability for long-distance dispersal increased in more isolated patches in the highlands whereas the number of species with short-distance dispersal increased in isolated patches in the lowlands (Aparicio et al. 2008). In some cases the effects of fragmentation on pollination and dispersal are plant-species specific and are contingent on the animal biota involved (Guariguata et al. 2002). For example, the fallen seeds of *Dipteryx panamensis* (Fabaceae) were dispersed more quickly by small scatter-hoarding rodents in forest fragments than in intact forest in and around La Selva, Costa Rica; fallen seeds of *Carapa quianensis* (Meliaceae), in contrast, were dispersed more quickly in intact forest than in forest fragments.

Edge effects and the type of vegetation matrix surrounding fragments also play a significant role in determining the types of mutualisms that survive fragmentation. In a study in the Atlantic forest of northeastern Brazil, secondary habitats and edges of fragments contained an assemblage of trees exhibiting reduced diversity of pollination systems, a higher abundance of reproductive traits associated with pollination by generalist diurnal vectors, and an increased abundance of hermaphroditic trees. From this, Lopes et al. (2009) concluded that narrow forest corridors and small fragments will become increasingly dominated by edge-affected habitats that will no longer contain the full complement of tree life history diversity and its attendant mutualists. Finally, climate change itself presents a potentially severe threat to biodiversity and mutualisms in fragmented habitats (see more below on climate change). As rates of environmental change accelerate due to increased CO_2 in the atmosphere, many species will be required to disperse rapidly through fragmented landscapes and large-scale corridors in order to keep pace with the changing climate. In many cases, standard dispersal mechanisms may not be able to respond rapidly enough to these changes (Pearson and Dawson 2005).

Birds may be particularly susceptible to habitat fragmentation because forest is the primary habitat for the majority of the avifauna with restricted geographic ranges—species that are especially prone to extinction (Oostra et al. 2008). Many of these bird species are important pollinators and seed dispersers in tropical habitats. For example, in a fragmented landscape in southern Costa Rica, nonforest hummingbirds occurred less frequently in fragments than in intact habitat, but fragments still supported a mixture of forest-interior and canopy-dwelling hummingbird species along with a diverse group of hummingbird-pollinated plants (Borgella et al. 2001). In contrast, in a species-rich scrub forest in Western Australia, the honeyeater pollinator community showed no significant response to fragment size (Yates et al. 2007a). Outcrossing rates of one bird-pollinated shrub were not significantly correlated with plant population or fragment size. However, mating in small populations occurred between much smaller numbers of plants, which could affect population fitness in subsequent generations. Further observations on these same populations showed a strong positive correlation between the number of seeds produced per fruit and population and fragment size (Yates et al. 2007b), indicating that fragment size does matter in terms of plant reproductive success. Borgella et al. (2001) reached the same conclusion in their study.

With regard to seed dispersal, numbers of species of frugivorous birds (and some primates) in forest fragments in montane Tanzania declined with decreasing fragment size. In addition, recruitment of seedlings and juveniles of 31 animal-dispersed tree species was more than three times greater in continuous forest and large forest fragments than in small fragments (Cordeiro and Howe 2001). Interestingly, recruitment of wind- and gravity-dispersed trees of the forest interior was unaffected by fragmentation in this region. A similar study in Los Tuxtlas, Mexico, demonstrated that habitat disturbance influences avian visitation patterns to fruiting trees; this may in turn affect recruitment patterns in some tree species, although the results were not consistent among the tree species studied (Graham et al. 2002). This result led the authors to suggest that it may be difficult to generalize about the effects of forest fragmentation on assemblages of frugivorous birds. For large fruit-eating birds such as keel-billed toucans, guans, and some hornbills, fragmentation may not have large effects on their seed dispersal abilities because they appear to be able to move through degraded habitats and along corridors of vegetation connecting fragments (Graham 2001; Pizo 2004; Raman and Mudappa 2003). Nonetheless, for many birds, habitat reduction and forest edges have substantial effects on fruit selection, which will ultimately affect plant fitness in forest fragments (e.g., Galetti et al. 2003).

Bat pollinators and dispersers appear to respond quite differently than their avian counterparts to habitat fragmentation. It has been shown in Veracruz, Mexico, that a combination of continuous forest, forest fragments, and agricultural habitats helps to conserve a diverse assemblage of bat species in the local landscape (Estrada and Coates-Estrada 2002). Similarly, in the Atlantic forests of Paraguay, species richness of bats was highest in partly deforested landscapes, whereas evenness was greatest in forested habitat; the highest diversity of bats occurred in landscapes comprising moderately fragmented forest habitat (Gorresen and Willig 2004). Nonetheless, despite the abundance and diversity of bats in disturbed habits, the effects of fragmentation on pollination and seed dispersal services by bats are still unclear. In Australian tropical rain forests, the common blossom bat (*Syconycteris australis*) is a significant pollinator and transports large quantities of pollen among flowers in forest fragments, but the quality (i.e., the effect of geographic genetic distance) of this pollen is not known (Law and Lean 1999). In Mexico, pollination and the reproductive success of the bombacaceous tree *Ceiba grandiflora* in dry forest habitats was negatively affected by habitat disruption. The effective pollinators of this species (*Glossophaga soricina* and *Musonycteris harrisoni*) visited flowers of trees in disturbed habitats significantly less often than trees in undisturbed habitats (Quesada et al. 2003). However, in a broader study of trees in the same plant family in Mexico and Costa Rica, the effects of forest fragmentation on bat pollinators, plant reproductive success, and mating patterns varied depending on the particular plant species, which suggests that the effects of forest fragmentation may differentially affect flowering patterns, bat foraging behavior, and plant self-incompatibility systems in these trees (Quesada et al. 2004).

In studies of bats as fruit and seed dispersers, Klingbeil and Willig (2009) demonstrated a guild-specific and scale-dependent response of bats to fragmented Amazonian rain forests. Abundance and richness of bats, including frugivorous species, were higher in moderately fragmented forest than in continuous forest. In contrast, large frugivores accounted for a higher proportion of total captures in continuous forest than in forest fragments, whereas small frugivores showed the opposite pattern in undisturbed forest and fragmented agricultural land in Peten, Guatemala (Schulze et al. 2000). These authors concluded that the relative abundances of large frugivores, which feed on large fruits of mature forest trees, and small frugivores, which feed on small-fruited plants occurring in early succession, are an indicator of forest disturbance. In Veracruz, Mexico, higher abundance and diversity of frugivorous bats were recorded for riparian sites than for isolated fruiting trees in pastures in a fragmented tropical landscape; the abundance of these bats decreased with distance from the nearest forest fragment (Galindo-Gonzalez and Sosa 2003). Although these few studies on bats are not necessarily consistent in the effects of fragmentation on bat diversity and abundance, they do suggest that bats may have a less severe response to habitat degradation than birds. In a comparison of bat- and bird-generated seed rain at isolated fruiting trees in pastures in fragmented tropical rain forest in Veracruz, Galindo-Gonzalez et al. (2000) found that seed diversity was similar between day and night seed captures and that the contribution of birds and bats to seed rain differed among seasons. Nonetheless, in this region both birds and bats are important seed dispersers of both pioneer and primary species in pastures; their dispersal activities help to connect forest fragments and maintain plant diversity in fragmented tropical forests. This result is likely to have broad generality throughout the tropics.

In contrast to bats, other small mammals appear to be significantly affected by habitat disturbance and fragmentation. In a study in the Atlantic forests of Brazil, total abundance and alpha diversity of small mammal populations were lower in small and medium-sized fragments than in large fragments and continuous forest (Pardini et al. 2005). Moreover, isolated fragments showed lower diversity and abundance compared to connected fragments, which highlights the importance of corridors for buffering the effects of habitat fragmentation on small mammals in tropical landscapes. For mammal-dispersed woody plants in tropical dry forests, the number of species declined with decreasing forest cover in a number of reserves in Central America (Gillespie 1999). Mammal-dispersed plants were rarest in the smallest fragments, perhaps as a result of the loss of dispersal vectors or because of other life history characteristics of the plants. For two Amazonian rodents (*Myoprocta acouchy* and *Dasyprocta leporina*), which are the most important dispersers of several large-seeded tree species, the larger species (*D. leporina*) was initially less affected by forest fragmentation than the smaller one, and continued fragmentation of Amazonian forests will most likely have strong negative consequences for the smaller species (Jorge 2008). In a comparison in a temperate fragmented forest of South America, it was found that the seeds of the mistletoe *Tristerix corymbosus* (Loranthaceae) are dispersed solely by the endemic marsupial *Dromiciops gliroides* (Rodríguez-Cabal et al. 2007). Fragmentation negatively affected marsupial abundance, fruit removal, seed dispersal, and seedling recruitment, and local extirpation of *D. gliroides* resulted in the complete disruption of mistletoe seed dispersal. Thus, the effects of forest fragmentation on this dispersal mutualism have clear demographic consequences for the survival of mistletoe populations.

Forest fragmentation can also have a profound effect on the behavior and ecology of primates, including their seed dispersal services. For example, forest fragmentation in the Orinoco Basin of Colombia has led to the local extinction of certain ateline monkeys (e.g., species of *Lagothrix* and *Ateles*) that originally inhabited the lowlands at the base of the Andes. Their absence has had negative effects on local plant populations because atelines play important roles as seed dispersers in these forests, especially for large-seeded plants (Stevenson and Aldana 2008). Ateline extinctions have apparently resulted in reduced species diversity in local plant communities in this region. While many other studies on the effects of habitat fragmentation on primates exist (e.g., reviewed in Arroyo-Rodríguez and Dias 2010; Harcourt and Doherty 2005; Isabirye-Basuta and Lwanga 2008; Onderdonk and Chapman 2000), none of these have explicitly focused on the effect of fragmentation on seed dispersal by primates. Nonetheless, because we know that forest fragments generally contain a reduced subset of the local or regional primate fauna, smaller group sizes per species, a lower biomass of large tree species favored as food sources by primates, and primates that sometimes move among and out of patches to find food, we can make the following predictions. (1) Large-seeded, primate-dispersed trees that occur in forest patches characterized by the long-term absence of primates will experience severely limited seed dispersal (compared with the same species in intact forests). (2) Both dispersal and recruitment limitation (see chap. 4) will be greater in patches that do contain primates compared with intact forest. (3) Long-distance seed dispersal mediated by primates will still occur for trees located in forest fragments, but it will be much less frequent than in intact forest. (4) Seeds of forest trees deposited in agricultural fields or other nonforest habitats by primates will have lower probabilities of producing new recruits than those dispersed in intact forest. Overall, we expect that seed dispersal mutualisms between primates and their food plants will be disrupted by habitat fragmentation, as is likely to be the case in many other tropical vertebrate-plant mutualisms.

Study of the effect of habitat fragmentation on the structure and stability of mutualistic networks is in its infancy (reviewed by Gonzalez et al. 2011). As discussed above, we expect fragments to contain a subset of animals and plants that co-occur in intact habitats, but the extent to which this threatens the integrity of mutualistic networks is still an open question. Indeed, even delimiting the spatial scope of these networks in a fragmented landscape can be problematic. High mobility on the part of many pollinators and seed dispersers (chaps. 7 and 8) will help to mitigate to some extent the effects of fragmentation on plant reproductive success in habitat patches. Nonetheless, the persistence of many species, especially ecological specialists among both animals and plants, and the stability of mutualistic networks in the long run are likely to decrease with increasing habitat fragmentation. There can be no doubt that fragmentation will accelerate the loss of species and their interactions. The question then becomes, how many species and which interactions?

INTRODUCED AND INVASIVE SPECIES

When they are introduced into new habitats, birds and mammals can act as competitors with or, more frequently, as predators on native bird and mammal nectarivores and frugivores. As described in the example from New Zealand at the beginning of this chapter, an introduced plant species can also disrupt vertebrate-plant mutualisms by competing with native plant species for vertebrate services. How far-ranging these alterations may be and how resilient the interacting species are to such disruptions remains to be determined (Traveset and Richardson 2006). Here we discuss what is known about the effects of introduced species, primarily animals, on vertebrate pollinator and seed dispersal mutualisms. Most of these studies occur on islands where introductions can have dramatic effects. Less is known about the effect of animal introductions in mainland ecosystems.

One of the best documented cases of an alien animal species that has had a wide-ranging affect on the native flora and fauna is the brown tree snake (*Boiga irregularis*) on the Pacific island of Guam (Mortensen et al. 2008). In a study of two bird-pollinated tree species native to Guam—*Bruguiera gymnorrhiza* (a mangrove tree in the Rhizophoraceae) and *Erythrina variegata* (a forest tree in the Fabaceae)—flower visitation and seed set were both significantly higher on Saipan, where the tree snake is absent, than on Guam, where the invader has caused severe losses of native vertebrates. The authors concluded that the bird-pollinated tree species were highly dependent on avian visitors for reproduction and that the decimation of flower-visiting birds by this snake has severely disrupted mutualistic interactions, as we described for two other island systems above. Each of these examples demonstrates the cascading effects of introduced predators on both partners of a vertebrate-plant mutualism.

In a more general and widespread fashion, invasive species that become integrated into networks of interacting mutualists may significantly alter the web structure that has evolved within a community. Aizen et al. (2008) analyzed the extent of mutual dependencies between interacting species (primarily insects and plants) in forests of the southern Andes as well as on several oceanic islands with respect to the pervasiveness of different alien species. They found that weaker, more asymmetrical mutualisms were present in highly invaded webs whereas mutualisms tended to persist better in less-invaded webs. The presence of aliens did not alter overall network connectivity, but connections were transferred from generalist native species to "super-generalist" alien species after invasion. They concluded that the introduction of alien species may leave native species open to new ecological and evolutionary dynamics with respect to their mutualistic partners.

In contrast to the study conducted in the Andes, a survey of plant-pollinator networks on the tropical island of Mauritius showed that newly introduced plant invaders had a relatively low impact on visitation rates to native plant species by both native invertebrate and vertebrate pollinators, suggesting that these invaders offered little direct competition for pollinators with native plant species (Kaiser-Bunbury et al. 2009). Conversely, the introduction of honey bees on the island of New Caledonia, which contains an evolutionarily unique flora, appears to have had a significant effect, including a change in patterns of gene flow, on a number of native plant species that were originally pollinated by native short-tongued bees (Kato and Kawakita 2004). Finally, nonpollinating introduced vertebrate species may also affect local plant-pollinator interactions. In one case, there is strong evidence that the introduction of domestic cattle has significantly modified the structure of local mutualistic networks. A comparison of plant-pollinator interaction networks in native forest sites with and without domestic cattle in montane Argentina demonstrated that the effect of cattle on the network structure was due primarily to the modification of only a few specific but common interactions, which resulted in a significant disruption of critical components of the overall network (Vazquéz and Simberloff 2003). Although these studies were not focused solely on vertebrate pollinators, they nonetheless indicate that introduced exotic species have the potential to significantly disrupt plant-pollinator mutualisms.

Before we discuss the impact of invasive species with respect to specific partners of the mutualistic interactions, it is important to note that vertebrates, especially fruit and seed dispersers, have played a major role in the propagation of invasive plant species in new environments. This issue is becoming an area of particular importance in weed and environmental management (Buckley et al. 2006). Moreover, as fragmentation of landscapes becomes increasing common in both temperate and tropical habitats, many invasive plants and their native dispersers readily use these disturbed environments and fragment edges to increase the dispersal of nonnative species. Where invasive plants are an important part of the diet of native frugivores, a conflict will inevitably arise between the control of the invasive plants and the conservation of populations of these frugivores, especially in cases in which other environmental threats have already reduced populations of native fruit-producing species. For example, in eastern Australia the

Bitou bush, *Chrysanthemoides monilifera* (Asteraceae), is an invasive weed of coastal habitats that produces fleshy bird- and mammal-dispersed fruits. This species, along with other nonnative plant invaders, has substantially altered the temporal pattern of fruit availability in the coastal vegetation of this region. Characteristics of its fruits as well as its seasonal pattern of fruit production are particularly attractive to native frugivores and have contributed to its successful spread as an invasive (Gosper 2004).

Mutualisms between plants and birds, both as flower pollinators and as seed dispersers, have received more attention regarding the effects of invasive species than plants and mammals. In an Australian study of bird pollination that focused on the use of exotic and native nectar resources by native species of birds, it was found that native plant genera such as *Banksia* and *Grevillea* produced significantly higher volumes of nectar than such exotic genera as *Camellia* and *Hibiscus* (French et al. 2005). Banksias also produced a significantly higher sugar reward per floral unit than the other three genera. The three most common local meliphagid nectarivores—the red wattlebird (*Anthochaera carunculata*), the little wattlebird (*Anthochaera chrysoptera*), and the noisy miner (*Manorina melanocephala*)—all preferred flowers of *Banksia* and *Grevillea* and spent significantly more time in flowers of *Banksia* than in those of any other genus. Therefore, contrary to expectations, the native genera of plants were not only a more valuable food source than the exotic genera, but they were also the preferred flowers for these nectarivorous birds (French et al. 2005).

In contrast, invasive alien plants and insects may also impede natural regeneration of native plant species by altering plant-animal interactions such as pollination. In Mauritius, the pollination ecology of a rare endemic cauliflorous tree, *Syzygium mamillatum*, was compared in a restored forest (all alien plant species removed) and an adjacent unrestored area (degraded by alien plants; Kaiser et al. 2008). Flowers of this tree were visited only by generalist bird species. Results indicated that fruit set and the number of seeds per fruit were lower in the restored forest than in the unrestored forest as a result of lower bird visitation rates in the restored area. In addition to differences in bird visitation rates, the difference in reproductive performance of *S. mamillatum* between the two localities was also caused by differences in the attack rates of insect herbivores on flower buds, indicating that multiple interspecific interactions may have compounding effects in habitats altered by invasive species.

Introduced insects also have direct effects on the reproductive ecology of bird-pollinated native plant species. The effects of introduced honey bees on the nectar-feeding activity of two endemic nectarivorous birds—the grey white-eye, *Zosterops borbonicus mauritianus*, and the olive white-eye, *Z. chloronothos*—at two endemic flowering trees, *Sideroxylon cinereum* and *S. puberulum* (Sapotaceae), was studied on Mauritius by Hansen et al. (2002). Results indicated that the introduced bees interfered with the interactions of endemic bird and plant species by reducing nectar levels and forcing birds to forage elsewhere. The authors suggested that native plant-pollinator mutualisms in island ecosystems may be especially vulnerable to disruption by introduced honey bees.

With regard to frugivorous birds and invasive species, most studies have concentrated on the effects of frugivores on the dispersal of introduced plant species. It is commonly assumed that exotic fruiting trees in degraded areas are attractive to frugivorous birds and may become centers of regeneration for invasive species. Supporting this is a study of the frugivore assemblage and seed rain/seedling establishment of exotic guava trees (*Psidium guajava*) in farmland adjacent to native forests in Kenya (Berens et al. 2008). Results indicated that 40 species of frugivorous birds visited guava trees, that 100% and 82% of the seed and seedling species found under guava crowns, respectively, were animal dispersed, and that the majority of these species were late-successional native forest species. Furthermore, the abundance of frugivorous shrubland birds, animal-dispersed seeds, and late-successional seeds at or under guava trees increased, rather than decreased, with increasing distance from primary forest. Therefore, even though guavas are an exotic species, these trees may have a positive effect on forest regeneration and may prove valuable for management plans concerning forest restoration in this area.

In contrast, Cordeiro et al. (2004) demonstrated that native bird dispersers also had a significant effect on the range expansion of an introduced tree species. Their study examined whether several generalist avian frugivores facilitated the invasion of the exotic early successional tree *Maesopsis eminii* (Rhamnaceae) in the East Usambara Mountains, Tanzania. Hornbills were shown to disperse more than 26 times more seeds of this species than monkeys and more than three times as many seeds as turacos per visit and were thus considered the most important disperser of this tree. They concluded that the extensive invasion of *M. eminii* in the East Usambara Mountains was enhanced in both speed and spatial scale by the silvery-cheeked hornbill, an extremely effective dispersal agent of this introduced tree.

The Hawaiian Islands have lost nearly all their native seed dispersers but have gained many frugivorous birds and fleshy-fruited plants through introductions. In this fragile ecosystem, introduced birds may not only aid invasions of exotic plants but may also be the sole dispersers of native plants. In a study including both native- and exotic-dominated forests, Foster and Robinson (2007) showed that introduced species of birds were the primary dispersers of native seeds into exotic-dominated forests, which may have enabled six native understory plant species to become reestablished. Introduced birds also dispersed seeds of two exotic plants into native forest habitats, but dispersal was localized and establishment of these exotics was minimal. Without suitable native dispersers, most common understory plants in Hawaiian rainforests now depend on introduced birds for dispersal, and these introduced species may actually facilitate perpetuation, and perhaps in some cases restoration, of native forests. However, the authors emphasized that restoration of native forests in Hawaii through seed dispersal by introduced birds depends on the existence of native forests to provide a source of local plant seeds. In contrast, on Mauritius the introduced red-whiskered bulbul (*Pycnonotus jocose*, Pycnonotidae), which is an effective disperser of many fleshy-fruited species, frequently moves from invaded and degraded habitats into less-disturbed native forests, thus potentially acting as a mediator of continued plant invasion into these areas. This exotic dispersal agent has been a major factor in the continued reinvasion of exotic plants such as *Clidemia hirta* (Melastomataceae) and *Ligustrum robustum* (Oleaceae) into secondary and restored conservation management areas. The initial invasion of native forests by exotic plants on Mauritius may not have happened as rapidly without efficient avian seed dispersers such as the red-whiskered bulbul (Linnebjerg et al. 2009).

Invasive insects may also have an impact on vertebrate-plant dispersal mutualisms. On Christmas Island in the Indian Ocean, Davis et al. (2010) have suggested that the invasion and formation of high-density supercolonies by the yellow crazy ant, *Anoplolepis gracilipes*, may severely disrupt frugivory and seed dispersal by endemic birds. This invasive ant, whose high densities are sustained through a mutualism with introduced scale insects, rapidly reduces fruit handling times by endemic island birds and may therefore reduce seed dispersal by these frugivores. Although additional studies need to be done on the effects of this introduced insect, this study complements the other investigations discussed above by showing that any disruption of mutualistic partners in a community network may have cascading effects on other members of the network.

Although investigations are limited on the effects of invasive species on mammal-plant interactions, it is expected that introduced plants as well as introduced vertebrates will significantly alter these mutualisms. At least one study has shown that the introduced brushtail possum (*Trichosurus vulpecula*; Phalangeridae) into New Zealand has contributed to the dispersal of seeds of 17% of the total species in a successional forest landscape (Dungan et al. 2002). As in the case of many introduced frugivorous birds, these possums have not only increased the spread of invasive weeds, but their seed dispersal behavior has also provided conservation benefits by accelerating forest regeneration in native vegetation. Because of the decrease in numbers of large-gaped native birds as fruit disperser over the last century, exotic possums may now be the only dispersal agent for large-seeded native species in many New Zealand habitats.

PATHOGENS AND DISEASES

The introduction of new parasites and pathogens to populations of vertebrates often accompanies human activities such as habitat degradation and the release of nonnative species. Primates living in forest fragments, for example, sometimes incur higher parasite loads because of contact with humans (or their feces) and their domestic pets (Arroyo-Rodríguez and Dias 2010; Gillespie and Chapman 2008; Mbora and McPeek 2009). Conversely, humans can also be infected with new pathogens as a result of increased contact with wildlife (e.g., Wolfe et al. 2005). But, except in the case of Hawaiian honeycreepers (Fringillidae, Drepanidinae), the impact that pathogens actually have on vertebrate-plant mutualisms is poorly known.

Hawaiian honeycreepers are a classic example of the adaptive radiation of a clade of birds on an isolated oceanic archipelago. In a period of about 4 Ma, these birds evolved into a substantial array of feeding niches, including nectarivory, occupied by about 20 genera and 50 species. Seventeen species and subspecies are now extinct, another 14 species are endangered, and only three species still have robust populations (Atkinson and LaPointe 2009; Spiegel et al. 2006). The nectar feeders of tribe Drepanidini included six genera and eight species (of which about five are now extinct) that evolved close ecological relationships with a number of native flowers, especially those in the Campanulaceae (Lobelioideae). Although causes of extinction or population loss include habitat degradation and predation by introduced mammals such as rats, dogs, and mongooses, avian malaria and birdpox, which were introduced into the archipelago in the late nineteenth or early twentieth centuries with the release of exotic birds, are thought to have played a major role in these extinctions. Populations of birds living below an elevation of about 1,700 m have been especially hard-hit by malaria. As a result of these avian extinctions plus habitat degradation and herbivory by feral goats, about 30 species of lobeliads have gone extinct in the past 100 years (Cox and Elmqvist 2000; Smith et al. 1995). In some places, remaining plant population densities are too low to support viable populations of their avian pollinators—an example of an "extinction vortex." Smith et al. (1995) reported that as a result of low lobelioid flower densities as well as the extinction of the 'o'o (*Moho nobilis*, Meliphagidae), which was behaviorally dominant to the i'iwi at ohia flowers, the nectar-feeding i'iwi (*Vestiaria coccinea*) has switched to feeding on flowers of ohia trees (*Metrosideros polymorpha*; Myrtaceae) with significant evolutionary results. Whereas its bill was originally strongly curved for feeding in the curved, tubular flowers of lobelioids (Speith 1966), current populations of *V. coccinea* have slightly shorter, straighter bills, presumably as a result of selection for feeding on the corolla-less flowers of *M. polymorpha*.

This example has several general messages. First, it illustrates the concept of cascading ecological consequences (i.e., an extinction vortex) associated with the extinction of mutualist partners. The extinction of nectar-feeding drepanidids presumably led to reduced reproductive success of their (co-evolved) food plants that led to their near or actual extinction and a significant change in the composition of the understories of Hawaiian forests. Second, it indicates how the extinction of one pollinator (the 'o'o) can create a new ecological opportunity for another pollinator (the i'iwi) that helped to prevent its extinction. Third, it emphasizes the importance of adaptability in the face of ecological change. If *V. coccinea* had not been able to successfully extract nectar from ohia flowers, it might have also gone extinct. Adaptive flexibility is a hallmark of this clade of birds (Lovette et al. 2002). Finally, this example shows us the precarious nature of strong ecological and evolutionary associations. Mutualisms such as pollination and frugivory are always vulnerable to disruption once one or more key partners are lost.

BUSHMEAT HUNTING

As indicated above, bushmeat hunting is a major threat to large-bodied frugivorous birds and mammals. The extraction of protein from forests via hunting is widespread in the tropics and is often unsustainable. It is more intense in Africa and Asia than in South America because of higher human population densities. In the forest belt of Central Africa, for example, Fa and Brown (2009) reported that about 4 million tons of dressed bushmeat are extracted annually and that 55% of all forest mammals are being hunted there. In contrast, Peres and Palacios (2007) reported that about 0.16 million tons of wild game are eaten annually in the Brazilian Amazon, where about 28% of all forest mammals are hunted. In regions where hunting has been long-standing, tropical forests are often devoid of large terrestrial herbivore/frugivores such as elephants, tragulids, cervids, and scatter-hoarding rodents as well as medium-to-large species of terrestrial and arboreal primates. Large fruit bats and birds such as guans, curassows, and hornbills are also hunted in many tropical areas (Kinnaird and O'Brien 2007; Mickleburgh et al. 2009; Trail 2007).

Since these birds and mammals are often the sole dispersers of large-seeded plants (i.e., seed length > 15 mm), the question becomes, how badly has bushmeat hunting disrupted seed dispersal mutualisms? We addressed this question in some detail in chapter 4 and will only give a few additional examples here. Brodie et al. (2009) studied the effects of variation in the density of three species of mammalian frugivores (lar gibbons, muntjac, and sambar deer) on seed removal and seedling density of the canopy tree *Choerospondias axillaris* (Anacardiaceae) in four national parks in Thailand. Three of these parks were affected by hunting, but the fourth was effectively protected. They found that the proportion of undispersed seeds decreased and the seed density in potential germination sites, especially light gaps, increased with increasing density of the three frugivores across these parks. They used population matrix simulation models (see chap. 4) to determine the effect of reduced seed dispersal on future population growth of this tree. Their results indicated that reduced dispersal slightly reduced population growth rates, but compared with the effects of juvenile and adult survival rates, the elasticity of decreased seed dispersal (i.e., the relative effect of this factor on a population's λ compared with other life history factors; chap. 4) had little effect on long-term population growth. As we discussed in chapter 4, owing to their low adult mortality rates, the extinction of tropical trees will occur long after their mammalian or avian dispersers have disappeared. The long time lag between loss of dispersal agents and the extinction of vertebrate-dispersed tree species creates an "extinction debt." All plants whose key dispersers or pollinators have been extirpated will incur this debt and are living on borrowed time. They will repay this debt on their death.

Additional recent studies of the effects of hunting on rates of seed dispersal in tropical trees include those of Holbrook and Loiselle (2009), Nuñez-Iturri et al. (2008), and Vanthomme et al. (2010). Holbrook and Loiselle (2009) studied rates of visitation and seed removal by frugivores in the tree *Virola flexuosa* (Myristicaceae) at one hunted and one unhunted site in lowland Ecuador. At the site where monkeys and toucans, which are the most important dispersers of this tree, were hunted, the proportion of seeds removed and frugivore visitation rates were reduced significantly. The remaining frugivorous birds at the hunted site included barbets, cotingids, and thrushes—species that do not swallow large *Virola* seeds but instead drop them under the canopies of fruiting trees. Hunting clearly reduces effective dispersal in this system. Nuñez-Iturri et al. (2008) compared the species richness and density of seedlings and juveniles of trees dispersed by medium-to-large primates in transects away from fruiting trees at three hunted and three unhunted sites in lowland Peru. In forests in which primates have been depleted, species richness of seedlings of trees dispersed by monkeys was reduced by 46% and the frequency of seedlings of abiotically dispersed species increased by 284% compared with unhunted forests. These researchers concluded that hunting results in severe recruitment limitation that will eventually (on a time scale of decades to centuries) lead to reduced densities of primate-dispersed trees and less food for primates. Hunting thus appears to lead to an extinction vortex, in which both frugivores and their food plants may be doomed to extinction. Finally, Vanthomme et al. (2010) censused diurnal frugivorous birds and mammals and compared seed sizes and the diversity and density of vertebrate-dispersed seedlings in transects around five species of primate-dispersed trees at one hunted and one unhunted forest site in the Central African Republic. As expected, animal densities were generally lower (including some extirpations), and mean seed lengths and diversity and density of seedlings of large-seeded species were lower at the hunted sites.

In summary, current evidence indicates that hunting has a significant negative effect on the seed dispersal services provided by frugivorous birds and mammals. As discussed in chapter 4, the long-term effects of hunting are less easy to predict but likely include decreased densities of vertebrates and their food plants. In the case of long-lived forest trees, these effects will occur over long periods of time so that the immediate botanical effects of extirpating frugivorous birds and mammals will not be obvious. Given this time lag (and its extinction debt), skeptics could say that it is too early to say that all is doomed for certain fruit-frugivore mutualisms because the jury is still out. We disagree with this point of view and contend that we cannot wait until the extinction debt has been repaid to reach unassailable conclusions. By the time sufficient evidence has accumulated to show that the high levels of bushmeat hunting of today will have effects on forest or other habitat dynamics that last for centuries (e,g., Nuñez-Iturri et al. 2008), it will be far too late to save substantial numbers of vertebrate frugivores and their food plants. We know that bushmeat hunting is unsustainable in much of the tropics today (Corlett 2007a; Fa and Brown 2009). If it continues to be unregulated, there will be dire consequences for both plants and animals involved in the frugivory mutualism. Bushmeat hunting throughout the tropics must be regulated and/or its effects mitigated via alternative forms of protein.

CLIMATE CHANGE

During the last 2 decades of the twentieth century, concern with environmental deterioration and the loss of biodiversity soared as habitats were being degraded at enormously increased rates due to the activities of human populations. As the twenty-first century dawned, a more all-encompassing environmental threat emerged with the realization that steadily rising global temperatures, primarily as a result of increased levels of atmospheric CO_2 from the burning of fossil fuels and tropical forests, would have a profound effect not only on biodiversity, but on human civilization as well. The future effects of climate change on the functioning of ecosystems have now eclipsed all other environmental concerns so that biodiversity loss and the threats to ecosystem performance are predominantly a function of increasing global temperature. To assess any current threat to biodiversity, we must consider the vulnerability and sensitivity of particular species or set of interacting species to climate change, the level of their exposure to environmental change, and the adaptive capacity of populations to cope with these changes (Dawson et al. 2011; Kiers et al. 2010).

All of the other threats to vertebrate-plant mutualisms as described in the first part of this chapter are either exacerbated or exceeded by the enormity of climate change effects. Within this century, these effects have been predicted to include an increase in mean global air temperature of 2–3°C, a rise in mean sea level of at least 0.5 m (and probably much more), significant shifts in the locations of habitats, particularly at high elevations and high latitudes, and a substantial increase in the extinction rates of plants and animals (but see He and Hubbell [2011] for a cautionary view about this).

Perhaps one of the most important things we have understood as global change has transformed our concern for biodiversity conservation is that historical change as reflected in the fossil record has much to tell us about the magnitude of these problems, the similarity of changes that the earth has gone through in the past, and what may happen in the future (e.g., Benton 2010; Erwin 2009; Ezard et al. 2011; Finarelli and Badgley 2010; Willis et al. 2010). It is quite difficult, if not impossible, to determine how global climate change will affect vertebrate-plant mutualisms discussed in this book (see Kiers et al. [2010] for an overview of this problem). We will therefore provide a brief analysis of past changes in vertebrate and plant interactions that may serve as a guide to and provide some insights into how climate change might affect these mutualisms in the future.

From a historical perspective, climate change has been the norm, not the exception, for most of the earth's history. This can be clearly seen in a plot of estimated global air temperatures throughout the Cenozoic Era (fig. 6. 1). The earth was much warmer during the Paleocene-Eocene Thermal Maximum than at any time since, and global air temperatures have declined steadily since the Mid-Miocene Climatic Optimum. As we describe in chapter 6, Cenozoic temperature changes, along with associated changes in precipitation, have caused the distributions of humid tropical forests to expand and shrink, which, in turn, has had a significant effect on the diversification and distribution of tropical plants and animals.

We used data on the geological ages of families of nectar- and fruit-eating birds and mammals (table 1.1) plus the temperature data found in figure 6.1 to address the question of what effect, if any, climate change has had on the diversification rates of these animals during the Cenozoic. Although diversification rates are usually calculated as the difference between rates of speciation and extinction at the species level (e.g., Mittlebach et al. 2007; Ricklefs 2010), we will use this term here to mean time of family origin. Thus, our primary question can be restated as: What effect has climate change (i.e., change in global air temperature) had on timing of the origins of families of mutualistic birds and mammals and their food plants? Have these families evolved primarily under conditions of elevated or reduced temperatures relative to climatic conditions before or after their origins? Or have these origins occurred independently of climatic conditions? Note that climate change is only one of several possible abiotic factors that are known to affect rates of biological diversification (e.g., Benton 2010; Finarelli and Badgley 2010). Tectonic events, including plate movements, orogeny, and volcanism, are also known to be important abiotic drivers, as we discussed in chapter 9. Thus, we would ultimately like to know the importance of climate change relative to that of various tectonic events as an abiotic driver of biological diversification.

To address our main question, we tallied the number of origins of families of nectar- or fruit-eating birds and mammals in 10 Ma time blocks throughout the Cenozoic. Results (fig. 10.4, birds and mammals combined) indicate that there are two peaks in the distribution of nectarivore origins (at 50–59 Ma and 30–39 Ma) and a single peak for frugivore origins (at 20–29 Ma). Next, we divided the Cenozoic Era temperature curve (fig. 6.1) into segments based on whether temperatures were higher, lower, or stable relative to temperature changes before and after particular time blocks (see col. 2 of table 10.3). We then compared the observed and expected number of origins of families of nectarivores and frugivores (birds and mammals

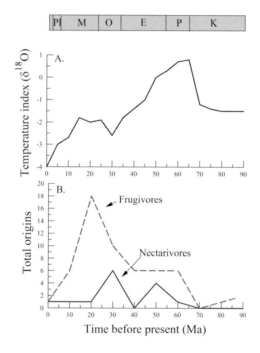

Figure 10.4. Estimate of global temperature during the Cenozoic (*A*) and distributions of time of origin of families of vertebrate nectarivores and frugivores (birds and mammals combined) in 10 Ma time blocks (*B*); based on data in table 1.1. Data for (*A*) come from Blois and Hadly (2009). Geological ages (at top): *Pl* = Pliocene; *M* = Miocene; *O* = Oligocene; *E* = Eocene; *P* = Paleocene; *K* = Cretaceous.

Table 10.3. The Times of Origin of Families of Avian and Mammalian Mutualists as a Function of Relative Temperature Changes during the Cenozoic Era

		Number of Family Origins					
		Nectarivores			Frugivores		
Relative Temperature Condition	Time Periods* (Ma)	Birds	Mammals	Total (Expected Origins)	Birds	Mammals	Total (Expected Origins)
Elevated	59–45; 26–15 (25)	3	2	5 (5.04)	14	16	30 (22.3)
Reduced	45–34; 15–0 (26)	7	0	7 (5.28)	5	9	14 (23.3)
Stable	34–26 (8)	0	0	0 (1.68)	7	2	9 (7.42)

Note. See fig. 6.1 and text for further discussion.
*Total Ma given in parentheses following time periods.

combined) occurring in each of the three segments by chi-square (table 10.3). Results indicated that nectarivore origins were independent of direction of temperature change ($\chi^2 = 2.24$, df = 2, $P \gg 0.05$) but that more frugivore origins occurred during periods of elevated temperatures than expected and fewer origins occurred during periods of reduced temperature ($\chi^2 = 6.71$, df = 2, 0.05 > P > 0.025). From this simple analysis, we tentatively conclude that (1) the origins of families of nectarivores appear to be independent of Cenozoic temperature changes (although the small number of nectarivore families limits our ability to detect a significant trend) and (2) the origins of families of frugivores are not independent of Cenozoic temperature changes; these origins were more common during periods of elevated temperatures than expected. It is important to note that the peak in the origin of families of vertebrate frugivores does not coincide with highest temperatures in the Cenozoic (i.e., with the Paleocene-Eocene Thermal Maximum). Instead, this peak is associated with a period of increasing temperature in the Late Oligocene and Early Miocene, which suggests that relative temperatures were more important than absolute temperatures in the diversification of modern families of fruit-eating birds and mammals (as well as mammals in general; Stadler 2011).

Rates of biological diversification are driven by both abiotic and biotic factors. In addition to competition and predation, biotic factors for verte-brate mutualists include new food resources resulting from the evolution of new families of food plants. However, as we discussed in chapter 6, most of the major families of angiosperms that provide nectar or fruit for birds and mammals evolved much earlier (e.g., in the Late Cretaceous) than their modern mutualists. Thus, there is not a close temporal association between the origins of families of vertebrate-pollinated or -dispersed plant families and families of their modern vertebrate mutualists. This does not preclude the possibility, however, that the origins of families of mutualistic birds and mammals involved coradiations with their food plants (chap. 5). That is, although origins of certain angiosperm families and their modern verte-brate mutualists were not coeval, perhaps major radiations of these plants did coincide with the origins of their mutualists. Some support for this idea comes from the evolutionary history of the cactus family, Cactaceae, and its phyllostomid bat pollinators. According to the phylogenetic analysis of Ara-kaki et al. (2011), this New World family first evolved in the Late Eocene but didn't undergo a major radiation of its crown lineages, including tribes of columnar cacti that are both pollinated and dispersed by bats, until the mid-Miocene in association with cooler and drier climatic conditions worldwide (fig. 10.5). The radiation of columnar cactus lineages also coincides with the first appearance of nectar-feeding phyllostomid bats (Datzman et al. 2010; fig. 10.5). If this cactus-bat example is general, then it supports the hypothesis that climate change can lead to an evolutionary response by plants, which in turn can favor an evolutionary response by vertebrate mutualists. We clearly need more detailed and dated plant and animal phylogenies in order to

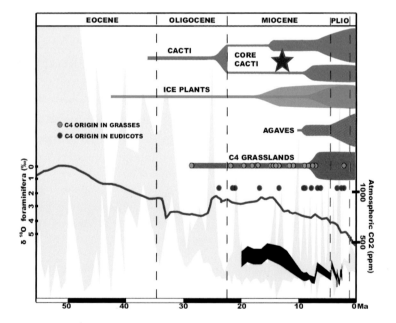

Figure 10.5. Evolutionary history of the Cactaceae and other families of arid-adapted plants with the time of origin of flower-visiting phyllostomid bats indicated by the *red star*. Based on Arakaki et al. (2011), with bat data from Datzmann et al. (2010).

determine the generality of this hypothesis. Data presented by Givnish et al. (unpubl. ms.) suggest that such a scenario also holds for the diversification of epiphytic bromeliads and their hummingbird pollinators during the mid-Miocene Andean uplift. Even if this hypothesis is generally true, however, it may not be applicable to the current period of climate change because of the speed with which this change is occurring. Coradiations likely occur on a time scale of thousands, if not millions, of years and far too slowly to respond quickly to changes that we are currently witnessing.

In conclusion, there can be no doubt that climate change has played an important role (in both a positive and negative sense) in the diversification of the earth's biota, and the evolution of vertebrate mutualists is no exception to this. The appearance of many modern families of vertebrate frugivores appears to have occurred during periods of elevated temperature in the latter half of the Cenozoic, perhaps in response to climate-induced expansions of tropical forests. Once they evolved, these vertebrates could have facilitated the radiation of their food plants via their seed dispersal behavior, including long-distance dispersal (e.g., Muellner et al. 2008; Viseshakul et al. 2011), but much more research is needed before these links are firmly established.

Whether the current global warming trend will have a net positive effect on frugivores and their food plants is difficult to say. This is because climate change is only one of many anthropogenically mediated global changes that are affecting biodiversity today. We suspect that evolutionary trends that held in the past have become void with the rise of modern *Homo sapiens.* As the tropical ecologist Ariel Lugo has said (in plenary addresses to the Association for Tropical Biology and Conservation in 2003 and 2006), we have now entered a new evolutionary era—the Anthropocene (also see *Economist* 2011). In this era, the past is not necessarily a mirror to the future of life on Earth. Like it or not, humans have become a major factor in the evolution of life on our planet.

Conservation of Vertebrate Pollination and Dispersal Mutualisms

What do vertebrate mutualists and their food plants need to avoid (premature) extinction as a result of human activities? Intact and extensive habitats is the easy answer. In the case of migratory mutualists, "extensive" can mean an extremely large area that often encompasses multiple political units. Many nectar- or fruit-eating birds fly thousands of kilometers annually between their wintering and breeding grounds, and a few phyllostomid and pteropodid bats fly nearly this far (Epstein et al. 2009; Fleming and Eby 2003; Greenberg and Marra 2005). However, most of the land encompassing migratory routes currently lacks conservation protection (e.g., Epstein et al. 2009; Fleming 2004b; Richter and Cumming 2008).Without serious conservation efforts, this easily identifiable goal will become more and more difficult to achieve with each passing year. In a world of increasingly fragmented habitats, shrinking wildlands, increasing temperatures, and ever-increasing human pressure, many of our ornaments of life will disappear within a few human generations unless we are proactive about their conservation. Given that these mutualisms occur in networks of interacting species embedded in larger sets of species in communities or habitats, their conservation clearly demands habitat or ecosystem conservation rather than simply species-based conservation. And it almost always involves in situ, rather than *ex situ*, conservation.

It is generally agreed that effective measures for conserving individual species or habitats and ecosystems involve a multi-pronged approach that includes environmental education, legislation, enforcement, and the involvement of local communities. Many international and local conservation agencies are involved in each of these activities, and we will not review this general topic here. Instead, we will focus on two topics that have special relevance to the conservation of nectar- and fruit-eating birds and mammals: pollinator and disperser protection campaigns and ecotourism.

POLLINATOR AND DISPERSER PROTECTION CAMPAIGNS

As a result of the United Nations Convention on Biological Diversity starting in 1992, a number of national, regional, and international efforts have been launched to increase awareness of the threats to pollinators and to protect them from further decline. The International Pollinators Initiative was established in 2000 at the fifth conference of the parties to the Convention on Biological Diversity as the International Initiative for the Conservation and Sustainable Use of Pollinators. This initiative to address the worldwide decline in pollinator diversity was followed by a plan of action, which was prepared by the Food and Agriculture Organization of the United Nations and the Convention on Biological Diversity secretariat and adopted in 2002 at sixth conference of the parties. The aim of the International Pollinators

Initiative is to promote coordinated worldwide action to monitor pollinator decline and identify its causes and its impact on pollination services; address the lack of taxonomic information on pollinators; assess the economic value of pollination and the economic impact of the decline of pollination services; and promote the conservation, restoration, and sustainable use of pollinator diversity in agriculture and other ecosystems. So far a number of reports have been prepared by the International Pollinators Initiative to further these aims (see www.internationalpollinatorsinitiative.org).

In addition to this international effort by the United Nations, the North American Pollinator Protection Campaign was established in 1999 as a private-public collaborative body to promote and protect pollinators in North America. The campaign is coordinated by the Pollinator Partnership, a 501(c)3 nonprofit organization that works with partners worldwide to further their goals. At present the campaign is made-up of more than 120 organizations that include scientists, researchers, conservationists, government officials, and volunteers who have organized major programs to protect pollinators, to raise awareness of pollinator-related issues by the public and governmental bodies, and to benefit the health of all species of pollinators, especially those most threatened by all of the factors described in the present chapter.

The North American Pollinator Protection Campaign's primary mission is "to encourage the health of resident and migratory pollinating animals in North America" by raising public awareness about the importance of pollinators for agriculture, ecosystem health, and food supplies; encouraging collaborative, working partnerships with federal, state, and local government entities and strengthening the network of associated organizations working on behalf of pollinators; promoting the conservation, protection and restoration of pollinator habitats; and documenting scientific, economic, and policy research on pollinators as well as developing and maintaining an international database of pollinator information. Similar efforts have now been instituted around the world, including the African Pollinator Initiative, the Brazilian Pollinator Initiative, the Colombian Pollinator Initiative, the European Pollinator Initiative, the French Pollinator Initiative, and the Oceania Pollinator Initiative.

Despite the diversity and breadth of these efforts to protect pollinators and their natural habitats, it is difficult to assess how effective these initiatives have been in preventing or at least slowing the decline of populations of pollinators. In addition, most of these efforts are aimed at insect pollinators, especially bees, because of their importance in agricultural systems. In fact, very little activity has been aimed at halting or slowing the loss of vertebrate pollinators. We know of no similar international programs that address the decline of vertebrate dispersal agents, though many agencies exist for the general protection of birds (e.g., US National Audubon Society, Royal Society for the Protection of Birds, etc.) and primates (e.g., Conservation International, Wildlife Conservation Society, World Wildlife Fund, etc.). Clearly, there is much room for further efforts specifically aimed at conserving the world's vertebrate pollinators and seed dispersers.

One agency that has taken considerable interest in the conservation status of vertebrate pollinators and seed dispersers is Bat Conservation International (www.batcon.org). Founded in 1982 by Dr. Merlin Tuttle and based in Austin, Texas, Bat Conservation International has a membership of about 14,000 people worldwide and has initiated or encouraged education and conservation efforts in many tropical countries around the world. Specifically, it has supported student research projects, many of which deal with phyllostomid and pteropodid bats, in 55 countries; its Global Grassroots Bat Conservation Fund supports local conservationists and researchers in countries such as Ghana, Kenya, Myanmar, and Vietnam; and its Wings across the Americas program focuses on conservation of two species of nectar-feeding bats in Mexico and the southwestern United States as well as conducting workshops on bat conservation throughout Latin America. In addition, it currently supports a special program on bats of the Philippines, which focuses on conservation, education, and research (Locke 2006; Waldien 2008). One of the major goals of this initiative is to set national priorities and to begin proactive conservation of this bat fauna, including at least 24 species of plant-visiting pteropodids (Heaney and Roberts 2009; D. Waldien, pers. comm.).

ECOTOURISM—BIODIVERSITY SAVIOR OR THREAT?

Ecotourism, or nature tourism as coined by Corlett (2009a), has become big business around the world in recent decades. For example, when one of us (Fleming) began working in Costa Rica in 1970, that country's main sources of income were coffee, bananas, and cattle. Within a decade or so, however, ecotourism had become one of its main income sources (currently second only to silicon chips), and this country continues to be a major destination for ecotourists, mostly from the United States. While staying in small-scale ecolodges in Costa Rica, guests can explore tropical forests where, depending on location, they can view manakins, macaws, toucans and araçaris, tanagers, quetzals and other trogons, three-wattled bellbirds and other cotingids, many species of hummingbirds, and several species of primates—myriads of the ornaments of life. Ecolodges are especially popular with bird watchers, and many well-run lodges cater specifically to this segment of the tourist industry. The world's avian ornaments surely are an important draw for these travelers.

In theory, ecotourism should be a win-win situation for a country: it offers visitors intact habitats full of flora and fauna while permitting countries to cash in on their natural resources without destroying them. Countries that attract ecotourists can parlay resource conservation into an important source of income. When it involves locally owned and operated small-scale lodges and local transportation systems, this income remains in the local economy. From a conservation standpoint, therefore, ecotourism can be a far better use of land and its natural resources than alternatives such as logging, farming, ranching, and mining.

In practice, however, ecotourism has its costs as well as its benefits. These costs include overuse of and damage to habitats caused by too many visitors; inadequate protection or misuse of the environment owing to too few park guards, poorly trained guides, and so on; an emphasis on profit over environmental protection by some tour companies; and heavy reliance on foreign capital and donors who do not necessarily respect a country's environmental ethics or laws (Dasenbrock 2002; Krüger 2005). As a final concern, ecotourism may not be a sustainable source of income in the long term because it relies significantly on a strong world economy (to generate ecotourists) and stable governments and safe conditions in destination countries (to attract ecotourists).

Ecotourism, of course, is only one of many options for conserving tropical biodiversity (Corlett 2009a; Ghazoul and Shell 2010). As Krüger (2005) indicates in his meta-analysis of the sustainability and conservation value of ecotourism, ecotourism is a major revenue producer in only a few countries (e.g., Costa Rica and Kenya), and of the world's 25 biodiversity hotspots identified by Myers et al. (2000), only six countries encompassing these localities are among the top 10 countries in terms of frequency of ecotourism case studies. Clearly, many areas of major conservation concern are not attracting ecotourists. Based on Krüger's (2005) analysis, major factors predicting whether countries are likely to have thriving ecotourism industries include proximity to major markets and the presence of species of charismatic birds and mammals (many of which are ornaments of life) or internationally recognized charismatic vertebrates (e.g., gorillas in Rwanda).

The Future of Vertebrate Pollination and Dispersal Mutualisms

We live in a world that continues to teem with many ornaments of life, especially in tropical ecosystems. One can take any number of well-run ecotours or walks through pristine as well as secondary forests to see beautiful nectar- or fruit-eating birds and mammals living with their plant hosts in apparently healthy habitats throughout the tropics (but which are decidedly depauperate in terms of large seed-dispersing birds and mammals; see Corlett 2007a). But what does the future hold for these systems? How many more generations of humans will be able to enjoy and appreciate these ecological interactions and these natural history phenomena? We suspect not many.

Our pessimism stems from two lifetimes of research by the authors in tropical ecosystems around the world. We have watched as forests have been butchered and reduced to low-productivity agricultural systems or, even worse, ecological deserts devoid of the life forms we expect to see in wet, ever-warm habitats. These observations are indeed discouraging and we do not need to dwell further on the threats to the ornaments of life that we have elaborately described in this volume. A paleontological colleague has stressed to us that if we are concerned with the future of the earth and the lineages of plants and animals that have evolved here, then we need not worry. The historical record testifies that our planet and its life forms will exist long after this current environmental onslaught has abated. However, he also has warned us that if we are worried about the future of our own species and the species we now find on Earth, then we should be extremely concerned (W. DiMichele, pers. comm.). And we are. As Davies et al. (2008, 11562) state in their analysis of extinction risk in mammals: "Diversity will almost certainly rebound after the current extinction event; however, it may be composed of species descended from a different . . . subset of lineages from those that dominate now, and humans will likely not be included among them." If nothing else, we hope that this book on the interactions between vertebrates and the plants they pollinate and disperse will help to inform our scientists, our citizens, and our policy makers that we have a responsibility to make sure that the ornaments of life will continue to prosper and flourish long into the future both for their own value and for the benefit and enjoyment of all.

Coda

In this book we have covered a wide range of topics dealing with the past, present, and future of interactions between nectar-feeding and fruit-eating birds and mammals and their food plants. In the 30–40 plus years that we have been studying these systems, we have seen tremendous advances in our understanding of many facets of these interactions. In the late 1960s and early 1970s, pollination and seed dispersal mutualisms were still being studied as natural history subjects without strong theoretical, evolutionary, or experimental underpinnings. One prominent theoretical ecologist was even questioning the importance of mutualisms in the dynamics of communities (May 1973). Most of our knowledge about tropical vertebrate mutualisms came from a few sites in Latin America, especially in Panama and Costa Rica.

As this book attests, all of this has changed in the last few decades. We now know (though this wasn't really news to field naturalists) that mutualisms play an extremely important role in the history of life; we have solid theoretical underpinnings for the study of pollination and dispersal mutualisms, which are now being studied experimentally everywhere; the geographic scope of mutualism studies now encompasses all tropical regions; and, owing to the surge of phylogenetic studies with the advent of molecular genetics and sophisticated computer-based analytical techniques, we can now study these mutualisms in an increasingly robust evolutionary framework. Although our basic understanding of the ecology and evolution of vertebrate-plant mutualisms has increased many-fold in recent decades, there is still much to be learned about them at all levels of biological organization, from genes through biogeographic realms, as we have repeatedly emphasized throughout this book. What we don't know about these mutualisms will keep "curious naturalists" busy for the foreseeable future. But, as we all know, we are in a race against time in this task because the habitats and organisms that we love to study are steadily disappearing as humans proclaim their dominion over land and sea in ever-increasing fashion. So the ultimate question concerning the fate of these mutualisms becomes: What will it take for our species to realize that biodiversity conservation in general, and conservation of vertebrate-plant mutualisms specifically, is one of the great environmental issues of our time? In writing about conserving biodiversity in tropical East Asia, Richard Corlett (Corlett 2009a, 209) said: "There is a small but growing minority of well-educated, mostly urban, ecologists and conservationists, both amateur and professional [in this region]. In between these groups, however, there is a high degree of ignorance about environmental problems in general and biodiversity conservation in particular, encompassing almost everyone from rural smallholders to urban civil servants and politicians. The extent to which this ignorance contributes to behavior that has harmful impacts on biodiversity is hard to judge, but combating ignorance is a crucial first step towards building the broad public support needed to sustain biodiversity conservation in the region." The same can be said for much of the world. It is our hope that this book will make an important contribution to combating ignorance and misunderstanding about the importance of a relatively small but nonetheless critical set of ecological and evolutionary interactions between plant-visiting vertebrates and their food plants. We can't imagine living in a world devoid of these ornaments of life.

Appendix 1

Overview of the Major Families of Bird and Mammal Pollinators and Seed Dispersers

Below are thumbnail sketches of many of the major families of nectar- and fruit-eating birds and mammals, as highlighted in table 1.1. These sketches indicate the major taxonomic subdivisions of each family, their species richness, and their major feeding adaptations. Taxonomic information comes from Cracraft et al. (2003; birds) and Wilson and Reeder (2005; mammals). Other information comes from Perrins and Middleton (1985) for birds and Fleagle (1988), Nowak (1991), and Rowe (1996) for mammals.

NECTAR-FEEDING BIRDS

FAMILY TROCHILIDAE (HUMMINGBIRDS).—This New World family contains the world's most specialized nectar-feeding birds. The 330 species occur in two subfamilies—hermits, Phaethornithinae (with six genera and about 34 species) and trochilines, Trochilinae (with 96 genera and about 294 species). Hermits are generally restricted to lowland forested habitats of tropical latitudes. Trochilines occur in a much wider array of habitats, elevations, and latitudes. Many species are altitudinal and/or latitudinal migrants, and 13 species breed in western North America as far north as southern Alaska. Trochilines, but not hermits, occur on most Caribbean islands. Hummingbirds are among the world's smallest birds (maximum mass about 20 g). Species-rich genera include *Phaethornis* (23 species), *Chlorostilbon* (14 species), and *Amazilia* (30 species). See examples in plate 1A–C, E.

FAMILY PSITTACIDAE, LORINAE.—This distinctive group of nectar- and pollen-feeding parrots contains 11 genera and about 54 species and occurs in

Indonesia, New Guinea, Australia, and the Pacific as far east as the Pitcairn Archipelago. Noisy and highly gregarious, these brilliantly colored birds attain their highest diversity in New Guinea, which contains 21 species. Owing to their erratic food supply, most lorikeets are nomadic and travel widely between food patches. They are small to medium-sized members of the parrot family (weight range, 20–240 g). See example in plate 1*D*.

FAMILY NECTARINIIDAE (SUNBIRDS).—This passerine family of 16 genera and 127 species is the Old World equivalent of hummingbirds. Maximum diversity occurs in Africa, which has 76 species. Sunbirds are substantially larger than hummingbirds (median mass = 21 g, compared to 10 g in hummers). Instead of hovering at flowers, they usually perch on or near flowers to feed. Like hummingbirds, sunbirds eat insects rather than pollen for protein, and many species are nomadic. See examples in plate 2*B, C*.

FAMILY MELIPHAGIDAE (HONEYEATERS).—This ecologically diverse family contains about 174 species in 44 genera. Greatest diversity occurs in Australia, which contains about 68 species. Small, straight-billed species tend to be mostly insectivorous, whereas larger, curve-billed species are nectar feeders. Arid-zone species are highly nomadic, whereas tropical forest dwellers are mostly sedentary. Weighing up to about 240 g, honeyeaters are among the largest "dedicated" nectar-feeding birds. See example in plate 2*A*.

FRUIT-EATING BIRDS

FAMILY CASUARIIDAE (CASSOWARIES).—These Australasian ratites are the world's largest avian frugivores (maximum height is 1.8 m and maximum mass is 44,000 g). The single genus contains three species, all found on New Guinea; one species lives in the rain forests of northeastern Australia. Solitary and wide ranging, these birds swallow fruits as large as papayas whole. Their fecal piles often contain a diverse array of seed sizes (from tiny fig seeds to large seeds of forest canopy trees). See example in plate 3*A*.

FAMILY COLUMBIDAE (PIGEONS AND DOVES).—Most members of this cosmopolitan family of about 300 species are primarily seed eaters. A clade of Australasian pigeons containing 126 species in 12 genera, however, is frugivorous and feeds heavily on fruits in the Lauraceae and Moraceae. Like the rest of the family, fruit pigeons are very strong fliers and have colonized many South Pacific islands as far east as the Pitcairns. In size, fruit pigeons range from 150 to 840 mm in length and weigh 49–810 g. Many pigeons, including the frugivores, migrate long distances seasonally. See example in plate 3*D*.

FAMILY TROGONIDAE (TROGONS).—Pantropical in distribution, New World and African species are much more frugivorous than Asian species. The family contains a total of 39 species in six genera. The largest Neotropical genus, *Trogon*, contains 15 species, and the Asian genus *Harpactes* contains 11 species. Many New World species feed heavily on fruits of the Lauraceae and undergo altitudinal migrations to track changes in the locations of fruit crops. These birds are medium-sized canopy frugivores (maximum mass is about 200g). See examples in plate 4*D, E*.

FAMILY BUCEROTIDAE (HORNBILLS).—Distributed in the Old World tropics as far east as New Guinea (one species), this ecologically and morphologically diverse family contains 49 species in 13 genera. Africa contains about half the species, and tropical Asia contains the other half. Most species have large (to enormous) bills. Small species are mostly insectivorous. Large terrestrial species in African grasslands are vertebrate predators. The forest-dwelling frugivores include many large species weighing up to 4,000 g. These gregarious birds are strong flyers and travel widely between fruit crops. See example in plate 3*E*.

FAMILY RAMPHASTIDAE (TOUCANS).—Superficially resembling hornbills, long-billed Neotropical toucans are classified in five genera containing a total of 38 species. Members of the largest genus, *Ramphastos*, occur in lowland forests, whereas the smaller toucanets and aracaris have montane distributions. Most species are moderately gregarious while foraging, and all are highly frugivorous. Maximum size (ca. 700 g) is far smaller than that of large hornbills. See examples in plate 4*B, C*.

FAMILY COTINGIDAE (COTINGAS).—This morphologically diverse Neotropical family of frugivores contains about 96 species in 33 genera. Snow (1982) did not advocate recognizing subfamilies. Mating systems are as diverse as the family's morphology. In addition to conventional monogamy, lek mating systems in which males advertise vocally and visually for mates in communal display areas are common. Most species feed in the canopies of tropical and subtropical forests. Sizes ranges from small (6 g) to relatively large (400 g). See example in plate 5*B*.

FAMILY PIPRIDAE (MANAKINS).—Closely related to the cotingids, manakins are more uniform in size (small: 10–25 g) and morphology. Found only on the Neotropical mainland (including Trinidad), this family contains about 48 species, classified in 13 genera of which *Pipra, Manacus*, and *Chiroxiphia* are the most species rich. Most species feed on fruit in the forest understory. Lek mating systems, often with very complex and bizarre display behaviors in males, are the rule in this family. See example in plate 5*C*.

FAMILY PTILONORHYNCHIDAE (BOWERBIRDS).—This Australasian family of sturdily built birds (size range, 70–230 g) contains 18 species, classified in eight genera. Nine species live in New Guinea, seven live in Australia, and two species occur in both areas. Most species are heavily frugivorous. Many species have promiscuous mating systems in which males build and display vocally at bowers of vegetation. These structures vary from relatively simple displays of large leaves on a cleared portion of the forest floor to "avenue" bowers of thin sticks to "maypole" bowers decorated with fruits and moss and finally to complex, hutlike structures. In promiscuous species, there is an inverse relationship between the gaudiness of a male's plumage and the complexity of its bower. Brightly plumaged species (e.g., satin or flamed bowerbirds) have simple bowers, whereas plain-plumaged species (e.g., striped or Vogelkop gardeners) build elaborate bowers.

FAMILY PARADISAEIDAE (BIRDS OF PARADISE).—Closely related to crows (and not to bowerbirds), this family of famously gaudy males is centered in New Guinea and surrounding islands. Only four of the 40 species (in 16 genera) occur in Australia. Most species live in wet forests, but some live in highland woodlands, lowland grasslands, or mangrove swamps. Most species are highly frugivorous, but some are insectivores, including the woodpecker-like sicklebills and riflebirds. As in bowerbirds, mating systems range from conventional monogamy to extreme polygyny, featuring males with extremely colorful plumages that sometimes include bizarre head or tail "wires." Polygamous species display visually and vocally in clumped communal leks or as widely dispersed singletons. Sizes range from about 40 to 300 g. See example in plate 5*E*.

FAMILY TURDIDAE (THRUSHES AND ROBINS).—Commonly classified as a subfamily of Muscicapidae (Turdinae), this cosmopolitan and ecologically diverse group contains about 1,655 species, classified in 24 genera. Sizes range from 8 to 220 g. Members of *Turdus* (60 species) represent a generalized ecological type that tends to be highly frugivorous. Highest thrush diversity occurs in Eurasia. Most species are insectivorous, but many often include fruit in their diet.

FAMILY STURNIDAE (STARLINGS).—Widely distributed in the Old World tropics and subtropics (but just barely reaching Australia), this family contains about 115 species, classified in 25 genera. Small to medium in size (40–220 g), many species have strong, straight bills for eating insects as well as fruit. The Brahminy starling (*Sturnus pagodarum*) has a brush-tipped tongue for collecting pollen and nectar. These birds are highly gregarious, and several species form communal sleeping roosts in trees.

FAMILY DICAEIDAE (FLOWERPECKERS).—This family of small, compact birds (weighing up to about 40 g) contains about 44 species, classified in two genera. Its geographic range includes southern Asia and Australasia, with greatest species richness in New Guinea. The largest genus is *Dicaeum* (35 species)—the mistletoe birds. These birds feed heavily on mistletoe fruits as well as fruits of many vines, trees, and shrubs. Monogamous during the breeding season, flowerpeckers often forage in flocks during the nonbreeding season.

FAMILY PYCNONOTIDAE (BULBULS).—Distributed in Africa, including Madagascar, and southern Asia, this family of about 118 species, classified in 22 genera has its greatest species richness in Africa. Relatively small in size (20–65 g), most species are nonmigratory forest dwellers. Many species are gregarious and forage in flocks with other species. Their diet includes both fruit and insects. Some species may be cooperative breeders (i.e., with nonbreeding helpers at the nest of a monogamous pair), but this is poorly documented. See example in plate 5D.

FAMILY EMBERIZIDAE, THRAUPINAE (TANAGERS).—This diverse group of often very colorful Neotropical fruit eaters contains about 202 species, classified in 50 genera (including the honeycreepers). Greatest species richness occurs in the lowland tropics, but the range of habits and elevations occupied by this group is extensive. Four tropical species migrate to North America to breed. Large genera include *Diglossa* (18 species), *Euphonia* (27 species, now classified in Fringillidae), and *Tangara* (49 species). Compactly built and relatively small in size (9–114 g), these birds feed on many kinds of berries produced by shrubs, vines, epiphytes (including mistletoes), and trees. Some species forage in flocks with other tanagers and species of insectivorous birds. See example in plate 5*F*.

NECTAR-FEEDING MAMMALS

FAMILY TARSIPEDIDAE (HONEY POSSUMS).—This marsupial family contains a single species, *Tarsipes spencerae*, distributed in southwestern Australia. Weighing about 10 g, it is highly specialized for visiting flowers of Proteaceae.

FAMILY PTEROPODIDAE (OLD WORLD FLYING FOXES).—This family of plant-visiting bats belongs to suborder Yinpterochiroptera (formerly known as Megachiroptera, the megabats) of order Chiroptera. It is widely distributed throughout the Old World tropics and subtropics, including islands in the South Pacific as far east as the Cook Islands. The family has traditionally been classified (e.g., Koopman 1993) in two subfamilies: Pteropodinae (flying foxes and their relatives) and Macroglossinae (blossom bats and their relatives), but recent molecular work (Kirsch et al. 1995) indicates that Macroglossinae is not monophyletic. The monotypic African genus *Megaloglossus* is not closely related to the other five nectar-feeding genera, which contain a total of 11 species that occur in the Asian tropics and Australasia. "Macroglossine" megabats are small (maximum mass about 80 g) flower visitors. Many nonmacroglossine megabats (e.g., species of *Pteropus*) are also frequent flower visitors. See example in plate 6*B*.

FAMILY PHYLLOSTOMIDAE (AMERICAN LEAF-NOSED BATS).—A member of suborder Yangochiroptera (formerly Microchiroptera, the microbats), this family of about 150 species is widely distributed in the American tropics and subtropics. It has traditionally been classified (e.g., Koopman 1993) in six subfamilies that basically reflect its broad ecological range: Phyllostominae (insectivores and carnivores), Glossophaginae (nectar bats), Carolliinae (*Piper*-eating fruit bats), Stenodermatinae (fig-eating fruit bats), Brachyphyllinae (West Indian fruit and flower bats), and Desmodontinae (vampire bats). Current taxonomic treatments, based on molecular as well as a host of more traditional characters (Wetterer et al. 2000, Baker et al. 2003), indicate that the family contains three closely related clades of nectar-feeding bats, sometimes given subfamilial status: Glossophaginae (with 10 genera and 22 species), Lonchophyllinae (with three genera and nine species), and Phyllonycterinae (with two genera and four species). Glossophagine bats occur throughout the American tropics and subtropics, including the Caribbean. The other two groups have more restricted distributions; Lonchophyllinae occurs in the mainland Neotropics and Phyllonycterinae occurs in the Greater Antilles, including the Bahamas. All of these bats are morphologically specialized for visiting flowers of a wide variety of families. Their size range is 8–30 g. Certain other phyllostomid bats (e.g., *Phyllostomus discolor*, Phyllostominae) and members of Carolliinae and Sternodermatinae are opportunistic flower visitors. See example in plate 6*A*.

FRUIT-EATING MAMMALS

FAMILY PTEROPODIDAE (FLYING FOXES).—Thirty-six genera containing about 160 species of megabats (the "pteropodinines") are primarily fruit eaters. Ten genera containing about 23 species occur in Africa; the rest occur in Southern Asia, Australasia, and the South Pacific. The genus *Pteropus* contains about 60 species and is widely distributed on islands in the Indian Ocean and the South Pacific but does not occur in Africa. Pteropodid fruit bats include the largest bats in the world (with a maximum mass of about 1,500 g). See examples in plate 6*D, F*.

FAMILY PHYLLOSTOMIDAE (AMERICAN LEAF-NOSED BATS).—Members of two traditional subfamilies—Carolliinae (with two genera and about 10 species) and Stenodermatinae (with 16 genera and about 60 species) are primarily frugivorous, as are two species of the West Indian genus *Brachyphylla* (Brachyphyllinae) and *Phyllostomus hastatus* (Phyllostominae). Carolliinine bats are almost exclusively mainland in their distribution as are most genera of stenodermatine bats. Exceptions to this include the Jamaican fruit bat, *Artibeus jamaicensis*, whose distribution includes the Neotropical mainland and the Caribbean, as well as an advanced clade of West Indian stenodermatines (*Ardops, Phyllops, Stenoderma*, etc.). Maximum size of these fruit bats is about 90 g. See examples in plate 6*C, E*.

FAMILY LEMURIDAE (LARGE LEMURS).—This family of arboreal Madagascan prosimians contains five genera and 19 species. These lemurs are medium-sized primates (maximum mass is about 4,500 g). Only species of *Lemur* and *Varecia* are highly frugivorous. *Lemur catta* is highly terrestrial, and *Varecia variegata* is both diurnal and nocturnal. See example in plate 7*A*.

FAMILY CEBIDAE (CAPUCHINS, MARMOSETS, AND TAMARINS).—This family of six genera and 56 species ranges from Honduras to northern Argentina. It includes small to medium-sized monkeys that lack prehensile tails. Larger cebids eat insects and vertebrates in addition to fruit; some marmosets and tamarins eat nectar and plant resins (gum). Squirrel monkeys (*Saimiri*) are highly gregarious. Night monkeys (*Aotus*) are sometimes included in this family. Social systems range from monogamous pairs (night monkeys), to polyandry (one female mates with several males, with males caring for babies), to polygynous multimale, multifemale groups. See example in plate 7*D*.

FAMILY ATELIDAE (SPIDER AND HOWLER MONKEYS).—This family of five genera and 24 species is distributed from southern Mexico to northern Argentina. A subfossil form (*Xenothrix*) is known from Jamaica. This family contains the largest New World monkeys (*Brachyteles*, weighing up to 12 kg), and its members are highly arboreal. Species in four genera (*Ateles*, *Brachyteles*, *Lagothrix*, and *Alouatta*) have prehensile tails. Spider monkeys are highly frugivorous and very mobile, whereas howlers are highly folivorous and sedentary. Social systems are based on polygyny, but group sizes are highly variable. See example in plate 7*B*.

FAMILY PITHECIDAE (SAKIS AND TITI MONKEYS).—This family of four genera and about 40 species occurs in South America. Its members are medium-sized (up to about 3,500 g) and highly arboreal. In addition to fruit, many species are seed predators and can crack hard seeds with their robust incisors and canines. Group size is variable and ranges from single adult male and female family groups to multimale, multifemale groups.

FAMILY CERCOPITHECIDAE (OLD WORLD MONKEYS).—This family of 22 genera and about 136 species is generally classified in two subfamilies with different dietary specializations: the frugivorous Cercopithecinae (with 12 genera and 77 species) and the herbivorous Colobinae (with 10 genera and 59 species). Frugivorous members of this family include mostly terrestrial species (e.g., *Papio*), terrestrial and arboreal species (e.g., *Macaca*), and highly arboreal species (e.g., *Cercopithecus*). Cercopithecines (maximum mass is about 54,000 g) tend to be significantly larger than cebids; no species has a truly prehensile tail. Several genera have cheek pouches in which they store fruits and seeds. Most genera of cercopithecines are African in distribution; only members of *Macaca* occur in Asia. *Cercopithecus* (with 25 species) is the largest African genus, and *Macaca* (with 22 species) is the largest Asian genus. Cercopithecines are gregarious and often live in multimale, multifemale groups. See example in plate 7C.

FAMILY HYLOBATIDAE (GIBBONS).—This family of lesser apes contains four genera with 16 species and is distributed only in Southeast Asia. These strictly arboreal primates are highly frugivorous. Maximum size is about 13,000 g. These apes live in monogamous family groups. See example in plate 7E.

FAMILY HOMINIDAE (GREAT APES).—The family of great apes contains four genera and seven species. Three taxa—chimpanzees, bonobos, and orangutans—are highly frugivorous. Weighing up to 90,000 g, *Pongo pygmaeus* is the world's largest arboreal frugivore. Chimpanzees and bonobos feed on the ground as well as in trees. Adult orangutans are solitary animals, but the other great apes are group dwellers. See example in plate 7F.

FAMILY PROCYONIDAE (RACCOONS).—This family of six genera and 15 species includes one Asian species (the lesser panda *Ailurus fulgen*, which feeds heavily on bamboo sprouts) and 14 New World species. Frugivores include raccoons (*Procyon*), ringtails (*Bassariscus*), coatis (*Nasua*), kinkajous (*Potos*), and olingos (*Bassaricyon*). Members of the latter two genera are strictly arboreal; the others are scansorial. The kinkajou has a prehensile tail. New World species are about the same size as cebid monkeys (maximum mass is about 11,000 g in *Procyon lotor*). Unlike most cebids, procyonids are nocturnal.

FAMILY VIVERRIDAE (CIVETS, GENETS, AND LINSANGS).—Members of only one subfamily—the palm civets, Paradoxurinae (with five genera and 10 species)—are frugivorous; the other 28 species (in three subfamilies) are mostly carnivorous or insectivorous. Nine of 10 species of palm civets occur in Southeast Asia; one species (*Nandinia binotata*) is African. Except for the terrestrial *Macrogalidia musschenbroeki*, palm civets are almost entirely arboreal. *Nandinia binotata* is highly frugivorous; the other species are more omnivorous. Maximum size is 14,000 g.

Appendix 2

Overview of the Major Families of Plants containing Species
That Are Pollinated or Dispersed by Birds or Mammals

Below are thumbnail sketches of the major families of plants providing food for nectar- and fruit-eating birds and mammals, as highlighted in table 1.3. These sketches indicate the major taxonomic subdivisions of each family, their species richness, their major flower or fruit adaptations, and some of the genera providing resources for vertebrates. Families were chosen based on their occurrence in the diets of more than one group of vertebrates. Plant information comes from APG III (APG 2009), Heywood et al. (2007) and Mabberley (1997). Dietary information comes from Buzato et al. (2000), Chapman et al. (2002), Dobat and Peikert-Holle (1985), Feinsinger (1976), Feinsinger et al. (1982), Galetti et al. (2000), Ganzhorn et al. (2004), Gautier-Hion et al. (1985), Gorchov et al. (1995), Kitamura et al. (2002), Kodric-Brown et al. (1984), Mickleburgh et al. (1992), Peres and Roosmalen (2002), Snow (1981), Snow and Snow (1980), Subramany and Radhamani (1993), and Terborgh (1983).

MAJOR FAMILIES PROVIDING NECTAR AND
POLLEN FOR TROPICAL BIRDS AND MAMMALS

AGAVACEAE (CENTURY PLANT FAMILY).—Distributed in arid and semi-arid habitats throughout tropical and subtropical parts of the world, this family contains about 637 species, classified in 23 genera. Most species are succulents, featuring a basal rosette of lance-shaped leaves and a central inflorescence bearing flowers arranged in racemes or panicles. Paniculate members of the American genus *Agave* (with about 300 species) are common members of xeric habitats and are bat pollinated within tropical latitudes. Hummingbirds (as well as bees) also pollinate various species of *Agave*. See an example in plate 9*F*.

BIGNONIACEAE (CATALPA FAMILY).—This cosmopolitan family containing about 800 species in 110 genera reaches its greatest diversity in South America. It contains many vines as well as shrubs and trees. Many species have large showy flowers that are often planted as ornamentals in tropical and subtropical countries. Vertebrate-pollinated genera include *Crescentia, Distictis, Enallagma, Kigelia, Oryoxlum*, and *Stereospermum* (bats), and *Fridericia, Jacaranda, Spathodea, and Tecoma* (birds). See an example in plate 9*B*.

BOMBACACEAE (BALSA FAMILY).—Closely related to the Malvaceae (mallow or cotton) and Sterculiaceae (chocolate) families (and included as a subfamily in Malvaceae in APG), this pantropical family of about 180 species (in 20 genera) of trees is most diverse in South America. Most species are deciduous and produce large, showy flowers during tropical dry seasons. Ecologically and/or economically important genera include *Adansonia* (baobabs), *Ceiba* (kapok or silk cotton), *Durio* (durian), and *Ochroma* (balsa). Species of these and other genera (e.g., *Bombax, Chorisia, Pachira*) are bat pollinated. Birds pollinate flowers of *Sprirotheca*. See an example in plate 9*D*.

BROMELIACEAE (PINEAPPLE FAMILY).—Except for one species (the West African *Pitcairnia feliciana*), this distinctive family of about 1,400 species (in 57 genera, classified in three subfamilies) occurs in the New World tropics and subtropics. Morphologically and physiologically adapted for xeric conditions, primitive members of this family are terrestrial with well-developed root systems, but most species are epiphytes. The bromeliads (subfamily Bromelioideae) include arboreal and terrestrial species that feature a leafy, water-filled tank and a central inflorescence with bright-colored flowers surrounded by equally colorful bracts. Many bromeliads are pollinated by hummingbirds (e.g., *Aechmea, Bilbergia, Guzmania, Pitcairnia, Tillandsia,* and *Vriesia*); a few (e.g., *Puya, Vriesia*) are pollinated by bats. See an example in plate 9*E*.

CACTACEAE (CACTUS FAMILY).—This family of spectacularly adapted xerophytes occurs in arid habitats throughout the New World. It contains about 2,000 species, classified in 87 genera occurring in three (or four) subfamilies (Pereskioideae [leafy cacti], Opuntioideae [prickly pears and chollas], and Cactoideae [columnar cacti]). The latter two subfamilies contain species that are either hummingbird or bat pollinated. Tribe Pachycereeae, centered in Mexico, contains huge columnars that are the dominant plants in their habitats. Most members of this tribe, whose genera include *Carnegiea* (saguaros), *Neobuxbaumia*, *Pachycereus* (cardons), and *Stenocereus* (organ pipes), are bat pollinated. With 34 species, *Pilosocereus* (tribe Cereeae) is the ecological equivalent of Mexican *Stenocereus* in arid parts of eastern and northern South America and the Caribbean. Cactus fruits are many-seeded berries that are eaten (and whose seeds are dispersed) by birds and mammals.

CAMPANULACEAE (BELLFLOWER FAMILY).—This cosmopolitan (except for tropical Africa) family containing about 2,380 species in about 84 genera is mostly north temperate in distribution. Most species are herbs. *Centropogon* and *Siphocampylus* (hummingbirds) and *Burmeistera* (bats) are three vertebrate-pollinated Neotropical genera. See an example in plate 8*C*.

CONVOLVULACEAE (MORNING GLORY FAMILY).—Cosmopolitan in distribution, this family of mostly woody or herbaceous climbers contains about 1,600 species classified in about 57 genera. Many genera produce large, bell-shaped flowers. The large genus *Ipomoea* contains species that are pollinated by birds and bats.

FABACEAE (PEA FAMILY).—This enormous cosmopolitan family contains about 19,400 species in 730 genera. Its three subfamilies (Mimosoideae, Caesalpinioideae, and Papilionoideae) are sometimes treated as separate families. Mimosoid flowers contain many stamens that resemble a shaving brush; caesalpinioid flowers have large lateral wings and fewer stamens; papilionoid flowers are butterfly-shaped with two ventral petals that are fused to form a keel. Most members of the first two subfamilies are trees or shrubs, whereas most members of the latter are herbs. Bird-pollinated plants include *Caesalpina*, *Delonix*, *Erythrina*, *Parkia*, *Peltophorum*, and *Sophora*. Bat-pollinated mimosoid genera include *Calliandra*, *Inga* (also hummingbird pollinated), and *Parkia*; caesalpiniod genera include *Bauhinia* and *Hymenaea*; papilionoid genera include *Castanospermum*, *Erythrina*, and *Mucuna*. See an example in plate 8*B*.

GESNERIACEAE (AFRICAN VIOLET FAMILY).—Pantropical in distribution (but uncommon in Africa), this family of mostly herbs and shrubs contains about 147 genera and over 3,870 species. Many Neotropical forms (e.g., in the genera *Columnea, Asteranthera, Kohleria, Nematanthus,* and *Sinningia*) are hummingbird-pollinated epiphytes. Some Neotropical forms (e.g., *Drymonia, Gesnaria,* and *Rhytidophyllum*) are bat pollinated. The Old World genus *Aeschyanthus* is bird pollinated.

HELICONIACEAE (HELICONIA FAMILY).—This family of about 200 species in a single genus is mostly Neotropical in distribution but also occurs in Southeast Asia. Flowers of these megaherbs are important nectar sources for hummingbirds (particularly hermits). See an example in plate 8*D*.

MALVACEAE (MALLOW FAMILY).—This cosmopolitan family contains over 1,000 species, classified in about 80 genera (not including Bombacaceae and Sterculiaceae); its growth habits include herbs, shrubs, and trees. Hummingbirds are major pollinators of *Abutilon, Malvaviscus,* and *Pavonia*. Bat-pollinated genera include *Abutilon* and *Wercklea*.

MUSACEAE (BANANA FAMILY).—This small family of paleotropical herbaceous monocots contains about 35 species in two genera. Greatest diversity in the primarily bat-pollinated genus *Musa* occurs in Southeast Asia and New Guinea. Sunbirds and honeyeaters also pollinate *Musa*s. See an example in plate 9*C*. Also see below.

MYRIACEAE (EUCALYPTUS FAMILY).—Pantropical in distribution, this family of shrubs and trees contains about 4,625 species in 131 genera. The large genus *Eucalyptus* (with more than 700 species), which is pollinated by a diverse array of birds and bats in Australasia, is very important ecologically and economically. *Syzygium* is another speciose vertebrate-pollinated taxon.

PROTEACEAE (BANKSIA FAMILY).—This family of trees and shrubs contains over 1,600 species in 80 genera and occurs in strongly seasonal habitats on the southern continents. Greatest diversity occurs in South Africa where sunbirds (Nectariniidae) are important pollinators. Honeyeaters (Meliphagidae), the honey possum (*Tarsipes*), and *Syconycteris* bats are important pollinators in Australia. Vertebrate-pollinated genera include *Banskia*, *Grevillea*, and *Protea*. See an example in plate 8*A*.

RUBIACEAE (COFFEE FAMILY).—This is a large (about 10,000 species in 600 genera), cosmopolitan family of trees and shrubs in tropical and subtropical habitats; temperate species tend to be herbs. Hummingbird-pollinated genera include *Hamelia*, *Isertia*, *Manettia*, and *Palicourea*. See below. See an example in plate 8*F*.

SOLANACEAE (POTATO FAMILY).—This cosmopolitan family of mostly herbs and a few shrubs and trees contains about 2,460 species in about 102 genera. Greatest diversity occurs in the Neotropics (about 40 genera) and Australia. Flower shape varies from round and flat (*Solanum*) to bell-shaped or tubular (*Nicotiana*). Bird-pollinated genera include *Acnistis*, *Cestrum*, and *Nicotiana*. Bat-pollinated genera include *Marke*a and *Trianaea*.

ZINGIBERACEAE (GINGER FAMILY).—This family of aromatic perennial understory herbs containing about 1,200 species in 50 genera is pantropical in distribution but has its greatest diversity in Indo-Malaysia. Important bird-pollinated genera include *Amomum* and *Plagiostachys*. Mammal-dispersed fruits occur in *Aframomum*. See an example in plate 12*A*.

MAJOR FAMILIES PROVIDING FRUIT FOR TROPICAL BIRDS AND MAMMALS

ANACARDIACEAE (CASHEW FAMILY).—This pantropical family contains about 985 species in about 70 genera. Growth habit includes trees, shrubs, and vines. Fruits tend to be relatively large and are usually single-seeded drupes. Birds disperse fruits of *Metopium*, *Rhus*, *Schinus*, and *Tapirira*. Bats and primates are often dispersers of seeds produced by *Anacardium*, *Mangifera*, and *Spondias*. Primates disperse seeds of *Antrocaryon*, *Choerospondias*, *Pseudospondias*, and *Tricoscypha*.

ANNONACEAE (SOURSOP FAMILY).—This pantropical family of trees and shrubs contains about 2,200 species in 129 genera. Its greatest diversity occurs in Old World tropical forests. Its relatively large fruit are aggregates of berries that are eaten by birds, bats, squirrels, and primates. Common genera occurring in the diets of birds include *Annona*, *Cananga*, *Cyath-*

ostemma, Dasymaschalon, Desmos, Polyalthia, and *Rollinia.* Genera in the diets of mammals include *Annona, Cananga,* and *Polyalthia* (bats) and *Alphonsia, Annona, Anonidium, Desmos, Guatteria, Oxandra, Polyalthia, Rollinia, Uvaria, Uvariopsis,* and *Xylopia* (primates).

APOCYNACEAE (OLEANDER FAMILY).—Members of this pantropical family, which contains about 4,550 species in 415 genera, are mostly trees, shrubs, and lianas. Fruits tend to occur in pairs. Bird-dispersed genera include *Alstonia, Rauvolfia, Stemmadenia,* and *Tabernaemontana.* Genera in the diets of bats include *Cerbera, Ochrosia,* and *Rauvolfia.* Primates eat fruits of *Bonafousia, Cylindropsis, Funtumia, Landolphia,* and *Rauvolfia*

ARACEAE (AROIDS).—This cosmopolitan family of fleshy herbaceous monocots includes about 4,025 species in 106 genera. Fruits are berries containing one to many seeds. Bird-dispersed genera include *Anthurium, Epiphyllum, Monstera,* and *Philodendron.* Bat-dispersed genera include *Anthurium* and *Philodendron.* Primates eat fruits of *Anthurium, Heteropsis,* and *Syngonium.*

ARECACEAE (PALM FAMILY).—This archetypical tropical family contains about 2,361 species in 189 genera. Palms are far more diverse in the Neotropics (about 64 genera and 857 species) and Southeast Asia (about 97 genera and 1,385 species) than Africa (about 16 genera and 116 species). Fruits are one-seeded berries or drupes. Many genera are dispersed by birds and mammals, including squirrels and other rodents (e.g., Echimyidae and Dasyproctidae in the New World).

BURSERACEAE (GUMBO LIMBO FAMILY).—This pantropical family of trees and shrubs contains about 550 species in 18 genera. It is closely related to the Anacardiaceae; both families contain resin ducts in many of their tissues. The fruit is a drupe or a capsule. Bird-dispersed genera include *Canarium, Commiphora, Dacroydes, Protium, Santiria,* and *Trattinickia.* Mammal-dispersed genera include *Bursera, Canarium, Dacroydes, Protium, Santiria,* and *Tetragastris.*

CELASTRACEAE (SPINDLE TREE FAMILY).—This family has a cosmopolitan distribution, but most species occur in the tropics or subtropics. The 1,300 species of trees and shrubs are classified in about 89 genera. Fleshy-fruited taxa produce berries or drupes. Bird-dispersed genera include *Bhesa, Cassine, Celastrus, Elaeodendron,* and *Maytenus*; mammal-dispersed genera include *Cassine* (bats) and *Maytenus* (primates).

CHRYSOBALANACEAE (COCOPLUM FAMILY).—This pantropical family of trees and shrubs contains about 460 species in 17 genera. Fruits are either dry or fleshy drupes. Birds and bats eat the fruit of *Parinari*; other fruit eaten by mammals include *Couepia, Hirtella, Licania,* and *Parinari* (primates).

CLUSIACEAE (MANGOSTEEN FAMILY).—This cosmopolitan family of trees and shrubs contains about 595 species in 14 genera. Fruit types include capsules as well as berry- or drupelike fruits. Vertebrate-dispersed genera include *Calophyllum* (birds, bats, and primates), *Clusia* (birds), *Garcinia* (bats, primates), *Mammea* (bats, primates), *Rheedia* (primates), *Symphonia* (primates), and *Vismia* (birds, bats).

COMBRETACEAE (TERMINALIA FAMILY).—This pantropical family of trees, shrubs, and lianas contains about 500 species in 20 genera. Fruits of many forest species tend to be fleshy and are dispersed by birds (*Terminalia*) or bats and primates (*Terminalia*, others). Flowers of *Combretum* are pollinated by hummingbirds and primates in the New World tropics.

CONVOLVULACEAE (MORNING GLORY FAMILY).—As mentioned above, some flowers of this family are vertebrate-pollinated. A few genera produce fleshy fruits that are eaten by birds (*Erycibe*) and primates (*Dicranostyles, Erycibe, Maripa*).

DILLENIACEAE (DILLENIAS).—Pantropical in distribution (but with only one genus in Africa), this family of trees, shrubs, and lianas contains about 300 species in 12 genera. Its berry-like fruit contain arillate seeds that are eaten by birds (e.g., *Davilla, Hibbertia, Lecistema*) and primates (*Doliocarpus*).

EBENACEAE (EBONY FAMILY).—Greatest diversity of this family of small trees (about 548 species in four genera of which *Diospyros* contains nearly all of the species) is concentrated in Southeast Asia and Africa, but it also occurs in the New World tropics and subtropics. Its fruits are berries that are eaten by birds, bats, and primates (*Diospyros*).

EUPHORBIACEAE (SPURGE FAMILY).—This cosmopolitan family contains over 5,970 species in about 220 genera. Its greatest concentration of species occurs in South America and tropical Asia. Fruits are either nonfleshy schizocarps or fleshy drupes. Genera dispersed by vertebrates include *Alchornea, Antidesma, Bridelia, Drypetes, Macaranga, Sapium,* and *Uapaca* (birds), *Balakata, Drypetes, Omphalea,* and *Uapaca* (primates), and *Bridelia, Sapium,* and *Uapaca* (bats and primates).

FLACOURTIACEAE–NOW PART OF SALICACEAE (WEST INDIAN BOX-WOOD FAMILY).—This pantropical family of trees and shrubs contains about 1,250 species in 89 genera. Fruit types are varied and include berries and drupes. Vertebrate-dispersed genera include *Casearia* (birds, primates), *Flacourtia* and *Hasseltia* (birds), and *Lindackeria, Ludia,* and *Mayna* (primates).

ICACINACEAE (NO COMMON FAMILY NAME).—With greatest diversity concentrated in Malaysia, this pantropical family of trees, shrubs, and lianas contains about 149 species in 24 genera. Fruits are usually drupes. Vertebrate-dispersed genera include *Apodytes, Citronella,* and *Gomphandra* (birds) and *Calatola, Discophera,* and *Leretia* (primates).

LAURACEAE (AVOCADO FAMILY).—Primarily tropical and subtropical in distribution, this family of about 2,500 species in 50 genera reaches its greatest diversity in lowland and montane rain forests of the Neotropics and Southeast Asia. Most species are trees or shrubs, and fruits are berry- or drupelike. Many fruits have lipid-rich pulp. Birds are the main dispersers of seeds of such genera as *Beilschmiedia, Cinnamomum, Cryptocarya, Litsea, Ocotea,* and *Persea.* Primates eat fruits of *Beilschmiedia, Cinnamomum, Cryptocarya, Licaria, Nectandra,* and *Ocotea.* See an example in plate 10*A*.

FABACEAE (PEA FAMILY).—Although most fruits of this huge family are woody pods (legumes) with dry seeds, some species produce pulp-covered seeds that attract vertebrate frugivores. Genera that are vertebrate-dispersed include *Acacia* (birds, bats), *Cassia* (primates), *Inga* (birds, primates), *Le-*

cointea (primates), *Parkia* (birds), *Peltogyne* (primates), and *Tamarindus* (bats). See an example in plate 11*A*.

MELASTOMATACEAE (MELASTOME FAMILY).—Pantropical in distribution, this family of about 5,005 species in 188 genera reaches its greatest diversity in the Neotropics. Most species are shrubs or small trees. Fruits are many-seeded berries in most genera. Vertebrate-dispersed genera include *Clidemia* (birds, bats, primates), *Conostegia* (birds), *Leandra* (birds), *Melastoma* (birds, bats), *Memecylon* (birds, primates), *Miconia* (birds, primates), and *Mouriri* (primates).

MELIACEAE (MAHOGANY FAMILY).—This pantropical family of trees and shrubs contains about 621 species in 52 genera. Among its various fruit types are berries and drupes. Seeds produced by members of subfamily Meliodeae have a fleshy aril or fleshy sarcotesta; those of subfamily Swietenioideae are dry and winged. Vertebrate-dispersed genera include *Aglaia* (birds), *Azadirachta* (bats), *Chisocheton* (birds?), *Dysoxylum* (birds, bats), *Guarea* (birds, primates), *Melia* (birds), and *Trichilia* (birds, bats, and primates).

MENISPERMACEAE (CURARE FAMILY).—Most members of this pantropical family (about 420 species in 70 genera) are lianas. Fruits are drupelike. Vertebrate-dispersed genera include *Cissampelos, Diploclisia, Fibraurea, Hyserpa,* and *Tinospora* (birds) and *Anomospermum, Chondrodendron, Curarea, Diploclisia,* and *Odontocarya* (primates).

MORACEAE (FIG FAMILY).—This pantropical family of ecologically important trees and shrubs contains about 1,100 species in 38 genera; the genus *Ficus* contains at least 700 species. Infructescences are variable in form but are invariably fleshy; they are highly sought after by birds and mammals. Vertebrate-dispersed genera include *Artocarpus* (birds and mammals), *Brosimum* (birds, bats, and primates), *Castilloa* (birds), *Chlorophora* (birds and mammals), *Ficus* (birds and mammals), *Maclura* (birds), *Musanga* (primates), *Pourouma* (primates), *Pseudomelia* (primates), and *Trophis* (birds, primates). See examples in plate 13*A, C*.

MYRISTICACEAE (NUTMEG FAMILY).—This family of trees occurs throughout the tropics and contains about 475 species in 20 genera. Greatest diversity occurs in New Guinea and the Amazon. The fruit is a dehiscent berry with an arillate seed. Old World *Myristica* and New World *Virola* are primarily bird dispersed. Bats eat *Pycnanthus* fruit. Primates eat fruits of *Coelocaryon, Osteophloem, Otoba,* and *Pycnanthus.* See an example in plate 10C.

MYRTACEAE (EUCALYPTUS FAMILY).—Fruits of this family, which is one of the uncommon examples of a family whose flowers are often vertebrate pollinated and whose fruits are often vertebrate dispersed, include berries, drupes, capsules, and nuts. Fleshy fruits occur in subfamily Myrtoideae; dry fruits occur in subfamily Leptospermoideae. Vertebrate-dispersed genera include *Acmena* (birds), *Ardisia* (birds, primates), *Calyptranthes* (primates), *Eugenia* (birds, bats, and primates), *Myrcia* (birds, primates), *Psidium* (birds, bats), and *Syzygium* (birds, bats).

OLACACEAE (AFRICAN WALNUT FAMILY).—This family of about 103 species in 14 genera is pantropical, but most species occur in the Old World. Fruits are drupes that are dispersed by birds (*Heisteria*) and primates (*Heisteria, Minquartia, Ptychopetalum,* and *Strombosia*).

PANDANACEAE (SCREW PINE FAMILY).—Restricted to coastal or marshy areas of the Old World tropics and subtropics, this family of trees, shrubs, and climbers contains about 885 species in four genera. Fruits are berries or multilocular drupes. Bats and primates are the major vertebrate seed dispersers.

PIPERACEAE (BLACK PEPPER FAMILY).—This pantropical family (about 3,600 species in five genera) of primitive angiosperms reaches its greatest diversity in the Neotropics, which has about 2,000 species of *Piper.* Most Neotropical forms are shrubs or treelets; most Old World forms are vines. The spikelike infructescences containing many small drupes are dispersed primarily by bats in the New World and by birds in the Old World. See an example in plate 13D.

ROSACEAE (ROSE FAMILY).—This cosmopolitan family of about 2,830 species, classified in 95 genera, reaches its greatest diversity in north temperate regions. All plant growth forms except aquatics are found in this family. Fruit types are diverse and have been used to delimit natural groups. Fleshy fruited taxa occur in three subfamilies: Maloideae, Prunoideae, and Rosoideae. Birds are major dispersers of genera such as *Hiratella*, *Prunus*, and *Rubus*. Bats eat fruit of *Eriobotrya* and *Prunus*.

RUBIACEAE (COFFEE FAMILY).—This large pantropical family contains both vertebrate-pollinated flowers and vertebrate-dispersed seeds. Fruit types are diverse and include berries, drupes, and schizocarps. Birds are important dispersers of many genera of understory shrubs (e.g., *Canthium*, *Hamelia*, *Henrietta*, *Morinda*, *Palicourea*, and *Psychotria*). Bats eat fruit of *Guettarda*, *Morinda*, *Nauclea*, and *Randia*. Primates eat fruit of *Alibertia*, *Anthocephalis*, *Canthium*, *Genipa*, *Nauclea*, and *Randia*.

RUTACEAE (CITRUS FAMILY).—This pantropical family of trees and shrubs, which contains about 1,815 species in 161 genera, reaches its greatest diversity in Australia and South Africa. Fruits are berries. Vertebrate-dispersed genera include *Acronychia*, *Clausena*, *Fagara*, *Teclea*, and *Zanthoxylum* (birds) and *Fagara*, *Teclea*, and *Toddalia* (primates). Bats also eat *Acronychia* fruit.

SAPINDACEAE (AKEE FAMILY).—Distributed throughout the tropics and subtropics, this family contains about 1,580 species in 135 genera. It contains many species of lianas as well as trees and shrubs. Fruit types include berries, drupes, and schizocarps with arillate seeds. Vertebrate-dispersed genera include *Allophyllus*, *Cupania*, *Sapindus*, and *Tristriopsis* (birds); *Cupaniopsis*, *Dimocarpus*, and *Nephelium* (bats), and *Allophyllus*, *Dimocarpus*, *Nephelium*, *Pancovia*, *Paullinia*, and *Tinopsis* (primates).

SAPOTACEAE (CHICLE FAMILY).—This pantropical family of trees contains about 975 species in up to 53 genera. Its fruit are berries. Different genera are dispersed by birds (e.g., *Bumelia*, *Chrysophyllum*, *Manilkara*, *Mimusops*, *Sideroxylon*), bats (*Chrysophyllum*, *Manilkara*, *Mimusops*, *Planchonella*, and *Sideroxylon*), and primates (*Chrysophyllum*, *Gambeya*, *Manilkara*, *Mimusops*, and *Pouteria*).

SIMAROUBACEAE (QUASSIA FAMILY).—Tropical and subtropical in distribution, this relatively small family of trees and shrubs contains about 95 species in 19 genera. Fruits include samaras, capsules, and schizocarps. Vertebrate-dispersed genera include *Eurycoma* (birds), *Picramnia* (birds, primates), *Odyendea* (birds), and *Simarouba* (primates).

SOLANACEAE (POTATO FAMILY).—Although mainly a family of herbaceous plants, the berries of a few species of shrubs and trees are dispersed by birds (*Acnistus, Capsicum, Cestrum, Lycium,* and *Solanum*) and bats (*Solanum* and *Cestrum*). See an example in plate 10*B*.

STERCULIACEAE—NOW PART OF MALVACEAE (CHOCOLATE FAMILY).— This pantropical family of trees, shrubs, and lianas contains about 700 species, classified in 60 genera. It is closely related to (and now included in) Malvaceae, along with Bombacaeae. Most taxa produce dry, capsular fruits, but a few produce berry-like, dehiscent fruit. Vertebrate-dispersed genera include *Guazuma* (birds, primates), *Sterculia* (birds, primates), and *Theobroma* (primates).

TILIACEAE (LINDEN FAMILY).—This relatively small family (about 400 species in 41 genera) of trees and shrubs has a cosmopolitan distribution. Tropical genera that are vertebrate dispersed include *Apeiba* (primates) and *Grewia* (birds, primates).

ULMACEAE—MANY TAXA NOW IN CANNABACEAE (HACKBERRY FAMILY).— This cosmopolitan family of trees and shrubs contains about 2,000 species in 16 genera. Its diverse fruits include nuts, samaras, and drupes. Vertebrate-dispersed genera include *Ampelocera* (primates), *Celtis* (birds, bats, and primates), *Gironniera* (birds), and *Trema* (birds).

URTICACEAE (NETTLE FAMILY).—This cosmopolitan family (which here includes Cecropiaceae) contains diverse growth habits among its 2,625 species (in about 54 genera). Some taxa produce berry-like or otherwise fleshy infructescences. Vertebrate-dispersed genera include *Cecropia* (eaten by many volant or arboreal birds and mammals), *Dendrocnide* (birds, bats), *Musanga* (birds, bats), and *Urera* (birds, primates).

VERBENACEAE (TEAK FAMILY).—This pantropical family of diverse growth habits contains over 1,100 species in about 30 genera. The most common fruit type is a drupe. Vertebrate-dispersed genera include *Duranta, Gmelina, Lantana, Lippia,* and *Vitex* (birds), *Faradaya, Premna,* and *Vitex* (bats), and *Lanatana* and *Vitex* (primates).

References

Ackerly, D. D. 2003. Community assembly, niche conservatism, and adaptive evolution in changing environments. International Journal of Plant Sciences 164:S165–S184.

———. 2009. Conservatism and diversification of plant functional traits: evolutionary rates versus phylogenetic signal. Proceedings of the National Academy of Sciences United States of America 106:19699–706.

Ackermann, M., and M. Weigend. 2006. Nectar, floral morphology and pollination syndrome in Loasaceae subfam. Loasoideae (Cornales). Annals of Botany 98:503–14.

Adams, C. D. 1972. Flowering plants of Jamaica. University of the West Indies, Mona, Jamaica.

Adler, L. S. 2000. The ecological significance of toxic nectar. Oikos 91:409–20.

Aguilar, R., L. Ashworth, L. Galetto, and M. A. Aizen. 2006. Plant reproductive susceptibility to habitat fragmentation: review and synthesis through a meta-analysis. Ecology Letters 9:968–80.

Ahmed, S., S. G. Compton, R. K. Butlin, and P. M. Gilmartin. 2009. Wind-borne insects mediate directional pollen transfer between desert fig trees 160 kilometers apart. Proceedings of the National Academy of Sciences of the United States of America 106:20342–47.

Aiba, S., and K. Kitayama. 1999. Structure, composition and species diversity in an altitude-substrate matrix of rain forest tree communities on Mount Kinabalu, Borneo. Plant Ecology 140:139–57.

Aiello, L., and C. Dean. 1990. An introduction to human evolutionary anatomy. Academic Press, London,.

Aizen, M. A., and L. D. Harder. 2007. Expanding the limits of the pollen-limitation concept: effects of pollen quantity and quality. Ecology 88:271–81.

Aizen, M. A., C. L. Morales, and J. M. Morales. 2008. Invasive mutualists erode native pollination webs. PLoS Biology 6:396–403.

Aizen, M. A., and D. P. Vazquez. 2006. Flowering phenologies of hummingbird plants from the temperate forest of southern South America: is there evidence of competitive displacement? Ecography 29:357–66.

Alados, C. L., T. Navarro, B. Komac, V. Pascual, and M. Rietkerk. 2010. Dispersal abilities and spatial patterns in fragmented landscapes. Biological Journal of the Linnean Society 100:935–47.

Albert, A. Y. K., and D. Schluter. 2005. Selection and the origin of species. Current Biology 15:R283–R288.

Alcantara, J. M., and P. J. Rey. 2003. Conflicting selection pressures on seed size: evolutionary ecology of fruit size in a bird-dispersed tree, *Olea europaea*. Journal of Evolutionary Biology 16:1168–76.

Alcantara, S., and L. G. Lohmann. 2010. Evolution of floral morphology and pollination system in Bignonieae (Bignoniaceae). American Journal of Botany 97:782–96.

Alexander, J. M., C. Kueffer, C. C. Daehler, P. J. Edwards, A. Pauchard, T. Seipel, and M. Consortium. 2011. Assemby of nonnative floras along elevational gradients explained by directional ecological filtering. Proceedings of the National Academy of Sciences of the United States of America 108:656–61.

Almeida-Neto, M., F. Campassi, M. Galetti, P. Jordano, and A. Oliveira. 2008. Vertebrate dispersal syndromes along the Atlantic forest: broad-scale patterns and macroecological correlates. Global Ecology and Biogeography 17:503–13.

Alonso, D., R. S. Etienne, and A. J. McKane. 2006. The merits of neutral theory. Trends in Ecology and Evolution 21:451–57.

Altmann, S. A. 1988. The monkey and the fig. American Scientist 77:256–63.

Altringham, J. D. 1996. Bats: biology and behaviour. Oxford University Press, Oxford.

Altshuler, D. L. 2001. Ultraviolet reflectance in fruits, ambient light composition and fruit removal in a tropical forest. Evolutionary Ecology Research 3:767–78.

Altshuler, D. L., F. G. Stiles, and R. Dudley. 2004. Of hummingbirds and helicopters: hovering costs, competitive ability, and foraging strategies. American Naturalist 163:16–25.

Alvarez-Buylla, E. R., and A. A. Garay. 1994. Population genetic structure of *Cecropia obtusifolia*, a tropical pioneer tree species. Evolution 48:437–53.

Amico, G. C., R. Vidal-Russell, and D. L. Nickrent. 2007. Phylogenetic relationships and ecological speciation in the mistletoe *Tristerix* (Loranthaceae): the influence of pollinators, dispersers, and hosts. American Journal of Botany 94:558–67.

Andersen, K. 1912. Catalogue of the Chiroptera in the collection of the British Museum. 2nd ed. Trustees of the British Museum (Natural History), London.

Anderson, B., W. W. Cole, and S. C. H. Barrett. 2005. Specialized bird perch aids cross-pollination. Nature 435:41.

Anderson, B., and S. D. Johnson. 2008. The geographical mosaic of coevolution in a plant-pollinator mutualism. Evolution 62:220–25.

Anderson, C. L., and T. Janßen. 2009. Monocots. Pages 203–12 in S. B. Hedges and S. Kumar, eds. The timetree of life. Oxford University Press, Oxford.

Anderson, E. F. 2001. The cactus family. Timber Press, Portland, OR.

Anderson, J. T., J. S. Rojas, and A. S. Flecker. 2009. High-quality seed dispersal by fruit-eating fishes in Amazonian floodplain habitats. Oecologia 161:279–90.

Anderson, S. H., D. Kelly, J. J. Ladley, S. Molloy, and J. Terry. 2011. Cascading effects of bird functional extinction reduce pollination and plant density. Science 331:1068–71.

Andriafidison, D., R. A. Andrianaivoarivelo, O. R. Ramilijaona, M. R. Razanahoera, J. MacKinnon, R. K. B. Jenkins, and P. A. Racey. 2006. Nectarivory by endemic Malagasy fruit bats during the dry season. Biotropica 38:85–90.

Anthony, N. M., M. Johnson-Bawe, K. Jeffery, S. L. Clifford, K. A. Abernethy, C. E. G. Tutin, S. A. Lahm, et al. 2007. The role of Pleistocene refugia and rivers in shaping gorilla genetic diversity in central Africa. Proceedings of the National Academy of Sciences of the United States of America 104:20432–36.

Antonelli, A., and I. Sanmartin. 2011. Why are there so many plant species in the Neotropics? Taxon 60:403–14.

Aparicio, A., R. G. Albaladejo, M. A. Olalla-Tarraga, L. F. Carrillo, and M. A. Rodriguez. 2008. Dispersal potentials determine responses of woody plant species richness to environmental factors in fragmented Mediterranean landscapes. Forest Ecology and Management 255:2894–906.

APG III. 2009. An update of the Angiosperm Phylogeny Group classification for the orders and families of flowering plants: APG III. Botanical Journal of the Linnean Society 161:105–21.

Arakaki, M., P.-A. Christin, R. Nyffeler, A. Lendel, U. Eggli, R. M. Ogburn, E. Spriggs, M. J. Moore, and E. J. Edwards. 2011. Contemporaneous and recent radiations of the world's major succulent plant lineages. Proceedings of the National Academy of Sciences of the United States of America 108:8379–84.

Arizmendi, M. D., A. Valiente-Banuet, A. Rojas-Martinez, and P. Davila-Aranda. 2002. Columnar cacti and the diets of nectar-feeding bats. Pages 264–82 in T. H. Fleming and A. Valiente-Banuet, eds. Columnar cacti and their mutualists: ecology, evolution, and conservation. University of Arizona Press, Tucson.

Armbruster, W. S., Y.-B. Gong, and S.-Q. Huang. 2011. Are pollination "syndromes" predictive? Asian *Dalechampia* fit Neotropical models. American Naturalist 178:135–43.

Armbruster, W. S., and N. Muchhala. 2009. Associations between floral specialization and species diversity: cause, effect, or correlation? Evolutionary Ecology 23:159–79.

Armstrong, J. A. 1979. Biotic pollination mechanisms in the Australian flora—a review. New Zealand Journal of Botany 17:467–508.

Arroyo-Rodríguez, V., and P. A. D. Dias. 2010. Effects of habitat fragmentation and disturbance on howler monkeys: a review. American Journal of Primatology 72:1–16.

Ascorra, C. F., S. Solari, and D. E. Wilson. 1996. Diversidad y ecología de los quirópteros en Pakitza. Pages 593–611 in D. E. Wilson and A. Sandoval, eds. Manú: The biodiversity of southeastern Peru. Smithsonian Institution, Washington DC.

Ashman, T. L., T. M. Knight, J. A. Steets, P. Amarasekare, M. Burd, et al. 2004. Pollen limitation of plant reproduction: ecological and evolutionary causes and consequences. Ecology 85:2408–21.

Ashman, T. L., and C. Majetic. 2006. Genetic constraints on floral evolution: a review and evaluation of patterns. Heredity 96:343–52.

Ashton, P. S. 1998. Niche specificity among tropical trees: a question of scales. Pages 491–514 in D. M. Newbery, H. H. T. Prins, and N. D. Brown, eds. Dynamics of tropical communities. Blackwell Scientific Ltd., Oxford.

Ashton, P. S., et al. 2004. Floristics and vegetation of the forest dynamics plots. Pages 90–106 in E. Losos and E. G. Leigh Jr., eds. Tropical forest diversity and dynamism. University of Chicago Press, Chicago.

Astaras, C., and M. Waltert. 2010. What does seed handling by the drill tell us about the ecological services of terrestrial cercopithecines in African forests? Animal Conservation 13:568–78.

Atkinson, C. T., and D. A. LaPointe. 2009. Introduced avian diseases, climate change, and the future of Hawaiian honeycreepers. Journal of Avian Medicine and Surgery 23:53–63.

Audet, D., and D. W. Thomas. 1997. Facultative hypothermia as a thermoregulatory strategy in the phyllostomid bats, *Carollia perspicillata* and *Sturnira lilium*. Journal of Comparative Physiology B: Biochemical Systemic and Environmental Physiology 167:146–52.

August, P. V. 1981. Fig consumption and seed dispersal by *Artibeus jamaicensis* in the llanos of Venezuela. Biotropica 13:70–76.

Ayala-Berdon, J., J. E. Schondube, K. E. Stoner, N. Rodriguez-Pena, and C. Martínez del Rio. 2008. The intake responses of three species of leaf-nosed Neotropical bats. Journal of Comparative Physiology B: Biochemical Systemic and Environmental Physiology 178:477–85.

Azuma, H., M. Toyota, Y. Asakawa, T. Takaso, and H. Tobe. 2002. Floral scent chemistry of mangrove plants. Journal of Plant Research 115:47–53.

Badyaev, A. V., and G. E. Hill. 2003. Avian sexual dichromatism in relation to phylogeny and ecology. Annual Review of Ecology Evolution and Systematics 34:27–49.

Bagchi, R., P. A. Henrys, P. E. Brown, D. F. R. P. Burslem, P. J. Diggle, C. V. Savitri Gunatilleke, I. A. U. Nimal Gunatilleke, et al. 2011. Spatial patterns reveal negative density dependence and habitat associations in tropical trees. Ecology 92:1723–29.

Baillie, J. E. M., C. Hilton-Taylor, and S. N. Stuart, eds. 2004. 2004 IUCN Red List of threatened species: a global species assessment. IUCN Publications Service Unit, Cambridge.

Baker, H. G. 1959. Reproductive methods as factors in speciation in flowering plants. Cold Spring Harbor Symposium on Quantitative Biology 24:177–99.

———. 1961. The adaptation of flowering plants to nocturnal and crepuscular pollinators. Quarterly Review of Biology 36:64–73.

———. 1978. Chemical aspects of the pollination of woody plants in the tropics. Pages 57–82 in P. B. Tomlinson and M. Zimmerman, eds. Tropical trees as living systems. Cambridge University Press, Cambridge.

Baker, H. G., and I. Baker. 1982. Chemical constituents of nectar in relation to pollination mechanisms and phylogeny. Pages 131–71 in M. H. Niteki, ed. Biochemical aspects of evolutionary biology. University of Chicago Press, Chicago.

———. 1983. Floral nectar constituents in relation to pollinator type. Pages 117–41 in C. E. Jones and R. J. Little, eds. Handbook of experimental pollination biology. Van Nostrand Reinhold, New York.

Baker, H. G., I. Baker, and S. A. Hodges. 1998. Sugar composition of nectars and fruits consumed by birds and bats in the tropics and subtropics. Biotropica 30:559–86.

Baker, H. G., and B. J. Harris. 1957. The pollination of *Parkia* by bats and its attendant evolutionary problems. Evolution 11:449–60.

Baker, R. J., O. R. P. Bininda-Emonds, H. Mantilla-Meluk, C. A. Porter, and R. A. Van den Bussche. 2012. Molecular timescale of diversification of feeding strategy and morphology in New World leaf-nosed bats (Phyllostomidae): a phylogenetic perspective. Pages 385–409 in G. F. Gunnell and N. B. Simmons, eds. Evolutionary history of bats: fossils, molecules and morphology. Cambridge University Press, Cambridge.

Baker, R. J., S. R. Hoofer, C. A. Porter, and R. A. Van den Bussche. 2003. Diversification among New World leaf-nosed bats: an evolutionary hypothesis and classification inferred from digenomic congruence of DNA sequences. Occasional Papers, Museum of Texas Tech University 230:1–32.

Bakken, B. H., and P. Sabat. 2007. The mechanisms and ecology of water balance in hummingbirds. Pages 501–9 in R. McNeil and I. Lazo, eds. Proceedings of the eighth Neotropical ornitological congress: Maturín, Venezuela, 13–19 May 2007.

Balcomb, S. R., and C. A. Chapman. 2003. Bridging the seed dispersal gap: consequences of seed deposition for seedling recruitment in primate-tree interactions. Ecological Monographs 73:625–42.

Balcomb, S. R., C. A. Chapman, and R. Wrangham. 2000. Relationship between chimpanzee (*Pan troglodytes*) density and large, fleshy-fruit tree density: conservation implications. American Journal of Primatology 51:197–203.

Barabási, A.-L. 2009. Scale-free networks: a decade and beyond. Science 325:412–13.

Barclay, R. M. R., L. E. Barclay, and D. S. Jacobs. 2006. Deliberate insectivory by the fruit bat *Rousettus aegyptiacus*. Acta Chiropterologica 8:549–53.

Barker, F. K., A. Cibois, P. Schikler, J. Feinstein, and J. Cracraft. 2004. Phylogeny and diversification of the largest avian radiation. Proceedings of the National Academy of Sciences of the United States of America 101:1040–45.

Barlow, J., and C. A. Peres. 2006. Effects of single and recurrent wildfires on fruit production and large vertebrate abundance in a central Amazonian forest. Biodiversity and Conservation 15:985–1012.

Barraclough, T. G., P. H. Harvey, and S. Nee. 1995. Sexual selection and taxonomic diversity in passerine birds. Proceedings of the Royal Society of London Series B: Biological Sciences 259:211–15.

Barrett, S. C. H. 1998. The evolution of mating strategies in flowering plants. Trends in Plant Science 3:335–41.

———. 2003. Mating strategies in flowering plants: the outcrossing-selfing paradigm and beyond.

Philosophical Transactions of the Royal Society of London B: Biological Sciences 358:991–1004.

Barrett, S. C. H., L. D. Harder, and A. C. Worley. 1996. The comparative biology of pollination and mating in flowering plants. Philosophical Transactions of the Royal Society of London Series B: Biological Sciences 351:1271–80.

Barton, R. A., A. Purvis, and P. H. Harvey. 1995. Evolutionary radiation of visual and olfactory brain systems in primates, bats, and insectivores. Philosophical Transactions of the Royal Society of London B: Biological Sciences 348:381–92.

Bascompte, J. 2009a. Disentangling the web of life. Science 325:416–19.

———. 2009b. Mutualistic networks. Frontiers in Ecology and the Environment 7:429–36.

Bascompte, J., P. Jordano, and J. M. Olesen. 2006. Asymmetric coevolutionary networks facilitate biodiversity maintenance. Science 312:431–33.

Bastolla, U., M. A. Fortuna, A. Pascual-Garcia, A. Ferrera, B. Luque, and J. Bascompte. 2009. The architecture of mutualistic networks minimizes competition and increases biodiversity. Nature 458:1018–21.

Bateman, A. J. 1948. Intra-sexual selection in *Drosophila*. Heredity 2:349–68.

Baum, D. A., R. L. Small, and J. F. Wendel. 1998. Biogeography and floral evolution of baobabs (*Adansonia*, Bombacaceae) as inferred from multiple data sets. Systematic Biology 47:181–207.

Bawa, K. S. 1990. Plant-pollinator interactions in tropical rain forests. Annual Review of Ecology and Systematics 21:399–422.

———. 1992. Mating systems, genetic differentiation and speciation in tropical rain-forest plants. Biotropica 24:250–55.

Bawa, K. S., and P. A. Opler. 1975. Dioecism in tropical forest trees. Evolution 29:167–79.

Beattie, A. J., and L. Hughes. 2002. Ant-plant interactions. Pages 211–35 in C. M. Herrera and O. Pellmyr, eds. Plant-animal interactions, an evolutionary approach. Blackwell Scientific, Oxford.

Beehler, B. M. 1981. Ecological structuring of forest bird communities in New Guinea. Monographiae Biologicae 42:837–61.

———. 1983. Frugivory and polygamy in birds of paradise. Auk 100:1–12.

———. 1985. Adaptive significance of monogamy in the Trumpet Manucode *Manucodia keraudrenii* (Aves, Paradisaeidae). Ornithological Monographs 37:83–99.

———. 1989. The birds of paradise. Scientific American 261:116–23.

Beehler, B. M., T. K. Pratt, and D. A. Zimmerman. 1986. Birds of New Guinea. Princeton University Press, Princeton, NJ.

Beehler, B. M., and S. G. Pruett-Jones. 1983. Display dispersion and diet of birds of paradise, a comparison of nine species. Behavioural Ecology and Sociobiology 13:229–38.

Bell, C. D., D. E. Soltis, and P. S. Soltis. 2005. The age of the angiosperms: a molecular timescale without a clock. Evolution 59:1245–58.

Bell, H. L. 1982. A bird community of lowland rainforest in New Guinea. 1. Composition and density of the avifauna. Emu 82:24–41.

Benkman, C. W. 1999. The selection mosaic and diversifying coevolution between crossbills and lodgepole pine. American Naturalist 153:S75–S91.

Bennet, A. T. D., and M. Thery. 2007. Avian color vision and coloration: multidisciplinary evolutionary biology. American Naturalist 169:S1–S6.

Bennett, P. M., and P. H. Harvey. 1985. Relative brain size and ecology in birds. Journal of Zoology 207:151–69.

Bennett, P. M., and I. P. F. Owens. 2002. Evolutionary ecology of birds. Oxford University Press, Oxford.

Benton, M. J. 2010. The origins of modern biodiversity on land. Philosophical Transactions of the Royal Society B: Biological Sciences 365:3667–79.

Benzing, D. H. 2000. Bromeliaceae, profile of an adaptive radiation. Cambridge University Press, Cambridge.

Berens, D. G., N. Farwig, G. Schaab, and K. Boehning-Gaese. 2008. Exotic guavas are foci of forest regeneration in Kenyan farmland. Biotropica 40:104–12.

Beresford, P., F. K. Barker, P. G. Ryan, and T. M. Crowe. 2005. African endemics span the tree of songbirds (Passeri): molecular systematics of several evolutionary "enigmas." Proceedings of the Royal Society B: Biological Sciences 272:849–58.

Bernard, E. 2001. Vertical stratification of bat communities in primary forests of Central Amazon, Brazil. Journal of Tropical Ecology 17:115–26.

Bernard, E., A. Albernaz, and W. E. Magnusson. 2001. Bat species composition in three localities in the Amazon Basin. Studies on Neotropical Fauna and Environment 36:177–84.

Bininda-Emonds, O. R. P., M. Cardillo, K. E. Jones, R. D. MacPhee, R. M. D. Beck, R. Grenyer, S. A. Price, R. A. Vos, J. L. Gittleman, and A. Purvis. 2007. The delayed rise of present-day mammals. Nature 446:507–12.

Birkinshaw, C. R., and I. C. Colquhoun. 2003. Lemur food plants. Pages 1207–20 in S. M. Goodman and J. P. Benstead, eds. The natural history of Madagascar. University of Chicago Press, Chicago.

Birt, P., L. S. Hall, and G. C. Smith. 1997. Ecomorphology of the tongues of Australian Megachiroptera (Chiroptera: Pteropodidae). Australian Journal of Zoology 45:369–84.

Bissell, E. K., and P. K. Diggle. 2010. Modular genetic architecture of floral morphology in *Nicotiana*: quantitative genetic and comparative phenotypic approaches to floral integration. Journal of Evolutionary Biology 23:1744–58.

Blake, J. G., and B. A. Loiselle. 2000. Diversity of birds along an elevational gradient in the Cordillera Central, Costa Rica. Auk 117:663–86.

Blake, S., S. L. Deem, E. Mossimbo, F. Maisels, and P. Walsh. 2009. Forest elephants: tree planters of the Congo. Biotropica 41:459–68.

Bleher, B., and K. Böhning-Gaese. 2001. Consequences of frugivore diversity for seed dispersal, seedling establishment and the spatial pattern of seedlings and trees. Oecologia 129:385–94.

Bleiweiss, R. 1997. Covariation of sexual dichromatism and plumage colours in lekking and non-lekking birds: a comparative analysis. Evolutionary Ecology 11:217–35.

———. 1998a. Origin of hummingbird faunas. Biological Journal of the Linnean Society 65:77–97.

———. 1998b. Tempo and mode of hummingbird evolution. Biological Journal of the Linnean Society 65:63–76.

Blois, J. L., and E. A. Hadly. 2009. Mammalian response to Cenozoic climatic change. Annual Review of Earth and Planetary Sciences 37:181–208.

Bobrowiec, P. E. D., and R. Gribel. 2010. Effects of different secondary types on bat community composition in Central Amazonia, Brazil. Animal Conservation 13:204–16.

Bochenski, Z., and Z. M. Bochenski. 2008. An old world hummingbird from the Oligocene: a new fossil from Polish Carpathians. Journal of Ornithology 149:211–16.

Bofarull, A. M., A. A. Royo, M. H. Fernandez, E. Ortiz-Jaureguizar, and J. Morales. 2008. Influence of continental history on the ecological specialization and macroevolutionary processes in the mammalian assemblage of South America: differences between small and large mammals. BMC Evolutionary Biology 8: article 97.

Bolmgren, K., and O. Eriksson. 2005. Fleshy fruits—origins, niche shifts, and diversification. Oikos 109:255–72.

Bonaccorso, F., and B. K. McNab. 1997. Plasticity of energetics in blossom bats (Pteropodidae): impact on distribution. Journal of Mammalogy 78:1073–88.

Bonaccorso, F., J. Winkelmann, and D. Byrnes. 2005. Home range, territoriality, and flight time budgets in the black-bellied fruit bat, *Melonycteris melanops* (Pteropodidae). Journal of Mammalogy 86:931–36.

Bonaccorso, F. J. 1979. Foraging and reproductive ecology in a Panamanian bat community. Bulletin of the Florida State Museum, Biological Science 24:359–408.

———. 1998. Bats of Papua New Guinea. Conservation International, Washington, DC.

Bonaccorso, F. J., and T. J. Gush. 1987. An experimental study of the feeding behaviour and foraging strategies of phyllostomid bats. Journal of Animal Ecology 56:907–20.

Borgella, R., A. A. Snow, and T. A. Gavin. 2001. Species richness and pollen loads of hummingbirds using forest fragments in southern Costa Rica. Biotropica 33:90–109.

Borges, R. M. 1993. Figs, Malabar giant squirrels, and fruit shortages within two tropical Indian forests. Biotropica 25:183–90.

Borges, R. M., J. M. Bessiere, and M. Hossaert-McKey. 2008. The chemical ecology of seed dispersal in monoecious and dioecious figs. Functional Ecology 22:484–93.

Bornemann, A., R. D. Norris, O. Friedrich, B. Beckmann, S. Schouten, J. S. Sinninghe Damste, J. Vogel, P. Hofmann, and T. Wagner. 2008. Isotopic evidence for glaciation during the Cretaceous supergreenhouse. Science 319:189–92.

Bosque, C., and R. Calchi. 2003. Food choice by blue-gray tanagers in relation to protein content. Comparative Biochemistry and Physiology A: Molecular and Integrative Physiology 135:321–27.

Bosque, C., and A. Pacheco. 2000. Dietary nitrogen as a limiting nutrient in frugivorous birds. Revista Chilena de Historia Natural 73:441–50.

Boulter, S. L., R. L. Kitching, B. G. Howlett, and K. Goodall. 2005. Any which way will do—the pollination biology of a northern Australian rainforest canopy tree (*Syzygium sayeri;* Myrtaceae). Botanical Journal of the Linnean Society 149:69–84.

Bowie, R. C. K., J. Fjeldsa, S. J. Hackett, and T. M. Crowe. 2004. Systematics and biogeography of double-collared sunbirds from the Eastern Arc Mountains, Tanzania. Auk 121:660–81.

Boyle, W. A., and C. J. Conway. 2007. Why migrate? a test of the evolutionary precursor hypothesis. American Naturalist 169:344–59.

Boyles, J. G., P. M. Cryan, G. F. McCracken, and T. H. Kunz. 2011. Economic importance of bats in agriculture. Science 332:41–42.

Bradbury, J. W. 1977. Lek mating behavior in the hammer-headed bat. Zeitschrift fur Tierpsychologie 45:225–55.

———. 1981. The evolution of leks. Pages 138–73 in R. D. Alexander and D. W. Tinkle, eds. Natural selection and social behavior. Chiron Press, New York.

Bradshaw, H. D., and D. W. Schemske. 2003. Allele substitution at a flower colour locus produces a pollinator shift in monkeyflowers. Nature 426:176–78.

Brearley, F. Q., J. Proctor, Suriantata, L. Nagy, G. Dalrymple, and B. C. Voysey. 2007. Reproductive phenology over a 10-year period in a lowland evergreen rain forest of central Borneo. Journal of Ecology 95:828–39.

Bremer, B., and O. Eriksson. 1992. Evolution of fruit characteristics and dispersal modes in the tropical family Rubiaceae. Biological Journal of the Linnean Society 47:79–95.

Brice, A. T., K. H. Dahl, and C. R. Grau. 1989. Pollen digestibility by hummingbirds and psittacines. Condor 91:681–88.

Brice, A. T., and C. R. Grau. 1991. Protein requirements of Costa's hummingbirds *Calypte costae*. Physiological Zoology 64:611–26.

Brodie, E. D., B. J. Ridenhour, and E. D. I. Brodie. 2002. The evolutionary response of predators to dangerous prey: hotspots and coldspots in the geographic mosaic of coevolution between garter snakes and newts. Evolution 56:2067–82.

Brodie, J. F., O. E. Helmy, W. Y. Brockelman, and J. L. Maron. 2009. Bushmeat poaching reduces the seed dispersal and population growth rate of a mammal-dispersed tree. Ecological Applications 19:854–63.

Bronstein, J. L. 2001. The exploitation of mutualisms. Ecology Letters 4:277–87.

————. 2009. The evolution of facilitation and mutualism. Journal of Ecology 97:1160–70.

Bronstein, J. L., W. G. Wilson, and W. E. Morris. 2001. Ecological dynamics of mutualist/antagonist communities. American Naturalist 162:S24–S39.

Brosset, A. 1978. Liste des vertebres du Bassin de L'Ivindo (Republique Gabonaise)—poissons exceptes. Unpublished manuscript.

Brown, B. J., and R. J. Mitchell. 2001. Competition for pollination: effects of pollen of an invasive plant on seed set of a native congener. Oecologia 129:43–49.

Brown, D. R., and T. W. Sherry. 2006. Behavioral response of resident Jamaican birds to dry season food supplementation. Biotropica 38:91–99.

Brown, E. D., and M. J. G. Hopkins. 1996. How New Guinea rainforest flower resources vary in time and space: implications for nectarivorous birds. Australian Journal of Ecology 21:363–78.

Brown, J. H., W. A. Calder III, and A. Kodric-Brown. 1978. Correlates and consequences of body size in nectar-feeding birds. American Zoologist 18:687–700.

Brown, J. W., R. B. Payne, and D. P. Mindell. 2007. Nuclear DNA does not reconcile "rocks" and "clocks" in Neoaves: a comment on Ericson et al. Biology Letters 3:257–60.

Brugiere, D., J. P. Gautier, A. Moungazi, and A. Gautier-Hion. 2002. Primate diet and biomass in relation to vegetation composition and fruiting phenology in a rain forest in Gabon. International Journal of Primatology 23:999–1024.

Bruna, E. M. 2003. Are plant populations in fragmented habitats recruitment limited? tests with an Amazonian herb. Ecology 84:932–47.

Bruneau, A. 1996. Phylogenetic and biogeographical patterns in *Erythrina* (Leguminosae: Phaseoleae) as inferred from morphological and chloroplast DNA characters. Systematic Botany 21:587–605.

————. 1997. Evolution and homology of bird pollination syndromes in *Erythrina* (Leguminosae). American Journal of Botany 84:54–71.

Buckley, Y. M., S. Anderson, C. P. Catterall, R. T. Corlett, T. Engel, C. R. Gosper, R. Nathan, et al. 2006. Management of plant invasions mediated by frugivore interactions. Journal of Applied Ecology 43:848–57.

Burd, M. 1994. Bateman principle and plant reproduction—the role of pollen limitation in fruit and seed set. Botanical Review 60:83–139.

Burdon, J. J., and P. H. Thrall. 1999. Spatial and temporal patterns in coevolving plant and pathogen associations. American Naturalist 153:S15–S33.

Burger, W. C. 1981. Why are there so many kinds of flowering plants? Bioscience 31:572–81.

Burgess, N. D., and C. O. F. Mlingwa. 2000. Evidence for altitudinal migration of forest birds between montane Eastern Arc and lowland forests in East Africa. Ostrich 71:184–90.

Burkle, L. A., and R. Alarcón. 2011. The future of plant-pollinator diversity: understanding interaction networks across time, space, and global change. American Journal of Botany 98: 528–38.

Burnham, R. J., and K. R. Johnson. 2004. South American palaeobotany and the origins of Neotropical rainforests. Philosophical Transactions of the Royal Society of London Series B: Biological Sciences 359:1595–1610.

Burns, K. C., E. Cazetta, M. Galetti, A. Valido, and H. M. Schaefer. 2009. Geographic patterns in fruit colour diversity: do leaves constrain the colour of fleshy fruits? Oecologia 159:337–43.

Burns, K. C., and B. Lake. 2009. Fruit-frugivore interactions in two southern hemisphere forests: allometry, phylogeny and body size. Oikos 118:1901–7.

Burns, K. J., and K. Naoki. 2004. Molecular phylogenetics and biogeography of Neotropical tanagers in the genus *Tangara*. Molecular Phylogenetics and Evolution 32:838–54.

Buzato, S., M. Sazima, and I. Sazima. 2000. Hummingbird-pollinated floras at three Atlantic forest sites. Biotropica 32:824–41.

Byrne, M. M., and D. J. Levey. 1993. Removal of seeds from frugivore defecations by ants in a

Costa Rican rain forest. Pages 363–74 in T. H. Fleming and A. Estrada, eds. Frugivory and seed dispersal: ecological and evolutionary aspects. Kluwer Academic Publishers, Dordrecht.

Campbell, C. J., A. Fuentes, K. C. MacKinnon, M. Panger, and S. K. Bearder, eds. 2007. Primates in perspective. Oxford University Press, New York.

Campbell, C., S. Yang, R. Albert, and K. Shea. 2011. A network model for plant-pollinator community assembly. Proceedings of the National Academy of Sciences of the United States of America 108:197–202.

Carlo, T. A., J. A. Collazo, and M. J. Groom. 2004. Influences of fruit diversity and abundance on bird use of two shaded coffee plantations. Biotropica 36:602–14.

Carpenter, F. L., and R. E. Macmillen. 1976. Threshold model of feeding territoriality and test with a Hawaiian honeycreeper. Science 194:639–42.

Carpenter, F. L., D. C. Paton, and M. A. Hixon. 1983. Weight gain and adjustment of feeding territory size in migrant hummingbirds. Proceedings of the National Academy of Sciences of the United States of America 80:7259–63.

Carriere, S. M., M. Andre, P. Letourmy, I. Olivier, and D. B. McKey. 2002. Seed rain beneath remnant trees in a slash-and-burn agricultural system in southern Cameroon. Journal of Tropical Ecology 18:353–74.

Carson, W. P., J. T. Anderson, J. Leigh, E. G., and S. A. Schnitzer. 2008. Challenges associated with testing and falsifying the Janzen-Connell hypothesis: a review and critique. Pages 210–41 in W. P. Carson and S. A. Schnitzer, eds. Tropical forest community ecology. Wiley-Blackwell, Oxford.

Carthew, S. M., and R. L. Goldingay. 1997. Non-flying mammals as pollinators. Trends in Ecology and Evolution 12:104–8.

Castro-Luna, A. A., V. J. Sosa, and G. Castillo-Campos. 2007. Bat diversity and abundance associated with the degree of secondary succession in a tropical forest mosaic in southeastern Mexico. Animal Conservation 10:219–28.

Caswell, H. 1989. Matrix population models: construction, analysis, and interpretation. Sinauer Associates, Sunderland, MA.

Cavender-Bares, J., K. H. Kozak, P. V. A. Fine, and S. W. Kembel. 2009. The merging of community ecology and phylogenetic biology. Ecology Letters 12:693–715.

Cazetta, E., H. M. Schaefer, and M. Galetti. 2009. Why are fruits colorful? the relative importance of achromatic and chromatic contrasts for detection by birds. Evolutionary Ecology 23:233–44.

Chan, K. 2001. Partial migration in Australian landbirds: a review. Emu 101:281–92.

Chapman, C. A. 1989. Primate seed dispersal: the fate of dispersed seeds. Biotropica 21:148–54.

Chapman, C. A., and L. J. Chapman. 1996. Frugivory and the fate of dispersed and non-dispersed seeds of six African tree species. Journal of Tropical Ecology 12:491–504.

Chapman, C. A., L. J. Chapman, M. Cords, J. M. Gathua, A. Gautier-Hion, J. E. Lambert, K. Rode, C. E. G. Tutin, and L. J. T. White. 2002. Variation in the diets of *Cercopithecus* species: differences within forests, among forests, and across species. Pages 325–50 in M. E. Glenn and M. Cords, eds. The guenons: diversity and adaptation in African monkeys. Kluwer Academic, New York.

Chapman, C. A., and D. A. Onderdonk. 1998. Forests without primates: Primate/plant codependency. American Journal of Primatology 45:127–41.

Chapman, C. A., and S. E. Russo. 2007. Primate seed dispersal: linking behavioral ecology with forest community. Pages 510–25 in C. J. Campbell and others, eds. Primates in perspective. Oxford University Press, New York.

Chapman, C. A., R. W. Wrangham, L. J. Chapman, D. K. Kennard, and A. E. Zanne. 1999. Fruit and flower phenology at two sites in Kibale National Park, Uganda. Journal of Tropical Ecology 15:189–211.

Charles-Dominique, P. 1986. Inter-relations between frugivorous vertebrates and pioneer plants: *Cecropia*, birds, and bats in French Guyana. Pages 119–35 in A. Estrada and T. H. Fleming, eds. Frugivores and seed dispersal. Dr. W. Junk, Dordrecht.

———. 1993. Speciation and coevolution: an interpretation of frugivory phenomena. Pages 75–84 in T. H. Fleming and A. Estrada, eds. Frugivory and sed dispersal: ecological and evolutionary aspects. Kluwer Academic Publishers, Dordrecht.

Charlesworth, D. 1999. Theories of the evolution of dioecy. Pages 33–60 in M. A. Geber, T. E. Dawson, and L. F. Delph, eds. Gender and sexual dimorphism in flowering plants. Springer-Verlag, Berlin.

Charlesworth, D., and B. Charlesworth. 1987. Inbreeding depression and its evolutionary consequences. Annual Review of Ecology and Systematics 18:237–68.

Chave, J. 2008. Spatial variation in tree species composition across tropical forests: pattern and process. Pages 11–30 in W. P. Carson and S. A. Schnitzer, eds. Tropical forest community ecology. Wiley-Blackwell, Oxford.

Chaves, J. A., J. P. Pollinger, T. B. Smith, and G. Lebuhn. 2006. The role of geography and ecology in shaping the phylogeography of the speckled hummingbird (*Adelomyia melanogenys*) in Ecuador. Molecular Phylogenetics and Evolution 43:795–807.

Chaves-Campos, J. 2004. Elevational movements of large frugivorous birds and temporal variation in abundance of fruits along an elevational gradient. Ornitologia Neotropical 15:433–45.

Chaves-Campos, J., J. E. Arevalo, and M. Araya. 2003. Altitudinal movements and conservation of bare-necked Umbrellabird *Cephalopterus glabricollis* of the Tilaran Mountains, Costa Rica. Bird Conservation International 13:45–58.

Chazdon, R. L., S. Careaga, C. Webb, and O. Vargas. 2003. Community and phylogenetic structure of reproductive traits of woody species in wet tropical forests. Ecological Monographs 73:331–48.

Cheke, R. A., and C. F. Mann. 2001. Sunbirds, a guide to the sunbirds, flowerpeckers, spiderhunters, and sugarbirds of the world. Yale University Press, New Haven, CT.

Chen, X.-S., and Q.-L. Li. 2008. Sexual systems and ecological correlates in an azonal tropical forest, SW China. Biotropica 40:160–67.

Cheney, D. L. 1986. Interactions and relationships between groups. Pages 267–81 in B. B. Smuts, D. L. Cheney, R. M. Seyfarth, R. W. Wrangham, and T. T. Struhsaker, eds. Primate societies. University of Chicago Press, Chicago,.

Cheptou, P.-O., and F. Massoi. 2009. Pollination fluctuations drive evolutionary syndromes linking dispersal and mating system. American Naturalist 174:46–55.

Chittka, L., A. Shmida, N. Troje, and R. Menzel. 1994. Ultraviolet as a component of flower reflections, and the colour perception of Hymenoptera. Vision Research 34:1489–1508.

Chittka, L., and J. D. Thomson. 2001. Cognitive ecology of pollination: animal behavior and floral evolution. Cambridge University Press, New York.

Chivers, D. J., ed. 1980. Malayan forest primates. Plenum Press, New York.

Chivers, D. J., and C. M. Hladik. 1980. Morphology of the gastrointestinal tract in primates: comparison with other mammals in relation to diet. Journal of Morphology 166:337–86.

Chust, G., J. Chave, R. Condit, S. Aguilar, S. Lao, and R. Perez. 2006. Determinants and spatial modeling of tree beta-diversity in a tropical forest landscape in Panama. Journal of Vegetation Science 17:83–92.

Cipollini, M. L. 2000. Secondary metabolites of vertebrate-dispersed fruits: evidence for adaptive functions. Revista Chilena De Historia Natural 73:421–40.

Cipollini, M. L., and D. J. Levey. 1997. Why are some fruits toxic? glycoalkaloids in *Solanum* and fruit choice by vertebrates. Ecology 78:782–98.

Clark, C. J., J. R. Poulsen, B. M. Bolker, E. F. Connor, and V. T. Parker. 2005. Comparative seed shadows of bird-, monkey-, and wind-dispersed trees. Ecology 86:2684–94.

Clark, C. J., J. R. Poulsen, D. J. Levey, and C. W. Osenberg. 2007. Are plant populations seed limited? a critique and meta-analysis of seed addition experiments. American Naturalist 170:128–42.

Clark, C. J., J. R. Poulsen, and V. T. Parker. 2001. The role of arboreal seed dispersal groups on the seed rain of a lowland tropical forest. Biotropica 33:606–20.

Clark, D. B., and D. A. Clark. 1991. The impact of physical damage on canopy tree regeneration in tropical rain forest. Journal of Ecology 79:447–57.

Clark, D. B., M. W. Palmer, and D. A. Clark. 1999. Edaphic factors and the landscape scale distributions of tropical rain forest trees. Ecology 80:2662–75.

Clark, J. S., C. Fastie, G. Hurtt, S. T. Jackson, C. Johnson, G. A. King, M. Lewis, et al. 1998. Dispersal theory and interpretation of paleoecological records. Bioscience 48:13–24.

Classen, R. 1987. Morphological adaptations for bird pollination in *Nicolaia elatior* (Jack) Horan (Zingiberaceae). Garden Bulletin of Singapore 40:37–43.

Clausing, G., K. Meyer, and S. S. Renner. 2000. Correlations among fruit traits and evolution of different fruits within the Melastomataceae. Botanical Journal of the Linnean Society 133:303–26.

Clausing, G., and S. S. Renner. 2001. Molecular phylogenetics of Melastomataceae and Memecylaceae: implications for character evolution. American Journal of Botany 88:486–98.

Cleary, D. F. R., and A. Priadjati. 2005. Vegetation responses to burning in a rain forest in Borneo. Plant Ecology 177:145–63.

Clutton-Brock, T. H. 1989. Mammalian mating systems. Proceedings of the Royal Society B: Biological Sciences 236:338–72.

Cocucci, A. A., and A. N. Sersic. 1998. Evidence of rodent pollination in *Cajophora coronata* (Loasaceae). Plant Systematics and Evolution 211:113–28.

Coe, M. J., D. L. Dilcher, J. G. Farlow, D. M. Jarzen, and D. A. Russell. 1987. Dinosaurs and land plants. Pages 225–58 in E. M. Friis, W. G. Chaloner, and P. R. Crane, eds. The origins of angiosperms and their biological consequences. Cambridge University Press, Cambridge.

Cohen, D., and A. Shmida. 1993. The evolution of flower display and reward. Evolutionary Biology 27:197–243.

Colinvaux, P. A., G. Irion, M. E. Rasanen, M. B. Bush, and J. de Mello. 2001. A paradigm to be discarded: geological and paleoecological data falsify the Haffer & Prance refuge hypothesis of Amazonian speciation. Amazoniana-Limnologia et Oecologia Regionalis Systemae Fluminis Amazonas 16:609–46.

Collevatti, R. G., M. E. C. Amaral, and F. S. Lopes. 1997. Role of pollinators in seed set and a test of pollen limitation hypothesis in the tropical weed *Triumfetta semitriloba* (Tiliaceae). Revista de Biologia Tropical 45:1401–7.

Collevatti, R. G., D. Grattapaglia, and J. D. Hay. 2001. Population genetic structure of the endangered tropical tree species *Caryocar brasiliense*, based on variability at microsatellite loci. Molecular Ecology 10:349–56.

Collins, B. G. 2008. Nectar intake and foraging efficiency: responses of honeyeaters and hummingbirds to variations in floral environments. Auk 125:574–87.

Collins, B. G., and D. C. Paton. 1989. Consequences of differences in body mass, wing length, and leg morphology for nectar-feeding birds. Australian Journal of Ecology 14:269–89.

Condit, R. 1995. Research in large, long-term tropical forest plots. Trends in Ecology and Evolution 10:18–22.

Condit, R., S. P. Hubbell, J. V. Lafrankie, R. Sukumar, N. Manokaran, R. B. Foster, and P. S. Ashton. 1996. Species-area and species-individual relationships for tropical trees: a comparison of three 50-ha plots. Journal of Ecology 84:549–62.

Condit, R., N. Pitman, E. G. Leigh, J. Chave, J. Terborgh, R. B. Foster, P. Nunez, et al. 2002. Beta-diversity in tropical forest trees. Science 295:666–69.

Connell, J. H. 1971. On the role of natural enemies in preventing competitive exclusion in some marine animals and in rain forest trees. Pages 298–310 in P. J. Boer and G. Gradwell, eds. Dynamics of populations. PUDOC, Wageningen, The Netherlands.

Connell, J. H., and P. T. Green. 2000. Seedling dynamics over thirty-two years in a tropical rain forest tree. Ecology 81:568–84.

Connell, J. H., and R. O. Slayter. 1977. Mechanisms of succession in natural communities and their role in community stability and organization. American Naturalist 111:1119–44.

Coombe, B. 1976. The development of fleshy fruits. Annual Review of Plant Physiology 27:507–28.

Cordeiro, N. J., and H. F. Howe. 2001. Low recruitment of trees dispersed by animals in African forest fragments. Conservation Biology 15:1733–41.

Cordeiro, N. J., D. A. G. Patrick, B. Munisi, and V. Gupta. 2004. Role of dispersal in the invasion of an exotic tree in an East African submontane forest. Journal of Tropical Ecology 20:449–57.

Cordell, G. A., and O. E. Araujo. 1993. Capsaicin: identification, nomenclature, and pharmacotherapy. Annals of Pharmacotherapy 27:330–36.

Corlett, R. 2004. Flower visitors and pollination in the Oriental (Indomalayan) region. Biological Reviews 79:497–532.

———. 1998. Frugivory and seed dispersal by vertebrates in the Oriental (Indomalayan) region. Biological Reviews 73:413–48.

———. 2002. Frugivory and seed dispersal in degraded tropical East Asian landscapes. Pages 451–65 in D. J. Levey, W. R. Silva, and M. Galetti, eds. Seed dispersal and frugivory: ecology, evolution, and conservation. CABI, New York.

———. 2007a. The impact of hunting on the mammalian fauna of tropical Asian forests. Biotropica 39:292–303.

———. 2007b. What's so special about Asian tropical forests? Current Science 93:1551–57.

———. 2009a. The ecology of tropical East Asia. Oxford University Press, Oxford.

———. 2009b. Seed dispersal distances and plant migration potential in tropical East Asia. Biotropica 41:592–98.

———. 2011. How to be a frugivore (in a changing world). Acta Oecologica 37:674–81.

Corlett, R. T., and J. V. LaFrankie. 1998. Potential impacts of climate change on tropical Asian forests through an influence on phenology. Climatic Change 39:439–53.

Corlett, R. T., and R. B. Primack. 2011. Tropical rain forests: an ecological and biogeographic comparison. 2nd ed. Wilely-Blackwell, New York.

Correa, S. B., K. O. Winemiller, H. Lopez-Fernandez, and M. Galetti. 2007. Evolutionary perspectives on seed consumption and dispersal by fishes. Bioscience 57:748–56.

Correll, D. S., and H. B. Correll. 1982. Flora of the Bahama Archipelago. A.R.G. Gantner Verlag, Vaduz, Lichtenstein.

Cortes-Ortiz, L., E. Bermingham, C. Rico, E. Rodriguez-Luna, I. Sampaio, and M. Ruiz-Garcia. 2003. Molecular systematics and biogeography of the Neotropical monkey genus, Alouatta. Molecular Phylogenetics and Evolution 26:64–81.

Courts, S. E. 1998. Dietary strategies of Old World Fruit Bats (Megachiroptera, Pteropodidae): how do they obtain sufficient protein? Mammal Review 28:185–93.

Cox, P., T. Elmqvist, E. Pierson, and W. Rainey. 1991. Flying foxes as strong interactors in South Pacific island ecosystems: a conservation hypothesis. Conservation Biology 5:448–54.

Cox, P. A., and T. Elmqvist. 2000. Pollinator extinction in the Pacific islands. Conservation Biology 14:1237–39.

Coyne, J. A., and H. A. Orr. 2004. Speciation. Sinauer Associates, Inc., Sunderland, MA.

Cracraft, J. 2001. Avian evolution, Gondwana biogeography and the Cretaceous-Tertiary mass extinction event. Proceedings of the Royal Society B: Biological Sciences 268:459–69.

Cracraft, J., and F. K. Barker. 2009. Passerine birds (Passeriformes). Pages 423–31 in S. B. Hedges and S. Kumar, eds. The timetree of life. Oxford University Press, Oxford.

Cracraft, J., F. K. Barker, M. J. Braun, J. Harshman, G. J. Dyke, and others. 2004. Phylogenetic relationships among modern birds (Neornithes). Pages 468–89 in J. Cracraft and M. J. Donoghue, eds. Assembling the tree of life. Oxford University Press, Oxford.

Cracraft, J., F. K. Barker, and A. Cibois. 2003. Avian higher-level phylogenetics and the Howard and Moore checklist of birds. Pages 16–21 in E. C. Dickinson, ed. The Howard and Moore complete checklist of the birds of the world. Princeton University Press, Princeton, NJ.

Cramer, J. M., R. Mesquita, and G. B. Williamson. 2007. Forest fragmentation differentially affects seed dispersal of large and small-seeded tropical trees. Biological Conservation 137:415–23.

Crepet, W. L., and E. M. Friis. 1987. The evolution of insect pollination in angiosperms. Pages 181–201 in E. M. Friis, W. G. Chaloner, and P. R. Crane, eds. The origins of angiosperms and their biological consequences. Cambridge University Press, Cambridge.

Crepet, W. L., K. C. Nixon, and M. A. Gandolfo. 2004. Fossil evidence and phylogeny: the age of major angiosperm clades based on mesofossil and macrofossil evidence from cretaceous deposits. American Journal of Botany 91:1666–82.

Cresswell, J. E. 2003. Towards the theory of pollinator-mediated gene flow. Philosophical Transactions of the Royal Society of London Series B: Biological Sciences 358:1005–8.

Crisp, M. D., M. T. K. Arroyo, L. G. Cook, M. A. Gandolfo, G. J. Jordan, M. S. McGlone, P. H. Weston, M. Westoby, P. Wilf, and H. P. Linder. 2009. Phylogenetic biome conservatism on a global scale. Nature 458:754–58.

Cristoffer, C. 1987. Body size differences between New World and Old World, arboreal, tropical vertebrates: causes and consequences. Journal of Biogeography 14:165–72.

Cristoffer, C., and C. A. Peres. 2003. Elephants versus butterflies: the ecological role of large herbivores in the evolutionary history of two tropical worlds. Journal of Biogeography 30:1357–80.

Croat, T. B. 1978. Flora of Barro Colorado Island. Stanford University Press, Stanford, CA.

Crome, F. H. J. 1975. The ecology of fruit pigeons in tropical northern Queensland. Australian Wildlife Research 2:155–85.

Crome, F. H. J., and A. K. Irvine. 1986. "Two bob each way": the pollination and breeding system of the Australian rain forest tree *Syzygium cormiflorum*. Biotropica 18:115–25.

Cronk, Q., and I. Ojeda. 2008. Bird-pollinated flowers in an evolutionary and molecular context. Journal of Experimental Botany 59:715–27.

Cruden, R. W. 1977. Pollen-ovule ratios: a conservative indicator of breeding systems in flowering plants. Evolution 31:32–46.

———. 2000. Pollen grains: why so many? Plant Systematics and Evolution 222:143–65.

———. 2009. Pollen grain size, stigma depth, and style length: the relationship revisited. Plant Systematics and Evolution 278:223–38.

Cruden, R. W., S. M. Hermann, and S. Peterson. 1983. Patterns of nectar production and plant-pollinator coevolution. Pages 80–125 in B. Bentley and T. Elias, eds. The biology of nectaries. Columbia University Press, New York.

Cunningham, S. A. 1995. Ecological constraints on fruit initiation by *Calyptrogyne ghiesbreghtiana* (Arecaceae): floral herbivory, pollen availability, and visitation by pollinating bats. American Journal of Botany 82:1527–36.

———. 1996. Pollen supply limits fruit initiation by a rain forest understorey palm. Journal of Ecology 84:185–94.

Curran, L. M., and M. Leighton. 2000. Vertebrate responses to spatiotemporal variation in seed production of mast-fruiting Dipterocarpaceae. Ecological Monographs 70:101–28.

Dacosta, J. M., and J. Klicka. 2008. The Great American Interchange in birds: a phylogenetic perspective with the genus *Trogon*. Molecular Ecology 17:1328–43.

Daily, G. C. 1997. Introduction: what are ecosystem services? Pages 1–10 in G. C. Daily, ed. Nature's services, societal dependence on natural ecosystems. Island Press, Washington, DC.

Dalgleish, E. 1999. Effectiveness of invertebrate and vertebrate pollinators and the influence of pollen limitation and inflorescence position on follicle production of *Banksia aemula* (Family Proteaceae). Australian Journal of Botany 47:553–62.

Dalling, J. W., H. C. Muller-Landau, S. J. Wright, and S. P. Hubbell. 2002. Role of dispersal in the recruitment limitation of Neotropical pioneer species. Journal of Ecology 90:714–27.

Daniel, M. J. 1976. Feeding by the short-tailed bat (*Mysticina tuberculata*) on fruit and possibly nectar. New Zealand Journal of Zoology 3:391–98.

Dar, S., M. D. Arizmendi, and A. Valiente-Banuet. 2006. Diurnal and nocturnal pollination of *Marginatocereus marginatus* (Pachycereeae: Cactaceae) in Central Mexico. Annals of Botany 97:423–27.

D'Arcy, W. G. 1991. The Solanaceae since 1976, with a review of its biogeography. Pages 75–137 in J. G. Hawkes, R. N. Lester, M. Nee, and N. Estrada, eds. Solanaceae III: taxonomy, chemistry, evolution. Royal Botanic Garden, Kew, Richmond.

Darwin, C. 1884. The different forms of flowers on plants of the same species. 2nd ed. J. Murray, London.

Dasenbrock, J. 2002. The pros and cons of ecotourism in Costa Rica. TED Case Studies 648. American University, Washington , DC

Datzmann, T., O. v, Helversen, and F. Mayer. 2010. Evolution of nectarivory in phyllostomid bats (Phyllostomidae Gray, 1825, Chiroptera: Mammalia). BMC Evolutionary Biology 10, article 165.

Davalos, L. M. 2009. Earth history and the evolution of Caribbean bats. Pages 96–115 in T. H. Fleming and P. A. Racey, eds. Island bats: evolution, ecology, and conservation. University of Chicago Press, Chicago.

Davidar, P., and E. S. Morton. 1986. The relationship between fruit crop sizes and fruit removal rates by birds. Ecology 67:262–65.

Davidar, P., B. Rajaopal, D. Mohandass, J.-P. Puyravaud, R. Condit, S. J. Wright, and E. G. Leigh Jr. 2007. The effect of climatic gradients, topographic variation and species traits on the beta diversity of rain forest trees. Global Ecology and Biogeography 16:510–18.

Davies, S. J., and P. S. Ashton. 1999. Phenology and fecundity in 11 sympatric pioneer species of *Macaranga* (Euphorbiaceae) in Borneo. American Journal of Botany 86:1786–95.

Davies, T. J., T. G. Barraclough, M. W. Chase, P. S. Soltis, D. E. Soltis, and V. Savolainen. 2004. Darwin's abominable mystery: insights from a supertree of the angiosperms. Proceedings of the National Academy of Sciences of the United States of America 101:1904–9.

Davies, T. J., S. A. Fritz, R. Grenyer, C. D. L. Orme, J. Bielby, O. R. P. Bininda-Emonds, M. Cardillo, et al. 2008. Phylogenetic trees and the future of mammalian biodiversity. Proceedings of the National Academy of Sciences of the United States of America 105:11556–63.

Davis, L. 1972. A field guide to the birds of Mexico and Central America. University of Texas Press, Austin.

Davis, N. E., D. J. O'Dowd, R. MacNally, and P. T. Green. 2010. Invasive ants disrupt frugivory by endemic island birds. Biology Letters 6:85–88.

Dawson, T. P., S. T. Jackson, J. I. House, I. C. Prentice, and G. M. Mace. 2011. Beyond predictions: biodiversity conservation in a changing climate. Science 332:53–58.

De Figueiredo, R. A., and E. Perin. 1995. Germination ecology of *Ficus luschnathiana* drupelets after bird and bat ingestion. Acta Oecologica: International Journal of Ecology 16:71–75.

Degen, B., and D. W. Roubik. 2004. Effects of animal pollination on pollen dispersal, selfing, and effective population size of tropical trees: a simulation study. Biotropica 36:165–79.

de Kloet, R. S., and S. R. de Kloet. 2005. The evolution of the spindlin gene in birds: sequence analysis of an intron of the spindlin W and Z genes reveals four major divisions of the Psittaciformes. Molecular Phylogenetics and Evolution 36:706–21.

de los Monteros, A. E. 2000. Higher-level phylogeny of Trogoniformes. Molecular Phylogenetics and Evolution 14:20–34.

de Queiroz, A. 2002. Contingent predictability in evolution: key traits and diversification. Systematic Biology 51:917–29.

Delacour, J. T. 1947. Birds of Malaysia. MacMillan Co., New York.

Delacour, J. T., and E. Mayr. 1946. Birds of the Philippines. Macmillan Co., New York.

Delorme, M., and D. W. Thomas. 1996. Nitrogen and energy requirements of the short tailed fruit bat (*Carollia perspicillata*): fruit bats are not nitrogen constrained. Journal of Comparative Physiology B: Biochemical Systemic and Environmental Physiology 166:427–34.

———. 1999. Comparative analysis of the digestive efficiency and nitrogen and energy requirements of the phyllostomid fruit-bat (*Artibeus jamaicensis*) and the pteropodid fruit-bat (*Rousettus aegyptiacus*). Journal of Comparative Physiology B: Biochemical Systemic and Environmental Physiology 169:123–32.

Desalegn, W., and C. Beierkuhnlein. 2010. Plant species and growth form richness along altitudinal gradients in the southwest Ethiopian highlands. Journal of Vegetation Science 21:617–26.

DeSwardt, D. H. 1993. Factors affecting the densities of nectarivores in Protea-Roupelliae woodland. Ostrich 64:172–77.

Devy, M. S., and P. Davidar. 2003. Pollination systems of trees in Kakaachi, a mid-elevation wet evergreen forest in Westen Ghats, India. American Journal of Botany 90:650–57.

DeWalt, S. J., S. K. Maliakal, and J. S. Denslow. 2003. Changes in vegetation structure and composition along a tropical forest chronosequence: implications for wildlife. Forest Ecology and Management 182:139–51.

DeWalt, S. J., S. A. Schnitzer, and J. S. Denslow. 2000. Density and diversity of lianas along a chronosequence in a central Panamanian lowland forest. Journal of Tropical Ecology 16:1–19.

Dick, C. W., K. Abdul-Salim, and E. Bermingham. 2003. Molecular systematic analysis reveals cryptic tertiary diversification of a widespread tropical rain forest tree. American Naturalist 162:691–703.

Dick, C. W., E. Bermingham, M. R. Lemes, and R. Gribel. 2007. Extreme long-distance dispersal of the lowland tropical rainforest tree *Ceiba pentandra* L. (Malvaceae) in Africa and the Neotropics. Molecular Ecology 14:3039–49.

Dick, C. W., and W. J. Kress. 2009. Dissecting tropical plant diversity with forest plots and a molecular toolkit. Bioscience 59:745–55.

Dickinson, E. C. 2003. The Howard and Moore complete checklist of birds of the world. Princeton University Press, Princeton, NJ.

DiFiore, A., and C. J. Campbell. 2007. The atelines: variation in ecology, behavior, and social organization. Pages 155–85 in C. J. Campbell, A. Fuentes, K. C. MacKinnon, M. Panger, and S. K. Bearder, eds. Primates in perspective. Oxford University Press, New York.

Dinerstein, E., and C. M. Wemmer. 1988. Fruits rhinoceros eat—dispersal of *Trewia nudiflora* (Euphorbiaceae) in lowland Nepal. Ecology 69:1768–74.

Dirzo, R., and E. Mendoza. 2007. Size-related differential seed predation in a heavily defaunated Neotropical rain forest. Biotropica 39:355–62.

Dobat, K., and T. Peikert-Holle. 1985. Bluten und fledermause. Verlag Waldemar Kramer, Frankfurt am Main.

Dobzhansky, T. 1950. Evolution in the tropics. American Scientist 38:209–21.

Dodd, M. E., J. Silvertown, and M. W. Chase. 1999. Phylogenetic analysis of trait evolution and species diversity variation among angiosperm families. Evolution 53:732–44.

Doherty, P. F., G. Sorci, J. A. Royle, J. E. Hines, J. D. Nichols, and T. Boulinier. 2003. Sexual selection affects local extinction and turnover in bird communities. Proceedings of the National Academy of Sciences of the United States of America 100:5858–62.

Dominguez, C. A., C. A. Abarca, L. E. Eguiarte, and F. Molina-Freaner. 2005. Local genetic differentiation among populations of the mass-flowering tropical shrub *Erythroxylum havanense* (Erythroxylaceae). New Phytologist 166:663–72.

Dominy, N. J. 2004. Fruits, fingers, and fermentation: the sensory cues available to foraging primates. Integrative and Comparative Biology 44:295–303.

Dominy, N. J., P. A. Garber, J. C. Bicca-Marques, and M. Azevedo-Lopes. 2003. Do female tamarins use visual cues to detect fruit rewards more successfully than do males? Animal Behaviour 66:829–37.

Dominy, N. J., P. W. Lucas, D. Osorio, and N. Yamashita. 2001. The sensory ecology of primate food perception. Evolutionary Anthropology 10:171–86.

Donatti, C. I., P. R. Guimaraes, M. Galetti, M. A. Pizo, F. M. D. Marquitti, and R. Dirzo. 2011. Analysis of a hyper-diverse seed dispersal network: modularity and underlying mechanisms. Ecology Letters 14:773–81.

Donoghue, P. C. J., and M. J. Benton. 2007. Rocks and clocks: calibrating the tree of life using fossils and molecules. Trends in Ecology and Evolution 22:424–31.

Doucet, S. M., D. J. Mennill, and G. E. Hill. 2007. The evolution of signal design in manakin plumage ornaments. American Naturalist 169:S62–S80.

Doyle, J. A., H. Sauquet, T. Scharaschkin, and A. Le Thomas. 2004. Phylogeny, molecular and fossil dating, and biogeographic history of Annonaceae and Myristicaceae (Magnoliales). International Journal of Plant Sciences 165:S55–S67.

Driskell, A. C., and L. Christidis. 2004. Phylogeny and evolution of the Australo-Papuan honeyeaters (Passeriformes, Meliphagidae). Molecular Phylogenetics and Evolution 31:943–60.

Dudley, R. 2004. Ethanol, fruit ripening, and the historical origins of human alcoholism in primate frugivory. Integrative and Comparative Biology 44:315–23.

Dudley, R., and Y. Winter. 2002. Hovering flight mechanics of Neotropical flower bats (Phyllostomidae : Glossophaginae) in normodense and hypodense gas mixtures. Journal of Experimental Biology 205:3669–77.

Duminil, J., S. Fineschi, A. Hampe, P. Jordano, D. Salvini, G. G. Vendramin, and R. J. Petit. 2007. Can population genetic structure be predicted from life-history traits? American Naturalist 169:662–72.

Dumont, E. R. 1997. Cranial shape in fruit, nectar, and exudate feeders: implications for interpreting the fossil record. American Journal of Physical Anthropology 102:187–202.

———. 1999. The effect of food hardness on feeding behaviour in frugivorous bats (Phyllostomidae): an experimental study. Journal of Zoology 248:219–29.

———. 2003. Bats and fruit: an ecomophological approach. Pages 398–429 in T. H. Kunz and M. B. Fenton, eds. Bat ecology. University of Chicago Press, Chicago.

———. 2004. Patterns of diversity in cranial shape among plant-visiting bats. Acta Chiropterologica 6:59–74.

———. 2007. Feeding mechanisms in bats: variation within the constraints of flight. Integrative and Comparative Biology 47:137–46.

Dumont, E. R., L. M. Davalos, A. Goldberg, S. E. Santana, K. Rex, and C. C. Voigt. 2011. Morphological innovation, diversification and invasion of a new adaptive zone. Proceedings of the Royal Society B: Biological Sciences 279:1797–1805.

Dumont, E. R., and R. O'Neal. 2004. Food hardness and feeding behavior in Old World fruit bats (Pteropodidae). Journal of Mammalogy 85:8–14.

Dungan, R. J., M. J. O'Cain, M. L. Lopez, and D. A. Norton. 2002. Contribution by possums to seed rain and subsequent seed germination in successional vegetation, Canterbury, New Zealand. New Zealand Journal of Ecology 26:121–27.

Dunn, R. R., N. C. Harris, R. K. Colwell, L. P. Koh, and N. S. Sodhi. 2009. The sixth mass coextinction: are most endangered species parasites and mutualists? Proceedings of the Royal Society B: Biological Sciences 276:3037–45.

Dunning, J. B., Jr., ed. 1993. CRC handbook of avian body masses. CRC Press, Boca Raton, FL.

Dunphy, B. K., J. L. Hamrick, and J. Schwagerl. 2004. A comparison of direct and indirect measures of gene flow in the bat-pollinated tree *Hymenaea courbaril* in the dry forest life zone of southwestern Puerto Rico. International Journal of Plant Sciences 165:427–36.

Duran, S. M., and G. H. Kattan. 2005. A test of the utility of exotic tree plantations for understory birds and food resources in the Colombian Andes. Biotropica 37:129–35.

Eberhard, J. R., and E. Bermingham. 2005. Phylogeny and comparative biogeography of *Pionopsitta* parrots and *Pteroglossus* toucans. Molecular Phylogenetics and Evolution 36:288–304.

Eby, P. 1991. Seasonal movements of grey-headed flying foxes, *Pteropus poliocephalus*, from two maternity camps in northern New South Wales. Wildlife Research 18:547–59.

Economist, The. 2011. Welcome to the Anthropocene. Economist, May 28, 2011, 11.

Edwards, E. J., R. Nyffeler, and M. J. Donoghue. 2005. Basal cactus phylogeny: implications of *Pereskia* (Cactaceae) paraphyly for the transition to the cactus life form. American Journal of Botany 92:1177–88.

Edwards, S. V., W. B. Jennings, and A. M. Shedlock. 2005. Phylogenetics of modern birds in the era of genomics. Proceedings of the Royal Society B: Biological Sciences 272:979–92.

Edwards, W. 2006. Plants reward seed dispersers in proportion to their effort: the relationship between pulp mass and seed mass in vertebrate dispersed plants. Evolutionary Ecology 20:365–76.

Eeley, H. A. C., and M. J. Lawes. 1999. Large-scale patterns of species richness and species range size in anthropoid primates. Pages 191–219 in J. G. Fleagle, C. H. Janson, and K. E. Reed, eds. Primate communities. Cambridge University Press, Cambridge.

Egas, M., M. W. Sabelis, and U. Dieckmann. 2005. Evolution of specialization and ecological character displacement of herbivores along a gradient of plant quality. Evolution 59:507–20.

Ehrlen, J., and O. Eriksson. 1993. Toxicity in fleshy fruits—a nonadaptive trait? Oikos 66:107–13.

Ehrlich, P. R., and P. H. Raven. 1964. Butterflies and plants: a study in coevolution. Evolution 18:586–608.

Eisenberg, J. F. 1981. The mammalian radiations: an analysis of trends in evolution, adaptation, and behavior. University of Chicago Press, Chicago.

Eisenberg, J. F., and D. E. Wilson. 1978. Relative brain size and feeding strategies in the Chiroptera. Evolution 32:740–51.

Elangovan, V., E. Y. S. Priya, and G. Marimuthu. 2006. Olfactory discrimination ability of the short-nosed fruit bat *Cynopterus sphinx*. Acta Chiropterologica 8:247–53.

Emerson, B. C., and R. G. Gillespie. 2008. Phylogenetic analysis of community assembly and structure over space and time. Trends in Ecology and Evolution 23:619–30.

Emmons, L. H. 1999. Of mice and monkeys: primates as predictors of mammal community richness. Pages 171–88 in J. G. Fleagle, C. Janson, and K. Reed, eds. Primate communities. Cambridge University Press, Cambridge.

Endress, P. K. 1994. Diversity and evolutionary biology of tropical flowers. Cambridge University Press, Cambridge.

———. 2011. Evolutionary diversification of the flowers in angiosperms. American Journal of Botany 98:370–96.

Engelbrecht, B. M. J., L. S. Comita, R. Condit, T. A. Kursar, M. T. Tyree, B. L. Turner, and S. P. Hubbell. 2007. Drought sensitivity shapes species distribution patterns in tropical forests. Nature 447:80–83.

Enquist, B. J., J. P. Haskell, and B. H. Tiffney. 2002. General patterns of taxonomic and biomass partitioning in extant and fossil plant communities. Nature 419:610–13.

Epstein, J. H., K. J. Olival, J. R. C. Pulliam, C. Smith, J. Westrum, T. Hughes, A. P. Dobson, et al. 2009. *Pteropus vampyrus*, a hunted migratory species with a multinational home-range and a need for regional management. Journal of Applied Ecology 46:991–1002.

Ericson, P. G. P., C. L. Anderson, T. Britton, A. Elzanowski, U. S. Johansson, M. Kallersjo, J. I. Ohlson, T. J. Parsons, D. Zuccon, and G. Mayr. 2006. Diversification of Neoaves: integration of molecular sequence data and fossils. Biology Letters 2:543–47.

Ericson, P. G. P., C. L. Anderson, and G. Mayr. 2007. Hangin' on to our rocks 'n clocks: a reply to Brown et al. Biology Letters 3:260–61.

Eriksson, O. 2008. Evolution of seed size and biotic seed dispersal in angiosperms: Paleoecological and neoecological evidence. International Journal of Plant Sciences 169:863–70.

Eriksson, O., and B. Bremer. 1991. Fruit characteristics, life forms, and species richness in the plant family Rubiaceae. American Naturalist 138:751–61.

Eriksson, O., E. M. Friis, and P. Lofgren. 2000. Seed size, fruit size, and dispersal systems in angiosperms from the early Cretaceous to the late Tertiary. American Naturalist 156:47–58.

Erkens, R. H. J., L. W. Chatrou, J. W. Maas, T. van der Niet, and V. Savolainen. 2007. A rapid diversification of rainforest trees (*Guatteria*; Annonaceae) following dispersal from Central into South America. Molecular Phylogenetics and Evolution 44:399–411.

Erwin, D. H. 2009. Climate as a driver of evolutionary change. Current Biology 19:R575–R583.

Esparza-Olguin, L., T. Valverde, and M. C. Mandujano. 2005. Comparative demographic analysis of three *Neobuxbaumia* species (Cactaceae) with differing degree of rarity. Population Ecology 47:229–45.

Estes, J. A., J. Terborgh, J. S. Brashares, M. E. Power, J. Berger, and others. 2011. Trophic downgrading of planet Earth. Science 333:301–6.

Estrada, A., and R. Coates-Estrada. 1986. Frugivory by howling monkeys (*Alouatta palliata*) at Los Tuxtlas, Mexico: dispersal and fate of seeds. Pages 93–104 in A. Estrada and T. H. Fleming, eds. Frugivores and seed dispersal. Dr. W. Junk Publishers, Dordrecht.

———. 2002. Bats in continuous forest, forest fragments and in an agricultural mosaic habitat-island at Los Tuxtlas, Mexico. Biological Conservation 103:237–45.

Ezard, T. H. G., T. Aze, P. N. Pearson, and A. Purvis. 2011. Interplay between changing climate and species' ecology drives macroevolutionary dynamics. Science 332:349–51.

Ezcurra, C. 2002. Phylogeny, morphology, and biogeography of *Chuquiraga*, an Andean-Patagonian genus of Asteraceae-Bamadesioideae. Botanical Review 68:153–70.

Fa, J. E., and D. Brown. 2009. Impacts of hunting on mammals in African tropical moist forests: a review and synthesis. Mammal Review 39:231–64.

Fa, J. E., and C. A. Peres. 2001. Game vertebrate extraction in African and Neotropical forests: an intercontinental comparison. Pages 203–41 in J. D. Reynolds, G. M. Mace, K. H. Redford, and J. G. Robinson, eds. Conservation of exploited species. Cambridge University Press, Cambridge.

Fabre, P.-H., A. Rodrigues, and E. J. P. Douzery. 2009. Patterns of macroevolution among Primates inferred from a supermatrix of mitochondrial and nuclear DNA. Molecular Phylogenetics and Evolution 53:808–25.

Faegri, K., and L. van der Pijl. 1979. The principles of pollination ecology. Pergamon Press, Oxford.

Fain, M. G., and P. Houde. 2004. Parallel radiations in the primary clades of birds. Evolution 58:2558–73.

Feare, C., and A. Craig. 1999. Starlings and mynas. Princeton University Press, Princeton, NJ.

Federov, A. A. 1966. The structure of the tropical rain forest and speciation in the humid tropics. Journal of Ecology 54:1–11.

Feinsinger, P. 1976. Organization of a tropical guild of nectarivorous birds. Ecological Monographs 46:257–91.

———. 1980. Asynchronous migration patterns and the coexistence of tropical hummingbirds. Pages 411–19 in A. Keast and E. S. Morton, eds. Migrant birds in the Neotropics. Smithsonian Institution Press, Washington, DC.

———. 1983. Coevolution and pollination. Pages 282–310 in D. J. Futuyma and M. Slatkin, eds. Coevolution. Sinauer Associates Inc., Sunderland, MA.

———. 1987. Approaches to nectarivore-plant interactions in the New World. Revista Chilena de Historia Natural 60:285–319.

Feinsinger, P., J. H. Beach, Y. B. Linhart, W. H. Busby, and K. G. Murray. 1987. Disturbance, pollinator predictability, and pollination success among Costa Rican cloud forest plants. Ecology 68:1294–1305.

Feinsinger, P., W. H. Busby, K. G. Murray, J. H. Beach, W. Z. Pounds, and Y. B. Linhart. 1988. Mixed support for spatial heterogeneity in species interactions—hummingbirds in a tropical disturbance mosaic. American Naturalist 131:33–57.

Feinsinger, P., and R. K. Colwell. 1978. Community organization among Neotropical nectar-feeding birds. American Zoologist 18:779–95.

Feinsinger, P., R. K. Colwell, J. Terborgh, and S. Budd. 1979. Elevation and the morphology, flight energetics, and foraging ecology of tropical hummingbirds. American Naturalist 113: 481–97.

Feinsinger, P., K. G. Murray, S. Kinsman, and W. H. Busby. 1986. Floral neighborhood and pollination success in 4 hummingbird-pollinated cloud forest plant species. Ecology 67:449–64.

Feinsinger, P., L. A. Swarm, and J. A. Wolfe. 1985. Nectar-feeding birds on Trinidad and Tobago—comparison of diverse and depauperate guilds. Ecological Monographs 55:1–28.

Feinsinger, P., H. M. Tiebout, and B. E. Young. 1991. Do tropical bird-pollinated plants exhibit density-dependent interactions—field experiments. Ecology 72:1953–63.

Fenster, C. B., W. S. Armbruster, P. Wilson, M. R. Dudash, and J. D. Thomson. 2004. Pollination syndromes and floral specialization. Annual Review of Ecology and Systematics 35:375–403.

Ferriere, R., J. L. Bronstein, S. Rinaldi, R. Law, and M. Gauduchon. 2002. Cheating and the evolutionary stability of mutualisms. Proceedings of the Royal Society B: Biological Sciences 269:773–80.

Finarelli, J. A., and C. Badgley. 2010. Diversity dynamics of Miocene mammals in relation to the history of tectonism and climate. Proceedings of the Royal Society B: Biological Sciences 277:2721–26.

Fine, P. V. A., D. C. Daly, G. V. Munoz, I. Mesones, and K. M. Cameron. 2005. The contribution of edaphic heterogeneity to the evolution and diversity of Burseraceae trees in the western Amazon. Evolution 59:1464–78.

Fink, R. D., C. A. Morrison, R. A. Zahawi, and K. D. Holl. 2009. Patch size and tree species influence the number and duration of bird visits in forest restoration plots in southern Costa Rica. Restoration Ecology 17:479–86.

Fitzpatrick, B. M., and M. Turelli. 2006. The geography of mammalian speciation: mixed signals from phylogenies and range maps. Evolution 60:601–15.

Fjeldsa, J., and R. C. K. Bowie. 2008. New perspectives on the origin and diversification of Africa's forest avifauna. African Journal of Ecology 46:235–47.

Fjeldsa, J., and C. Rahbek. 2006. Diversification of tanagers, a species rich bird group, from lowlands to montane regions of South America. Integrative and Comparative Biology 46:72–81.

Fleagle, J. G. 1988. Primate adaptation and evolution. Academic Press, Inc., San Diego, CA.

Fleagle, J. G., C. Janson, and K. Reed, eds. 1999. Primate communities. Cambridge University Press, Cambridge.

Fleagle, J. G., and R. A. Mittermeier. 1980. Locomotor behavior, body size and comparative ecology of seven Suriname monkeys. American Journal of Physical Anthropology 52:301–22.

Fleagle, J. G., and K. E. Reed. 1999. Phylogenetic and temporal perspectives on primate ecology. Pages 92–115 in J. G. Fleagle, C. H. Janson, and K. E. Reed, eds. Primate communities. Cambridge University Press, Cambridge.

Fleming, P. A., D. A. Gray, and S. W. Nicolson. 2004. Osmoregulatory response to acute diet change in an avian nectarivore: rapid rehydration following water shortage. Comparative Biochemistry and Physiology A: Molecular and Integrative Physiology 138:321–26.

Fleming, P. A., and S. W. Nicolson. 2002. How important is the relationship between *Protea humiflora* (Proteaceae) and its non-flying mammal pollinators? Oecologia 132:361–68.

———. 2003. Osmoregulation in an avian nectarivore, the whitebellied sunbird *Nectarinia talatala:* response to extremes of diet concentration. Journal of Experimental Biology 206:1845–54.

Fleming, P. A., S. Xie, K. Napier, T. J. McWhorter, and S. W. Nicolson. 2008. Nectar concentration affects sugar preferences in two Australian honeyeaters and a lorikeet. Functional Ecology 22:599–605.

Fleming, T. H. 1981. Fecundity, fruiting pattern, and seed dispersal in *Piper amalago* (Piperaceae), a bat-dispersed tropical shrub. Oecologia 51:42–46.

———. 1985. Coexistence of five sympatric *Piper* (Piperaceae) species in a tropical dry forest. Ecology 66:688–700.

———. 1986a. Opportunism versus specialization: the evolution of feeding strategies in frugivorous bats. Pages 105–18 in A. Estrada and T. H. Fleming, eds. Frugivores and seed dispersal. Dr. W. Junk Publishers, Dordrecht.

———. 1986b. The structure of Neotropical bat communities: a preliminary analysis. Revista Chilena de Historia Natural 59:135–50.

———. 1988. The short-tailed fruit bat, a study in plant-animal interactions. University of Chicago Press, Chicago.

———. 1991a. Fruiting plant-frugivore mutualism: the evolutionary theater and the ecological play. Pages 119–44 in P. W. Price, T. M. Lewinsohn, G. W. Fernandes, and W. W. Benson, eds. Plant-animal interactions: evolutionary ecology in tropical and temperate regions. J. Wiley and Sons, New York.

———. 1991b. The relationship between body size, diet, and habitat use in frugivorous bats, genus *Carollia* (Phyllostomidae). Journal of Mammalogy 72:493–501.

———. 1992. How do fruit- and nectar-feeding birds and mammals track their food resources? Pages 355–91 in M. D. Hunter, T. Ohgushi, and P. W. Price, eds. Resource distributions and plant-animal interactions. Academic Press, Orlando, FL.

———. 1995. Pollination and frugivory in phyllostomid bats of arid regions. Marmosiana 1:87–93.

———. 2002. The pollination biology of Sonoran Desert columnar cacti. Pages 207–24 in T. H. Fleming and A. Valiente-Banuet, eds. Columnar cacti and their mutualists: evolution, ecology, and conservation. University of Arizona Press, Tucson.

———. 2004a. Dispersal ecology of Neotropical *Piper* shrubs and treelets. Pages 58–77 in L. A. Dyer and A. D. N. Palmer, eds. *Piper:* a model genus for studies of phytochemistry, ecology, and evolution. Kluwer Academic/Plenum Publishers, New York.

———. 2004b. Nectar corridors: migration and the annual cycle of lesser long-nosed bats. Pages 23–42 in G. P. Nabhan, ed. Conserving migratory pollinators and nectar corridors in western North America. University of Arizona Press, Tucson, Arizona.

———. 2005. The relationship between species richness of vertebrate mutualists and their food plants in tropical and subtropical communities differs among hemispheres. Oikos 111: 556–62.

———. 2006. Reproductive consequences of early flowering in organ pipe cactus, *Stenocereus thurberi*. International Journal of Plant Sciences 167:473–81.

Fleming, T. H., R. L. Breitwisch, and G. W. Whitesides. 1987. Patterns of tropical vertebrate frugivore diversity. Annual Review of Ecology and Systematics 18:91–109.

Fleming, T. H., C. K. Geiselman, and W. J. Kress. 2009. The evolution of bat pollination: a phylogenetic perspective. Annals of Botany 104:1017–43.

Fleming, T. H., and P. Eby. 2003. Ecology of bat migration. Pages 156–208 in T. H. Kunz and M. B. Fenton, eds. Bat ecology. University of Chicago Press, Chicago.

Fleming, T. H., and E. R. Heithaus. 1981. Frugivorous bats, seed shadows, and the structure of tropical forests. Biotropica 13 (supplement):45–53.

Fleming, T. H., E. R. Heithaus, and W. B. Sawyer. 1977. An experimental analysis of the food location behavior of frugivorous bats. Ecology 58:619–27.

Fleming, T. H., and J. N. Holland. 1998. The evolution of obligate mutualisms: the senita cactus and senita moth. Oecologia 114:368–75.

Fleming, T. H., and W. J. Kress. 2011. A brief history of fruits and frugivores. Acta Oecologica 37:521–30.

Fleming, T. H., S. Maurice, S. L. Buchmann, and M. D. Tuttle. 1994. Reproductive biology and relative male and female fitness in a trioecious cactus, *Pachycereus pringlei* (Cactaceae). American Journal of Botany 81:858–67.

Fleming, T. H., and N. Muchhala. 2008. Nectar-feeding bird and bat niches in two worlds: pantropical comparisons of vertebrate pollination systems. Journal of Biogeography 35:764–80.

Fleming, T. H., N. Muchhala, and J. F. Ornelas. 2005. New World nectar-feeding vertebrates: community patterns and processes. Pages 161–84 in V. Sanchez-Cordero and R. A. Medellin, eds. Contribuciones Mastozoologicas en Homenaje a Bernardo Villa. Instituto de Biologia y Instituto de Ecologia, Universidad Nacional Autonoma de Mexico, Mexico, DF.

Fleming, T. H., and J. Nassar. 2002. Population biology of the lesser long-nosed bat, *Leptonycteris curasoae*, in Mexico and northern South America. Pages 283–305 in T. H. Fleming and A. Valiente-Banuet, eds. Columnar cacti and their mutualists: evolution, ecology, and conservation. University of Arizona Press, Tucson.

Fleming, T. H., C. T. Sahley, J. N. Holland, J. D. Nason, and J. L. Hamrick. 2001. Sonoran Desert columnar cacti and the evolution of generalized pollination systems. Ecological Monographs 71:511–30.

Fleming, T. H., and V. J. Sosa. 1994. Effects of nectarivorous and frugivorous mammals on reproductive success of plants. Journal of Mammalogy 75:845–51.

Fleming, T. H., M. D. Tuttle, and M. A. Horner. 1996. Pollination biology and the relative importance of nocturnal and diurnal pollinators in three species of Sonoran Desert columnar cacti. Southwestern Naturalist 41:257–69.

Fleming, T. H., and C. F. Williams. 1990. Phenology, seed dispersal, and recruitment in *Cecropia peltata* (Moraceae) in Costa Rican tropical dry forest. Journal of Tropical Ecology 6:163–78.

Fleming, T. H., C. F. Williams, F. J. Bonaccorso, and L. H. Herbst. 1985. Phenology, seed dispersal, and colonization in *Muntingia calabura*, a Neotropical pioneer tree. American Journal of Botany 72:383–91.

Florchinger, M., J. Braun, K. Bohning-Gaese, and H. M. Schaefer. 2010. Fruit size, crop mass, and plant height explain differential fruit choice of primates and birds. Oecologia 164:151–61.

Fogden, M. P. 1976. A census of a bird community in tropical rain forest in Sarawak. Sarawak Museum Journal 24:252–57.

———. 2000. Appendix 9: birds of the Monteverde area. Pages 541–52 in N. M. Nadkarni and N. T. Wheelwright, eds. Monteverde, ecology and conservation of a tropical cloud forest. Oxford University Press, New York.

Ford, H. A. 1985. A synthesis of the foraging ecology and behaviour of birds in eucalypt forests and woodlands. Pages 249–54 in A. Keast, H. F. Recher, H. A. Ford, and D. Saunders, eds. Birds of eucalypt forests and woodlands: ecology, conservation, and management. Royal Australian Ornithologists Union and Surrey and Sons, Sydney.

Ford, H. A., and D. C. Paton. 1977. The comparative ecology of ten species of honeyeaters in South Australia. Australian Journal of Ecology 2:399–407.

Ford, H. A., D. C. Paton, and N. Forde. 1979. Birds as pollinators of Australian plants. New Zealand Journal of Botany 17:309–19.

Forget, P. M., J. E. Lambert, P. E. Hulme, and S. B. Vander Wall, eds. 2005. Seed fate: predation, dispersal and seedling establishment. CABI, Wallingford, UK.

Forget, P. M., and D. G. Wenny. 2005. How to elucidate seed fate? a review of methods used to study seed removal and secondary seed dispersal. Pages 379–94 in P. M. Forget et al., eds. Seed fate. CABI, Wallingford, UK.

Forget, P.-M., A. J. Dennis, S. J. Mazer, P. A. Jansen, S. Kitamura, J. E. Lambert, and D. A. West-cott. 2007. Seed allometry and disperser assemblages in tropical rainforests: a comparison of four floras on different continents. Pages 5–36 in A. J. Dennis, E. W. Schupp, R. J. Green, and D. A. Westcott, eds. Seed dispersal, theory and its application in a changing world. CABI, Wallingford, UK.

Forman, G. L., C. J. Phillips, and C. S. Rouk. 1979. Alimentary tract. Pages 205–27 in R. J. Baker, J. K. Jones Jr., and D. C. Carter, eds. Biology of bats of the New World family Phyllostomidae. Pt. 3. Special Publications of the Museum Texas Tech University, no. 16, Lubbock.

Foster, J. T., and S. K. Robinson. 2007. Introduced birds and the fate of Hawaiian rainforests. Conservation Biology 21:1248–57.

Foster, M. S. 1990. Factors influencing bird foraging preferences among conspecific fruit-trees. Condor 92:844–54.

Foster, R. B. 1982. Famine on Barro Colorado Island. Pages 201–12 in E. G. Leigh, A. S. Rand, and D. M. Windsor, eds. The ecology of a tropical forest, seasonal rhythms and long-term changes. Smithsonian Institution Press, Washington, DC.

Foster, R. B., J. Arce, and T. S. Wachter. 1986. Dispersal and the sequential plant communities in Amazonian Peru floodplain. Pages 357–70 in A. Estrada and T. H. Fleming, eds. Frugivores and seed dispersal. Dr. W. Junk Publishers, Dordrecht.

Foster, S. A. 1986. On the adaptive value of large seeds for tropical moist forest trees: a review and synthesis. Botanical Review 52:260–99.

Foster, S. A., and C. H. Janson. 1985. The relationship between seed size and establishment conditions in tropical woody plants. Ecology 66:773–80.

Fragoso, J. M. V. 1997. Tapir-generated seed shadows: scale-dependent patchiness in the Amazon rain forest. Journal of Ecology 85:519–29.

Franceschinelli, E. V., and K. Bawa. 2000. The effect of ecological factors on the mating system of a South American shrub species (*Helicteres brevispira*). Heredity 84:116–23.

Frankie, G. W., H. G. Baker, and P. A. Opler. 1974. Comparative phenological studies of trees in tropical wet and dry forests in the lowlands of Costa Rica. Journal of Ecology 62:881–919.

Franklin, D. C., and C. S. Bacht. 2006. Assessing intraspecific phenological synchrony in zoochorous trees from the monsoon forests of northern Australia. Journal of Tropical Ecology 22:419–29.

Franklin, D. C., and R. A. Noske. 1999. Birds and nectar in a monsoonal woodland: correlations at three spatio-temporal scales. Emu 99:15–28.

———. 2000. Nectar sources used by birds in monsoonal north-western Australia: a regional survey. Australian Journal of Botany 48:461–74.

Franklin, J. 2003. Regeneration and growth of pioneer and shade-tolerant rain forest trees in Tonga. New Zealand Journal of Botany 41:669–84.

Freeman, P. W. 1988. Frugivorous and animalivorous bats (Microchiroptera): dental and cranial adaptations. Biological Journal of the Linnean Society 33:249–72.

———. 1995. Nectarivorous feeding mechanisms in bats. Biological Journal of the Linnean Society 56:439–63.

———. 1998. Form, function, and evolution of skulls and teeth of bats. Pages 140–56 in T. H. Kunz and P. A. Racey, eds. Bat biology and conservation. Smithsonian Institution Press, Washington, DC.

———. 2000. Macroevolution in Microchiroptera: recoupling morphology and ecology with phylogeny. Evolutionary Ecology Research 2:317–35.

French, A. R., and T. B. Smith. 2005. Importance of body size in determining dominance hierarchies among diverse tropical frugivores. Biotropica 37:96–101.

French, K., R. Major, and K. Hely. 2005. Use of native and exotic garden plants by suburban nectarivorous birds. Biological Conservation 12:545–59.

Frick, W. F., P. A. Heady, and J. P. Hayes. 2009. Facultative nectar-feeding behavior in a gleaning insectivorous bat (*Antrozous pallidus*). Journal of Mammalogy 90:1157–64.

Frith, C. B., and B. M. Beehler. 1998. The birds of paradise. Oxford University Press, Oxford.

Frolich, M. W., and M. W. Chase. 2007. After a dozen years of progress the origin of angiosperms is still a great mystery. Nature 450:1184–89.

Frost, S. K., and P. G. H. Frost. 1981. Sunbird pollination of *Strelitzia nicolai*. Oecologia 49:379–84.

Fuchs, E. J., J. A. Lobo, and M. Quesada. 2003. Effects of forest fragmentation and flowering phenology on the reproductive success and mating patterns of the tropical dry forest tree *Pachira quinata*. Conservation Biology 17:149–57.

Fumero-Caban, J. J., and E. J. Melendez-Ackerman. 2007. Relative pollination effectiveness of floral visitors of *Pitcairnia angustifolia* (Bromeliaceae). American Journal of Botany 94: 419–24.

Funch, L. S., R. Funch, and G. M. Barroso. 2002. Phenology of gallery and montane forest in the Chapada Diamantina, Bahia, Brazil. Biotropica 34:40–50.

Funk, D. J., K. E. Filchak, and J. L. Feder. 2002. Herbivorous insects: model systems for the comparative study of speciation ecology. Genetica 116:251–67.

Futuyma, D. J., and M. Slatkin, eds. 1983. Coevolution. Sinauer Associates, Sunderland, MA.

Gage, M. J. G., G. A. Parker, S. Nylin, and C. Wiklund. 2002. Sexual selection and speciation in mammals, butterflies and spiders. Proceedings of the Royal Society B: Biological Sciences 269:2309–16.

Galeano, G., S. A. Suarez, and H. Balslev. 1998. Vascular plant species count in a wet forest in the Choco area on the Pacific coast of Colombia. Biodiversity and Conservation 7: 1563–75.

Galetti, M., C. P. Alves-Costa, and E. Cazetta. 2003. Effects of forest fragmentation, anthropogenic edges and fruit color on the consumption of ornithocoric fruits. Biological Conservation 111:269–73.

Galindo-Gonzalez, J., S. Guevara, and V. J. Sosa. 2000. Bat- and bird-generated seed rains at isolated trees in pastures in a tropical forest. Conservation Biology 14:1693–1703.

Galindo-Gonzalez, J., and V. J. Sosa. 2003. Frugivorous bats in isolated trees and riparian vegetation associated with human-made pastures in a fragmented tropical landscape. Southwestern Naturalist 48:579–89.

Ganesh, T., and P. Davidar. 2001. Dispersal modes of tree species in the wet forests of southern Western Ghats. Current Science 80:394–99.

Ganzhorn, J. U., P. C. Wright, and J. Ratsimbazafy. 1999. Primate communities: Madagascar. Pages 75–89 in J. G. Fleagle, C. H. Janson, and K. E. Reed, eds. Primate communities. Cambridge University Press, Cambridge.

Garcia-Moreno, J., P. Arctander, and J. Fjeldsa. 1999. Strong diversification at the treeline among *Metallura* hummingbirds. Auk 116:702–11.

Gartrell, B. D. 2000. The nutritional, morphologic, and physiologic bases of nectarivory in Australian birds. Journal of Avian Medicine and Surgery 14:85–94.

Gautier-Hion, A., J.-M. Duplantier, R. Quris, F. Feer, C. Sourd, J.-P. Decoux, G. Dubost, et al. 1985. Fruit characters as a basis of fruit choice and seed dispersal in a tropical forest vertebrate community. Oecologia 65:324–37.

Gautier-Hion, A., and G. Michaloud. 1989. Are figs always keystone resources for tropical vertebrate frugivores? a test in Gabon. Ecology 70:1826–30.

Geerts, S., and A. Pauw. 2009. Hyper-specialization for long-billed bird pollination in a guild of South African plants: the Malachite Sunbird pollination syndrome. South African Journal of Botany 75:699–706.

Gentry, A. H. 1974. Flowering phenology and diversity in tropical Bignoniaceae. Biotropica 6:64–68.

———. 1982. Neotropical floristic diversity: phytogeographical connections between Central and South America, Pleistocene climatic fluctuations, or an accident of the Andean orogeny? Annals of the Missouri Botanical Garden 69:557–93.

———. 1988. Tree species richness of upper Amazonian forests. Proceedings of the National Academy of Sciences of the United States of America 85:156–59.

———. 1992. Tropical forest biodiversity—distributional patterns and their conservational significance. Oikos 63:19–28.

Gentry, A. H., and C. Dodson. 1987. Contribution of nontrees to species richness of a tropical rain-forest. Biotropica 19:149–56.

Gentry, A. H., and L. H. Emmons. 1987. Geographical variation in fertility, phenology, and composition of the understory of Neotropical forests. Biotropica 19:216–27.

Ghazoul, J. 2006. Floral diversity and the facilitation of pollination. Journal of Ecology 94: 295–304.

Ghazoul, J., and D. Shell. 2010. Tropical rain forest ecology, diversity, and conservation. Oxford University Press, Oxford.

Giannini, N. P., and N. B. Simmons. 2005. Conflict and congruence in a combined DNA-morphology analysis of megachiropteran bat relationships (Mammalia: Chiroptera: Pteropodidae). Cladistics 21:411–37.

Gideon, O. 1996. Systematics and evolution of the genus *Tapeinochilos* Miq. (Costaceae: Zingiberales). Ph.D. diss. James Cook University, Townsville, Australia.

Gilbert, L. E. 1980. Food web organization and the conservation of Neotropical diversity. Pages 11–33 in M. E. Soule and B. A. Wilcox, eds. Conservation biology. Sinauer, Sunderland, MA.

Gill, F. B. 1990. Ornithology. W. H. Freeman and Co., New York.

Gill, F. B., and L. L. Wolf. 1975a. Economics of feeding territoriality in the golden-winged sunbird. Ecology 56:333–45.

———. 1975b. Foraging strategies and energetics of east African sunbirds at mistletoe flowers. American Naturalist 109:491–510.

Gillespie, T. R., and C. A. Chapman. 2008. Forest fragmentation, the decline of an endangered primate, and changes in host-parasite interactions relative to an unfragmented forest. American Journal of Primatology 70:222–30.

Gillespie, T. W. 1999. Life history characteristics and rarity of woody plants in tropical dry forest fragments of Central America. Journal of Tropical Ecology 15:637–49.

Givinish, T. J. 1998. Adaptive plant evolution on islands: classical patterns, molecular data, new insights. Pages 281–304 in P. R. Grant, ed. Evolution on islands. Oxford University Press, Oxford.

———. 1999. On the causes of gradients in tropical tree diversity. Journal of Ecology 87:193–210.

Givnish, T. J., M. H. J. Barfuss, B. Van Ee, R. Riina, K. Schulte, R. Horres, P. A. Gonsiska, et al. 2011. Phylogeny, adaptive radiation, and historical biogeography in Bromeliaceae: insights from an eight-locus plastid phylogeny. American Journal of Botany 98:872–95.

Givnish, T. J., K. C. Millam, T. M. Evans, J. C. Hall, J. C. Pires, P. E. Berry, and K. J. Sytsma. 2004. Ancient vicariance or recent long-distance dispersal? inferences about phylogeny and South American–African disjunctions in Rapateaceae and Bromeliaceae based on ndhF sequence data. International Journal of Plant Sciences 165:S35–S54.

Givnish, T. J., J. C. Pires, S. W. Graham, M. A. McPherson, L. M. Prince, T. B. Patterson, H. S. Rai, et al. 2005. Repeated evolution of net venation and fleshy fruits among monocots in shaded habitats confirms a priori predictions: evidence from an ndhF phylogeny. Proceedings of the Royal Society B: Biological Sciences 272:1481–90.

Glanz, W. E. 1990. Neotropical mammal densities: how unusual is the community on Barro Colorado Island, Panama? Pages 287–313 in A. H. Gentry, ed. Four Neotropical forests. Yale University Press, New Haven, CT.

Godinez-Alvarez, H., and P. Jordano. 2007. An empirical approach to analysing the demographic consequences of seed dispersal by frugivores. Pages 391–406 in A. J. Dennis, E. W. Schupp, R. J. Green, and D. A. Westcott, eds. Seed dispersal: theory and its application in a changing world. CABI, Wallingford, UK.

Godinez-Alvarez, H., and A. Valiente-Banuet. 2004. Demography of the columnar cactus *Neobuxbaumia macrocephala:* a comparative approach using population projection matrices. Plant Ecology 174:109–18.

Godinez-Alvarez, H., A. Valiente-Banuet, and A. Rojas-Martinez. 2002. The role of seed dispersers in the population dynamics of the columnar cactus *Neobuxbaumia tetetzo*. Ecology 83:2617–29.

Goldsmith, T. H. 2006. What birds see. Scientific American 295:68–75.

Gomes, L. G. L., V. Ostra, V. Nijman, A. M. Cleef, and M. Kappelle. 2008. Tolerance of frugivorous birds to habitat disturbance in a tropical cloud forest. Biological Conservation 14:860–71.

Gomez, J. M., F. Perfectti, J. Bosch, and J. P. M. Camacho. 2009. A geographic selection mosaic in a generalized plant-pollinator-herbivore system. Ecological Monographs 79:245–263.

Gomez, J. M., M. Verdú, and F. Perfectti. 2010. Ecological interactions are evolutionarily conserved across the entire tree of life. Nature 465:918–22.

Gomez-Pompa, A., T. C. Whitmore, and M. Hadley, eds. 1991. Rain forest regeneration and management. United Nations Educational Scientific and Cultural Organization, Paris.

Gonzales, R. S., N. R. Ingle, D. A. Lagunzad, and T. Nakashizuka. 2009. Seed dispersal by birds and bats in lowland Philippine forest successional area. Biotropica 41:452–58.

Gonzalez, A., B. Rayfield, and Z. Lindo. 2011. The disentangled bank: how loss of habitat fragments and disassembles ecological networks. American Journal of Botany 98:503–16.

Good-Avila, S. V., V. Souza, B. S. Gaut, and L. E. Eguiarte. 2006. Timing and rate of speciation in *Agave* (Agavaceae). Proceedings of the National Academy of Sciences of the United States of America 103:9124–29.

Goodman, R. S., and J. U. Ganzhorn. 1998. Rarity of figs (*Ficus*) on Madagascar and its relationship to a depauperate frugivore fauna. Revue d' Ecologie: La Terre et la Vie 52:321–29.

Goodwin, D. 1983. Pigeons and doves of the world. 3rd ed. Comstock Publishing Associates, Ithaca, NY.

Gorchov, D. L., F. Cornejo, C. F. Ascorra, and M. Jaramillo. 1995. Dietary overlap between frugivorous birds and bats in the Peruvian Amazon. Oikos 74:235–50.

Gorresen, P. M., and M. R. Willig. 2004. Landscape responses of bats to habitat fragmentation in Atlantic forest of Paraguay. Journal of Mammalogy 85:688–97.

Gosper, C. R. 2004. Fruit characteristics of invasive bitou bush, *Chrysanthemoides monilifera* (Asteraceae), and a comparison with co-occurring native plant species. Australian Journal of Botany 52:223–30.

Gould, E. H. 1978. Foraging behavior of Malaysian nectar-feeding bats. Biotropica 10:184–93.

Goulding, M. 1980. The fishes and the forest: explorations in Amazonian natural history. University of California Press, Berkeley.

Graham, A. 2009. The Andes: A geological overview from a biological perspective. Annals of the Missouri Botanical Garden 96:371–85.

———. 2010. A natural history of the New World: the ecology and evolution of plants in the Americas. University of Chicago Press, Chicago.

———. 2011. The age and diversification of terrestrial New World ecosystems through Cretaceous and Cenozoic time. American Journal of Botany 98:336–51.

Graham, C., J. E. Martinez-Leyva, and L. Cruz-Paredes. 2002. Use of fruiting trees by birds in continuous forest and riparian forest remnants in Los Tuxtlas, Veracruz, Mexico. Biotropica 34:589–97.

Graham, C. H. 2001. Factors influencing movement patterns of keel-billed toucans in a fragmented tropical landscape in southern Mexico. Conservation Biology 15:1789–98.

Graham, C. H., and P. V. A. Fine. 2008. Phylogenetic beta diversity: linking ecological and evolutionary processes across space in time. Ecology Letters 11:1265–77.

Graham, C. H., J. L. Parra, C. Rahbek, and J. A. McGuire. 2009. Phylogenetic structure in tropical hummingbird communities. Proceedings of the National Academy of Sciences of the United States of America 106:19673–78.

Graham, G. L. 1990. Bats versus birds—comparisons among Peruvian volant vertebrate faunas along an elevational gradient. Journal of Biogeography 17:657–68.

Gravendeel, B., A. Smithson, F. J. W. Slik, and A. Schuiteman. 2004. Epiphytism and pollinator specialization: drivers for orchid diversity? Philosophical Transactions of the Royal Society of London B 359:1523–35.

Greenberg, R., and P. P. Marra, eds. 2005. Birds of two worlds, the ecology and evolution of migration. Johns Hopkins University Press, Baltimore, MD.

Gribel, R., and J. D. Hay. 1993. Pollination ecology of *Caryocar brasiliense* (Caryocaraceae) in Central Brazil cerrado vegetation. Journal of Tropical Ecology 9:199–211.

Griffiths, T. A. 1978. Muscular and vascular adaptations for nectar-feeding in the glossophagine bats *Monophyllus* and *Glossophaga*. Journal of Mammalogy 59:414–18.

———. 1982. Systematics of the New World nectar-feeding bats (Mammalia, Phyllostomidae), based on the morphology of the hyoid and lingual regions. American Museum Novitates 2742:1–45.

Gross, C. L. 2005. A comparison of the sexual systems in the trees from the Australian tropics with other tropical biomes—more monoecy but why? American Journal of Botany 92:907–19.

Gross-Camp, N., and B. A. Kaplin. 2005. Chimpanzee (*Pan troglodytes*) seed dispersal in an afromontane forest: microhabitat influences on the postdispersal fate of large seeds. Biotropica 37:641–49.

Grubb, P. J. 1998. Seeds and fruits of tropical rainforest plants: interpretation of the range in seed size, degree of defence and flesh/seed quotients. Pages 1–24 in D. M. Newbury, H. H. T. Prins, and N. D. Brown, eds. Dynamics of tropical communities. Blackwell Scientific, Oxford.

Grytnes, J. A., and J. H. Beaman. 2006. Elevational species richness patterns for vascular plants on Mount Kinabalu, Borneo. Journal of Biogeography 33:1838–49.

Grytnes, J. A., and O. R. Vetaas. 2002. Species richness and altitude: a comparison between null models and interpolated plant species richness along the Himalayan altitudinal gradient, Nepal. American Naturalist 159:294–304.

Guariguata, M. R. 2000. Seed and seedling ecology of tree species in Neotropical secondary forests: management implications. Ecological Applications 10:145–56.

Guariguata, M. R., H. Arias-LeClaire, and G. Jones. 2002. Tree seed fate in a logged and fragmented forest landscape, northeastern Costa Rica. Biotropica 34:405–15.

Guevara, S., J. Laborde, and G. Sanchez-Rios. 2004. Rain forest regeneration beneath the canopy of fig trees isolated in pastures of Los Tuxtlas, Mexico. Biotropica 36:99–108.

Guimaraes, P. R., M. Galetti, and P. Jordano. 2008. Seed dispersal anachronisms: rethinking the fruits extinct megafauna ate. PLoS One 3:e1745.

Guimaraes, P. R., P. Jordano, and J. N. Thompson. 2011. Evolution and coevolution in mutualistic networks. Ecology Letters 14:877–85.

Gupta, A. K., and D. J. Chivers. 1999. Biomass and use of resources in south and south-east Asian primate communities. Pages 38–54 in J. G. Fleagle, C. H. Janson, and K. E. Reed, eds. Primate communities. Cambridge University Press, Cambridge.

Haffer, J., and G. T. Prance. 2001. Climatic forcing of evolution in Amazonia during the Cenozoic: on the refuge theory of biotic differentiation. Amazonia 16:579–605.

Hall, J. B., and M. D. Swaine. 1976. Classification and ecology of closed-canopy forest in Ghana. Journal of Ecology 64:913–51.

———. 1981. Distribution and ecology of vascular plants in a tropical rain forest: forest vegetation in Ghana. Dr. W. Junk Publishers, The Hague.

Hall, L. S., and G. Richards. 2000. Flying foxes: fruit and blossom bats of Australia. Krieger Publishing, Malabar, FL.

Hamann, A. 2004. Flowering and fruiting phenology of a Philippine submontane rain forest: climatic factors as proximate and ultimate causes. Journal of Ecology 92:24–31.

Hamilton, W. J., III, and K. E. F. Watt. 1970. Refuging. Annual Review of Ecology and Systematics 1:263–86.

Hammond, D. S., and V. K. Brown. 1998. Disturbance, phenology and life-history characteristics: factors influencing distance/density-dependent attack on tropical seeds and seedlings. Pages 51–78 in D. M. Newbury, H. H. T. Prins, and N. D. Brown, eds. Dynamics of tropical communities. Blackwell Scientific, Oxford.

Hamrick, J. L., and M. J. Godt. 1996. Effects of life history traits on genetic diversity in plant species. Philosophical Transactions of the Royal Society of London Series B: Biological Sciences 351:1291–98.

Hamrick, J. L., M. J. Godt, and S. L. Sherman-Broyles. 1992. Factors influencing levels of genetic diversity in woody plant species. New Forests 6:95–124.

Hamrick, J. L., and M. D. Loveless. 1986. The influence of seed dispersal mechanisms on the genetic structure of plant populations. Pages 211–23 in A. Estrada and T. H. Fleming, eds. Frugivores and seed dispersal. Dr. W. Junk, Dordrecht.

Hamrick, J. L., D. A. Murawski, and J. D. Nason. 1993. The influence of seed dispersal mechanisms on the genetic structure of tropical tree populations. Pages 281–97 in T. H. Fleming and A. Estrada, eds. Frugivory and seed dispersal: ecological and evolutionary aspects. Kluwer Academic, Dordrecht.

Hamrick, J. L., J. D. Nason, T. H. Fleming, and J. Nassar. 2002. Genetic diversity in columnar cacti. Pages 122–33 in T. H. Fleming and A. Valiente-Banuet, eds. Columnar cacti and their mutualists: evolution, ecology, and conservation. University of Arizona Press, Tucson.

Handley, C. O., Jr, D. E. Wilson, and A. L. Gardner. 1991. Demography and natural history of the common fruit bat, *Artibeus jamaicensis*, on Barro Colorado Island, Panama. Smithsonian Contributions to Zoology 511:1–173.

Hanley, M. E., M. Franco, S. Pichon, B. Darvilll, and D. Goulson. 2008. Breeding system, pollinator choice and variation in pollen quality in British herbaceous plants. Functional Ecology 22:592–98.

Hansen, D. M., J. M. Olesen, and C. G. Jones. 2002. Trees, birds and bees in Mauritius: exploitative competition between introduced honey bees and endemic nectarivorous birds? Journal of Biogeography 29:721–34.

Hanski, I. 2010. The theories of island biogeography and metapopulation dynamics. Pages 186–213 in J. B. Losos and R. E. Ricklefs, eds. The theory of island biogeography revisited. Princeton University Press, Princeton, NJ.

Harcourt, A. H., and D. A. Doherty. 2005. Species-area relationships of primates in tropical forest fragments: a global analysis. Journal of Applied Ecology 42:630–37.

Harder, L. D., and S. D. Johnson. 2009. Darwin's beautiful contrivences: evolutionary and functional evidence for floral adaptation. New Phytologist 183:530–45.

Hardy, O. J., and B. Sonke. 2004. Spatial pattern analysis of tree species distribution in a tropical rain forest of Cameroon: assessing the role of limited dispersal and niche differentiation. Forest Ecology and Management 197:191–202.

Harms, K. E., S. J. Wright, O. Calderon, A. Hernandez, and E. A. Herre. 2000. Pervasive density-dependent recruitment enhances seedling diversity in a tropical forest. Nature 404:493–95.

Harper, J. T., P. H. Lovell, and K. G. Moore. 1970. The shapes and sizes of seeds. Annual Review of Ecology and Systematics 1:327–56.

Harrison, R. D. 2005. Figs and the diversity of tropical rainforests. Bioscience 55:1053–64.

Hart, N. S., and D. M. Hunt. 2007. Avian visual pigments: characteristics, spectral tuning, and evolution. American Naturalist 169:S7–S26.

Harvey, P. H., and M. D. Pagel. 1991. The comparative method in evolutionary biology. Oxford University Press, Oxford.

Harvey, P. H., R. D. Martin, and T. H. Clutton-Brock. 1986. Life histories in comparative perspective. Pages 181–196 in B. B. Smuts, D. L. Cheney, R. M. Seyfarth, R. W. Wrangham, and T. T. Struhsaker, eds. Primate societies. University of Chicago Press, Chicago.

He, F. L., and S. P. Hubbell. 2011. Species-area relationships always overestimate extinction rates from habitat loss. Nature 473:368–71.

Heaney, L. R., P. D. Heideman, E. A. Rickart, R. C. B. Utzurrum, and J. S. H. Klompen. 1989. Elevational zonation of mammals in the central Philippines. Journal of Tropical Ecology 5:259–80.

Heaney, L. R., and T. E. Roberts. 2009. New perspectives on the long-term biogeographic dynamics and conservation of Philippine fruit bats. Pages 17–58 in T. H. Fleming and P. A. Racey, eds. Island bats: evolution, ecology, and conservation. University of Chicago Press, Chicago.

Heard, S. B., and G. H. Cox. 2007. The shapes of phylogenetic trees of clades, faunas, and local assemblages: exploring spatial pattern in differential diversification. American Naturalist 169:E107–E118.

Hedges, S. B., and S. Kumar, eds. 2009. The timetree of life. Oxford University Press, Oxford.

Heinrich, B., and P. H. Raven. 1972. Energetics and pollination. Science 176:597–602.

Heithaus, E. R. 1982. Coevolution between bats and plants. Pages 327–67 in T. H. Kunz, ed. Ecology of bats. Plenum Press, New York.

Heithaus, E. R., T. H. Fleming, and P. A. Opler. 1975. Patterns of foraging and resource utilization in seven species of bats in a seasonal tropical forest. Ecology 56:841–54.

Heithaus, E. R., P. A. Opler, and H. G. Baker. 1974. Bat activity and pollination of *Bauhinia pauletia*: plant-pollinator coevolution. Ecology 55:412–19.

Heithaus, E. R., E. Stashko, and P. K. Anderson. 1982. Cumulative effects of plant-animal interactions on seed production by *Bauhinia ungulata*, a Neotropical legume. Ecology 63:1294–1302.

Helversen, O. von 1993. Adaptations of flowers to the pollination by glossophagine bats. Pages 41–59 in W. F. A. Barthlott, ed. Plant-animal interactions in tropical environments. Museum Alexander Koenig, Bonn.

Helversen, O. von, and D. von Helversen. 1999. Acoustic guide in a bat-pollinated flower. Nature 398:759–60.

Helversen, O. von, L. Winkler, and H. J. Bestmann. 2000. Sulphur-containing "perfumes" attract flower-visiting bats. Journal of Comparative Physiology A 186:143–53.

Helversen, O. von, and Y. Winter. 2003. Glossophagine bats and their flowers: cost and benefit for flower and pollinator. Pages 346–97 in T. H. Kunz and M. B. Fenton, eds. Bat ecology. University of Chicago Press, Chicago.

Henkel, T. W., J. R. Mayor, and L. P. Woolley. 2005. Mast fruiting and seedling survival of the ectomycorrhizal, monodominant *Dicymbe corymbosa* (Caesalpiniaceae) in Guyana. New Phytologist 167:543–56.

Henry, M., and S. Jouard. 2007. Effect of bat exclusion on patterns of seed rain in tropical rain forest in French Guiana. Biotropica 39:510–18.

Henry, M., and E. K. V. Kalko. 2007. Foraging strategy and breeding constraints of *Rhinophylla pumilio* (Phyllostomidae) in the Amazon lowlands. Journal of Mammalogy 88:81–93.

Herbst, L. H. 1985. The role of nitrogen from fruit pulp in the nutrition of a frugivorous bat, *Carollia perspicillata*. Biotropica 18:39–44.

Herre, E. A., K. C. Jander, and C. A. Machado. 2008. Evolutionary ecology of figs and their associates: recent progress and outstanding puzzles. Annual Review of Ecology Evolution and Systematics 39:439–58.

Herrera, C. M. 1985. Determinants of plant-animal coevolution: the case of mutualistic dispersal of seeds by vertebrates. Oikos 44:132–41.

———. 1986. Vertebrate-dispersed plants: why they don't behave the way they should. Pages 5–18 in A. Estrada and T. H. Fleming, eds. Frugivores and seed dispersal. Dr. W. Junk Publishers, Dordrecht.

———. 1989. Seed dispersal by animals: a role in angiosperm diversification? American Naturalist 133:309–22.

———. 2002. Seed dispersal by vertebrates. Pages 185–208 in C. M. Herrera and O. Pellmyr, eds. Plant-animal interactions: an evolutionary approach. Blackwell Scientific, Oxford.

Herrera, C. M., P. Jordano, L. Lopezsoria, and J. A. Amat. 1994. Recruitment of a mast-fruiting, bird-dispersed tree—bridging frugivore activity and seedling establishment. Ecological Monographs 64:315–44.

Herrera, L. G. 1999. Preferences for different sugars in Neotropical nectarivorous and frugivorous bats. Journal of Mammalogy 80:683–88.

Herrera, L. G., and C. Martínez del Rio. 1998. Pollen digestion by New World bats: effects of processing time and feeding habits. Ecology 79:2828–38.

Herrera, L. G., and C. A. Mancina. 2008. Sucrose hydrolysis does not limit food intake by Pallas's long-tongued bats. Physiological and Biochemical Zoology 81:119–24.

Heywood, J. S., and T. H. Fleming. 1986. Patterns of allozyme variation in three Costa Rican species of *Piper*. Biotropica 18:208–13.

Heywood, V. H., R. K. Brummitt, A. Culham, and O. Seberg. 2007. Flowering plant families of the world. Firefly Books, Ontario.

Hladik, C. M., D. J. Chivers, and P. Pasquet. 1999. On diet and gut size in non-human primates and humans: is there a relationship to brain size? discussion and criticism. Current Anthropology 40:695–97.

Hockey, P. A. R. 2000. Patterns and correlates of bird migrations in sub-saharan Africa. Emu 100:401–17.

Hodgkison, R., M. Ayasse, E. K. V. Kalko, C. Haeberlein, S. Schulz, W. A. W. Mustapha, A. Zubaid, and T. H. Kunz. 2007. Chemical ecology of fruit bat foraging behavior in relation to the fruit odors of two species of paleotropical bat-dispersed figs (*Ficus hispida* and *Ficus scortechinii*). Journal of Chemical Ecology 33:2097–2110.

Hodgkison, R., S. T. Balding, Z. Akbar, and T. H. Kunz. 2003. Fruit bats (Chiroptera) as seed dispersers and pollinators in a lowland Malaysian rain forest. Biotropica 35:491–502.

Hodgkison, R., S. T. Balding, A. Zubaid, and T. H. Kunz. 2004a. Temporal variation in the relative abundance of fruit bats (Megachiroptera: Pteropodidae) in relation to the availability of food in a lowland Malaysian rain forest. Biotropica 36:522–33.

———. 2004b. Habitat structure, wing morphology, and the vertical stratification of Malaysian fruit bats (Megachiroptera: Pteropodidae). Journal of Tropical Ecology 20:667–73.

Hoffmann, F. G., and R. J. Baker. 2001. Systematics of bats of the genus *Glossophaga* (Chiroptera: Phyllostomidae) and phylogeography in *G. soricina* based on the cytochrome-b gene. Journal of Mammalogy 82:1092–1101.

———. 2003. Comparative phylogeography of short-tailed bats (*Carollia*: Phyllostomidae). Molecular Ecology 12:3403–14.

Hoffmann, M., C. Hilton-Taylor, A. Angulo, M. Bohm, T. M. Brooks, S. H. M. Butchart, and others. 2010. The impact of conservation on the status of the world's vertebrates. Science 330:1503–1509.

Holbrook, K. M., and B. A. Loiselle. 2009. Dispersal in a Neotropical tree, *Virola flexuosa* (Myristicaceae): Does hunting of large vertebrates limit seed removal? Ecology 90:1449–1455.

Holbrook, K. M., and T. B. Smith. 2000. Seed dispersal and movement patterns in two species of *Ceratogymna* hornbills in a West African tropical lowland forest. Oecologia 125:249–257.

Holbrook, K. M., T. B. Smith, and B. D. Hardesty. 2002. Implications of long-distance movements of frugivorous rain forest hornbills. Ecography 25:745–749.

Hoorn, C., F. P. Wesselingh, H. Ter Steege, M. A. Bermudez, A. Mora, J. Sevink, I. Sanmartin, et al. 2010. Amazonia through time: Andean uplift, climate change, landscape evolution, and biodiversity. Science 330:927–31.

Hopper, S. D., and A. H. Burbidge. 1986. Speciation and bird-pollinated plants in southwestern Australia. Pages 20–31 in H. A. Ford and D. C. Paton, eds. The dynamic partnership: birds and plants in Southern Australia. SA Government Printer, Adelaide.

Horner, M. A., T. H. Fleming, and C. T. Sahley. 1998. Foraging behaviour and energetics of a nectar-feeding bat, *Leptonycteris curasoae* (Chiroptera: Phyllostomidae). Journal of Zoology 244:575–86.

Horvitz, C. C., M. A. Pizo, B. Bello y Bello, J. LeCorff, and R. Dirzo. 2002. Are plant species that need gaps for recruitment more attractive to seed-dispersing birds and ants than other species? Pages 145–59 in D. J. Levey, W. R. Silva, and M. Galetti, eds. Seed dispersal and frugivory: ecology, evolution, and conservation. CABI, Wallingford, UK.

Hosner, P. A., F. H. Sheldon, H. C. Lim, and R. G. Moyle. 2010. Phylogeny and biogeography of the Asian trogons (Aves: Trogoniformes) inferred from nuclear and mitochondrial DNA sequences. Molecular Phylogenetics and Evolution 57:1219–25.

Howe, H. F. 1980. Monkey dispersal and waste of a Neotropical fruit. Ecology 61:944–59.

———. 1981. Dispersal of a Neotropical nutmeg (*Virola sebifera*) by birds. Auk 98:88–98.

———. 1986. Seed dispersal by fruit-eating birds and mammals. Pages 123–89 in D. R. Murray, ed. Seed dispersal. Academic Press, Sydney.

———. 1989. Scatter- and clump-dispersal by birds and mammals: implications for seedling demography. Oecologia 79:417–26.

———. 1993a. Aspects of variation in a Neotropical seed dispersal system. Vegetatio 107/108: 149–62.

———. 1993b. Specialized and generalized dispersal systems: where does "the paradigm" stand? Vegetatio 107/108:3–13.

Howe, H. F., and G. F. Estabrook. 1977. On intraspecific competition for avian dispersers in tropical trees. American Naturalist 111:817–32.

Howe, H. F., and M. N. Mitiri. 2004. When seed dispersal matters. Bioscience 54:651–60.

Howe, H. F., E. W. Schupp, and L. C. Westley. 1985. Early consequences of seed dispersal for a Neotropical tree (*Virola surinamensis*). Ecology 66:781–91.

Howe, H. F., and J. Smalllwood. 1982. Ecology of seed dispersal. Annual Review of Ecology and Systematics 13:201–28.

Howe, H. F., and G. A. Vande Kerckhove. 1981. Removal of nutmeg (*Virola surinamensis*) crops by birds. Ecology 62:1093–1106.

Howe, H. F., and L. C. Westley. 1988. Ecological relationships of plants and animals. Oxford University Press, New York.

Howell, D. J. 1974. Bats and pollen: physiological aspects of the syndrome of chiropterophily. Comparative Biochemistry and Physiology A 48:263–76.

———. 1979. Flock foraging in nectar-feeding bats: advantages to the bats and to the host plants. American Naturalist 114:23–49.

Howell, D. J., and B. S. Roth. 1981. Sexual reproduction in agaves: the benefits of bats: cost of semelparous advertising. Ecology 62:3–7.

Hu, S., D. L. Dilcher, D. M. Jarzen, and D. W. Taylor. 2008. Early steps of angiosperm-pollinator coevolution. Proceedings of the National Academy of Sciences of the United States of America 105:240–45.

Hubbell, S. P. 2001. The unified neutral theory of biodiversity and biogeography. Princeton University Press, Princeton, NJ.

———. 2008. Approaching ecological complexity from the perspective of symmteric neutral theory. Pages 143–59 in W. P. Carson and S. A. Schnitzer, eds. Tropical forest community ecology. Wiley-Blackwell, Oxford.

———. 2010. Neutral theory and the theory of island biogeography. Pages 264–92 in J. B. Losos and R. E. Ricklefs, eds. The theory of island biogeography revisited. Princeton University Press, Princeton, NJ.

Hubbell, S. P., F. He, R. Condit, L. Borda-de-Agua, J. Kellner, and H. ter Steege. 2008. How many tree species are there in the Amazon and how many of them will go extinct? Proceedings of the National Academy of Sciences of the United States of America 105:11498–504.

Hughes, C. E., and R. Eastwood. 2006. Island radiation on a continental scale: exceptional rates of plant diversification after uplift of the Andes. Proceedings of the National Academy of Sciences of the United States of America 103:10334–39.

Hulme, P. E., and C. W. Benkman. 2002. Granivory. Pages 132–54 in C. M. Herrera and O. Pellmyr, eds. Plant-animal interactions. Blackwell Scientific, Oxford.

Hutcheon, J. M., J. W. Kirsch, and T. Garland. 2002. A comparative analysis of brain size in relation to foraging ecology and phylogeny in the Chiroptera. Brain Behavior and Evolution 60:165–80.

Hyatt, L. A., M. S. Rosenberg, T. G. Howard, G. Bole, W. Fang, J. Anastasia, K. Brown, R. Grella, J. P. Kurdziel, and J. Gurevitch. 2005. The distance dependence prediction of the Janzen-Connell hypothesis: a meta-analysis. Oikos 103:590–602.

Ingle, N. R. 2003. Seed dispersal by wind, birds, and bats between Philippine montane rainforest and successional vegetation. Oecologia 134:251–61.

Ippolito, A., and J. E. Armstrong. 1993. Floral biology of *Hornstedtia scottiana* (Zingiberaceae) in a lowland rain-forest of Australia. Biotropica 25:281–89.

Irestedt, M., K. A. Jonsson, J. Fjeldsa, L. Christidis, and P. G. P. Ericson. 2009. An unexpectedly long history of sexual selection in birds-of-paradise. BMC Evolutionary Biology 9: article 235.

Irwin, A. J., J. L. Hamrick, M. J. W. Godt, and P. E. Smouse. 2003. A multiyear estimate of the effective pollen donor pool for *Albizia julibrissin*. Heredity 90:187–94.

Irwin, M. P. S. 1999. The genus *Nectarinia* and the evolution and diversification of sunbirds: an Afrotropical perspective. Honeyguide 45:45–58.

Isaac, N. J. B., K. E. Jones, J. L. Gittleman, and A. Purvis. 2005. Correlates of species richness in mammals: body size, life history, and ecology. American Naturalist 165:600–607.

Isabirye-Basuta, G. M., and J. S. Lwanga. 2008. Primate populations and their interactions with changing habitats. International Journal of Primatology 29:35–48.

Isler, M. L., and P. R. Isler. 1999. The tanagers, natural history, distribution, identification. Smithsonian Institution Press, Washington, DC.

Itino, T., M. Kato, and M. Hotta. 1991. Pollination ecology of the two wild bananas, *Musa acuminata* subsp *halabanensis* and *M. salaccensis*—chiropterophily and ornithophily. Biotropica 23:151–58.

Iwaniuk, A. N., K. M. Dean, and J. E. Nelson. 2005. Interspecific allometry of the brain and brain regions in parrots (Psittaciformes): comparisons with other birds and primates. Brain Behavior and Evolution 65:40–59.

Jacobs, B. F. 2004. Palaeobotanical studies from tropical Africa: relevance to the evolution of forest, woodland and savannah biomes. Philosophical Transactions of the Royal Society of London Series B: Biological Sciences 359:1573–83.

Jacobs, G. H., and J. Nathans. 2009. The evolution of primate color vision. Scientific American 300:56–63.

Janson, C. H., and C. A. Chapman. 1999. Resources and primate community structure. Pages 237–67 in J. G. Fleagle, C. H. Janson, and K. E. Reed, eds. Primate communities. Cambridge Unversity Press, Cambridge.

Janson, C. H., J. Terborgh, and L. Emmons. 1981. Non-flying mammals as pollination agents in the Amazonian forest. Biotropica 13:1–6.

Janzen, D. H. 1970. Herbivores and the number of tree species in tropical forests. American Naturalist 104:501–28.

———. 1977. Promising directions of study in tropical animal-plant interactions. Annals of the Missouri Botanical Garden 64:706–36.

———. 1979. How to be a fig. Annual Review of Ecology and Systematics 10:13–51.

———. 1980. When is it coevolution? Evolution 34:611–12.

———. 1983a. Seed and pollen dispersal by animals: convergence in the ecology of contamination and sloppy harvest. Biological Journal of the Linnean Society 20:103–13.

———. 1983b. Dispersal of seeds by vertebrate guts. Pages 232–62 in D. J. Futuyma and M. Slatkin, eds. Coevolution. Sinauer Associates Inc., Sunderland, MA.

Jaramillo, C., D. Ochoa, L. Contreras, M. Pagani, H. Carvajal-Ortiz, L. M. Pratt, S. Krishnan, et al. 2010. Effects of rapid global warming at the Paleocene-Eocene boundary on Neotropical vegetation. Science 330:957–61.

Jaramillo, C., M. J. Rueda, and G. Mora. 2006. Cenozoic plant diversity in the Neotropics. Science 311:1893–96.

Jaramillo, M. A., and R. Callejas. 2004. Current perspectives on the classification and phylogenetics of the genus *Piper* L. Pages 179–98 in L. A. Dyer and A. D. N. Palmer, eds. *Piper:* a model genus for studies of phytochemistry, ecology, and evolution. Kluwer Academic/Plenum Publishers, New York.

Jetz, W., and C. Rahbek. 2001. Geometric constraints explain much of the species richness pattern in African birds. Proceedings of the National Academy of Sciences of the United States of America 98:5661–66.

Johansson, U. S., J. Fjeldsa, and R. C. K. Bowie. 2008. Phylogenetic relationships within Passerida (Aves: Passeriformes): a review and a new molecular phylogeny based on three nuclear intron markers. Molecular Phylogenetics and Evolution 48:858–76.

John, R., et al. 2007. Soil nutrients influence spatial distributions of tropical tree species. Proceedings of the National Academy of Sciences of the United States of America 104:864–69.

Johnsgard, P. A. 1997. The hummingbirds of North America. 2nd ed. Smithsonian Institution Press, Washington, DC.

Johnson, S. D. 1996. Pollination, adaptation and speciation in the Cape flora of South Africa. Taxon 45:59–66.

Johnson, S. D., and W. J. Bond. 1997. Evidence for widespread pollen limitation of fruiting success in Cape wildflowers. Oecologia 109:530–34.

Johnson, S. D., A. L. Hargreaves, and M. Brown. 2006. Dark, bitter-tasting nectar functions as a filter of flower visitors in a bird-pollinated plant. Ecology 87:2709–16.

Johnson, S. D., H. P. Linder, and K. E. Steiner. 1998. Phylogeny and radiation of pollination systems in *Disa* (Orchidaceae). American Journal of Botany 85:402–11.

Johnson, S. D., and S. W. Nicolson. 2008. Evolutionary associations between nectar properties and specificity in bird pollination systems. Biology Letters 4:49–52.

Johnson, S. D., A. Pauw, and J. Midgley. 2001. Rodent pollination in the African lily *Massonia depressa* (Hyacinthaceae). American Journal of Botany 88:1768–73.

Johnson, S. D., and K. E. Steiner. 2003. Specialized pollination systems in southern Africa. South African Journal of Science 99:345–48.

Jones, F. A., and H. C. Muller-Landau. 2008. Measuring long-distance seed dispersal in complex natural environments: an evaluation and integration of classical and genetic methods. Journal of Ecology 96:642–52.

Jones, K. E., O. R. P. Bininda-Emonds, and J. L. Gittleman. 2005. Bats, clocks, and rocks: diversification patterns in Chiroptera. Evolution 59:2243–55.

Jonsson, K. A., P.-H. Fabre, R. E. Ricklefs, and J. Fjeldsa. 2011. Major global radiation of corvoid birds originated in the proto-Papuan archipelago. Proceedings of the National Academy of Sciences of the United States of America 108:2328–33.

Jonsson, K. A., and J. Fjeldsa. 2006. A phylogenetic supertree of oscine passerine birds (Aves: Passeri). Zoologica Scripta 35:149–86.

Jordan, F. 2009. Keystone species and food webs. Philosophical Transactions of the Royal Society B: Biological Sciences 364:1733–41.

Jordano, P. 1987. Patterns of mutualistic interactions in pollination and seed dispersal: connectance, dependence asymmetries, and coevolution. American Naturalist 129:657–77.

———. 1992. Fruits and frugivory. Pages 105–56 in M. Fenner, ed Seeds: the ecology of regeneration in plant communities. CABI, Wallingford, UK.

———. 1995. Angiosperm fleshy fruits and seed dispersers—a comparative analysis of adaptation and constraints in plant-animal interactions. American Naturalist 145:163–91.

Jordano, P., J. Bascompte, and J. M. Olesen. 2003. Invariant properties in coevolutionary networks of plant-animal interactions. Ecology Letters 6:69–81.

Jordano, P., C. Garcia, J. A. Godoy, and J. L. Garcia-Castano. 2007. Differential contribution of frugivores to complex seed dispersal patterns. Proceedings of the National Academy of Sciences of the United States of America 104:3278–82.

Jorge, M. L. S. P. 2008. Effects of forest fragmentation on two sister genera of Amazonian rodents (*Myoprocta acouchy* and *Dasyprocta leporina*). Biological Conservation 141:617–23.

Jost, L. 2007. Partitioning diversity into independent alpha and beta components. Ecology 88:2427–39.

Jousselin, E., J. Y. Rasplus, and F. Kjellberg. 2003. Convergence and coevolution in a mutualism: evidence from a molecular phylogeny of *Ficus*. Evolution 57:1255–69.

Joy, J. B., and B. J. Crespi. 2007. Adaptive radiation of gall-inducing insects within a single host-plant species. Evolution 61:784–95.

Kaiser, C. N., D. M. Hansen, and C. B. Muller. 2008. Habitat structure affects reproductive success of the rare endemic tree *Syzygium mamillatum* (Myrtaceae) in restored and unrestored sites in Mauritius. Biotropica 40:86–94.

Kaiser-Bunbury, C. N., J. Memmott, and C. B. Muller. 2009. Community structure of pollination webs of Mauritian heathland habitats. Perspectives in Plant Ecology Evolution and Systematics 11:241–54.

Kalko, E. K. V., and M. A. Condon. 1998. Echolocation, olfaction and fruit display: how bats find fruit of flagellichorous cucurbits. Functional Ecology 12:364–72.

Kalko, E. K. V., and C. O. Handley. 2001. Neotropical bats in the canopy: diversity, community structure, and implications for conservation. Plant Ecology 153:319–33.

Kalko, E. K. V., E. A. Herre, and C. O. Handley, Jr. 1996. The relation of fig fruit syndromes to fruit-eating bats in the New and Old World tropics. Journal of Biogeography 23:565–76.

Kalmar, A., and D. J. Currie. 2007. A unified model of avian species richness on islands and continents. Ecology 88:1309–21.

Kappeler, P. M., and E. W. Heymann. 1996. Nonconvergence in the evolution of primate life history and socio-ecology. Biological Journal of the Linnean Society 59:297–326.

Kappeler, P. M., and C. P. Van Schaik. 2002. Evolution of primate social systems. International Journal of Primatology 23:707–40.

Kappelman, J., D. T. Rasmussen, W. J. Sanders, M. Feseha, T. Bown, P. Copeland, J. Crabaugh, et al. 2003. Oligocene mammals from Ethiopia and faunal exchange between Afro-Arabia and Eurasia. Nature 426:549–52.

Karp, D. S., G. Ziv, P. R. Ehrlich, and G. C. Daily. 2011. Resilience and stability in bird guilds across tropical countryside. Proceedings of the National Academy of Sciences of the United States of America 108:21134–39.

Karr, J. R. 1976. Within-habitat avian diversity in African and Neotropical lowland habitats. Ecological Monographs 46:457–81.

———. 1980. Geographical variation in the avifaunas of tropical forest undergrowth. Auk 97: 283–98.

Karr, J. R., S. K. Robinson, J. G. Blake, and R. O. Bierregaard, Jr. 1990. Birds of four Neotropical forests. Pages 237–69 in A. H. Gentry, ed. Four Neotropical forests. Yale University Press, New Haven, CT.

Kato, M. 1996. Plant-pollinator interactions in the understory of a lowland mixed dipterocarp forest in Sarawak. American Journal of Botany 83:732–43.

Kato, M., T. Itino, and T. Nagamitsu. 1993. Melittophily and ornithophily of long-tubed flowers in Zingiberaceae and Gesneriaceae in West Sumatra. Tropics 2:129–42.

Kato, M., and A. Kawakita. 2004. Plant-pollinator interactions in New Caledonia influenced by introduced honey bees. American Journal of Botany 91:1814–27.

Kaufman, S. R., P. E. Smouse, and E. R. Alvarez-Buylla. 1998. Pollen-mediated gene flow and differential male reproductive success in a tropical pioneer tree, *Cecropia obtusifolia* Bertol. (Moraceae): a paternity analysis. Heredity 81:164–73.

Kawecki, T. J. 1998. Red queen meets Santa Rosalia: arms races and the evolution of host specialization in organisms with parasitic lifestyles. American Naturalist 152:635–51.

Kay, K. M., P. A. Reeves, R. G. Olmstead, and D. W. Schemske. 2005. Rapid speciation and the evolution of hummingbird pollination in Neotropical *Costus* subgenus *Costus* (Costaceae): evidence from nrDNA ITS and ETS sequences. American Journal of Botany 92:1899–1910.

Kay, K. M., and D. W. Schemske. 2003. Pollinator assemblages and visitation rates for 11 species of Neotropical *Costus* (Costaceae). Biotropica 35:198–207.

———. 2005 Natural selection reinforces speciation in a radiation of Neotropical rainforest plants. Evolution 62:2628–42.

Kay, K. M., C. Voelckel, J. Y. Yang, K. M. Hufford, D. D. Kaska, and S. A. Hodges. 2006. Floral characters and species diversification. Pages 311–25 in L. D. Harder and S. C. H. Barrett, eds. Ecology and evolution of flowers. Oxford University Press, New York.

Kay, R. F. 1984. On the use of anatomical features to infer foraging behavior in extinct primates. Pages 21–53 in P. S. Rodman and J. G. H. Cant, eds. Adaptations for foraging in nonhuman primates: contributions to an organismal biology of prosimians. Columbia University Press, New York.

Kearns, C. A., and D. W. Inouye. 1993. Techniques for pollination biologists. University of Colorado Press, Niwot.

Kearns, C. A., D. W. Inouye, and N. M. Waser. 1998. Endangered mutualisms: the conservation of plant-pollinator interactions. Annual Review of Ecology and Systematics 29:83–112.

Keast, A. 1968. Seasonal movements in the Australian honeyeaters (Meliphagidae) and their ecological significance. Emu 67:159–209.

Keitt, T. H. 2009. Habitat conversion, extinction thresholds, and pollination services in agroecosystems. Ecological Applications 19:1561–73.

Kelly, C. K., M. G. Bowler, O. Pybus, and P. H. Harvey. 2008. Phylogeny, niches, and relative abundance in natural communities. Ecology 89:962–70.

Kelm, D. H., J. Schaer, S. Ortmann, G. Wibbelt, J. R. Speakman, and C. C. Voigt. 2008. Efficiency of facultative frugivory in the nectar-feeding bat *Glossophaga commissarisi*: the quality of fruits as an alternative food source. Journal of Comparative Physiology B: Biochemical Systemic and Environmental Physiology 178:985–96.

Kemp, A. 1995. The hornbills, Bucerotiformes. Oxford University Press, Oxford.

Kessler, D., K. Gase, and I. T. Baldwin. 2008. Field experiments with transformed plants reveal the sense of floral scents. Science 321:1200–1202.

Kessler, M. 1999. Plant species richness and endemism during natural landslide succession in a perhumid montane forest in the Bolivian Andes. Ecotropica 5:123–36.

———. 2000. Elevational gradients in species richness and endemism of selected plant groups in the central Bolivian Andes. Plant Ecology 149:181–93.

———. 2001. Patterns of diversity and range size of selected plant groups along an elevational transect in the Bolivian Andes. Biodiversity and Conservation 10:1897–1921.

———. 2002. Environmental patterns and ecological correlates of range size among bromeliad communities of Andean forests in Bolivia. Botanical Review 68:100–127.

Kiers, E. T., T. M. Palmer, A. R. Ives, J. F. Bruno, and J. L. Bronstein. 2010. Mutualisms in a changing world: an evolutionary perspective. Ecology Letters 13:1459–74.

Kiester, A. R., R. Lande, and D. W. Schemske. 1984. Models of coevolution and speciation in plants and their pollinators. American Naturalist 124:220–43.

Kikvidze, Z., and R. M. Callaway. 2009. Ecological facilitation may drive major evolutionary transitions. Bioscience 59:399–404.

Kim, W., T. Gilet, and J. W. M. Bush. 2011. Optimal concentrations in nectar feeding. Proceedings of the National Academy of Sciences of the United States of America 108:16618–21.

Kimura, K., T. Yumoto, and K. Kikuzawa. 2001. Fruiting phenology of fleshy-fruited plants and seasonal dynamics of frugivorous birds in four vegetation zones on Mt. Kinabalu, Borneo. Journal of Tropical Ecology 17:833–58.

Kinnaird, M. F. 1998. Evidence for effective seed dispersal by the Sulawesi red-knobbed hornbill, *Aceros cassidix*. Biotropica 30:50–55.

Kinnaird, M. F., and T. G. O'Brien. 2007. The ecology and conservation of Asian hornbills, farmers of the forest. University of Chicago Press, Chicago.

Kinnaird, M. F., T. G. O'Brien, and S. Suryadi. 1996. Population fluctuation in Sulawesi red-knobbed hornbills: tracking figs in space and time. Auk 113:431–40.

Kirsch, J. A. W., T. F. Flannery, M. S. Springer, and F. J. Lapointe. 1995. Phylogeny of the Pteropodidae (Mammalia, Chiroptera) based on DNA hybridization, with evidence for bat monophyly. Australian Journal of Zoology 43:395–428.

Kirschel, A. N. G., H. Slabbekoorn, D. T. Blumstein, R. E. Cohen, S. R. de Kort, W. Buermann, and T. B. Smith. 2011. Testing alternative hypotheses for evolutionary diversification in an African songbird: rainforest refugia versus ecological gradients. Evolution 65:3162–74.

Kissling, W. D., K. Böhning-Gaese, and W. Jetz. 2009. The global distribution of frugivory in birds. Global Ecology and Biogeography 18:150–62.

Kissling, W. D., C. H. Sekercioglu, and W. Jetz. 2011. Bird dietary guild richness across latitudes, environments and biogeographic regions. Global Ecology and Biogeography 21:328–40.

Kitamura, S., S. Suzuki, T. Yumoto, P. Chuailua, P. Poonswad, N. Noma, T. Maruhashi, and C. Suckasam. 2005. A botanical inventory of a tropical seasonal forest in Khao Yai National Park, Thailand: implications for fruit-frugivore interactions. Biodiversity and Conservation 14:1241–62.

Kleizen, C., J. Midgley, and S. D. Johnson. 2008. Pollination systems of *Colchicum* (Colchicaceae) in Southern Africa: evidence for rodent pollination. Annals of Botany 102:747–55.

Klingbeil, B. T., and M. R. Willig. 2009. Guild-specific responses of bats to landscape composition and configuration in fragmented Amazonian rainforest. Journal of Applied Ecology 46:203–13.

Knapp, S. 2002. Tobacco to tomatoes: a phylogenetic perspective on fruit diversity in the Solanaceae. Journal of Experimental Botany 53:2001–22.

———. 2010. On "various contrivances": pollination, phylogeny, and flower form in the Solanaceae. Philosophical Transactions of the Royal Society B: Biological Sciences 365:449–60.

Knight, T. M., J. A. Steets, J. C. Vamosi, S. J. Mazer, M. Burd, D. R. Campbell, M. R. Dudash, M. O. Johnston, R. J. Mitchell, and T. L. Ashman. 2005. Pollen limitation of plant reproduction: pattern and processes. Annual Review of Ecology and Systematics 36:467–97.

Knudsen, J. T., R. Eriksson, J. Gershenzon, and B. Stahl. 2006. Diversity and distribution of floral scent. Botanical Review 72:1–120.

Knudsen, J. T., L. Tollsten, I. Groth, G. Bergstrom, and R. A. Raguso. 2004. Trends in floral scent chemistry in pollination syndromes: floral scent composition in hummingbird-pollinated taxa. Botanical Journal of the Linnean Society 146:191–99.

Koffi, K. G., O. J. Hardy, C. Doumenge, C. Cruaud, and M. Heuertz. 2011. Diversity gradients and phylogeographic patterns in *Santiria trimera* (Burseraceae), a widespread African tree typical of mature rainforests. American Journal of Botany 98:254–64.

Koh, L. P., R. R. Dunn, N. S. Sodhi, R. K. Colwell, H. C. Proctor, and V. S. Smith. 2004. Species coextinctions and the biodiversity crisis. Science 305:1632–34.

Koopman, K. F. 1981. The distributional patterns of New World nectar-feeding bats. Annals of the Missouri Botanical Gardens 68:352–69.

———. 1993. Order Chiroptera. Pages 137–241 in D. E. Wilson and D. M. Reeder, eds. Mammal species of the world. Smithsonian Institution Press, Washington, DC.

Korine, C., Z. Arad, and A. Arieli. 1996. Nitrogen and energy balance of the fruit bat *Rousettus aegyptiacus* on natural fruit diets. Physiological Zoology 69:618–34.

Korine, C., and E. K. V. Kalko. 2005. Fruit detection and discrimination by small fruit-eating bats (Phyllostomidae): echolocation call design and olfaction. Behavioral Ecology and Sociobiology 59:12–23.

Korine, C., E. K. V. Kalko, and E. A. Herre. 2000. Fruit characteristics and factors affecting fruit removal in a Panamanian community of strangler figs. Oecologia 123:560–68.

Korine, C., I. Vatnick, I. G. van Tets, and B. Pinshow. 2006. The influence of ambient temperature and the energy and protein content of food on nitrogenous excretion in the Egyptian fruit bat (*Rousettus aegyptiacus*). Physiological and Biochemical Zoology 79:957–64.

Kraft, N. J. B., L. S. Comita, J. M. Chase, N. J. Sanders, N. G. Swenson, T. O. Crist, J. C. Stegen, et al. 2011. Disentangling the drivers of β diversity along latitudinal and elevational gradients. Science 333:1755–58.

Kraft, N. J. B., R. Valencia, and D. D. Ackerly. 2008. Functional traits and niche-based tree community assembly in an Amazonian forest. Science 322:580–82.

Kreft, H., N. Koster, W. Kuper, J. Nieder, and W. Barthlott. 2004. Diversity and biogeography of vascular epiphytes in Western Amazonia, Yasuni, Ecuador. Journal of Biogeography 31: 1463–76.

Kress, W. J. 1983a. Self-incompatability in Central American *Heliconia*. Evolution 37:735–44.

———. 1983b. Crossability barriers in Neotropical *Heliconia*. Annals of Botany 52:131–47.

———. 1985. Bat pollination of an Old World Heliconia. Biotropica 17:302–8.

———. 1986. The phylogeny and classification of the Zingiberales. American Journal of Botany 73:744–45.

———. 1990. The phylogeny and classification of the Zingiberales. Annals of the Missouri Botanical Garden 77:698–721.

Kress, W. J., and J. H. Beach. 1994. Flowering plant reproductive systems. Pages 161–82 in

L. A. McDade, K. S. Bawa, H. A. Hespenheide, and G. S. Hartshorn, eds. La Selva: ecology and natural history of a Neotropical rain forest. University of Chicago Press, Chicago.

Kress, W. J., D. L. Erickson, F. A. Jones, N. G. Swenson, R. Perez, O. Sanjur, and E. Bermingham. 2009. Plant DNA barcodes and a community phylogeny of a tropical forest dynamics plot in Panama. Proceedings of the National Academy of Sciences of the United States of America 106:18621–26.

Kress, W. J., L. M. Prince, W. J. Hahn, and E. A. Zimmer. 2001. Unraveling the evolutionary radiation of the families of the Zingiberales using morphological and molecular evidence. Systematic Biology 50:926–44.

Kress, W. J., G. E. Schatz, M. Andrianifahanana, and H. S. Morland. 1994. Pollination of *Ravenala madagascariensis* (Strelitziaceae) by lemurs: evidence for an archaic coevolutionary system? American Journal of Botany 81:542–51.

Kress, W. J., and C. D. Specht. 2005. Between Cancer and Capricorn: phylogeny, evolution, and ecology of tropical Zingiberales. Pages 459–78 in I. Friis and H. Balslev, eds. Proceedings of a symposium on plant diversity and complexity patterns—local, regional and global dimensions. Biologiske Skrifter, Royal Danish Academy of Sciences and Letters, Copenhagen.

Kress, W. J., and D. E. Stone. 1993. Morphology and floral biology of *Phenakospermum* (Strelitziaceae), an arborescent herb of the Neotropics. Biotropica 25:290–300.

Krömer, T., M. Kessler, G. Lohaus, and A. N. Schmidt-Lebuhn. 2008. Nectar sugar composition and concentration in relation to pollination syndromes in Bromeliaceae. Plant Biology 10:502–11.

Krüger, O. 2005. The role of ecotourism in conservation: panacea or Pandora's box? Biodiversity and Conservation 14:579–600.

Kunz, B. K., and K. E. Linsenmair. 2007. Changes in baboon feeding behavior: maturity-dependent fruit and seed size selection within a food plant species. International Journal of Primatology 28:819–35.

———. 2010. Fruit traits in baboon diet: a comparison with plant species characteristics in West Africa. Biotropica 42:363–71.

Kunz, T. H., E. Braun de Torrez, D. Bauer, T. A. Lobova, and T. H. Fleming. 2011. Ecosystem services provided by bats. Annals of the New York Academy of Sciences 1223:1–38.

Kunz, T. H., and C. A. Diaz. 1995. Folivory in fruit-eating bats, with new evidence from *Artibeus jamaicensis* (Chiroptera, Phyllostomidae). Biotropica 27:106–20.

Kunz, T. H., and K. A. Ingalls. 1994. Folivory in bats—an adaptation derived from frugivory. Functional Ecology 8:665–68.

Kusmierski, R., G. Borgia, A. Uy, and R. H. Crozier. 1997. Labile evolution of display traits in bowerbirds indicates reduced effects of phylogenetic constraints. Proceedings of the Royal Society B: Biological Sciences 264:307–11.

Labandeira, C. C. 2002. The history of associations between plants and animals. Pages 26–74 in C. M. Herrera and O. Pellmyr, eds. Plant-animal interactions: an evolutionary approach. Blackwell Scientific, Oxford.

Lack, D. 1968. Ecological adaptations for breeding in birds. Methuen and Co., London.

LaFrankie, J. V., P. S. Ashton, G. B. Chuyong, L. Co, R. Condit, S. J. Davies, R. B. Foster, et al. 2006. Contrasting structure and composition of the understory in species-rich tropical rain forests. Ecology 87:2298–2305.

Laman, T. G. 1996. *Ficus* seed shadows in a Bornean rain forest. Oecologia 107:347–55.

Lambert, F. R., and A. G. Marshall. 1991. Keystone characteristics of bird-dispersed *Ficus* in a Malaysian lowland rain-forest. Journal of Ecology 79:793–809.

Lambert, J. 1998. Primate digestion: interactions between anatomy, physiology, and feeding ecology. Evolutionary Anthropology 7:8–20.

Lambert, J. E., and C. A. Chapman. 2005. The fate of primate-dispersed seeds: deposition pattern, dispersal distance and implications for conservation. Pages 137–50 in P. M. Forget, ed. Seed fate. CABI, Wallingford.

Langrand, O. 1990. Guide to the birds of Madagascar. Yale University Press, New Haven, CT.

Larsen, P. A., S. R. Hoofer, M. C. Bozeman, S. C. Pedersen, H. H. Genoways, C. J. Phillips, D. E. Pumo, and R. J. Baker. 2007. Phylogenetics and phylogeography of the *Artibeus jamaicensis* complex based on cytochrome-b DNA sequences. Journal of Mammalogy 88:712–27.

Larson, B. M. H., and S. C. H. Barrett. 1999. The ecology of pollen limitation in buzz-pollinated *Rhexia virginica* (Melastomataceae). Journal of Ecology 87:371–81.

———. 2000. A comparative analysis of pollen limitation in flowering plants. Biological Journal of the Linnean Society 69:503–20.

Larson, D. L. 1996. Seed dispersal by specialist versus generalist foragers: the plant's perspective. Oikos 76:113–20.

Laska, M. 1990. Olfactory sensitivity to food odor components in the short-tailed fruit bat, *Carollia perspicillata* (Phyllostomatidae, Chiroptera). Journal of Comparative Physiology A 166:395–99.

Laska, M., D. Hofelmann, D. Huber, and M. Schumacher. 2006. The frequency of occurrence of acyclic monoterpene alcohols in the chemical environment does not determine olfactory sensitivity in nonhuman primates. Journal of Chemical Ecology 32:1317–31.

Laurance, W. F., and J. Bieregaard, R. O., eds. 1997. Tropical forest remnents: ecology, management, and conservation of fragmented communities. University of Chicago Press, Chicago.

Laurance, W. F., and C. A. Peres, eds. 2006. Emerging threats to tropical forests. University of Chicago Press, Chicago.

Lavin, M., B. P. Schrire, G. Lewis, R. T. Pennington, A. Delgado-Salinas, M. Thulin, C. E. Hughes, A. B. Matos, and M. F. Wojciechowski. 2004. Metacommunity process rather than continental tectonic history better explains geographically structured phylogenies in legumes. Philosophical Transactions of the Royal Society of London Series B: Biological Sciences 359:1509–22.

Law, B. S. 1992a. Physiological factors affecting pollen use by Queensland blossom bats (*Syconycteris australis*). Functional Ecology 6:257–64.

———. 1992b. The maintenance nitrogen requirements of the Queensland blossom bat (*Syconycteris australis*) on a sugar pollen diet—is nitrogen a limiting resource. Physiological Zoology 65:634–48.

———. 1994. *Banksia* nectar and pollen—dietary items affecting the abundance of the common blossom bat, *Syconycteris australis*, in southeastern Australia. Australian Journal of Ecology 19:425–34.

———. 1995. The effect of energy supplementation on the local abundance of the common blossom bat, *Syconycteris australis*, in south-eastern Australia. Oikos 72:42–50.

———. 1996. Residency and site fidelity of marked populations of the common blossom bat *Syconycteris australis* in relation to the availability of *Banksia* inflorescences in New South Wales, Australia. Oikos 77:447–58.

Law, B. S., and M. Lean. 1999. Common blossom bats (*Syconycteris australis*) as pollinators in fragmented Australian tropical rainforest. Biological Conservation 91:201–12.

Lawrence, D. 2004. Erosion of tree diversity during 200 years of shifting cultivation in Bornean rain forest. Ecological Applications 14:1855–69.

Lawrence, D. C., and J. Mogea. 1996. A preliminary analysis of tree diversity under shifting cultivation north of Gunung Palung National Park. Tropical Biodiversity 3:297–319.

Lazaro, A., R. Lundgren, and O. Totland. 2009. Co-flowering neighbors influence the diversity and identity of pollinator groups visiting plant species. Oikos 118:691–702.

Lee, D. 2007. Nature's palette: the science of plant color. University of Chicago Press, Chicago.

Lefebvre, L., S. M. Reader, and D. Sol. 2004. Brains, innovations and evolution in birds and primates. Brain Behavior and Evolution 63:233–46.

Leibold, M. A., E. P. Economo, and P. Peres-Neto. 2010. Metacommunity phylogenetics: separating the roles of environmental filters and historical biogeography. Ecology Letters 13:1290–99.

Leighton, M. 1986. Hornbill social dispersion: variations on a monogamous theme. Pages 108–30 in D. I. Rubenstein and R. Wrangham, eds. Ecological aspects of social evolution. Princeton University Press, Princeton, NJ.

Leighton, M., and D. R. Leighton. 1983. Vertebrate response to fruiting seasonality within a Bornean rain forest. Pages 181–96 in S. L. Sutton, T. C. Whitmore, and A. C. Chadwick, eds. Tropical rain forest: ecology and management. Blackwell Scientific, Oxford.

Leishman, M. R., I. J. Wright, A. T. Moles, and M. Westoby. 2000. The evolutionary ecology of seed size. Pages 31–57 in M. Fenner, ed. Seeds: the ecology of regeneration in plant communities. CABI, Wallingford, UK.

Lemke, T. O. 1984. Foraging ecology of the long-nosed bat, *Glossophaga soricina*, with respect to resource availability. Ecology 65:538–48.

Letcher, S. G. 2010. Phylogenetic structure of angiosperm communities during tropical forest succession. Proceedings of the Royal Society B: Biological Sciences 277:97–104.

Letcher, S. G., and R. L. Chazdon. 2009. Rapid recovery of biomass, species richness, and species composition in a forest chronosequence in northeastern Costa Rica. Biotropica 41:608–17.

Levey, D. J. 1987. Seed size and fruit-handling techniques of avian frugivores. American Naturalist 129:471–85.

———. 1988. Spatial and temporal variation in Costa Rican fruit and fruit-eating bird abundance. Ecological Monographs 58:251–69.

———. 2004. The evolutionary ecology of ethanol production and alcoholism. Integrative and Comparative Biology 44:284–89.

Levey, D. J., and M. M. Byrne. 1993. Complex ant plant interactions—rain-forest ants as secondary dispersers and postdispersal seed predators. Ecology 74:1802–12.

Levey, D. J., and C. Martínez del Rio. 2001. It takes guts (and more) to eat fruit: Lessons from avian nutritional ecology. Auk 118:819–31.

Levey, D. J., and W. H. Karasov. 1992. Digestive modulation in a seasonal frugivore, the American robin (*Turdus migratorius*). American Journal of Physiology 262:G711–G718.

Levey, D. J., and F. G. Stiles. 1992. Evolutionary precursors of long-distance migration: resource availability and movement patterns in Neotropical landbirds. American Naturalist 140: 447–76.

Levey, D. J., J. Tewksbury, and B. M. Bolker. 2008. Modelling long-distance seed dispersal in heterogeneous landscapes. Journal of Ecology 96:599–608.

Levey, D. J., J. J. Tewksbury, M. L. Cipollini, and T. A. Carlo. 2006. A field test of the directed deterrence hypothesis in two species of wild chili. Oecologia 150:61–68.

Levin, D. A. 2006. Ancient dispersals, propagule pressure, and species selection in flowering plants. Systematic Botany 31:443–48.

Levin, S. A., H. C. Muller-Landau, R. Nathan, and J. Chave. 2003. The ecology and evolution of seed dispersal: a theoretical perspective. Annual Review of Ecology and Systematics 34: 575–603.

Levine, J. M., and D. J. Murrell. 2003. The community-level consequences of seed dispersal. Annual Review of Ecology and Systematics 34:549–73.

Ley, A. C., and R. Classen-Bockhoff. 2009. Pollination syndromes in African Marantaceae. Annals of Botany 104:41–56.

Li, Q.-L., Z.-F. Zu, W. J. Kress, Y.-M. Xia, L. Zhang, X.-B. Deng, J.-Y. Gao, and Z.-L. Bai. 2001. Pollination: flexible style that encourages outcrossing. Nature 410:432.

Lieberman, D., M. Lieberman, R. Peralta, and G. S. Hartshorn. 1996. Tropical forest structure and composition on a large-scale altitudinal gradient. Journal of Ecology 84:137–52.

Liebsch, D., M. C. M. Marques, and R. Goldenberg. 2008. How long does the Atlantic rain forest take to recover after a disturbance? Changes in species composition and ecological features during secondary succession. Biological Conservation 141:1717–25.

Lim, H. C., M. A. Rahman, S. L. H. Lim, R. G. Moyle, and F. H. Sheldon. 2010. Revisiting Wallace's haunt: coalescent simulations and comparative niche modeling reveal historical mechanisms that promoted avian population divergence in the Malay Archipelago. Evolution 65:321–34.

Linhart, Y. B. 1973. Ecological and behavioral determinants of pollen dispersal in hummingbird-pollinated *Heliconia*. American Naturalist 107:511–23.

Linhart, Y. B., W. H. Busby, J. H. Beach, and P. Feinsinger. 1987. Forager behavior, pollen dispersal, and inbreeding in 2 species of hummingbird-pollinated plants. Evolution 41:679–82.

Linnebjerg, J. F., D. M. Hansen, and J. M. Olesen. 2009. Gut passage effect of the introduced red-whiskered bulbul (*Pycnonotus jocosus*) on germination of invasive plant species in Mauritius. Austral Ecology 34:272–77.

Liu, A. Z., D. Z. Li, and H. Wang. 2001. Pollination ecology of a pioneer species: *Musa itinerans* (Musaceae) in Xishuangbanna, South Yunnan, China. Acta Botanica Sinica 43:319–22.

Liu, A. Z., D. Z. Li, H. Wang, and W. J. Kress. 2002. Ornithophilous and chiropterophilous pollination in *Musa itinerans* (Musaceae), a pioneer species in tropical rain forests of Yunnan, southwestern China. Biotropica 34:254–60.

Lively, C. M. 1999. Migration, virulence, and the geographic mosaic of adaptation by parasites. American Naturalist 153:S34–S47.

Lobo, J. A., M. Quesada, and K. E. Stoner. 2005. Effects of pollination by bats on the mating system of *Ceiba pentandra* (Bombacaceae) populations in two tropical life zones in Costa Rica. American Journal of Botany 92:370–76.

Lobova, T. A., C. K. Geiselman, and S. Mori. 2009. Seed dispersal by bats in the neotropics. New York Botanical Garden, Bronx.

Locke, R. 2006. A treasure trove of fruit bats. Bats 24:1–7.

Loiselle, B. A. 1990. Seeds in droppings of tropical fruit-eating birds: importance of considering seed composition. Oecologia 82:494–500.

Loiselle, B. A., and J. G. Blake. 1993. Spatial-distribution of understory fruit-eating birds and fruiting plants in a Neotropical lowland wet forest. Vegetatio 108:177–89.

———. 1994. Annual variation in birds and plants of a tropical second-growth woodland. Condor 96:368–80.

———. 2002. Potential consequences of extinction of frugivorous birds for shrubs of a tropical wet forest. Pages 397–406 in D. J. Levey, W. R. Silva, and M. Galetti, eds. Seed dispersal and frugivory: ecology, evolution and conservation. CABI, Wallingford, UK.

Loiselle, B. A., V. L. Sork, and C. Graham. 1995. Comparison of genetic variation in bird-dispersed shrubs of a tropical wet forest. Biotropica 27:487–94.

Lomáscolo, S. B., D. J. Levey, R. T. Kimball, B. M. Bolker, and H. T. Alborn. 2010. Dispersers shape fruit diversity in *Ficus* (Moraceae). Proceedings of the National Academy of Sciences of the United States of America 107:14668–72.

Lomáscolo, S. B., and H. M. Schaefer. 2010. Signal convergence in fruits: a result of selection by frugivores? Journal of Evolutionary Biology 23:614–24.

Lomáscolo, S. B., P. Speranza, and R. T. Kimball. 2008. Correlated evolution of fig size and color supports the dispersal syndromes hypothesis. Oecologia 156:783–96.

Lomolino, M. V., and L. R. Heaney. 2004. Frontiers of biogeography: new directions in the geography of nature. Sinauer Associates, Inc., Sunderland, MA.

Lomolino, M. V., B. R. Riddle, and J. H. Brown. 2006. Biogeography. 3rd ed. Sinauer Associates, Inc., Sunderland, MA.

Lopes, A. V., L. C. Girao, B. A. Santos, C. A. Peres, and M. Tabarelli. 2009. Long-term erosion of tree reproductive trait diversity in edge-dominated Atlantic forest fragments. Biological Conservation 142:1154–65.

Lord, J. M. 2004. Frugivore gape size and the evolution of fruit size and shape in southern hemisphere floras. Austral Ecology 29:430–36.

Lorts, C. M., T. Briggeman, and T. Sang. 2008. Evolution of fruit types and seed dispersal: a phylogenetic and ecological snapshot. Journal of Systematics and Evolution 46:396–404.

Losos, E., and E. G. Leigh, Jr. 2004. Tropical forest diversity and dynamics: findings from a large-scale plot network. University of Chicago Press, Chicago.

Losos, J. B. 2010. Adaptive radiation, ecological opportunity, and evolutionary determinism. American Naturalist 175:623–39.

———. 2011. Convergence, adaptation, and constraint. Evolution 65:1827–40.

Losos, J. B., and R. E. Ricklefs. 2009. Adaptation and diversification on islands. Nature 457: 830–36.

Lotz, C. N., and J. E. Schondube. 2006. Sugar preferences in nectar- and fruit-eating birds: behavioral patterns and physiological causes. Biotropica 38:3–15.

Louchart, A., N. Tourment, J. Carrier, T. Roux, and C. Mourer-Chauvire. 2008. Hummingbird with modern feathering: an exceptionally well-preserved Oligocene fossil from southern France. Naturwissenschaften 95:171–75.

Loveless, M. D. 2002. Genetic diversity and differentiation in tropical trees. Pages 3–30 in B. Degen, M. D. Loveless, and A. Kremer, eds. Modelling and experimental research on genetic processes in tropical and temperate forests. Embrapa Amazonia Oriental, Belem, Brazil.

Loveless, M. D., and J. L. Hamrick. 1984. Ecological determinants of genetic structure in plant populations. Annual Review of Ecology and Systematics 15:65–95.

Lovett, J. C. 1996. Elevational and latitudinal changes in tree associations and diversity in the Eastern Arc Mountains of Tanzania. Journal of Tropical Ecology 12:629–50.

Lovette, I. J., E. Bermingham, and R. E. Ricklefs. 2002. Clade-specific morphological diversification and adaptive radiation in Hawaiian songbirds. Proceedings of the Royal Society of London Series B: Biological Sciences 269:37–42.

Lovette, I. J., and D. R. Rubenstein. 2007. A comprehensive molecular phylogeny of the starlings (Aves: Sturnidae) and mockingbirds (Aves: Mimidae): congruent mtDNA and nuclear trees for a cosmopolitan avian radiation. Molecular Phylogenetics and Evolution 44:1031–56.

Lucas, P. W., N. J. Dominy, P. Riba-Hernandez, K. E. Stoner, N. Yamashita, E. Loria-Calderon, W. Petersen-Pereira, et al. 2003. Evolution and function of routine trichromatic vision in primates. Evolution 57:2636–43.

Luck, G. W., G. C. Daily, and P. R. Ehrlich. 2003. Population diversity and ecosystem services. Trends in Ecology and Evolution 18:331–36.

Luckow, M., and H. C. F. Hopkins. 1995. A cladistic analysis of *Parkia* (Leguminosae: Mimosoideae). American Journal of Botany 82:1300–1320.

Lugo, A. E., and J. L. Frangi. 1993. Fruit fall in the Luquillo Experimental Forest, Puerto Rico. Biotropica 25:73–84.

Lumpkin, H. A., and W. A. Boyle. 2009. Effects of forest age on fruit composition and removal in tropical bird-dispersed understorey trees. Journal of Tropical Ecology 25:515–22.

Luteyn, J. L. 2002. Diversity, adaptation, and endemism in Neotropical Ericaceae: biogeographical patterns in the Vaccinieae. Botanical Review 68:55–87.

Lwanga, J. S. 2003. Forest succession in Kibale National Park, Uganda: implications for forest restoration and management. African Journal of Ecology 41:9–22.

Maas, P. J. M. 1977. *Renealmia* (Zingiberaceae-Zingiberoideae) Costoideae (additions)(Zingiberaceae). New York Botanical Gardens, New York.

Mabberley, D. J. 1997. The plant-book. Cambridge University Press, Cambridge.

MacArthur, R. H. 1972. Geographical ecology: patterns in the distribution of species. Harper & Row, New York.

Machado, C. A., E. Jousselin, F. Kjellberg, S. G. Compton, and E. A. Herre. 2001. Phylogenetic relationships, historical biogeography, and character evolution of fig-pollinating wasps. Proceedings of the Royal Society B: Biological Sciences 268:685–94.

Machado, C. A., N. Robbins, M. T. P. Gilbert, and E. A. Herre. 2005. Critical review of host specificity and its coevolutionary implications in the fig/fig-wasp mutualism. Proceedings of the National Academy of Sciences of the United States of America 102:6558–65.

Machado, I. C., and A. V. Lopes. 2004. Floral traits and pollination systems in the Caatinga, a Brazilian tropical dry forest. Annals of Botany 94:365–76.

Mack, A. L. 1993. The sizes of vertebrate-dispersed fruits—a Neotropical- Paleotropical comparison. American Naturalist 142:840–56.

———. 1995. Distance and non-randomness of seed dispersal by the dwarf cassowary, *Casuarius bennetti*. Ecography 18:286–95.

MacNally, R., and J. M. McGoldrick. 1997. Landscape dynamics of bird communities in relation to mass flowering in some eucalypt forests of central Victoria, Australia. Journal of Avian Biology 28:171–83.

Maddison, W., and D. Maddison. 2007. Mesquite, version 2.5.

Malcolm, J. R. 1990. Estimates of mammalian densities in continuous forest north of Manaus. Pages 339–57 in A. H. Gentry, ed. Four Neotropical forests. Yale University Press, New Haven, CT.

Malizia, L. R. 2001. Seasonal fluctuations of birds, fruits, and flowers in a subtropical forest of Argentina. Condor 103:45–61.

Mariot, A., L. C. Di Stasi, and M. S. dos Reis. 2002. Genetic diversity in natural populations of *Piper cernuum*. Journal of Heredity 93:365–69.

Marquis, R. J. 2004. Biogeography of Neotropical *Piper*. Pages 78–96 in L. A. Dyer and A. D. N. Palmer, eds. *Piper*: a model genus for studies of phytochemistry, ecology, and evolution. Kluwer Academic, New York.

Marroig, G., and J. M. Cheverud. 2005. Size as a line of least evolutionary resistance: diet and adaptive morphological radiation in New World monkeys. Evolution 59:1128–42.

Marshall, L. G. 1988. Land mammals and the Great American Interchange. American Scientist 76:380–88.

Martén-Rodríguez, S., A. Almares-Castro, and C. B. Fenster. 2009. Evaluation of pollination syndromes in Antillean Gesneriaceae: evidence for bat, hummingbird and generalized flowers. Journal of Ecology 97:348–59.

Martén Rodríguez, S., and C. B. Fenster. 2010. Pollen limitation and reproductive assurance in Antillean Gesnerieae: a specialist vs. generalist comparison. Ecology 91:155–65.

Martén-Rodríguez, S., C. B. Fenster, I. Agnarsson, L. E. Skog, and E. A. Zimmer. 2010. Evolutionary breakdown of pollination specialization in a Caribbean plant radiation. New Phytologist 188:403–17.

Martén-Rodríguez, S., W. J. Kress, E. Temeles, E. Meléndez-Ackerman. 2011. Plant-pollinator interactions and floral convergence in two species of *Heliconia* from the Caribbean Antilles. Oecologia 167:1075–83.

Martin, R. D., C. Soligo, and S. Tavare. 2007. Primate origins: implications of a Cretaceous ancestry. Folia Primatologica 78:277–96.

Martínez del Rio, C. 1990. Dietary, phylogenetic, and ecological correlates of intestinal sucrase and maltase activity in birds. Physiological Zoology 63:987–1011.

Martínez del Rio, C., M. Hourdequin, A. Silva, and R. Medel. 1995. The influence of cactus size and previous infection on bird deposition of mistletoe seeds. Australian Journal of Ecology 20:571–76.

Martínez del Rio, C., and W. H. Karasov. 1990. Digestion strategies in nectar-and fruit-eating birds and the sugar composition of plant rewards. American Naturalist 136:618–37.

Martínez del Rio, C., and C. Restrepo. 1993. Ecological and behavioral consequences of digestion in frugivorous animals. Vegetatio 107/108:205–16.

Martínez del Rio, C., J. E. Schondube, T. J. McWhorter, and L. G. Herrera. 2001. Intake responses in nectar feeding birds: digestive and metabolic causes, osmoregulatory consequences, and coevolutionary effects. American Zoologist 41:902–15.

Martínez, I., D. Garcia, and J. R. Obeso. 2007. Allometric allocation in fruit and seed packaging conditions the conflict among selective pressures on seed size. Evolutionary Ecology 21:517–33.

Martins, M. M. 2006. Comparative seed dispersal effectiveness of sympatric *Alouatta guariba* and *Brachyteles arachnoides* in southeastern Brazil. Biotropica 38:57–63.

Marussich, W. A., and C. A. Machado. 2007. Host-specificity and coevolution among pollinating and nonpollinating New World fig wasps. Molecular Ecology 16:1925–46.

Masters, J. C., M. J. de Wit, and R. J. Asher. 2006. Reconciling the origins of Africa, India and Madagascar with vertebrate dispersal scenarios. Folia Primatologica 77:399–418.

May, R. M. 1973. Stability and complexity in model ecosystems. Princeton University Press, Princeton, NJ.

Mayfield, M. M., D. Ackerly, and G. C. Daily. 2006. The diversity and conservation of plant reproductive and dispersal functional traits in human-dominated tropical landscapes. Journal of Ecology 94:522–36.

Mayr, E. 1995. Species, classification, and evolution. Pages 3–12 in R. Arai, M. Kato, and Y. Doi, eds. Biodiversity and evolution. National Science Museum Foundation, Tokyo.

Mayr, G. 2004. Old World fossil record of modern-type hummingbirds. Science 304:861–64.

———. 2005a. Fossil hummingbirds in the Old World. Biologist 52:12–16.

———. 2005b. The Paleogene fossil record of birds in Europe. Biological Reviews 80:515–42.

———. 2007. New specimens of the early Oligocene old world hummingbird *Eurotrochilus inexpectatus*. Journal of Ornithology 148:105–11.

———. 2011. Metaves, Mirandornithes, Strisores and other novelties—a critical review of the higher-level phylogeny of neornithine birds. Journal of Zoological Systematics and Evolutionary Research 49:58–76.

Mazer, S. J., and N. T. Wheelwright. 1993. Fruit size and shape: allometry at different taxonomic levels in bird-dispersed plants. Evolutionary Ecology 7:556–75.

Mbora, D. N. M., and M. A. McPeek. 2009. Endangered species in small habitat patches can possess high genetic diversity: the case of the Tana River red colobus and mangabey. Conservation Genetics 11:1725–35.

McCain, C. M. 2007. Area and mammalian elevational diversity. Ecology 88:76–86.

———. 2009. Global analysis of bird elevational diversity. Global Ecology and Biogeography 18:346–60.

McConkey, K. R. 2000. Primary seed shadow generated by gibbons in the rain forests of Barito Ulu, central Borneo. American Journal of Primatology 52:13–29.

McConkey, K. R., F. Aldy, A. Ario, and D. J. Chivers. 2002. Selection of fruit by gibbons (*Hylobates muelleri x agilis*) in the rain forests of Central Borneo. International Journal of Primatology 23:123–145.

McConkey, K. R., and W. Y. Brockelman. 2011. Nonredundancy in the dispersal network of a generalist tropical forest tree. Ecology 92:1492–1502.

McConkey, K. R., and D. R. Drake. 2002. Extinct pigeons and declining bat populations: are large seeds still being dispersed in the Tropical Pacific? Pages 381–95 in D. J. Levey, W. R. Silva, and M. Galetti, eds. Seed dispersal and frugivory: ecology, evolution and conservation. CABI, Wallingford.

———. 2006. Flying foxes cease to function as seed dispersers long before they become rare. Ecology 87:271–76.

McCoy, M. 1990. Pollination of two eucalypts by flying-foxes in northern Australia. Pages 33–37 in Flying fox workshop proceedings. New South Wales Department of Agriculture and Fisheries, Sydney.

McCracken, G. F., and R. B. Bradbury. 1981. Social organization and kinship in the polygynous bat *Phyllostomus hastatus*. Behavioral Ecology and Sociobiology 8:11–34.

McCracken, G. F., and G. S. Wilkinson. 2000. Bat mating systems. Pages 321–62 in E. G. Crichton and P. H. Krutzsch, eds. Reproductive biology of bats. Academic Press, San Diego, CA.

McDade, L. A. 1983. Long-tailed hermit hummingbird visits to inflorescence color morphs of *Heliconia irrasa*. Condor 85:360–64.

———. 1992. Pollinator relationships, biogeography, and phylogenetics. Bioscience 42:21–26.

McGuire, J. A., C. C. Witt, D. L. Altshuler, and J. V. Remsen, Jr. 2007. Phylogenetic systematics and biogeography of hummingbirds: Bayesian and maximum likelihood analyses of partitioned data and selection of an appropriate partitioning strategy. Systematic Biology 56:837–56.

McKey, D. 1975. The ecology of coevolved seed dispersal systems. Pages 159–91 in L. E. Gilbert and P. H. Raven, eds. Coevolution of animals and plants. University of Texas Press, Austin.

McNab, B. K. 1982. Evolutionary alternatives in the physiological ecology of bats. Pages 151–200 in T. H. Kunz, ed Ecology of bats. Plenum Press, New York.

———. 2002. The physiological ecology of vertebrates. Cornell University Press, Ithaca, NY.

———. 2003. Standard energetics of phyllostomid bats: the inadequacies of phylogenetic-contrast analyses. Comparative Biochemistry and Physiology A: Molecular and Integrative Physiology 135A:357–68.

McPeek, M. A. 2008. The ecological dynamics of clade diversification and community assembly. American Naturalist 172:E270–E284.

Medellin, R. A., and O. Gaona. 1999. Seed dispersal by bats and birds in forest and disturbed habitats of Chiapas, Mexico. Biotropica 31:478–85.

Medina, A., C. A. Harvey, D. S. Merlo, S. Vilchez, and B. Hernandez. 2007. Bat diversity and movement in an agricultural landscape in Matiguas, Nicaragua. Biotropica 39:120–28.

Mello, M. A. R., F. M. D. Marquitti, P. R. Guimaraes, E. K. V. Kalko, P. Jordano, and M. A. M. de Aguiar. 2011a. The modularity of seed dispersal: differences in structure and robustness between bat- and bird-fruit networks. Oecologia 167:131–40.

———. 2011b. The missing part of seed dispersal networks: structure and robustness of bat-fruit interactions. PLoS One 6: e17395.

Meredith, R. W., J. E. Janecka, J. Gatesy, O. A. Ryder, C. A. Fisher, E. C. Teeling, A. Goodbla, et al. 2011. Impacts of the Cretaceous terrestrial revolution and K/Pg extinction on mammal diversification. Science 334:521–24.

Mesquita, R. C. G., K. Ickes, G. Ganade, and G. B. Williamson. 2001. Alternative successional pathways in the Amazon Basin. Journal of Ecology 89:528–37.

Michalski, S. G., and W. Durka. 2009. Pollination mode and life form strongly affect the relation between mating system and pollen to ovule ratios. New Phytologist 183:470–79.

Mickleburgh, S., K. Waylen, and P. Racey. 2009. Bats as bushmeat: a global review. Oryx 43:217–34.

Mickleburgh, S. P., A. M. Hutson, and P. A. Racey, eds. 1992. Old World fruit bats, an action plan for their conservation. International Union for the Conservation of Nature and Natural Resources, Gland, Switzerland.

Miller, J. S., and P. K. Diggle. 2007. Correlated evolution of fruit size and sexual expression in andromonoecious *Solanum* sections *Acanthophora* and *Lasiocarpa* (Solanaceae). American Journal of Botany 94:1706–15.

Miller, K. G., M. A. Kominz, J. V. Browning, J. D. Wright, G. S. Mountain, M. E. Katz, P. J. Sugarman, B. S. Cramer, N. Christie-Blick, and S. F. Pekar. 2005. The Phanerozoic record of global sea-level change. Science 310:1293–98.

Miller, R. S. 1985. Why hummingbirds hover. Auk 102:722–26.

Milton, K. 1993. Diet and primate evolution. Scientific American 269:86–93.

Milton, K., J. Giacalone, S. J. Wright, and G. Stockmayer. 2005. Do frugivore population fluctuations reflect fruit production? evidence from Panama. Pages 5–35 in J. L. Dew and J. P. Boubli, eds. Tropical fruits and frugivores: the search for strong interactors. Springer, Dordrecht.

Mittelbach, G. G., D. W. Schemske, H. V. Cornell, A. P. Allen, J. M. Brown, M. B. Bush, S. P. Harrison, et al. 2007. Evolution and the latitudinal diversity gradient: speciation, extinction and biogeography. Ecology Letters 10:315–31.

Moegenburg, S. M., and D. J. Levey. 2003. Do frugivores respond to fruit harvest? an experimental study of short-term responses. Ecology 84:2600–2612.

Moermond, T. C., and J. S. Denslow. 1985. Neotropical avian frugivores: patterns of behavior, morphology, and nutrition. Ornithological Monographs 36:865–97.

Molina-Freaner, F., and L. E. Eguiarte. 2003. The pollination biology of two paniculate *Agaves* (Agavaceae) from northwestern Mexico: contrasting roles of bats as pollinators. American Journal of Botany 90:1016–24.

Molina-Freaner, F., A. Rojas-Martinez, T. H. Fleming, and A. Valiente-Banuet. 2004. Pollination biology of the columnar cactus *Pachycereus pecten-aboriginum* in north-western Mexico. Journal of Arid Environments 56:117–27.

Moller, A. P., and J. J. Cuervo. 1998. Speciation and feather ornamentation in birds. Evolution 52:859–69.

Momose, K., R. Ishii, S. Sakai, and T. Inoue. 1998. Plant reproductive intervals and pollinators in the aseasonal tropics: a new model. Proceedings of the Royal Society of London Series B: Biological Sciences 265:2333–39.

Moraes, M. L. T., P. Y. Kageyama, and A. M. Sebbenn. 2007. Mating system in small fragmented populations and isolated trees of *Hymenaea stigonocarpa*. Scientia Forestalis 74:75–86.

Moraes, M. L. T., and A. M. Sebbenn. 2011. Pollen dispersal between isolated trees in the Brazilian savannah: a case study of the Neotropical tree *Hymenaea stigonocarpa*. Biotropica 43:192–99.

Moran, C., and C. P. Catterall. 2010. Can functional traits predict ecological interactions? a case study using rain forest frugivores and plants in Australia. Biotropica 42:318–26.

Mori, S., and G. T. Prance. 1987. Phenology. Chap. 9 of The Lecythidaceae of a lowland Neotropical forest: La Fumee Mountain, French Guiana. Memoirs of the New York Botanical Gardens 44:124–36.

Moritz, C., J. L. Patton, C. J. Schneider, and T. B. Smith. 2000. Diversification of rainforest faunas: an integrated molecular approach. Annual Review of Ecology and Systematics 31:533–63.

Morley, R. J. 2000. Origin and evolution of tropical forests. Wiley, New York.

———. 2003. Interplate dispersal routes for megathermal angiosperms. Perspectives in Plant Ecology Evolution and Systematics 6:5–20.

———. 2007. Cretaceous and Tertiary climate change and the past distribution of megathermal rainforests. Pages 1–32 in M. B. Bush and J. R. Flenley, eds. Tropical rainforest responses to climate change. Springer, New York.

Morley, R. J., and C. W. Dick. 2003. Missing fossils, molecular clocks, and the origin of the Melastomataceae. American Journal of Botany 90:1638–44.

Morris, W. F., R. A. Hufbauer, A. A. Agrawal, J. D. Bever, V. A. Borowicz, G. S. Gilbert, J. L. Maron, et al. 2007. Direct and interactive effects of enemies and mutualists on plant performance: a meta-analysis. Ecology 88:1021–29.

Morrison, D. W. 1978a. Foraging ecology and energetics of the frugivorous bat *Artibeus jamaicensis*. Ecology 59:716–23.

———. 1978b. Influence of habitat on the foraging distances of the fruit bat, *Artibeus jamaicensis*. Journal of Mammalogy 59:622–24.

———. 1980a. Foraging and day-roosting dynamics of canopy fruit bats in Panama. Journal of Mammalogy 61:20–29.

———. 1980b. Efficiency of food utilization by fruit bats. Oecologia 45:270–73.

Morrow, E. H., and T. E. Pitcher. 2003. Sexual selection and the risk of extinction in birds. Proceedings of the Royal Society of London Series B: Biological Sciences 270:1793–99.

Morrow, E. H., T. E. Pitcher, and G. Arnqvist. 2003. No evidence that sexual selection is an "engine of speciation" in birds. Ecology Letters 6:228–34.

Mortensen, H. S., Y. L. Dupont, and J. M. Olesen. 2008. A snake in paradise: disturbance of plant reproduction following extirpation of bird flower-visitors on Guam. Biological Conservation 141:2146–54.

Morton, E. S. 1973. On the evolutionary advantages and disadvantages of fruit eating in tropical birds. American Naturalist 107:8–22.

Moyle, R. G. 2004. Phylogenetics of barbets (Aves: Piciformes) based on nuclear and mitrochondrial DNA sequence data. Molecular Phylogenetics and Evolution 30:187–200.

———. 2005. Phylogeny and biogeographical history of Trogoniformes, a pantropical bird order. Biological Journal of the Linnean Society 84:725–38.

Moyle, R. G., and B. D. Marks. 2006. Phylogenetic relationships of the bulbuls (Aves: Pycnonotidae) based on mitochondrial and nuclear DNA sequence data. Molecular Phylogenetics and Evolution 40:687–95.

Muchhala, N. 2003. Exploring the boundary between pollination syndromes: bats and hummingbirds as pollinators of *Burmeistera cyclostigmata* and *B. tenuiflora* (Campanulaceae). Oecologia 134:373–80.

———. 2006a. The pollination biology of *Burmeistera* (Campanulaceae): specialization and syndromes. American Journal of Botany 93:1081–89.

———. 2006b. Nectar bat stows huge tongue in its rib cage. Nature 444:701–2.

Muchhala, N., A. Caiza, J. C. Vizuete, and J. D. Thomson. 2009. A generalized pollination system in the tropics: bats, birds and *Aphelandra acanthus*. Annals of Botany 103:1481–87.

Muchhala, N., and P. Jarrín-V. 2002. Flower visitation by bats in cloud forests of western Ecuador. Biotropica 34:387–95.

Muchhala, N., and J. D. Thomson. 2009. Going to great lengths: selection for long corolla tubes in an extremely specialized bat-flower mutualism. Proceedings of the Royal Society B: Biological Sciences 276: 2147–52.

———. 2010. Fur versus feathers: pollen delivery by bats and hummingbirds and consequences for pollen production. American Naturalist 175:717–26.

Muellner, A. N., C. M. Pannell, A. Coleman, and M. W. Chase. 2008. The origin and evolution of Indomalesian, Australasian and Pacific island biotas: insights from Aglaieae (Meliaceae, Sapindales). Journal of Biogeography 35:1769–89.

Muller, B., S. M. Goodman, and L. Peichl. 2007. Cone photoreceptor diversity in the retinas of fruit bats (Megachiroptera). Brain Behavior and Evolution 70:90–104.

Muller-Landau, H. C. 2007. Predicting the long-term effects of hunting on plant species composition and diversity in tropical forests. Biotropica 39:372–84.

———. 2010. The tolerance-fecundity trade-off and the maintenance of diversity in seed size. Proceedings of the National Academy of Sciences of the United States of America 107:4242–47.

Muller-Landau, H. C., S. J. Wright, O. Calderon, R. Condit, and S. P. Hubbell. 2008. Interspecific variation in primary seed dispersal in a tropical forest. Journal of Ecology 96:653–67.

Muller-Landau, H. C., S. J. Wright, O. Calderon, S. P. Hubbell, and R. B. Foster. 2002. Assessing recruitment limitation: concepts, methods and case-studies from a tropical forest. Pages 35–53 in D. J. Levey, W. R. Silva, and M. Galetti, eds. Seed dispersal and frugivory: ecology, evolution and conservation. CABI, Wallingford.

Munzbergova, Z., and T. Herben. 2005. Seed, dispersal, microsite, habitat and recruitment limitation: Identification of terms and concepts in studies of limitations. Oecologia 145:1–8.

Murawski, D. A. 1995. Reproductive biology and genetics from a canopy perspective. Pages

457–93 in M. D. Lowman and N. M. Nadkarni, eds. Forest canopies. Academic Press, San Diego, CA.

Murcia, C., and P. Feinsinger. 1996. Interspecific pollen loss by hummingbirds visiting flower mixtures: effects of floral architecture. Ecology 77:550–60.

Murray, K. G. 1988. Avian seed dispersal of three Neotropical gap-dependent plants. Ecological Monographs 58:271–98.

———. 2000. Fruiting phenologies of pioneer plants, constraints imposed by flowering phenology, disturbance regime, and disperser migration patterns. Pages 283–286 in N. M. Nadkarni and N. T. Wheelwright, eds. Monteverde.

Murray, K. L., and T. H. Fleming. 2008. Social structure and mating system of the buffy flower bat, *Erophylla sezekorni* (Chiroptera: Phyllostomidae). Journal of Mammalogy 89:1391–1400.

Muscarella, R., and T. H. Fleming. 2007. The role of frugivorous bats in tropical forest succession. Biological Reviews 82:573–90.

Myers, N., R. A. Mittermeier, C. G. Mittermeier, G. A. B. da Fonseca, and J. Kent. 2000. Biodiversity hotspots for conservation priorities. Nature 403:853–58.

Nabhan, G. P., and S. L. Buchmann. 1997. Services provided by pollinators. Pages 133–50 in G. C. Daily, ed. Nature's services, societal dependence on natural ecosystems. Island Press, Washington, DC.

Nagy, K. A., I. A. Girard, and T. K. Brown. 1999. Energetics of free-ranging mammals, reptiles, and birds. Annual Review of Nutrition 19:247–77.

Namoff, S., Q. Luke, F. Jimenez, A. Veloz, C. Lewis, V. Sosa, M. Maunder, and J. Francisco-Ortega. 2010. Phylogenetic analyses of nucleotide sequences confirm a unique plant intercontinental disjunction between tropical Africa, the Caribbean, and the Hawaiian Islands. Journal of Plant Research 123:57–65.

Naranjo, M. E., C. Rengifo, and P. J. Soriano. 2003. Effect of ingestion by bats and birds on seed germination of *Stenocereus griseus* and *Subpilocereus repandus* (Cactaceae). Journal of Tropical Ecology 19:19–25.

Nason, J. D., E. A. Herre, and J. L. Hamrick. 1998. The breeding structure of a tropical keystone plant resource. Nature 391:685–87.

Nassar, J., N. Ramirez, and O. Linares. 1997. Comparative pollination biology of Venezuelan columnar cacti and the role of nectar-feeding bats in their sexual reproduction. American Journal of Botany 84:918–27.

Nassar, J. M., J. L. Hamrick, and T. H. Fleming. 2001. Genetic variation and population structure of the mixed-mating cactus, *Melocactus curvispinus* (Cactaceae). Heredity 87:69–79.

Nassar, J. M., M. V. Salazar, A. Quintero, K. E. Stoner, M. Gomez, A. Cabrera, and K. Jaffe. 2008. Seasonal sebaceous patch in the nectar-feeding bats *Leptonycteris curasoae* and *L. yerbabuenae* (Phyllostomidae: Glossophaginae): phenological, histological, and preliminary chemical characterization. Zoology 111:363–76.

Nathan, R. 2006. Long-distance dispersal of plants. Science 313:786–88.

Nathan, R., and R. Casagrandi. 2004. A simple mechanistic model of seed dispersal, predation and plant establishment: Janzen-Connell and beyond. Journal of Ecology 92:733–46.

Nathan, R., and H. C. Muller-Landau. 2000. Spatial patterns of seed dispersal, their determinants and consequences for recruitment. Trends in Ecology and Evolution 15:278–85.

Nathan, R., F. M. Schurr, O. Spiegel, O. Steinitz, A. Trakhtenbrot, and A. Tsoar. 2008. Mechanisms of long-distance seed dispersal. Trends in Ecology and Evolution 23:638–47.

Nattero, J., A. N. Sersic, and A. A. Cocucci. 2010. Patterns of contemporary phenotypic selection and flower integration in the hummingbird-pollinated *Nicotiana glauca* between populations with different flower-pollinator combinations. Oikos 119:852–63.

Nealen, P. M., and R. E. Ricklefs. 2001. Early diversification of the avian brain: body relationship. Journal of Zoology 253:391–404.

Ne'eman, G., A. Jurgens, L. Newstrom-Lloyd, S. G. Potts, and A. Dafni. 2010. A framework for comparing pollinator performance: effectiveness and efficiency. Biological Reviews 85:435–51.

Nelson, S. L., D. V. Masters, S. R. Humphrey, and T. H. Kunz. 2005. Fruit choice and calcium block use by Tongan fruit bats in American Samoa. Journal of Mammalogy 86:1205–9.

Nemeth, M. B., and N. L. Smith-Huerta. 2002. Effects of pollen load composition and deposition pattern on pollen performance in *Clarkia unguiculata* (Onagraceae). International Journal of Plant Sciences 163:795–802.

Newbury, D. M., G. B. Chuyong, and L. Zimmermann. 2006. Mast fruiting of large ectomycorrhizal African rain forest trees: importance of dry season intensity, and the resource-limitation hypothesis. New Phytologist 170:561–79.

Newbury, D. M., N. Songwe, and G. B. Chuyong. 1998. Phenology and dynamics of an African rainforest at Korup, Cameroon. Pages 267–308 in D. M. Newbury, H. H. T. Prins, and N. D. Brown, eds. Dynamics of tropical communities. Blackwell, Oxford.

Newstrom, L., and A. Robertson. 2005. Progress in understanding pollination systems in New Zealand. New Zealand Journal of Botany 43:1–59.

Newstrom, L. E., G. W. Frankie, and H. G. Baker. 1994. A new classification for plant phenology based on flowering patterns in lowland tropical rain-forest trees at La Selva, Costa Rica. Biotropica 26:141–59.

Nicolay, C. W., and Y. Winter. 2006. Performance analysis as a tool for understanding the ecological morphology of flower-visiting bats. Pages 131–44 in A. Zubaid, G. F. McCracken, and T. H. Kunz, eds. Functional and evolutionary ecology of bats. Oxford University Press, New York.

Nicolson, S. W., and P. A. Fleming. 2003. Nectar as food for birds: the physiological consequences of drinking dilute sugar solutions. Plant Systematics and Evolution 238:139–53.

Niklas, K. J. 1994. Plant allometry: the scaling of form and process. University of Chicago Press, Chicago.

Nishihara, H., S. Maruyama, and N. Okada. 2009. Retroposon analysis and recent geological data suggest near-simultaneous divergence of the three superorders of mammals. Proceedings of the National Academy of Sciences of the United States of America 106:5235–40.

Norberg, U. M., and J. M. V. Rayner. 1987. Ecological morphology and flight in bats (Mammalia; Chiroptera): wing adaptations, flight performance, foraging strategy, and echolocation. Philosophical Transactions of the Royal Society of London Series B: Biological Sciences 316:335–427.

Norden, N., J. Chave, P. Belbenoit, A. Caubere, P. Chatelet, P.-M. Forget, and C. Thebaud. 2007. Mast fruiting is a frequent strategy in woody species of eastern South America. PLoS One 2:e1079.

Norghauer, J. M., J. R. Malcolm, B. L. Zimmerman, and J. M. Felfili. 2006. An experimental test of density- and distant-dependent recruitment of mahogany (*Swietenia macrophylla*) in southeastern Amazonia. Oecologia 148:437–46.

Norman, J. A., F. E. Rheindt, D. L. Rowe, and L. Christidis. 2007. Speciation dynamics in the Australo-Papuan *Meliphaga* honeyeaters. Molecular Phylogenetics and Evolution 42:80–91.

Nowak, R. M. 1991. Walker's mammals of the world. 5th ed. Johns Hopkins University Press, Baltimore, MD.

———. 1994. Walker's bats of the world. Johns Hopkins University Press, Baltimore, MD.

Numata, S., M. Yasuda, T. Okuda, N. Kachi, and N. S. M. Noor. 2003. Temporal and spatial patterns of mass flowerings on the Malay Peninsula. American Journal of Botany 90:1025–31.

Nuñez-Iturri, G., and H. F. Howe. 2007. Bushmeat and the fate of trees with seeds dispersed by large primates in a lowland rain forest in western Amazonia. Biotropica 39:348–54.

Nuñez-Iturri, G., O. Olsson, and H. F. Howe. 2008. Hunting reduces recruitment of primate-dispersed trees in Amazonian Peru. Biological Conservation 141:1536–46.

Nur, N. 1976. Studies on pollination in Musaceae. Annals of Botany 40:167–77.

Nyári, A. S., A. T. Peterson, N. H. Rice, and R. G. Moyle. 2009. Phylogenetic relationships of flowerpeckers (Aves: Dicaeidae): novel insights into the evolution of a tropical passerine clade. Molecular Phylogenetics and Evolution 53:613–19.

Nybom, H. 2004. Comparison of different nuclear DNA markers for estimating intraspecific genetic diversity in plants. Molecular Ecology 13:1143–55.

Nyffeler, R. 2002. Phylogenetic relationships in the cactus family (Cactaceae) based on evidence from trnK/matK and trnL-trnF sequences. American Journal of Botany 89:312–26.

O'Gorman, E. J., and M. C. Emmerson. 2009. Perturbations to trophic interactions and the stability of complex food webs. Proceedings of the National Academy of Sciences of the United States of America 106:13393–98.

Ohlson, J. I. 2007. A molecular phylogeny of the cotingas (Aves: Cotingidae). Molecular Phylogenetics and Evolution 42:25–37.

Ohsawa, M. 1995. Latitudinal comparison of altitudinal changes in forest structure, leaf-type, and species richness in humid monsoon Asia. Vegetatio 121:3–10.

O'Leary, M. A., J. I. Bloch, J. J. Flynn, T. J. Gaudin, A. Giallombardo, N. P. Giannini, et al. 2013. The placental mammal ancestor and the post-K-Pg radiation of placentals. Science 339:662–67.

Olesen, J. M., and A. Valido. 2003. Lizards as pollinators and seed dispersers: an island phenomenon. Trends in Ecology and Evolution 18:177–81.

Ollerton, J., R. Alarcon, N. M. Waser, and others. 2009. A global test of the pollination syndrome hypothesis. Annals of Botany 103: 1471–80.

Olson, V. A., and I. P. F. Owens. 2005. Interspecific variation in the use of carotenoid-based coloration in birds: diet, life history and phylogeny. Journal of Evolutionary Biology 18:1534–46.

Onderdonk, D. A., and C. A. Chapman. 2000. Coping with forest fragmentation: the primates of Kibale National Park, Uganda. International Journal of Primatology 21:587–611.

Oostra, V., L. G. L. Gomes, and V. Nijman. 2008. Implications of deforestation for the abundance of restricted-range bird species in a Costa Rican cloud-forest. Bird Conservation International 18:11–19.

Opler, P. A. 1983. Nectar production in a tropical ecosystem. Pages 30–79 in B. Bentley and T. Elias, eds. The biology of nectaries. Columbia University Press, New York.

Opler, P. A., G. W. Frankie, and H. G. Baker. 1980. Comparative phenological studies of treelet and shrub species in tropical wet and dry forests in the lowlands of Costa Rica. Journal of Ecology 68:167–88.

Ordano, M., J. Fornoni, K. Boege, and C. A. Dominguez. 2008. The adaptive value of phenotypic floral integration. New Phytologist 179:1183–92.

Ornelas, J. F. 1994. Serrate tomia—an adaptation for nectar robbing in hummingbirds. Auk 111:703–10.

Ornelas, J. F., A. Ordano, J. de Nova, M. E. Quintero, and T. Garland. 2007. Phylogenetic analysis of interspecific variation in nectar of hummingbird-visited plants. Journal of Evolutionary Biology 20:1904–17.

Ortega-Baes, P., M. Saravia, S. Suhring, H. Godinez-Alvarez, and M. Zamar. 2011. Reproductive biology of *Echinopsis terscheckii* (Cactaceae): the role of nocturnal and diurnal pollinators. Plant Biology 13:33–40.

Ortiz-Pulido, R., J. Laborde, and S. Guevara. 2000. Fruit-eating habits of birds in a fragmented landscape: implications for seed dispersal. Biotropica 32:473–88.

Owens, I. P. F., P. M. Bennett, and P. H. Harvey. 1999. Species richness among birds: body size, life history, sexual selection or ecology? Proceedings of the Royal Society of London Series B: Biological Sciences 266:933–39.

Pabón-Mora, N., and A. Litt. 2011. Comparative anatomical and developmental analysis of dry and fleshy fruits of Solanaceae. American Journal of Botany 98:1415–36.

Pacheco, V., and E. Vivar. 1996. Annotated checklist of the non-flying mammals at Pakitza, Manu

Reserve Zone, Manu National Park, Peru. Pages 577–92 in D. E. Wilson and A. Sandoval, eds. Manu, the biodiversity of southeastern Peru. Smithsonian Institution, Washington, DC.

Pacini, E., M. Nepi, and J. L. Vesprini. 2003. Nectar biodiversity: a short review. Plant Systematics and Evolution 238:7–21.

Pagel, M. D. 1999. Inferring the historical patterns of biological evolution. Nature 401: 877–84.

Palmer, T. M., M. L. Stanton, and T. P. Young. 2003. Competition and coexistence: exploring mechanisms that restrict and maintain diversity within mutualist guilds. American Naturalist 162:S63–S79.

Parachnowitsch, A. L., and A. Kessler. 2010. Pollinators exert natural selection on flower size and floral display in *Penstemon digitalis*. New Phytologist 188:393–402.

Pardini, R., S. M. de Souza, R. Braga-Neto, and J. P. Metzger. 2005. The role of forest structure, fragment size and corridors in maintaining small mammal abundance and diversity in an Atlantic forest landscape. Biological Conservation 124:253–266.

Pardini, E. A., and J. L. Hamrick. 2007. Hierarchical patterns of paternity within crowns of *Albizia julibrissin* (Fabaceae). American Journal of Botany 94:111–18.

Parmentier, I., Y. Malhi, B. Senterre, R. J. Whittaker, A. Alonso, M. P. B. Balinga, A. Bakayoko, et al. 2007. The odd man out? might climate explain the lower tree alpha-diversity of African rain forests relative to Amazonian rain forests? Journal of Ecology 95:1058–71.

Parra, J. L., J. A. McGuire, and C. H. Graham. 2010. Incorporating clade identity in analyses of phylogenetic community structure: an example with hummingbirds. American Naturalist 176:573–87.

Patel, A. 1997. Phenological patterns of *Ficus* in relation to other forest trees in southern India. Journal of Tropical Ecology 13:681–95.

Patel, S., J. D. Weckstein, J. S. L. Patane, J. M. Bates, and A. Aleixo. 2011. Temporal and spatial diversification of *Pteroglossus* araçaris (AVES: Ramphastidae) in the neotropics: constant rate of diversification does not support an increase in radiation during the Pleistocene. Molecular Phylogenetics and Evolution 58:105–15.

Paton, D. C. 1985. Food supply, population structure, and behavior of New Holland honeyeaters *Phylidonyris novaehollandiae* in woodland near Horsham, Victoria. Pages 219–30 in A. Keast, H. F. Recher, H. A. Ford, and D. A. Saunders, eds. Birds of eucalypt forests and woodlands: ecology, conservation, management. Royal Australian Ornithologists Union and Surrey Beatty & Sons, Sydney.

———. 1986. Honeyeaters and their plants in south-eastern Australia. Pages 9–19 in H. A. Ford and D. C. Paton, eds. The dynamic partnership: birds and plants in southern Australia. South Australia Government Printer, Adelaide.

Paton, D. C., and B. G. Collins. 1989. Bills and tongues of nectar-feeding birds: a review of morphology, function and performance, with intercontinental comparisons. Australian Journal of Ecology 14:473–506.

Paton, D. C., and H. A. Ford. 1977. Pollination by birds of native plants in Australia. Emu 77: 73–85.

Pausas, J. G., and M. Verdu. 2010. The jungle of methods for evaluating phenotypic and phylogenetic structure of communities. Bioscience 60:614–25.

Pauw, A., and J. A. Hawkins. 2011. Reconstruction of historical pollination rates reveals linked declines of pollinators and plants. Oikos 120:344–49.

Payne, J., C. M. Francis, and K. Phillips. 1985. A field guide to the mammals of Borneo. The Sabah Society, Sabah, Malaysia.

Pearson, D. L. 1977. A pantropical comparison of bird community structure on six lowland forest sites. Condor 79:232–44.

———. 1982. Historical factors and bird species richness. Pages 441–52 in G. T. Prance, ed. Biological diversification in the tropics. Columbia University Press, New York.

Pearson, R. G., and T. P. Dawson. 2005. Long-distance plant dispersal and habitat fragmentation: identifying conservation targets for spatial landscape planning under climate change. Biological Conservation 123:389–401.

Pedersen, L. B., and W. J. Kress. 1999. Honeyeater (Meliphagidae) pollination and the floral biology of Polynesian *Heliconia* (Heliconiaceae). Plant Systematics and Evolution 216:1–21.

Pellmyr, O. 2002. Pollination by animals. Pages 157–84 in C. M. Herrera and O. Pellmyr, eds. Plant-animal interactions, an evolutionary approach. Blackwell Scientific, Oxford.

———. 2003. Yuccas, yucca moths, and coevolution: a review. Annals of the Missouri Botanical Garden 90:35–55.

Pennington, R. T., and C. W. Dick. 2004. The role of immigrants in the assembly of the South American rainforest flora. Philosophical Transactions of the Royal Society of London Series B: Biological Sciences 359:1611–22.

Pennington, R. T., M. Lavin, D. E. Prado, C. A. Pendry, S. K. Pell, and C. A. Butterworth. 2004. Historical climate change and speciation: Neotropical seasonally dry forest plants show patterns of both Tertiary and Quaternary diversification. Philosophical Transactions of the Royal Society of London Series B: Biological Sciences 359:515–37.

Pennington, R. T., M. Lavin, T. E. Sarkinen, G. P. Lewis, B. B. Klitgaard, and C. E. Hughes. 2010. Contrasting plant diversification histories within the Andean biodiversity hotspot. Proceedings of the National Academy of Sciences of the United States of America 107:13783–87.

Pereira, S. L., and A. J. Baker. 2004. Vicariant speciation of curassows (Aves, Cracidae): a hypothesis based on a mitochondrial DNA phylogeny. Auk 121:682–94.

Pereira, S. L., K. P. Johnson, D. H. Clayton, and A. J. Baker. 2007. Mitochondrial and nuclear DNA sequences support a Cretaceous origin of Columbiformes and a dispersal-driven radiation in the Paleogene. Systematic Biology 56:656–72.

Peres, C. A. 2000. Identifying keystone plant resources in tropical forests: the case of gums from *Parkia* pods. Journal of Tropical Ecology 16:287–317.

———. 2008. Soil fertility and arboreal mammal biomass in tropical forests. Pages 349–64 in W. P. Carson and S. A. Schnitzer, eds. Tropical forest community ecology. Wiley-Blackwell, Oxford.

Peres, C. A., and C. H. Janson. 1999. Species coexistence, distribution and environmental determinants of Neotropical primate richness. Pages 55–74 in J. G. Fleagle, C. H. Janson, and K. E. Reed, eds. Primate communities. Cambridge University Press, Cambridge.

Peres, C. A., and E. Palacios. 2007. Basin-wide effect of game harvest on vertebrate population densities in Amazonian forests: implications for animal-mediated seed dispersal. Biotropica 39:304–15.

Peres, C. A., and M. van Roosmalen. 2002. Primate frugivory in two species-rich Neotropical forests: implications for the demography of large-seeded plants in overhunted areas. Pages 407–21 in D. J. Levey, W. R. Silva, and M. Galetti, eds. Seed dispersal and frugivory: ecology, evolution and conservation. CABI, Wallingford.

Perez, F., M. T. K. Arroyo, and R. Medel. 2007. Phylogenetic analysis of floral integration in *Schizanthus* (Solanaceae): does pollination truly integrate corolla traits? Journal of Evolutionary Biology 20:1730–38.

Perez, F., M. T. K. Arroyo, R. Medel, and M. A. Hershkovitz. 2006. Ancestral reconstruction of flower morphology and pollination systems in *Schizanthus* (Solanaceae). American Journal of Botany 93:1029–38.

Perez-Nasser, N., L. E. Eguiarte, and D. Pinero. 1993. Mating system and genetic structure of the distylous tropical tree *Psychotria faxlucens* (Rubiaceae). American Journal of Botany 80:45–52.

Perret, M., A. Chautems, R. Spichiger, T. G. Barraclough, and V. Savolainen. 2007. The geographical pattern of speciation and floral diversification in the Neotropics: the tribe Sinningieae (Gesneriaceae) as a case study. Evolution 61:1641–60.

Perret, M., A. Chautems, R. Spichiger, G. Kite, and V. Savolainen. 2003. Systematics and evolution of tribe Sinningieae (Gesneriaceae): evidence from phylogenetic analyses of six plastid DNA regions and nuclear ncpGS. American Journal of Botany 90:445–60.

Perret, M., A. Chautems, R. Spichiger, M. Peixoto, and V. Savolainen. 2001. Nectar sugar composition in relation to pollination syndrome in Sinningieae (Gesneriaceae). Annals of Botany 87:267–73.

Perrins, C. M., and A. L. A. Middleton, eds. 1985. The encyclopedia of birds. Facts on File, New York.

Peters, S. L., J. R. Malcolm, and B. L. Zimmerman. 2006. Effects of selective logging on bat communities in the southeastern Amazon. Conservation Biology 20:1410–21.

Petit, S., and L. Pors. 1996. Survey of columnar cacti and carrying capacity for nectar feeding bats on Curaçao. Conservation Biology 10:769–75.

Pettersson, S., F. Ervik, and J. T. Knudsen. 2004. Floral scent of bat-pollinated species: West Africa vs. the New World. Biological Journal of the Linnean Society 82:161–68.

Pettersson, S., and J. T. Knudsen. 2001. Floral scent and nectar production in Parkia biglobosa Jacq. (Leguminosae: Mimosoideae). Botanical Journal of the Linnean Society 135:97–106.

Phillimore, A. B., R. P. Freckleton, C. D. L. Orme, and I. P. F. Owens. 2006. Ecology predicts large-scale patterns of phylogenetic diversification in birds. American Naturalist 168:220–29.

Phillips, C. J., G. W. Grimes, and G. L. Forman. 1977. Oral biology. Pages 121–246 in R. J. Baker, J.K. Jones, and D. C. Carter, eds. Biology of bats of the New World family Phyllostomatidae, vol. 2. Texas Tech University Press, Lubbock.

Phillips, O. L., P. N. Vargas, A. L. Monteagudo, A. P. Cruz, M. E. C. Zans, W. G. Sanchez, M. Yli-Halla, and S. Rose. 2003. Habitat association among Amazonian tree species: a landscape-scale approach. Journal of Ecology 91:757–75.

Piechowski, D., S. Dotterl, and G. Gottsberger. 2010. Pollination biology and floral scent chemistry of the Neotropical chiropterophilous Parkia pendula. Plant Biology 12:172–82.

Pitman, N. C. A., H. Mogollon, N. Davila, M. Rios, and others. 2008. Tree community change across 700 km of lowland Amazonian forest from the Andean foothills to Brazil. Biotropica 40:525–35.

Pitman, N. C. A., J. Terborgh, M. R. Silman, and P. Nuez. 1999. Tree species distributions in an upper Amazonian forest. Ecology 80:2651–61.

Pitman, N. C. A., J. W. Terborgh, M. R. Silman, P. Nunez, D. A. Neill, C. E. Ceron, W. A. Palacios, and M. Aulestia. 2001. Dominance and distribution of tree species in upper Amazonian terra firme forests. Ecology 82:2101–17.

———. 2002. A comparison of tree species diversity in two upper Amazonian forests. Ecology 83:3210–24.

Pizo, M. A. 2004. Frugivory and habitat use by fruit-eating birds in a fragmented landscape of southeast Brazil. Ornitologia Neotropical 15:117–26.

Pizzey, G. 1980. A field guide to the birds of Australia. Princeton University Press, Princeton, NJ.

Plana, V. 2004. Mechanisms and tempo of evolution in the African Guineo-Congolian rainforest. Philosophical Transactions of the Royal Society of London B 359:1585–94.

Platt, W. J., and J. H. Connell. 2003. Natural disturbances and directional replacement of species. Ecological Monographs 73:507–22.

Pontzer, H., D. A. Raichlen, R. W. Shumaker, C. Ocobock, and S. A. Wich. 2010. Metabolic adpatation for low energy throughput in orangatans. Proceedings of the National Academy of Sciences of the United States of America 107:14048–52.

Popp, M., V. Mirre, and C. Brochmann. 2011. A single Mid-Pleistocene long-distance dispersal by a bird can explain the extreme bipolar disjunction in crowberries (Empetrum). Proceedings of the National Academy of Sciences of the United States of America 108:6520–25.

Porcher, E., and R. Lande. 2005. The evolution of self-fertilization and inbreeding depression under pollen discounting and pollen limitation. Journal of Evolutionary Biology 18:497–508.

Potts, M. D., P. S. Ashton, L. S. Kaufman, and J. B. Plotkin. 2002. Habitat patterns in tropical rain forests: a comparison of 105 plots in Northwest Borneo. Ecology 83:2782–97.

Potts, S. G., J. C. Biesmeijer, C. Kremen, P. Neumann, O. Schweiger, and W. E. Kunin. 2010. Global pollinator declines: trends, impacts and drivers. Trends in Ecology and Evolution 25:345–53.

Poulin, B., S. J. Wright, G. Lefebvre, and O. Calderon. 1999. Interspecific synchrony and asynchrony in the fruiting phenologies of congeneric bird-dispersed plants in Panama. Journal of Tropical Ecology 15:213–27.

Poulsen, J. R., C. J. Clark, and T. B. Smith. 2001. Seed dispersal by a diurnal primate community in the Dja Reserve, Cameroon. Journal of Tropical Ecology 17:787–808.

Powell, G. V. N., and R. Bjork. 1995. Implications of intratropical migration on reserve design—a case-study using *Pharomachrus mocinno*. Conservation Biology 9:354–62.

———. 2004. Habitat linkages and the conservation of tropical biodiversity as indicated by seasonal migrations of three-wattled bellbirds. Conservation Biology 18:500–509.

Prather, L. A. 1999. The relative lability of floral vs. non-floral characters and a morphological phylogenetic analysis of *Cobaea* (Polemoniaceae). Botanical Journal of the Linnean Society 131:433–50.

Preston, J. C., L. C. Hileman, and P. Cubas. 2011. Reduce, reuse, and recycle: developmental evolution of trait diversification. American Journal of Botany 98:397–403.

Price, J. P., and W. L. Wagner. 2004. Speciation in Hawaiian angiosperm lineages: cause, consequence, and mode. Evolution 58:2185–2200.

Price, T. 1998. Sexual selection and natural selection in bird speciation. Philosophical Transactions of the Royal Society of London B 353:251–60.

———. 2007. Speciation in birds. Roberts and Co., Greenwood Village, CO.

Primack, R., and R. Corlett. 2005. Tropical rain forests: an ecological and biogeographical comparison. Blackwell Scientific, Malden, MA.

Primack, R. B. 1987. Relationships among flowers, fruits, and seeds. Annual Review of Ecology and Systematics 18:409–30.

Proctor, G. R. 1984. Flora of the Cayman Islands. Kew Bulletin Additional Series 11, London.

Proctor, M., P. Yeo, and A. Lack. 1996. The natural history of pollination. Timber Press, Portland, OR.

Prum, R. O. 1997. Phylogenetic tests of alternative intersexual selection mechanisms: trait macroevolution in a polygynous clade (Aves: Pipridae). American Naturalist 149:668–92.

Purvis, A. 1995. A composite estimate of primate phylogeny. Philosophical Transactions of the Royal Society of London Series B: Biological Sciences 348:405–21.

Purvis, A., S. Nee, and P. H. Harvey. 1995. Macroevolutionary inferences from primate phylogeny. Proceedings of the Royal Society of London B 260:329–33.

Pyke, C. P., R. Condit, A. Aguilar, and A. Hernandez. 2001. Floristic composition across a climatic gradient in a Neotropical lowland forest. Journal of Vegetation Science 12:533–66.

Pyke, G. H. 1980. The foraging behaviour of Australian honeyeaters: a review and some comparisons with hummingbirds. Australian Journal of Ecology 5:343–69.

———. 1984. Optimal foraging theory: a critical review. Annual Review of Ecology and Systematics 15:523–75.

Quesada, M., K. E. Stoner, J. A. Lobo, Y. Herrerias-Diego, C. Palacios-Guevara, M. A. Munguia-Rosas, K. A. O. Salazar, and V. Rosas-Guerrero. 2004. Effects of forest fragmentation on pollinator activity and consequences for plant reproductive success and mating patterns in bat-pollinated bombacaceous trees. Biotropica 36:131–38.

Quesada, M., K. E. Stoner, V. Rosas-Guerrero, C. Palacios-Guevara, and J. A. Lobo. 2003. Effects of habitat disruption on the activity of nectarivorous bats (Chiroptera: Phyllostomidae) in a dry tropical forest: implications for the reproductive success of the Neotropical tree *Ceiba grandiflora*. Oecologia 135:400–406.

Rabosky, D. L. 2009. Ecological limits on clade diversification in higher taxa. American Naturalist 173:662–74.

Raffaele, H., J. Wiley, O. Garrido, A. Keith, and J. Raffaele. 1998. A guide to the birds of the West Indies. Princeton University Press, Princeton, NJ.

Raguso, R. A. 2008. Wake up and smell the roses: the ecology and evolution of floral scent. Annual Review of Ecology Evolution and Systematics 39:549–70.

Rahbek, C. 1995. The elevational gradient of species richness: a uniform pattern? Ecography 18:200–205.

———. 1997. The relationship among area, elevation, and regional species richness in Neotropical birds. American Naturalist 149:875–902.

Rahbek, C., and G. R. Graves. 2000. Detection of macro-ecological patterns in South American hummingbirds is affected by spatial scale. Proceedings of the Royal Society of London B: Biological Sciences 267:2259–65.

Rainey, W. E., E. D. Pierson, T. Elmqvist, and P. A. Cox. 1995. The role of flying foxes (Pteropodidae) in oceanic island ecosystems of the Pacific. Pages 47–62 in P. A. Racey and S. Swift, eds. Ecology, evolution and behaviour of bats. Clarendon Press, Oxford.

Rakotomanana, H., T. Hino, M. Kanzaki, and H. Morioka. 2001. The role of the velvet asity *Philepitta castanea* in regeneration of understory shrubs in Madagascan rainforest. Ornithological Science 2:49–58.

Raman, T. R. S., and D. Mudappa. 2003. Correlates of hornbill distribution and abundance in rainforest fragments in the southern Western Ghats, India. Bird Conservation International 13:199–212.

Ratcliffe, J. M. 2009. Neuroecology and diet selection in phyllostomid bats. Behavioural Processes 80:247–51.

Rathcke, B. J. 2000. Hurricane causes resource and pollination limitation of fruit set in a bird-pollinated shrub. Ecology 81:1951–58.

Rausher, M. D. 2008. Evolutionary transitions in floral color. International Journal of Plant Sciences 169:7–21.

Raven, P. H., and D. I. Axelrod. 1974. Angiosperm biogeography and past continental movements. Annals of the Missouri Botanical Garden 61:539–673.

Reader, S. M., and K. N. Laland. 2002. Social intelligence, innovation, and enhanced brain size in primates. Proceedings of the National Academy of Sciences of the United States of America 99:4436–41.

Redford, K. H. 1992. The empty forest. Bioscience 42:412–22.

Rees, M., R. Condit, M. Crawley, S. Pacala, and D. Tilman. 2001. Long-term studies of vegetation dynamics. Science 293:650–55.

Regal, P. J. 1977. Ecology and evolution of flowering plant dominance. Science 196:622–29.

Regan, B. C., C. Julliot, B. Simmen, F. Vienot, P. Charles-Dominique, and J. D. Mollon. 2001. Fruits, foliage and the evolution of primate colour vision. Philosophical Transactions of the Royal Society of London Series B: Biological Sciences 356:229–83.

Reid, N. 1989. Dispersal of mistletoes by honeyeaters and flowerpeckers—components of seed dispersal quality. Ecology 70:137–45.

———. 1990. Mutualistic interdependence between mistletoes (*Amyema quandang*), and spiny-cheeked honeyeaters and mistletoebirds in an arid woodland. Australian Journal of Ecology 15:175–90.

Remis, M. J. 2006. The role of taste in food selection by African apes: implications for niche separation and overlap in tropical forests. Primates 47:56–64.

Renner, S. S. 2004. Plant dispersal across the tropical Atlantic by wind and sea currents. International Journal of Plant Sciences 164:S23–S33.

Renner, S. S., G. Clausing, and K. Meyer. 2001. Historical biogeography of Melastomataceae: the roles of Tertiary migration and long-distance dispersal. American Journal of Botany 88:1290–1300.

Renner, S. S., and R. E. Ricklefs. 1995. Dioecy and its correlates in the flowering plants. American Journal of Botany 82:596–606.

Rey, P. J., J. E. Gutierrez, J. Alcantara, and F. Valera. 1997. Fruit size in wild olives: implications for avian seed dispersal. Functional Ecology 11:611–18.

Reynolds, R. J., M. R. Dudash, and C. B. Fenster. 2010. Multiyear study of multivariate linear and nonlinear phenotypic selection on floral traits of hummingbird-pollinated *Silene virginica*. Evolution 64:358–69.

Rezende, E. L., P. Jordano, and J. Bascompte. 2007a. Effects of phenotypic complementarity and phylogeny on the nested structure of mutualistic networks. Oikos 116:1919–29.

Rezende, E. L., J. E. Lavabre, P. R. Guimaraes, P. Jordano, and J. Bascompte. 2007b. Non-random coextinctions in phylogenetically structured mutualistic networks. Nature 448:925–29.

Rhoades, D. F., and J. C. Bergdahl. 1981. Adaptive significance of toxic nectar. American Naturalist 117:798–803.

Ribas, C. C., R. G. Moyle, C. Y. Miyaki, and J. Cracraft. 2007. The assembly of montane biotas: linking Andean tectonics and climatic oscillations to independent regimes of diversification in *Pionus* parrots. Proceedings of the Royal Society B: Biological Sciences 274:2399–2408.

Richards, G. C. 1995. A review of ecological interactions of fruit bats in Australian ecosystems. Pages 79–96 in P. A. Racey and S. Swift, eds. Ecology, evolution and behaviour of bats. Clarendon Press, Oxford.

Richards, P. W. 1952. The tropical rain forest. Cambridge University Press, Cambridge.

Richardson, J. E., L. W. Chatrou, J. B. Mols, R. H. J. Erkens, and M. D. Pirie. 2004. Historical biogeography of two cosmopolitan families of flowering plants: Annonaceae and Rhamnaceae. Philosophical Transactions of the Royal Society of London Series B: Biological Sciences 359:1495–1508.

Richardson, K. C., and R. D. Wooller. 1986. The structures of the gastrointestinal tracts of honeyeaters and other small birds in relation to their diets. Australian Journal of Zoology 34:119–24.

———. 1990. Adaptations of the alimentary tracts of some Australian lorikeets to a diet of pollen and nectar. Australian Journal of Zoology 38:581–86.

Richter, H. V., and G. S. Cumming. 2006. Food availability and annual migration of the straw-colored fruit bat (*Eidolon helvum*). Journal of Zoology 268:35–44.

———. 2008. First application of satellite telemetry to track African straw-coloured fruit bat migration. Journal of Zoology 275:172–76.

Ricklefs, R. E. 1990. Ecology. 3rd ed. W. H. Freeman and Co., New York.

———. 2010. Evolutionary diversification, coevolution between populations and their antagonists, and the filling of niche space. Proceedings of the National Academy of Sciences of the United States of America 107:1265–72.

Ricklefs, R. E., and R. E. Latham. 1993. Global patterns of diversity in mangrove floras. Pages 215–29 in R. E. Ricklefs and D. Schluter, eds. Species diversity in ecological communities: historical and geographical perspectives. University of Chicago Press, Chicago.

Ricklefs, R. E., and S. S. Renner. 1994. Species richness within families of flowering plants. Evolution 48:1619–36.

Rico-Guevara, A., and M. A. Rubega. 2011. The hummingbird tongue is a fluid trap, not a capillary tube. Proceedings of the National Academy of Sciences of the United States of America 108:9356–60.

Ridley, H. N. 1930. The dispersal of plants throughout the world. Reeve, Ashford, UK.

Rieseberg, L. H., and J. H. Willis. 2007. Plant speciation. Science 317:910–14.

Ritland, K. 1989. Correlated matings in the partial selfer *Mimulus guttatus*. Evolution 43:848–59.

Ritz, C. M., L. Martins, R. Mecklenberg, V. Goremykin, and F. H. Hellwig. 2007. The molecular phylogeny of *Rebutia* (Cactaceae) and its allies demonstrates the influence of paleogeography on the evolution of South American mountain cacti. American Journal of Botany 94:1321–32.

Robinson, W. D., J. D. Brawn, and S. K. Robinson. 2000. Forest bird community structure in central Panama: Influence of spatial scale and biogeography. Ecological Monographs 70:209–35.

Rocha, M., S. V. Good-Avila, F. Molina-Freaner, H. T. Arita, A. Castillo, A. Garcia-Mendoza, A. Silva-Montellano, B. S. Gaut, V. Souza, and L. E. Eguiarte. 2006. Pollination biology and adaptive radiation of Agavaceae, with special emphasis on the genus *Agave*. Aliso 22:329–44.

Rocha, M., A. Valera, and L. E. Eguiarte. 2005. Reproductive ecology of five sympatric *Agave littaea* (Agavaceae) species in Central Mexico. American Journal of Botany 92:1330–41.

Rockwood, L. L. 1985. Seed weight as a function of life form, elevation and life zone in Neotropical forests. Biotropica 17:32–39.

Rodríguez-Cabal, M. A., M. A. Aizen, and A. J. Novaro. 2007. Habitat fragmentation disrupts a plant-disperser mutualism in the temperate forest of South America. Biological Conservation 139:195–202.

Rodríguez-Duran, A. 1995. Metabolic rates and themal conductance in four species of Neotropical bats roosting in hot caves. Comparative Biochemistry and Physiology A: Molecular and Integrative Physiology 110:347–55.

Rodríguez-Duran, A., and T. H. Kunz. 2001. Biogeography of West Indian bats: an ecological perspective. Pages 355–68 in C. A. Woods and F. E. Sergile, eds. Biogeography of the West Indies. CRC Press, Boca Raton, FL.

Rodríguez-Peña, N., K. E. Stoner, J. E. Schondube, J. Ayala-Berdon, C. M. Flores-Ortiz, and C. Martínez del Rio. 2007. Effects of sugar composition and concentration on food selection by Saussure's long-nosed bat (*Leptonycteris curasoae*) and the long-tongued bat (*Glossophaga soricina*). Journal of Mammalogy 88:1466–74.

Rojas, D., A. Vale, V. Ferrero, and L. Navarro. 2011. When did plants become important to leaf-nosed bats? diversification of feeding habits in the family Phyllostomidae. Molecular Ecology 20: 2217–28.

Root, R. B. 1967. The niche exploitation pattern of the blue-gray gnatcatcher. Ecological Monographs 37:317–50.

Rosas-Guerrero, V., M. Quesada, W. S. Armbruster, R. Perez-Barrales, and S. D. Smith. 2010. Influence of pollination specialization and breeding system on floral integration and phenotypic variation in *Ipomoea*. Evolution 65:350–64.

Rosenzweig, M. L. 1995. Species diversity in space and time. Cambridge University Press, Cambridge.

Roulston, T. H., and J. H. Cane. 2000. Pollen nutritional content and digestibility for animals. Plant Systematics and Evolution 222:187–209.

Roulston, T. H., J. H. Cane, and S. L. Buchmann. 2000. What governs protein content of pollen: pollinator preferences, pollen-pistil interactions, or phylogeny? Ecological Monographs 70:617–43.

Rourke, J., and D. Wiens. 1977. Convergent floral evolution in South African and Australian Proteaceae and its possible bearing on pollination by nonflying mammals. Annals of the Missouri Botanical Garden 64:1–17.

Rowe, N. 1996. The pictorial guide to the living primates. Pogonias Press, Charlestown, RI.

Rowe, T., T. H. Rich, P. Vickers-Rich, M. Springer, and M. O. Woodburne. 2008. The oldest platypus and its bearing on divergence timing of the platypus and echidna clades. Proceedings of the National Academy of Sciences of the United States of America 105:1238–42.

Roxburgh, L., and B. Pinshow. 2000. Nitrogen requirements of an old world nectarivore, the orange- tufted sunbird *Nectarinia osea*. Physiological and Biochemical Zoology 73:638–45.

Roy, M. S., J. C. Torres-Mura, and F. Herel. 1998. Evolution and history of hummingbirds (Aves: Trochilidae) from the Juan Fernandez Islands, Chile. Ibis 140:265–73.

Ruby, J., P. T. Nathan, J. Balasingh, and T. H. Kunz. 2000. Chemical composition of fruits and leaves eaten by short-nosed fruit bat, *Cynopterus sphinx*. Journal of Chemical Ecology 26:2825–41.

Russo, S. E. 2003. Responses of dispersal agents to tree and fruit traits in *Virola calophylla* (Myristicaceae): implications for selection. Oecologia 136:80–87.

Russo, S. E., and C. K. Augspurger. 2004. Aggregated seed dispersal by spider monkeys limits recruitment to clumped patterns in *Virola calophylla*. Ecology Letters 7:1058–67.

Russo, S. E., C. J. Campbell, J. L. Dew, P. R. Stevenson, and S. A. Suarez. 2005. A multi-forest comparison of dietary preferences and seed dispersal by *Ateles* spp. International Journal of Primatology 26:1017–37.

Russo, S. E., S. Portnoy, and C. K. Augspurger. 2006. Incorporating animal behavior into seed dispersal models: Implications for seed shadows. Ecology 87:3160–74.

Rust, J., H. Singh, R. S. Rana, T. McCann, L. Singh, K. Anderson, N. Sarkar, et al. 2010. Biogeographic and evolutionary implications of a diverse paleobiota in amber from the early Eocene of India. Proceedings of the National Academy of Sciences of the United States of America 107:18360–65.

Sahley, C. T. 1996. Bat and hummingbird pollination of an autotetraploid columnar cactus, *Weberbauerocereus weberbaueri* (Cactaceae). American Journal of Botany 83:1329–36.

Sahley, C. T., M. A. Horner, and T. H. Fleming. 1993. Flight speeds and mechanical power outputs of the nectar-feeding bat, *Leptonycteris curasoae* (Phyllostomidae: Glossophaginae). Journal of Mammalogy 74:594–600.

Sakai, S. 2000. Reproductive phenology of gingers in a lowland mixed dipterocarp forest in Borneo. Journal of Tropical Ecology 16:337–54.

———. 2002. General flowering in lowland mixed dipterocarp forests of South-east Asia. Biological Journal of the Linnean Society 75:233–47.

Sakai, S., R. D. Harrison, K. Momose, K. Kuraji, H. Nagamasu, T. Yasunari, L. Chong, and T. Nakashizuka. 2006. Irregular droughts trigger mass flowering in aseasonal tropical forests in Asia. American Journal of Botany 93:1134–39.

Sakai, S., and T. Inoue. 1999. A new pollination system: dung-beetle pollination discovered in *Orchidantha inouei* (Lowiaceae, Zingiberales) in Sarawak. American Journal of Botany 86:56–61.

Sakai, S., M. Kato, and T. Inoue. 1999. Three pollinator guilds and variation in floral characteristics of Bornean gingers (Zingiberaceae and Costaceae). American Journal of Botany 86:646–58.

Saldarriaga, J. G., and C. Uhl. 1991. Recovery of forest vegetation following slash-and-burn agriculture in the upper Rio Negro. Pages 303–12 in A. Gomez-Pompa, T. C. Whitmore, and M. Hadley, eds. Rain forest regeneration and management. United Nations Educational Scientific and Cultural Organization, Paris.

Saldarriaga, J. G., D. C. West, M. L. Tharp, and C. Uhl. 1988. Long-term chronosequence of forest succession in the upper Rio Negro of Colombia and Venezuela. Journal of Ecology 76:938–58.

Salewski, V., and B. Bruderer. 2007. The evolution of bird migration—a synthesis. Naturwissenschaften 94:268–79.

Sanchez, F., C. Korine, M. Steeghs, L. J. Laarhoven, S. M. Cristescu, F. J. M. Harren, R. Dudley, and B. Pinshow. 2006. Ethanol and methanol as possible odor cues for Egyptian fruit bats (*Rousettus aegyptiacus*). Journal of Chemical Ecology 32:1289–1300.

Sandring, S., and J. Ågren. 2009. Pollinator-mediated selection on floral display and flowering time in the perennial herb *Arabidopsis lyrata*. Evolution 63:1292–1300.

Sanitjan, S., and J. Chen. 2009. Habitat and fig characteristics influence the bird assemblage and network properties of fig trees from Xishuangbanna, South-West China. Journal of Tropical Ecology 25:161–70.

San Martin-Gajardo, I., and M. Sazima. 2005. Chiropterophily in Sinningieae (Gesneriaceae): *Sinningia brasiliensis* and *Paliavana prasinata* are bat-pollinated, but *P. sericiflora* is not—not yet? Annals of Botany 95:1097–1103.

Santana, S. E., and E. R. Dumont. 2009. Connecting behaviour and performance: the evolution of biting behaviour and bite performance in bats. Journal of Evolutionary Biology 22:2131–45.

Saracco, J. F., J. A. Collazo, and M. J. Groom. 2004. How do frugivores track resources? insights from spatial analyses of bird foraging in a tropical forest. Oecologia 139:235–45.

Sargent, R. D. 2004. Floral symmetry affects speciation rates in angiosperms. Proceedings of the Royal Society B: Biological Sciences 271:603–8.

Sargent, R. D., and D. D. Ackerly. 2008. Plant-pollinator interactions and the assembly of plant communities. Trends in Ecology and Evolution 23:123–30.

Sargent, R. D., and J. C. Vamosi. 2008. The influence of canopy position, pollinator syndrome, and region on evolutionary transitions in pollinator guild size. International Journal of Plant Sciences 169:39–47.

Sargent, S. 2000. Specialized seed dispersal: mistletoes and fruit-eating birds. Pages 288–89 in N. M. Nadkarni and N. T. Wheelwright, eds. Monteverde: ecology and conservation of a tropical cloud forest. Oxford University Press, New York

Sarkinen, T. E., M. F. Newman, P. J. M. Maas, H. Maas, A. D. Poulsen, D. J. Harris, J. E. Richardson, A. Clark, M. Hollingsworth, and R. T. Pennington. 2007. Recent oceanic long-distance dispersal and divergence in the amphi-Atlantic rain forest genus *Renealmia* L.f. (Zingiberaceae). Molecular Phylogenetics and Evolution 44:968–80.

Sazima, M., S. Buzato, and I. Sazima. 1999. Bat-pollinated flower assemblages and bat visitors at two Atlantic forest sites in Brazil. Annals of Botany 83:705–12.

Schaefer, H. M., D. J. Levey, V. Schaefer, and M. L. Avery. 2006. The role of chromatic and achromatic signals for fruit detection by birds. Behavioral Ecology 17:784–89.

Schaefer, H. M., K. McGraw, and C. Catoni. 2008. Birds use fruit colour as honest signal of dietary antioxidant rewards. Functional Ecology 22:303–10.

Schaefer, H. M., and V. Schaefer. 2007. The evolution of visual fruit signals: Concepts and constraints. Pages 59–77 in A. J. Dennis, E. W. Schupp, R. J. Green, and D. A. Westcott, eds. Seed dispersal, theory and its application in a changing world. CABI, Wallingford, U.K.

Schaefer, H. M., V. Schaefer, and D. J. Levey. 2004. How plant-animal interactions signal new insights in communication. Trends in Ecology and Evolution 19:577–84.

Schaefer, H. M., V. Schaefer, and M. Vorobyev. 2007. Are fruit colors adapted to consumer vision and birds equally efficient in detecting colorful signals? American Naturalist 169:S159–S169.

Schaefer, H. M., and V. Schmidt. 2004. Detectability and content as opposing signal characteristics in fruits. Proceedings of the Royal Society of London Series B: Biological Sciences 271:S370–S373.

Schaefer, H. M., V. Schmidt, and F. Bairlein. 2003. Discrimination abilities for nutrients: which difference matters for choosy birds and why? Animal Behaviour 65:531–41.

Schaefer, H. M., V. Schmidt, and J. Wesenberg. 2002. Vertical stratification and caloric content of the standing fruit crop in a tropical lowland forest. Biotropica 34:244–53.

Schemske, D. W. 1981. Floral convergence and pollinator sharing in two bee-pollinated tropical herbs. Ecology 62:946–54.

———. 2009. Biotic interactions and speciation in the tropics. Pages 219–39 in R. K. Butlin, J. R. Bridle, and D. Schluter, eds. Speciation and patterns of diversity. Cambridge University Press, Cambridge.

Schillaci, M. A. 2008. Primate mating systems and the evolution of neocortex size. Journal of Mammalogy 89:58–63.

Schleuning, M., N. Bluthgen, M. Florchinger, J. Braun, H. M. Schaefer, and K. Böhning-Gaese. 2011. Specialization and interaction strength in a tropical plant-frugivore network differ among forest strata. Ecology 92:26–36.

Schluter, D. 2000. The ecology of adaptive radiation. Oxford University Press, Oxford.

———. 2009. Evidence for ecological speciation and its alternative. Science 323:737–41.

Schmid, J., and J. R. Speakman. 2000. Daily energy expenditure of the grey mouse lemur (*Microcebus murinus*): a small primate that uses torpor. Journal of Comparative Physiology B: Biochemical Systemic and Environmental Physiology 170:633–41.

Schmid, S., V. S. Schmid, A. Zillikens, B. Harter-Marques, and J. Steiner. 2011. Bimodal pollination system of the bromeliad *Aechmea nudicaulis* involving hummingbirds and bees. Plant Biology 13:41–50.

Schmidt-Lebuhn, A. N., M. Kessler, and I. Hensen. 2007. Hummingbirds as drivers of plant speciation? Trends in Plant Science 12:329–31.

Schnitzer, S. A. 2005. A mechanistic explanation for global patterns of liana abundance and distribution. American Naturalist 166:262–76.

Schoener, T. W. 1971. Theory of feeding strategies. Annual Review of Ecology and Systematics 2:369–404.

Schondube, J. E., L. G. Herrera-M, and C. Martínez del Rio. 2001. Diet and the evolution of digestion and renal function in phyllostomid bats. Zoology-Analysis of Complex Systems 104:59–73.

Schreier, B. M., A. H. Harcourt, S. A. Coppeto, and M. F. Somi. 2009. Interspecific competition and niche separation in primates: a global analysis. Biotropica 41:283–91.

Schuchmann, K. L. 1999. Family Trochilidae (Hummingbirds). Pages 468–680 in Handbook of the birds of the world. Lynx Editions, Barcelona.

Schulenberg, T. S., D. F. Stotz, D. F. Lane, J. P. O'Neill, and T. A. Parker III. 2007. Birds of Peru. Princeton University Press, Princeton, NJ.

Schulte, P., L. Alegret, I. Arenillas, J. A. Arz, P. J. Barton, P. R. Bown, T. J. Bralower, et al. 2010. The Chicxulub asteroid impact and mass extinction at the Cretaceous-Paleogene boundary. Science 327:1214–18.

Schulze, M. D., N. E. Seavy, and D. F. Whitacre. 2000. A comparison of the phyllostomid bat assemblages in undisturbed Neotropical forest and in forest fragments of a slash-and-burn farming mosaic in Peten, Guatemala. Biotropica 32:174–84.

Schupp, E. W. 1993. Quantity, quality and the effectiveness of seed dispersal by animals. Pages 15–29 in T. H. Fleming and A. Estrada, eds. Frugivory and seed dispersal: ecological and evolutionary aspects. Kluwer Academic Publishers, Dordrecht.

Schupp, E. W., P. Jordano, and J. M. Gomez. 2010. Seed dispersal effectiveness revisited: a conceptual review. New Phytologist 188:333–53.

Schupp, E. W., T. Milleron, and S. E. Russo. 2002. Dissemination limitation and the origin and maintenance of species-rich tropical forests. Pages 19–33 in D. J. Levey, W. R. Silva, and M. Galetti, eds. Seed dispersal and frugivory: ecology, evolution and conservation. CABI, Wallingford.

Schweizer, M., O. Seehausen, M. Guntert, and S. T. Hertwig. 2010. The evolutionary diversification of parrots supports a taxon pulse model with multiple trans-oceanic dispersal events and local radiations. Molecular Phylogenetics and Evolution 54:984–94.

Seidler, T. G., and J. B. Plotkin. 2006. Seed dispersal and spatial pattern in tropical trees. PLoS Biology 4:2132–37.

Sekercioglu, C. H. 2006. Increasing awareness of avian ecological function. Trends in Ecology and Evolution 21:464–71.

———. 2011. Functional extinctions of bird pollinators cause plant declines. Science 331:1019–20.

Servat, G. P. 1996. An annotated list of birds of the Biolat Biological Station at Pakitza, Peru. Pages

555–75 in D. E. Wilson and A. Sandoval, eds. Manu, the biodiversity of southeastern Peru. Smithsonian Institution, Washington, DC.

Shanahan, M., and S. G. Compton. 2001. Vertical stratification of figs and fig-eaters in a Bornean lowland rain forest: how is the canopy different? Plant Ecology 153:121–32.

Shanahan, M., S. So, S. G. Compton, and R. Corlett. 2001. Fig-eating by vertebrate frugivores: a global review. Biological Reviews 76:529–72.

Shapcott, A. 1998. The patterns of genetic diversity in *Carpentaria acuminata* (Arecaceae), and rainforest history in northern Australia. Molecular Ecology 7:833–47.

———. 1999. Comparison of the population genetics and densities of five *Pinanga* palm species at Kuala Belalong, Brunei. Molecular Ecology 8:1641–54.

Shilton, L. A., J. D. Altringham, S. G. Compton, and R. J. Whittaker. 1999. Old World fruit bats can be long-distance seed dispersers through extended retention of viable seeds in the gut. Proceedings of the Royal Society of London Series B: Biological Sciences 266:219–23.

Shilton, L. A., and R. H. Whittaker. 2009. The role of pteropodid bats in re-establishing tropical forests on Krakatau. Pages 176–215 in T. H. Fleming and P. A. Racey, eds. Island bats: evolution, ecology, and conservation. University of Chicago Press, Chicago.

Shull, A. M. 1988. Endangered and threatened wildlife and plants: determination of endangered status of two long-nosed bats. Federal Register 53:38456–60.

Shultz, S., C. Opie, and Q. D. Atkinson. 2011. Stepwise evolution of stable sociality in primates. Nature 479:219–24.

Sibley, C. G., and J. E. Ahlquist. 1990. Phylogeny and classification of birds. Yale University Press, New Haven, CT.

Silva Taboada, G. 1979. Los murcielagos de Cuba. Editora de la Academia de Ciencias de Cuba, Havana.

Silvertown, J. W., M. Franco, I. Pisanty, and A. Mendoza. 1993. Comparative plant demography: relative importance of life-cycle components to the finite rate of increase in woody and herbaceous perennials. Journal of Ecology 81:465–76.

Simmons, N. B. 2005. Order Chiroptera. Pages 312–529 in D. E. Wilson and D. M. Reeder, eds. Mammal species of the world: a taxonomic and geographic reference. Johns Hopkins University Press, Baltimore, MD.

Simmons, N. B., and J. H. Geisler. 1998. Phylogenetic relationships of *Icaronycteris*, *Archaeonycteris*, *Hassianycteris*, and *Palaeochiropteryx* to extant bat lineages, with comments on the evolution of echolocation and foraging strategies in Microchiroptera. Bulletin of the American Museum of Natural History 235:1–182.

Simmons, N. B., K. L. Seymour, J. Habersetzer, and G. F. Gunnell. 2008. Primitive early Eocene bat from Wyoming and the evolution of flight and echolocation. Nature 451:818–22.

Simmons, N. B., and A. L. Wetterer. 2002. Phylogeny and convergence in cactophilic bats. Pages 87–121 in T. H. Fleming and A. Valiente-Banuet, eds. Columnar cacti and their mutualists: evolution, ecology, and conservation. University of Arizona Press, Tucson.

Simoes, C. G., and M. C. M. Marques. 2007. The role of sprouts in the restoration of Atlantic rainforest in southern Brazil. Restoration Ecology 15:53–59.

Simon, R., M. W. Holderied, C. U. Koch, and O. von Helversen. 2011. Floral acoustics: conspicuous echoes of a dish-shaped leaf attract bat pollinators. Science 333:631–33.

Sinclair, I., and P. Ryan. 2003. Birds of Africa south of the Sahara. Princeton University Press, Princeton, NJ.

Singaravelan, N., and G. Marimuthu. 2004. Nectar feeding and pollen carrying from *Ceiba pentandra* by pteropodid bats. Journal of Mammalogy 85:1–7.

Singer, R. B., and M. Sazima. 2000. The pollination of *Stenorrhynchos lanceolatus* (Aublet) L. C. Rich. (Orchidaceae: Spiranthinae) by hummingbirds in southeastern Brazil. Plant Systematics and Evolution 223:221–27.

Siol, M., S. I. Wright, and S. C. H. Barrett. 2010. The population genomics of plant adaptation. New Phytologist 188:313–32.

Slauson, L. A. 2000. Pollination biology of two chiropterophilous agaves in Arizona. American Journal of Botany 87:825–36.

Slik, J. W. F., A. D. Poulsen, P. S. Ashton, C. H. Cannon, K. A. O. Eichhorn, K. Kartawinata, I. Lanniari, et al. 2003. A floristic analysis of the lowland dipterocarp forests of Borneo. Journal of Biogeography 30:1517–31.

Slik, J. W. F., N. Raes, S. I. Aiba, F. Q. Brearley, C. H. Cannon, E. Meijaard, H. Nagamasu, et al. 2009. Environmental correlates for tropical tree diversity and distribution patterns in Borneo. Diversity and Distributions 15:523–32.

Slocum, M. G. 2001. How tree species differ as recruitment foci in a tropical pasture. Ecology 82:2547–59.

Smedmark, J. E. E., and A. A. Anderberg. 2007. Boreotropical migration explains hybridization between geographically distant lineages in the pantropical clade Sideroxyleae (Sapotaceae). American Journal of Botany 94:1491–1505.

Smith, A. P., and J. U. Ganzhorn. 1996. Convergence in community structure and dietary adaptation in Australian possums and gliders and Malagasy lemurs. Australian Journal of Ecology 21:31–46.

Smith, C. C., and S. D. Fretwell. 1974. The optimal balance between size and number of offspring. American Naturalist 108:499–506.

Smith, J. F. 2001. High species diversity in fleshy-fruited tropical understory plants. American Naturalist 157:646–53.

Smith, N. J. H. 1981. Man, fishes, and the Amazon. Columbia University Press, New York.

Smith, S. A., J. M. Beaulieu, and M. J. Donoghue. 2010. An uncorrelated relaxed-clock analysis suggests an earlier origin for flowering plants. Proceedings of the National Academy of Sciences of the United States of America 107:5897–5902.

Smith, S. D. 2010. Using phylogenetics to detect pollinator-mediated floral evolution. New Phytologist 188:354–63.

Smith, S. D., C. Ane, and D. A. Baum. 2008a. The role of pollinator shifts in the floral divergence of *Iochroma* (Solanaceae). Evolution 62:793–806.

Smith, S. D., S. J. Hall, P. R. Izquierdo, and D. A. Baum. 2008b. Comparative pollination biology of sympatric and allopatric Andean *Iochroma* (Solanaceae). Annals of the Missouri Botanical Garden 95:600–617.

Smith, T. B., R. Calsbeek, R. K. Wayne, K. H. Holder, D. Pires, and C. Bardeleben. 2005. Testing alternative mechanisms of evolutionary divergence in an African rain forest passerine bird. Journal of Evolutionary Biology 18:257–68.

Smith, T. B., L. A. Freed, J. K. Lepson, and J. H. Carothers. 1995. Evolutionary consequences of extinctions in populations of a Hawaiian honeycreeper. Conservation Biology 9:107–13.

Smuts, B. B., D. L. Cheney, R. M. Seyfarth, R. W. Wrangham, and T. T. Struhsaker, eds. 1986. Primate societies. University of Chicago Press, Chicago.

Smythies, B. E. 1968. The birds of Borneo. 2nd ed. Oliver and Boyd, Edinburgh.

Snow, D. W. 1962. The natural history of the oilbird (*Steatornis caripensis*) in Trinidad, W. I. Pt. 2. Population, breeding ecology, and food. Zoologica 47:199–221.

———. 1971. Evolutionary aspects of fruit-eating by birds. Ibis 113:194–202.

———. 1981. Tropical frugivorous birds and their food plants: a world survey. Biotropica 13:1–14.

———. 1982. The cotingas. Comstock Publishing Associates, Ithaca, NY.

Snow, D. W., and B. K. Snow. 1980. Relationships between hummingbirds and flowers in the Andes of Colombia. Bulletin of the British Museum of Natural History (Zoology) 38:105–39.

Sodhi, N. S., B. W. Brook, and C. J. A. Bradshaw. 2007. Tropical conservation biology. Blackwell Publishing, Oxford.

Soligo, C., and R. D. Martin. 2006. Adaptive origins of primates revisited. Journal of Human Evolution 50:414–30.

Solorzano, S., S. Castillo, T. Valverde, and L. Avila. 2000. Quetzal abundance in relation to fruit availability in a cloud forest in southeastern Mexico. Biotropica 32:523–32.

Soltis, D. E., S. A. Smith, N. Cellinese, K. J. Wurdack, D. C. Tank, S. F. Brockington, N. F. Refulio-Rodriguez, et al. 2011. Angiosperm phylogeny: 17 genes, 640 taxa. American Journal of Botany 98:704–30.

Soltis, D. E., P. S. Soltis, P. K. Endress, and M. W. Chase. 2005. Phylogeny and evolution of angiosperms. Sinauer Associates, Sunderland, MA.

Sosa, V. J., and T. H. Fleming. 2002. Why are columnar cacti associated with nurse plants? Pages 306–23 in T. H. Fleming and A. Valiente-Banuet, eds. Columnar cact and their mutualists: evolution, ecology, and conservation. University of Arizona Press, Tucson.

Speith, H. T. 1966. Hawaiian honeycreeper, *Vestaria coccinea* (Foster), feeding on lobeliad flowers, *Clermontia arborescens* (Mann) Hillebrand. American Naturalist 100:470–73.

Spencer, H. J., and T. H. Fleming. 1989. Roosting and foraging behavior of the Queensland tube-nosed bat, *Nyctimene robinsoni*: preliminary radio-tracking observations. Australian Wildlife Research 16:413–20.

Spiegel, C. S., P. J. Hart, B. L. Woodworth, E. J. Tweed, and J. J. LeBrun. 2006. Distribution and abundance of forest birds in low-altitude habitat on Hawai'i Island: evidence for range expansion of native species. Bird Conservation International 16:175–85.

Spoor, F., T. Garland, G. Krovitz, T. M. Ryan, M. T. Silcox, and A. Walker. 2007. The primate semicircular canal system and locomotion. Proceedings of the National Academy of Sciences of the United States of America 104:10808–12.

Spujt, R. W. 1994. A systematic treatment of fruit types. Memoirs of the New York Botanical Garden 70:1–181.

Stadler, T. 2011. Mammalian phylogeny reveals recent diversification rate shifts. Proceedings of the National Academy of Sciences of the United States of America 108:6187–92.

Staggemeier, V. G., J. A. F. Diniz-Filho, and L. P. C. Morellato. 2010. The shared influence of phylogeny and ecology on the reproductive patterns of Myrteae (Myrtaceae). Journal of Ecology 98:1409–21.

Stanley, M. C., and A. Lill. 2002. Importance of seed ingestion to an avian frugivore: an experimental approach to fruit choice based on seed load. Auk 119:175–84.

Stanley, S. M. 1979. Macroevolution. W. H. Freeman, San Francisco, CA.

———. 2005. Earth system history. 2nd ed. W. H. Freeman and Co., New York.

Steadman, D. W. 1997. The historic biogeography and community ecology of Polynesian pigeons and doves. Journal of Biogeography 24:737–53.

———. 2006. Extinction and biogeography of tropical Pacific birds. University of Chicago Press, Chicago.

Stebbins, G. L. 1970. Adaptive radiation of reproductive characteristics in angiosperms. Pt. 1: Pollination mechanisms. Annual Review of Ecology and Systematics 1:307–26.

———. 1974. Flowering plants: evolution above the species. Harvard University Press, Cambridge, MA.

Stebbins, G. L. 1989. Adaptive shifts toward hummingbird pollination. Pages 39–60 in J. H. Bock and Y. B. Linhart, eds. The evolutionary ecology of plants. Westview Press, Boulder, CO.

Stephens, D. W., and J. R. Krebs. 1986. Foraging theory. Princeton University Press, Princeton, NJ.

Stevenson, P. 2005. Potential keystone plant species for the frugivore community at Tinigua Park, Colombia. Pages 37–57 in J. L. Dew and J. P. Boubli, eds. Tropical fruits and frugivores: the search for strong interactors. Springer, Dordrecht.

Stevenson, P. R. 2001. The relationship between fruit production and primate abundance in Neotropical communities. Biological Journal of the Linnean Society 72:161–78.

Stevenson, P. R., and A. M. Aldana. 2008. Potential effects of ateline extinction and forest fragmentation on plant diversity and composition in the western Orinoco Basin, Colombia. International Journal of Primatology 29:365–77.

Stevenson, P. R., M. C. Castellanos, A. I. Cortes, and A. Link. 2008. Flowering patterns in a seasonal tropical lowland forest in western Amazonia. Biotropica 40:559–67.

Stiles, F. G. 1975. Ecology, flowering phenology, and hummingbird pollination of some Costa
Rican *Heliconia* species. Ecology 56:285–301.

———.1978a. Ecological and evolutionary implications of bird pollination. American Zoologist
18:715–27.

———.1978b. Temporal organization of flowering among hummingbird foodplants of a tropical
wet forest. Biotropica 10:194–210.

———. 1979. Notes on the natural history of *Heliconia* (Musaceae) in Costa Rica. Brenesia 15:
151–80.

———. 1980. The annual cycle in a tropical wet forest hummingbird community. Ibis 122:322–43.

———. 1981. Geographical aspects of bird-flower coevolution, with particular reference to Central
America. Annals of the Missouri Botanical Gardens 68:323–51.

———. 1983. Birds. Pages 502–30 in D. H. Janzen, ed. Costa Rican natural history. University of
Chicago Press, Chicago.

———. 2004. Phylogenetic constraints upon morphological and ecological adaptation in hummingbirds (Trochilidae): why are there no hermits in the paramo? Ornitologia Neotropical
15:191–98.

———. 2008. Ecomorphology and phylogeny of hummingbirds: divergence and convergence in
adaptations to high elevations. Ornitologia Neotropical 19:511–19.

Stiles, F. G., and D. J. Levey. 1994. Appendix 7: Birds of La Selva and vicinity. Pages 384–93 in
L. A. McDade, K. S. Bawa, H. A. Hespenheide, and G. S. Hartshorn, eds. La Selva: ecology and
natural history of a Neotropical rain forest. University of Chicago Press, Chicago.

Stiles, F. G., and L. Rosselli. 1993. Consumption of fruits of the Melastomataceae by birds—how
diffuse is coevolution? Vegetatio 107/108:57–73.

Stiles, F. G., and A. F. Skutch. 1989. A guide to the birds of Costa Rica. Cornell University Press,
Ithaca, NY.

Stiles, F. G., and L. L. Wolf. 1979. Ecology and evolution of lek mating behavior in the long-tailed
hermit hummingbird. Ornithological Monographs 27:1–78.

Stocker, G. C., and A. K. Irvine. 1983. Seed dispersal by cassowaries (*Casuarius casuarius*) in
North Queensland's rainforests. Biotropica 15:170–76.

Stoddard, M. C., and R. O. Prum. 2008. Evolution of avian plumage color in a tetrahedral color
space: a phylogenetic analysis of New World buntings. American Naturalist 171:755–76.

Stouffer, D. B., and J. Bascompte. 2011. Compartmentalization increases food-web persistence.
Proceedings of the National Academy of Sciences of the United States of America 108:
3648–52.

Strauss, S. Y., and A. R. Zangerl. 2002. Plant-insect interactions in terrestrial ecosystems. Pages
77–106 in C. M. Herrera and O. Pellmyr, eds. Plant-animal interactions. Blackwell Scientific,
Oxford.

Streisfeld, M. A., and M. D. Rausher. 2009. Genetic changes contributing to the parallel evolution
of red floral pigments among *Ipomoea* species. New Phytologist 183:751–63.

Stroo, A. 2000. Pollen morphological evolution in bat pollinated plants. Plant Systematics and
Evolution 222:225–42.

Stutchbury, B. J. M., and E. S. Morton. 2001. Behavioral ecology of tropical birds. Academic Press,
San Diego, CA.

Suarez, R. K., K. C. Welch, S. K. Hanna, and L. G. Herrera. 2009. Flight muscle enzymes and
metabolic flux rates during hovering flight of the nectar bat, *Glossophaga soricina:* further

evidence of convergence with hummingbirds. Comparative Biochemistry and Physiology A: Molecular and Integrative Physiology 153:136–40.

Sun, C., T. C. Moermond, and T. J. Givnish. 1997. Nutritional determinants of diet in three Turacos in a tropical montane forest. Auk 114:200–211.

Sussman, R. W. 1991. Primate origins and the evolution of angiosperms. American Journal of Primatology 23:209–23.

Suthers, R. A. 1970. Vision, olfaction, taste. Pages 265–309 in W. A. Wimsatt, ed. Biology of bats. Academic Press, New York.

Svenning, J. C., B. M. J. Engelbrecht, D. A. Kinner, T. A. Kursar, R. F. Stallard, and S. J. Wright. 2006. The relative roles of environment, history and local dispersal in controlling the distributions of common tree and shrub species in a tropical forest landscape, Panama. Journal of Tropical Ecology 22:575–86.

Svenning, J. C., and S. J. Wright. 2005. Seed limitation in a Panamanian forest. Journal of Ecology 93:853–62.

Swaine, M. D., and P. Becker. 1999. Woody life-form composition and association on rainfall and soil fertility gradients in Ghana. Plant Ecology 145:167–73.

Sytsma, K. J., A. Litt, M. L. Zjhra, J. C. Pires, M. Nepokroeff, E. Conti, J. Walker, and P. G. Wilson. 2004. Clades, clocks, and continents: historical and biogeographical analysis of Myrtaceae, Vochysiaceae, and relatives in the Southern Hemisphere. International Journal of Plant Sciences 165:S85–S105.

Sytsma, K. J., and R. W. Pippen. 1985. Morphology and pollination biology of an intersectional hybrid of *Costus*. Systematic Botany 10:353–62.

Tang, Z., A. Mukherjee, L. Sheng, M. Cao, B. Liang, R. T. Corlett, and S. Zhang. 2007. Effect of ingestion by two frugivorous bat species on the seed germination of *Ficus racemosa* and *F. hispida* (Moraceae). Journal of Tropical Ecology 23:125–27.

Tarr, C. L., and R. C. Fleischer. 1995. Evolutionary relationships of the Hawaiian honeycreepers (Aves, Drepanidinae). Pages 147–59 in W. L. Wagner and V. A. Funk, eds. Hawaiian biogeography. Smithsonian Institution Press, Washington, DC.

Tattersall, G. J., D. V. Andrade, and A. S. Abe. 2009. Heat exchange from the toucan bill reveals a controllable vascular thermal radiator. Science 325:468–70.

Teeling, E. C., M. S. Springer, O. Madsen, P. Bates, S. J. O'Brien, and W. J. Murphy. 2005. A molecular phylogeny for bats illuminates biogeography and the fossil record. Science 307:580–84.

Temeles, E. J., C. R. Koulouris, S. E. Sander, and W. J. Kress. 2009. Effect of flower shape and size on foraging performance and trade-offs in a tropical hummingbird. Ecology 90:1147–61.

Temeles, E. J., and W. J. Kress. 2003. Adaptation in a plant-hummingbird association. Science 300:630–33.

———. 2010. Mate choice and mate competition by a tropical hummingbird at a floral resource. Proceedings of the Royal Society B: Biological Sciences 277:1607–13.

Temeles, E. J., I. L. Pan, J. L. Brennan, and J. N. Horwitt. 2000. Evidence for ecological causation of sexual dimorphism in a hummingbird. Science 289:441–43.

Ter Steege, H., N. Pitman, D. Sabatier, H. Castellanos, P. Van der Hout, D. C. Daly, M. Silveira, et al. 2003. A spatial model of tree alpha-diversity and tree density for the Amazon. Biodiversity and Conservation 12:2255–77.

Terborgh, J. 1977. Bird species diversity on an Andean elevational gradient. Ecology 58:1007–19.

———. 1983. Five New World primates: a study in comparative ecology. Princeton University Press, Princeton, NJ.

———. 1986. Keystone plant resources in the tropical forests. Pages 330–44 in M. Soule, ed. Conservation biology: the science of scarcity and diversity. Sinauer Associates, Sunderland, MA.

Terborgh, J., P. Alvarez-Loayza, K. Dexter, F. Cornejo, and C. Carrasco. 2011. Decomposing dispersal limitation: limits on fecundity or seed distribution? Journal of Ecology 99:935–44.

Terborgh, J., and G. Nuñez-Iturri. 2006. Disperser-free tropical forests await an unhappy fate. Pages 241–52 in W. F. Laurance and C. A. Peres, eds. Emerging threats to tropical forests. University of Chicago Press, Chicago.

Terborgh, J., and C. P. Van Schaik. 1987. Convergence vs. non-convergence in primate communities. Pages 205–26 in J. H. R. Gee and P. S. Giller, eds. Organization of communities: past and present. Blackwell Scientific, Oxford.

Terborgh, J. W., S. K. Robinson, T. A. Parker III, C. A. Munn, and N. Pierpont. 1990. Structure and organization of an Amazonian forest bird community. Ecological Monographs 60:213–38.

Tewksbury, J. J. 2002. Fruits, frugivores and the evolutionary arms race. New Phytologist 156: 137–39.

Tewksbury, J. J., D. J. Levey, M. Huizinga, D. C. Haak, and A. Traveset. 2008. Costs and benefits of capsaicin-mediated control of gut retention in dispersers of wild chilies. Ecology 89:107–17.

Tewksbury, J. J., and G. P. Nabhan. 2001. Seed dispersal—directed deterrence by capsaicin in chillies. Nature 412:403–4.

Tewksbury, J. J., G. P. Nabhan, D. Norman, H. Suzan, J. D. Tuxill, and J. Donovan. 1999. In situ conservation of wild chilies and their biotic associates. Conservation Biology 13:98–107.

Thebault, E., and C. Fontaine. 2010. Stability of ecological communities and the architecture of mutualistic and trophic networks. Science 329:853–56.

Thies, W., and E. K. V. Kalko. 2004. Phenology of Neotropical pepper plants (Piperaceae) and their association with their main dispersers, two short-tailed fruit bats, *Carollia perspicillata* and *C. castanea* (Phyllostomidae). Oikos 104:362–76.

Thies, W., E. K. V. Kalko, and H.-U. Schnitzler. 1998. The roles of echolocation and olfaction in two Neotropical fruit-eating bats, *Carollia perspicillata* and *C. castanea*, feeding on *Piper*. Behavioral Ecology and Sociobiology 42:397–409.

Thomas, D. W. 1983. The annual migration of three species of West African fruit bats (Chiroptera: Pteropodidae). Canadian Journal of Zoology-Revue Canadienne De Zoologie 61:2266–72.

Thomas, G. H., C. D. L. Orme, R. G. Davies, V. A. Olson, P. M. Bennett, K. J. Gaston, I. P. F. Owens, and T. M. Blackburn. 2008. Regional variation in the historical components of global avian species richness. Global Ecology and Biogeography 17:340–51.

Thompson, J. N. 1994. The coevolutionary process. University of Chicago Press, Chicago.

———. 1997. Evaluating the dynamics of coevolution among geographically structured populations. Ecology 78:1619–23.

———. 1999. Specific hypotheses on the geographic mosaic theory of coevolution. American Naturalist 153:S1–S14.

———. 2005. The geographic mosaic of coevolution. University of Chicago Press, Chicago.

———. 2006. Mutualistic webs of species. Science 312:372–73.

———. 2009. The coevolving web of life. American Naturalist 173:125–40.

Thompson, J. N., and B. M. Cunningham. 2002. Geographic structure and dynamics of coevolutionary selection. Nature 417:735–38.

Thompson, J. N., and O. Pellmyr. 1992. Mutualism with pollinating parasites amid co-pollinators. Ecology 73:1780–91.

Thomson, J. D., and P. Wilson. 2008. Explaining evolutionary shifts between bee and hummingbird pollination: convergence, divergence, and directionality. International Journal of Plant Sciences 169:23–38.

Thorne, R. 2004. Tropical plant disjunctions: a personal reflection. International Journal of Plant Sciences 165:S137–S138.

Tiffney, B. H. 1984. Seed size, dispersal syndromes, and the rise of the angiosperms: evidence and hypothesis. Annals of the Missouri Botanical Garden 71:551–76.

———. 1986. Evolution of seed dispersal syndromes according to the fossil record. Pages 273–305 in D. R. Murray, ed. Seed dispersal. Academic Press, Sydney.

———. 2004. Vertebrate dispersal of seed plants through time. Annual Review of Ecology, Evolution, and Systematics 35:1–29.

Tiffney, B. H., and S. J. Mazer. 1995. Angiosperm growth habit, dispersal and diversification reconsidered. Evolutionary Ecology 9:93–117.

Tilman, D. 1982. Resource competition and community structure. Princeton University Press, Princeton, NJ.

Timm, R. M. 1994. Appendix 8: Mammals. Pages 394–98 in L. A. McDade, K. S. Bawa, H. A. Hespenheide, and G. S. Hartshorn, eds. 1994. La Selva: ecology and natural history of a Neotropical rain forest. University of Chicago Press, Chicago.

Timm, R. M., and R. K. LaVal. 2000. Appendix 10: Mammals of Monteverde. Pages 553–57 in N. M. Nadkarni and N. T. Wheelwright, eds. Monteverde: ecology and conservation of a tropical cloud forest. Oxford University Press, New York.

Toledo, M., and J. Salick. 2006. Secondary succession and indigenous management in semideciduous forest fallows of the Amazon basin. Biotropica 38:161–70.

Toriola, D. 1998. Fruiting of a 19-year old secondary forest in French Guiana. Journal of Tropical Ecology 14:373–79.

Trail, P. W. 2007. African hornbills: keystone species threatened by habitat loss, hunting and international trade. Ostrich 78:609–13.

Traill, L. W., M. L. M. Lim, N. S. Sodhi, and C. J. A. Bradshaw. 2010. Mechanisms driving change: altered species interactions and ecosystem function through global warming. Journal of Animal Ecology 79:937–47.

Traveset, A. 1999. La importancia de los mutualismos para la conservacion de la biodiversidad en ecosistemas insulares. Revista Chilena de Historia Natural 72:527–38.

Traveset, A., and D. M. Richardson. 2006. Biological invasions as disruptors of plant reproductive mutualisms. Trends in Ecology and Evolution 21:208–16.

Traveset, A., and N. Riera. 2005. Disruption of a plant-lizard seed dispersal system and its ecological effects on a threatened endemic plant in the Balearic Islands. Conservation Biology 19:421–31.

Tripp, E. A., and P. S. Manos. 2008. Is floral specialization an evolutionary dead-end? pollination system transitions in *Ruellia* (Acanthaceae). Evolution 62:1712–36.

Tsahar, E., C. Martínez del Rio, Z. Arad, J. P. Joy, and I. Izhaki. 2005a. Are the low protein requirements of nectarivorous birds the consequence of their sugary and watery diet? a test with an omnivore. Physiological and Biochemical Zoology 78:239–45.

Tsahar, E., C. Martínez del Rio, I. Izhaki, and Z. Arad. 2005b. Can birds be ammonotelic? nitrogen balance and excretion in two frugivores. Journal of Experimental Biology 208:1025–34.

Tschapka, M. 2004. Energy density patterns of nectar resources permit coexistence within a guild of Neotropical flower-visiting bats. Journal of Zoology 263:7–21.

Tschapka, M., S. Dressler, and O. v. Helversen. 2006. Bat visits to *Marcgravia pittieri* and notes on the inflorescence diversity within the genus *Marcgravia* (Marcgraviaceae). Flora 201:383–88.

Tscharntke, T., C. H. Sekercioglu, T. V. Dietsch, N. S. Sodhi, P. Hoehn, and J. M. Tylianakis. 2008. Landscape consraints on functional diversity of birds and insects in tropical agroecosystems. Ecology 89:944–51.

Tuljapurkar, S., C. C. Horvitz, and J. B. Pascarella. 2003. The many growth rates and elasticities of populations in random environments. American Naturalist 162:489–502.

Tuomisto, H., and K. Ruokolainen. 2006. Analyzing or explaining beta diversity? understanding the targets of different methods of analysis. Ecology 87:2697–2708.

Tuomisto, H., K. Ruokolainen, M. Aguilar, and A. Sarmiento. 2003a. Floristic patterns along a 43-km long transect in an Amazonian rain forest. Journal of Ecology 91:743–56.

Tuomisto, H., K. Ruokolainen, and M. Yli-Halla. 2003b. Dispersal, environment, and floristic variation of western Amazonian forests. Science 299:241–44.

Turnbull, L. A., M. J. Crawley, and M. Rees. 2000. Are plant populations seed-limited? a review of seed sowing experiments. Oikos 88:225–38.

Turner, I. M., and R. T. Corlett. 1996. The conservation value of small, isolated fragments of lowland tropical rain forest. Trends in Ecology and Evolution 11:330–33.

Tutin, C. E. G., R. M. Ham, L. J. T. White, and M. J. S. Harrison. 1997. The primate community of the Lope Reserve, Gabon: Diets, responses to fruit scarcity, and effects on biomass. American Journal of Primatology 42:1–24.

Tutin, C. E. G., and L. J. T. White. 1998. Primates, phenology and frugivory: present, past and future patterns in the Lope Reserve, Gabon. Pages 309–37 in D. M. Newbury, H. H. T. Prins, and N. D. Brown, eds. Dynamics of tropical communities. Blackwell Scientific, Oxford.

Uhl, C. 1987. Factors controlling succession following slash-and-burn agriculture in Amazonia. Journal of Ecology 75:377–407.

Uhl, C., K. Clark, H. Clark, and P. Murphy. 1981. Early plant succession after cutting and burning in the upper Rio Negro region of the Amazon basin. Journal of Ecology 69:631–49.

Uriarte, M., M. Anciaes, M. T. B. da Silva, P. Rubim, E. Johnson, and E. M. Bruna. 2011. Disentangling the drivers of reduced long-distance seed dispersal by birds in an experimentally fragmented landscape. Ecology 92:924–37.

Utzurrum, R. C. B. 1995. Feeding ecology of Philippine fruit bats: patterns of resource use and seed dispersal. Pages 63–77 in P. A. Racey and S. Swift, eds. Ecology, evolution, and behaviour of bats. Clarendon Press, Oxford.

Valencia, R., R. B. Foster, G. Villa, R. Condit, J. C. Svenning, C. Hernandez, K. Romoleroux, E. Losos, E. Magard, and H. Balslev. 2004. Tree species distributions and local habitat variation in the Amazon: large forest plot in eastern Ecuador. Journal of Ecology 92:214–29.

Valido, A., H. M. Schaefer, and P. Jordano. 2011. Colour, design and reward: phenotypic integration of fleshy fruit displays. Journal of Evolutionary Biology 24:751–60.

Valiente-Banuet, A., M. D. Arizmendi, A. Martinez-Rojas, and P. Davila. 1997. Pollination of two columnar cacti (*Neobuxbaumia mezcalensis* and *Neobuxbaumia macrocephala*) in the Tehuacan Valley, central Mexico. American Journal of Botany 84:452–55.

Valiente-Banuet, A., M. D. Arizmendi, A. Martinez-Rojas, and L. Dominguez-Canesco. 1996. Geographical and ecological correlates between columnar cacti and nectar-feeding bats in Mexico. Journal of Tropical Ecology 12:103–19.

Valiente-Banuet, A., and M. Verdú. 2007. Facilitation can increase the phylogenetic diversity of plant communities. Ecology Letters 10:1029–36.

Valiente-Banuet, A., F. Vite, and J. A. Zavala-Hurtado. 1991. Interaction between the cactus *Neobuxbaumia tetezo* and the nurse shrub *Mimosa lusiana*. Journal of Vegetation Science 2:11–14.

Vamosi, J. C., and S. M. Vamosi. 2010. Key innovations within a geographical context in flowering plants: towards resolving Darwin's abominable mystery. Ecology Letters 13:1270–79.

Vamosi, J. C., T. M. Knight, J. A. Steets, S. J. Mazer, M. Burd, and T. L. Ashman. 2006. Pollination decays in biodiversity hotspots. Proceedings of the National Academy of Sciences of the United States of America 103:956–61.

Vamosi, J. C., S. P. Otto, and S. C. H. Barrett. 2003. Phylogenetic analysis of the ecological correlates of dioecy in angiosperms. Journal of Evolutionary Biology 16:1006–18.

Vamosi, S. M., S. B. Heard, J. C. Vamosi, and C. O. Webb. 2009. Emerging patterns in the comparative analysis of phylogenetic community structure. Molecular Ecology 18:572–92.

van der Pijl, L. 1936. Fledermause und blumen. Flora 131:1–40.

———. 1982. Principles of dispersal in higher plants. 3rd ed. Springer-Verlag, Berlin.

Van Devender, T. R. 2002. Environmental history of the Sonoran Desert. Pages 3–24 in T. H. Fleming and A. Valiente-Banuet, eds. Columnar cacti and their mutualists: evolution, ecology, and conservation. University of Arizona Press, Tucson.

Van Schaik, C. 1986. Phenological changes in a Sumatran rain forest. Journal of Tropical Ecology 2:327–47.

Van Schaik, C. P., and K. R. Pfannes. 2005. Tropical climates and phenology: a primate perspective. Pages 23–54 in D. K. Brockman and C. P. Van Schaik, eds. Seasonality in primates: studies of living and extinct human and non-human primates. Cambridge University Press, Cambridge.

Van Schaik, C. P., J. Terborgh, and S. J. Wright. 1993. The phenology of tropical forests: adaptive significance and consequences for primary consumers. Annual Review of Ecology and Systematics 24:353–77.

van Tets, I. G., and A. J. Hulbert. 1999. A comparison of the nitrogen requirements of the eastern pygmy possum, *Cercartetus nanus*, on a pollen and on a mealworm diet. Physiological and Biochemical Zoology 72:127–37.

van Tets, I. G., and S. W. Nicolson. 2000. Pollen and nitrogen requirements of the lesser double-collared sunbird. Auk 117:826–30.

Van Tyne, J., and A. J. Berger. 1976. Fundamentals of ornithology. 2nd ed. John Wiley, New York.

Vanthomme, H., B. Belle, and P. M. Forget. 2010. Bushmeat hunting alters recruitment of large-seeded plant species in Central Africa. Biotropica 42:672–79.

Varassin, I. G., J. R. Trigo, and M. Sazima. 2001. The role of nectar production, flower pigments and odour in the pollination of four species of *Passiflora* (Passifloraceae) in south-eastern Brazil. Botanical Journal of the Linnean Society 136:139–52.

Vardon, M. J., P. S. Brocklehurst, J. C. Z. Woinarski, R. B. Cunningham, C. F. Donnelly, and C. R. Tidemann. 2001. Seasonal habitat use by flying-foxes, *Pteropus alecto* and *P. scapulatus* (Megachiroptera), in monsoonal Australia. Journal of Zoology 253:523–35.

Vaughan, T. N., J. M. Ryan, and N. J. Czaplewski. 2000. Mammalogy. 4th ed. Saunders College Publishing, Fort Worth, TX.

Vazquéz, D. P., and M. A. Aizen. 2004. Asymmetric specialization: a pervasive feature of plant-pollinator interactions. Ecology 85:1251–57.

Vazquéz, D. P., N. Bluthgen, L. Cagnolo, and N. P. Chacoff. 2009. Uniting pattern and process in plant-animal mutualistic networks: a review. Annals of Botany 103:1445–57.

Vazquéz, D. P., and D. Simberloff. 2003. Changes in interaction biodiversity induced by an introduced ungulate. Ecology Letters 6:1077–83.

Vazquez-Yanes, C., and A. Orozco-Segovia. 1993. Patterns of seed longevity and germination in the tropical rainforest. Annual Review of Ecology and Systematics 24:69–87.

Venable, D. L. 1996. Packaging and provisioning in plant reproduction. Philosophical Transactions of the Royal Society of London Series B: Biological Sciences 351:1319–29.

Verdú, M., P. J. Rey, J. M. Alcantara, G. Siles, and A. Valiente-Banuet. 2009. Phylogenetic signatures of facilitation and competition in successional communities. Journal of Ecology 97:1171–80.

Verdú, M., and A. Traveset. 2004. Bridging meta-analysis and the comparative method: a test of seed size effect on germination after frugivores' gut passage. Oecologia 138:414–18.

Viseshakul, N., W. Charoennitikul, S. Kitamura, A. Kemp, S. Thong-Aree, Y. Surapunpitak, P. Poonswad, and M. Ponglikitmongkol. 2011. A phylogeny of frugivorous horbills linked to the evolution of Indian plants within Asian rainforests. Journal of Evolutionary Biology 24:1533–45.

Visser, M. D., E. Jongejans, M. van Breugel, P. A. Zuidema, Y.-Y. Chen, A. R. Kassim, and H. de Kroon. 2011. Strict mast fruiting for a tropical dipterocarp tree: a demographic cost-benefit analysis of delayed reproduction and seed predation. Journal of Ecology 99:1033–44.

Voigt, C. C., and Y. Winter. 1999. Energetic cost of hovering flight in nectar-feeding bats (Phyllostomidae: Glossophaginae) and its scaling in moths, birds and bats. Journal of Comparative Physiology B: Biochemical Systemic and Environmental Physiology 169:38–48.

Voigt, F. A., B. Bleher, J. Fietz, J. U. Ganzhorn, D. Schwab, and K. Bohning-Gaese. 2004. A comparison of morphological and chemical fruit traits between two sites with different frugivore assemblages. Oecologia 141:94–104.

Vormisto, J., J. C. Svenning, P. Hall, and H. Balslev. 2004. Diversity and dominance in palm (Arecaceae) communities in terra firme forests in the western Amazon basin. Journal of Ecology 92:577–88.

Wadt, L. H. D., and P. Y. Kageyama. 2004. Genetic structure and mating system of *Piper hispidinervum*. Pesquisa Agropecuaria Brasileira 39:151–57.

Waites, A. R., and J. Ågren. 2004. Pollinator visitation, stigmatic pollen loads and among-population variation in seed set in *Lythrum salicaria*. Journal of Ecology 92:512–26.

Wake, D. B., M. H. Wake, and C. D. Specht. 2011. Homoplasy: from detecting pattern to determining process and mechanism of evolution. Science 331:1032–35.

Waldien, D. 2008. Meeting the challenge in the Philippines. Bats 26:4–7.

Walker, L. R., F. H. Landau, E. Velazquez, A. B. Shiels, and A. D. Sparrow. 2010. Early successional woody plants facilitate and ferns inhibit forest development on Puerto Rican landslides. Journal of Ecology 98:625–35.

Walsberg, G. E. 1975. Digestive adaptations of *Phainopepla nitens* associated with the eating of mistletoe berries. Condor 77:169–74.

Wang, B. C., and T. B. Smith. 2002. Closing the seed dispersal loop. Trends in Ecology and Evolution 17:379–85.

Wang, B. C., V. L. Sork, M. T. Leong, and T. B. Smith. 2007. Hunting of mammals reduces seed removal and dispersal of the Afrotropical tree *Antrocaryon klaineanum* (Anacardiaceae). Biotropica 39:340–47.

Ward, M., C. W. Dick, R. Gribel, and A. J. Lowe. 2005. To self, or not to self . . . a review of outcrossing and pollen-mediated gene flow in Neotropical trees. Heredity 95:246–54.

Waser, N. M. 1983. Competition for pollination and floral character differences among sympatric plant species: a review of evidence. Pages 277–93 in C. E. Jones and R. J. Little, eds. Handbook of experimental pollination ecology. Van Nostrand-Reinhold, New York.

Waser, N. M., L. Chittka, M. V. Price, N. M. Williams, and J. Ollerton. 1996. Generalization in pollination systems, and why it matters. Ecology 77:1043–60.

Weathers, W. W., D. C. Paton, and R. S. Seymour. 1996. Field metabolic rate and water flux of nectarivorous honeyeaters. Australian Journal of Zoology 44:445–60.

Webb, C. J., and K. S. Bawa. 1983. Pollen dispersal by hummingbirds and butterflies: a comparative study of two lowland tropical plants. Evolution 37:1258–70.

Webb, C. O. 2000. Exploring the phylogenetic structure of ecological communities: an example for rain forest trees. American Naturalist 156:145–55.

Webb, C. O., D. D. Ackerly, M. A. McPeek, and M. J. Donoghue. 2002. Phylogenies and community ecology. Annual Review of Ecology and Systematics 33:475–505.

Webb, C. O., C. H. Cannon, and S. J. Davies. 2008. Ecological organization, biogeography, and the phylogenetic structure of tropical forest tree communities. Pages 79–97 in W. P. Carson and S. A. Schnitzer, eds. Tropical forest community ecology. Wiley-Blackwell, Oxford.

Webb, C. O., and M. J. Donoghue. 2005. Phylomatic: tree assembly for applied phylogenetics. Molecular Ecology Notes 5:181–83.

Webb, C. O., and D. R. Peart. 2001. High seed dispersal rates in faunally intact tropical rain forest: theoretical and conservation implications. Ecology Letters 4:491–99.

Weiher, E., and P. A. Keddy, eds. 1999. Ecological assembly rules, perspectives, advances, retreats. Cambridge University Press, Cambridge.

Weir, J. T. 2006. Divergent timing and patterns of species accumulation in lowland and highland Neotropical birds. Evolution 60:842–55.

Weir, J. T., E. Bermingham, and D. Schluter. 2009. The great American interchange in birds. Proceedings of the National Academy of Sciences of the United States of America 106:21737–42.

Welch, K. C., Jr., D. L. Altshuler, and R. K. Suarez. 2007. Oxygen consumption rates in hovering hummingbirds reflect substrate-dependent differences in P/O ratios: carbohydrate as a "premium" fuel. Journal of Experimental Biology 210:2146–53.

Welch, K. C., Jr., B. H. Bakken, C. Martínez del Rio, and R. K. Suarez. 2006. Hummingbirds fuel hovering flight with newly-ingested sugar. Physiological and Biochemical Zoology 79:1082–87.

Wen, J., and S. M. Ickert-Bond. 2009. Evolution of the Madrean-Tethyan disjunctions and the North and South American amphitropical disjunctions. Journal of Systematics and Evolution 47:331–48.

Wendeln, M. C., J. R. Runkle, and E. K. V. Kalko. 2000. Nutritional values of 14 fig species and bat feeding preferences in Panama. Biotropica 32:489–501.

Wenny, D. G. 2000a. Seed dispersal of a high quality fruit by specialized frugivores: high quality dispersal? Biotropica 32:327–37.

———. 2000b. Seed dispersal, seed predation, and seedling recruitment of a Neotropical montane tree. Ecological Monographs 70:331–51.

———. 2001. Advantages of seed dispersal: a re-evaluation of directed dispersal. Evolutionary Ecology Research 3:51–74.

Wenny, D. G., and D. J. Levey. 1998. Directed seed dispersal by bellbirds in a tropical cloud forest. Proceedings of the National Academy of Sciences of the United States of America 95:6204–7.

Westcott, D. A., J. Bentrupperbaumer, M. G. Bradford, and A. McKeown. 2005a. Incorporating patterns of disperser behaviour into models of seed dispersal and its effects on estimated dispersal curves. Oecologia 146:57–67.

Westcott, D. A., M. G. Bradford, A. J. Dennis, and G. Lipsett-Moore. 2005b. Keystone fruit resources and Australia's tropical rain forests. Pages 237–60 in J. L. Dew and J. P. Boubli, eds. Tropical fruits and frugivores: the search for strong interactors. Springer: Dordrecht.

Westcott, D. A., and D. L. Graham. 2000. Patterns of movement and seed dispersal of a tropical frugivore. Oecologia 122:249–57.

Wester, P., and R. Classen-Bockhoff. 2007. Floral diversity and pollen transfer mechanisms in bird-pollinated *Salvia* species. Annals of Botany 100:401–21.

Westerkamp, C. 1990. Bird-flowers—hovering versus perching exploitation. Botanica Acta 103:366–71.

Westoby, M., D. S. Falster, A. T. Moles, P. A. Vesk, and I. J. Wright. 2002. Plant ecological strategies: some leading dimensions of variation between species. Annual Review of Ecology and Systematics 33:125–59.

Wetterer, A. L., M. V. Rockman, and N. B. Simmons. 2000. Phylogeny of phyllostomid bats (Mammalia: Chiroptera): data from diverse morphological systems, sex chromosomes, and restriction sites. Bulletin of the American Museum of Natural History 248:1–200.

Wheelwright, N. T. 1983. Fruits and the ecology of resplendent quetzals. Auk 100:286–301.

———. 1985. Fruit size, gape width, and the diets of fruit-eating birds. Ecology 66:808–18.

Wheelwright, N. T., W. A. Haber, K. G. Murray, and C. Guindon. 1984. Tropical fruit-eating birds and their food plants: a survey of Costa Rican lower montane forest. Biotropica 16:173–92.

Wheelwright, N. T., and C. H. Janson. 1985. Colors of fruit displays of bird-dispersed plants in 2 tropical forests. American Naturalist 126:777–99.

Wheelwright, N. T., and G. H. Orians. 1982. Seed dispersal by animals: contrasts with pollen dispersal, problems of terminology, and constraints on coevolution. American Naturalist 119:402–13.

Whelan, C. J., D. G. Wenny, and R. J. Marquis. 2008. Ecosystem services provided by birds. Annals of the New York Academy of Sciences 1134:25–60.

Whelan, R. J., and R. L. Goldingay. 1989. Factors affecting fruit-set in *Telopea speciosissima* (Proteaceae)—the importance of pollen limitation. Journal of Ecology 77:1123–34.

White, E., N. Tucker, N. Meyers, and J. Wilson. 2004. Seed dispersal to revegetated isolated rainforest patches in North Queensland. Forest Ecology and Management 192:409–26.

White, L. J. T. 1994. Patterns of fruit-fall phenology in the Lope Reserve, Gabon. Journal of Tropical Ecology 10:289–312.

Whitmore, T. C. 1998. An introduction to tropical rain forests. 2nd ed. Clarendon Press, Oxford.

Whitmore, T. C., and D. F. R. P. Burslem. 1998. Major disturbances in tropical rainforests. Pages 549–65 in D. M. Newbury, H. H. T. Prins, and N. D. Brown, eds. Dynamics of tropical communities. Blackwell Scientific, Oxford.

Whitney, K. D., M. K. Fogiel, A. M. Lamperti, K. M. Holbrook, D. J. Stauffer, B. D. Hardesty, V. T. Parker, and T. B. Smith. 1998. Seed dispersal by *Ceratogymna* hornbills in the Dja Reserve, Cameroon. Journal of Tropical Ecology 14:351–71.

Whittaker, R. H. 1972. Evolution and measurement of species diversity. Taxon 21:213–51.

———. 1977. Evolution of species diversity in land communities. Pages 1–67 in M. K. Hecht, W. C. Steere, and B. Wallace, eds. Evolutionary biology. Plenum Press, New York.

Whittaker, R. J., and S. H. Jones. 1994. The role of frugivorous bats and birds in the rebuilding of a tropical forest ecosystem, Krakatau, Indonesia. Journal of Biogeography 21:245–58.

Wible, J. R., G. W. Rougier, M. J. Novacek, and R. J. Asher. 2007. Cretaceous eutherians and Laurasian origin for placental mammals near the K/T boundary. Nature 447:1003–6.

Wich, S. A., R. Buij, and C. P. Van Schaik. 2004. Determinants of orangutan density in the dryland forests of the Leuser Ecosystem. Primates 45:177–82.

Wich, S. A., S. S. Utami-Atmoko, T. M. Setia, S. Djoyosudharmo, and M. L. Geurts. 2006. Dietary and energetic responses of *Pongo abelii* to fruit availability fluctuations. International Journal of Primatology 27:1535–50.

Wich, S. A., and C. P. Van Schaik. 2000. The impact of El Niño on mast fruiting in Sumatra and elsewhere in Malesia. Journal of Tropical Ecology 16:563–77.

Wiens, J. J., and M. J. Donoghue. 2004. Historical biogeography, ecology, and species richness. Trends in Ecology and Evolution 19:639–44.

Wiens, D., M. Renfree, and R. O. Wooller. 1979. Pollen loads of honey possums (*Tarsipes spenserae*) and non-flying mammal pollination in Southwestern Australia. Annals of the Missouri Botanical Garden 66:830–38.

Wiens, J. J., D. D. Ackerly, A. P. Allen, B. L. Anacker, L. B. Buckley, H. V. Cornell, E. I. Damschen, et al. 2010. Niche conservatism as an emerging principle in ecology and conservation biology. Ecology Letters 13:1310–24.

Wiens, J. J., Pyron, R.A., and D.S. Moen. 2011. Phylogenetic origins of local-scale diversity patterns and the causes of Amazonian megadiversity. Ecology Letters 14:643–52.

Wikstrom, N., V. Savolainen, and M. W. Chase. 2001. Evolution of the angiosperms: calibrating the family tree. Proceedings of the Royal Society B: Biological Sciences 268:2211–20.

Wildman, D. E., M. Uddin, J. C. Opazo, G. Liu, V. Lefort, S. Guindon, O. Gascuel, L. I. Grossman, R. Romero, and M. Goodman. 2007. Genomics, biogeography, and the diversification of placental mammals. Proceedings of the National Academy of Sciences of the United States of America 104:14395–400.

Wilf, P., N. R. Cuneo, K. R. Johnson, J. F. Hicks, S. L. Wing, and J. D. Obradovich. 2003. High plant diversity in Eocene South America: evidence from Patagonia. Science 300:122–25.

Wilf, P., K. R. Johnson, N. R. Cuneo, M. E. Smith, B. S. Singer, and M. A. Gandolfo. 2005. Eocene plant diversity at Laguna del Hunco and Rio Richileufu, Patagonia, Argentina. American Naturalist 165:634–50.

Williams, B. A., R. F. Kay, and E. C. Kirk. 2010. New perspectives on primate origins. Proceedings of the National Academy of Sciences of the United States of America 107:4797–4804.

Williams, P. H., H. M. de Klerk, and T. M. Crowe. 1999. Interpreting biogeographical boundaries among Afrotropical birds: spatial patterns in richness gradients and species replacement. Journal of Biogeography 26:459–74.

Williams, P. H., C. J. Humphries, and K. J. Gaston. 1994. Centers of seed-plant diversity—the family way. Proceedings of the Royal Society of London Series B: Biological Sciences 256:67–70.

Williams, R. J., B. A. Myers, D. Eamus, and G. A. Duff. 1999. Reproductive phenology of woody species in a north Australian tropical savanna. Biotropica 31:626–36.

Willig, M. R., D. M. Kaufman, and R. D. Stevens. 2003. Latitudinal gradients of biodiversity: pattern, process, scale, and synthesis. Annual Review of Ecology and Systematics 34:273–309.

Willig, M. R., S. J. Presley, C. P. Bloch, C. L. Hice, S. P. Yanoviak, M. M. Diaz, L. A. Chauca, V. Pacheco, and S. C. Weaver. 2007. Phyllostomid bats of lowland Amazonia: effects of habitat alteration on abundance. Biotropica 39:737–46.

Willis, K. J., K. D. Bennett, S. A. Bhagwat, and H. J. B. Birks. 2010. 4 degrees C and beyond: what did this mean for biodiversity in the past? Systematics and Biodiversity 8:3–9.

Willson, M. F. 1993. Dispersal mode, seed shadows, and colonization patterns. Vegetatio 107/108:261–80.

Willson, M. F., and N. Burley. 1983. Mate choice in plants: tactics, mechanisms, and consequences. Princeton University Press, Princeton, NJ.

Willson, M. F., A. K. Irvine, and N. G. Walsh. 1989. Vertebrate dispersal syndromes in some Australian and New Zealand plant communities, with geographic comparisons. Biotropica 21:133–47.

Willson, M. F., and C. J. Whelan. 1990. The evolution of fruit color in fleshy-fruited plants. American Naturalist 136:790–809.

Wilson, D. E. 1983. Checklist of mammals. Pages 443–47 in D. H. Janzen, ed. Costa Rican natural history. University of Chicago Press, Chicago.

Wilson, D. E., and D. M. Reeder, eds. 2005. Mammal species of the world: a taxonomic and geographic reference. 3rd ed. Johns Hopkins University Press, Baltimore, MD.

Wing, S. L., F. Herrera, C. A. Jaramillo, C. Gomez-Navarro, P. Wilf, and C. C. Labandeira. 2009. Late Paleocene fossils from the Cerrejon Formation, Colombia, are the earliest record of Neotropical rainforest. Proceedings of the National Academy of Sciences of the United States of America 106:18627–32.

Wing, S. L., and B. H. Tiffney. 1987. Interactions of angiosperms and herbivorous tetrapods through time. Pages 203–24 in E. M. Friis, W. G. Chaloner, and P. H. Crane, eds. The origins of angiosperms and their biological consequences. Cambridge University Press, Cambridge.

Winkelmann, J. R., F. J. Bonaccorso, E. E. Goedeke, and L. J. Ballock. 2003. Home range and territoriality in the least blossom bat, *Macroglossus minimus*, in Papua New Guinea. Journal of Mammalogy 84:561–70.

Winter, Y., J. Lopez, and O. v. Helversen. 2003. Ultraviolet vision in a bat. Nature 425:612–14.

Winter, Y., and O. v. Helversen. 2001. Bats as pollinators: foraging energetics and floral adaptations. Pages 148–70 in L. Chittka and J. D. Thomson, eds. Cognitive ecology of pollination. Cambridge University Press, Cambridge.

———. 2003. Operational tongue length in phyllostomid nectar-feeding bats. Journal of Mammalogy 84:886–96.

Witmer, M. C., and P. J. Van Soest. 1998. Contrasting digestive strategies of fruit-eating birds. Functional Ecology 12:728–41.

Wolf, L. L. 1975. Energy intake and expenditures in a nectar-feeding sunbird. Ecology 56:92–104.

Wolf, L. L., F. R. Hainsworth, and F. G. Stiles. 1972. Energetics of foraging: rate and efficiency of nectar extraction by hummingbirds. Science 176:1351–52.

Wolfe, N. D., P. Daszak, A. M. Kilpatrick, and D. S. Burke. 2005. Bushmeat hunting deforestation, and prediction of zoonoses emergence. Emerging Infectious Diseases 11:1822–27.

Wolff, D. 2006. Nectar sugar composition and volumes of 47 species of Gentianales from a southern Ecuadorian montane forest. Annals of Botany 97:767–77.

Woltmann, S. 2003. Bird community responses to disturbance in a forestry concession in lowland Bolivia. Biodiversity and Conservation 12:1921–36.

Wolton, R. J., P. A. Arak, H. C. J. Godfray, and R. P. Wilson. 1982. Ecological and behavioural studies of the Megachiroptera at Mount Nimba, Liberia, with notes on Microchiroptera. Mammalia 46:419–48.

Wong, M. 1986. Trophic organisation of understory birds in a Malaysian dipterocarp forest. Auk 103:100–116.

Wong, S. T., C. Servheen, L. Ambu, and A. Norhayati. 2005. Impacts of fruit production cycles on Malayan sun bears and bearded pigs in lowland tropical forest of Sabah, Malaysian Borneo. Journal of Tropical Ecology 21:627–39.

Woodburne, M. O., G. F. Gunnell, and R. K. Stucky. 2009. Climate directly influences Eocene mammal faunal dynamics in North America. Proceedings of the National Academy of Sciences of the United States of America 106:13399–403.

Wooller, F. L. S., and K. C. Richardson. 1988. Morphological relationships of passerine birds from Australia and New Guinea in relation to their diets. Zoological Journal of the Linnean Society 94:193–201.

Wooller, R. D., K. C. Richardson, and B. G. Collins. 1993. The relationship between nectar supply and the rate of capture of a nectar-dependent small marsupial *Tarsipes rostratus*. Journal of Zoology 229:651–58.

Wooller, R. D., K. C. Richardson, and C. M. Pagendham. 1988. The digestion of pollen by some Australian birds. Australian Journal of Zoology 36:357–62.

Worthington, A. H. 1989. Adaptations for avian frugivory: assimilation efficiency and gut transit time of *Manacus vitellinus* and *Pipra mentalis*. Oecologia 80:381–89.

Wrangham, R. W. 1986. Evolution of social structure. Pages 282–96 in B. B. Smuts, D. L. Cheney, R. M. Seyfarth, R. W. Wrangham, and T. T. Struhsaker, eds. Primate societies. University of Chicago Press, Chicago.

Wright, D. D. 1998. Fruit choice by the dwarf cassowary, *Casuarius bennetti*, over a three year period in Papua New Guinea. Ph.D. diss., University of Miami, Miami, FL.

Wright, I. J., D. D. Ackerly, F. Bongers, K. E. Harms, and others. 2007. Relationships among ecologically important dimensions of plant trait variation in seven Neotropical forests. Annals of Botany 99:1003–15.

Wright, S. 1965. The interpretation of population structure by F-statistics with special regard to systems of mating. Evolution 19:395–420.

Wright, S. J., and O. Calderon. 1995. Phylogenetic patterns among tropical flowering phenologies. Journal of Ecology 83:937–48.

Wright, S. J., C. Carrasco, O. Calderon, and S. Paton. 1999. The El Niño Southern Oscillation variable fruit production, and famine in a tropical forest. Ecology 80:1632–47.

Wright, S. J., A. Hernandez, and R. Condit. 2007. The bushmeat harvest alters seedling banks by favoring lianas, large seeds, and seeds dispersed by bats, birds, and wind. Biotropica 39:363–71.

Wunderle, J. M., L. M. P. Henriques, and M. R. Willig. 2006. Short-term responses of birds to forest gaps and understory: an assessment of reduced-impact logging in a lowland Amazon forest. Biotropica 38:235–55.

Wyatt, J. L., and M. R. Silman. 2004. Distance-dependence in two Amazonian palms: effects of spatial and temporal variation in seed predator communities. Oecologia 140:26–35.

Yates, C. J., D. J. Elliott, and M. Byrne. 2007a. Composition of the pollinator community, pollination and the mating system for a shrub in fragments of species rich kwongan in south-west Western Australia. Biodiversity and Conservation 16:1379–95.

Yates, C. J., D. J. Elliott, M. Byrne, D. J. Coates, and R. Fairman. 2007b. Seed production, germinability and seedling growth for a bird-pollinated shrub in fragments of kwongan in southwestern Australia. Biological Conservation 136:306–14.

Yoder, J. B., E. Clancey, S. Des Roches, J. M. Eastman, L. Gentry, W. Godsoe, T. J. Hagey, et al. 2010. Ecological opportunity and the origin of adaptive radiations. Journal of Evolutionary Biology 23:1581–96.

York, H. A., and S. A. Billings. 2009. Stable-isotope analysis of diets of short-tailed fruit bats (Chiroptera: Phyllostomidae: *Carollia*). Journal of Mammalogy 90:1469–77.

Young, K. R., C. U. Ulloa, J. L. Luteyn, and S. Knapp. 2002. Plant evolution and endemism in Andean South America: an introduction. Botanical Review 68:4–21.

Yumoto, T. 2000. Bird-pollination of three *Durio* species (Bombacaceae) in a tropical rainforest in Sarawak, Malaysia. American Journal of Botany 87:1181–88.

Yumoto, T., T. Itino, and H. Nagamasu. 1997. Pollination of hemiparasites (Loranthaceae) by spider hunters (Nectariniidae) in the canopy of a Bornean rain forest. Selbyana 18:51–60.

Zahwai, R. A., and C. K. Augspurger. 2006. Tropical forest restoration: tree islands as recruitment foci in degraded lands of Honduras. Ecological Applications 16:464–78.

Zeipel, H., von, and O. Eriksson. 2007. Fruit removal in the forest herb *Actaea spicata* depends on local context of fruits sharing the same dispersers. International Journal of Plant Sciences 168:855–60.

Zerega, N. J. C., W. L. Clement, S. L. Datwyler, and G. D. Weiblen. 2005. Biogeography and divergence times in the mulberry family (Moraceae). Molecular Phylogenetics and Evolution 37:402–16.

Ziegler, A. C. 2002. Hawaiian natural history, ecology, and evolution. University of Hawai'i Press, Honolulu.

Zimmerman, M., and G. H. Pyke. 1988. Reproduction in *Polemonium:* assessing the factors limiting seed set. American Naturalist 131:723–38.

Zuccon, D., R. Prys-Jones, P. C. Rasmussen, and P. G. P. Ericson. 2012. The phylogenetic relationships and generic limits of finches (Fringillidae). Molecular Phylogenetics and Evolution 62:581–96.

Species Index

Subject Index

adaptive radiation: and biotic interactions, 425–29; and environmental filtering, 423–25; and immigration/dispersal, 422–23; in regional biotas, 419–29; and speciation, 420–22. *See also* earth history; evolution

Andes: effect on animal speciation, 203, 204, 262–64, 430; effect on plant speciation, 194, 195, 262–64

angiosperm evolution: chronology, 228–331; congruence with the evolution of vertebrate mutualists, 255–57; and long-distance dispersal, 232–33, 257; ordinal phylogeny, 218; and vicariance, 231–33, 257. *See also* mutualisms

Anthropocene, 477

atelids (Atelidae), 371–72

bats: assembly of contemporary communities, 418; Caribbean phyllostomids, 418; color vision in, 309; evolution of body size and jaw length, 310–16; flower visiting in phyllostomids vs. pteropodids, 309; foraging behavior, 124–31, 324–25; fruit processing in, 369, 370; morphology of fruit eaters, 367–70; morphology of nectar-feeders, 303–16; and neotropical fruits, 336–38; network structure of, 58–61; physiology of fruit eaters, 375–78; physiology of nectar-feeders, 319–21; social and mating systems in, 324–25, 384–85. *See also* frugivores

benefits of seed dispersal: the colonization benefit, 150–52; the directed dispersal benefit, 152; the escape benefit, 145–50

biogeography: in angiosperms, 231–36; Asian tropical forests, 261; of Neotropical forests, 261–64; and species richness, 30–40. *See also* birds; mammals

biomass: of frugivores, 101–2; of fruits, 101; of nectar, 96–99; of nectarivores, 96–101

birds: biogeography, 245–48; chronology of family appearances, 244–45; foraging behavior in nectar-feeders, 125–31, 322–24; lek mating in, 323, 382–84; morphology of frugivores, 363–67; morphology of nectar-feeders, 296–303; phylogenetic distribution of frugivory, 241–44; phylogenetic distribution of nectarivory, 237–41; phylogeny, 237–45; physiology of frugivores, 375–78; physiology of nectar-feeders, 316–19; social systems, 323. *See also* frugivores; hummingbirds; *and specific families*

birds of paradise (Paradisaeidae), 55, 186, 202–3, 244, 247, 367, 383–84, 391

body size (mass): and elevation, 48; in functional groups, 401; of frugivores, 8; of nectarivores, 8, 70, 72; of pollinating insects, 18; and vertical stratification, 54

Bombacoideae (Malvaceae), 65, 118, 185, 191, 210, 256, 262, 280, 289, 292, 332

Bromeliaceae, 30, 37, 48, 65, 69, 194, 196–97, 222, 230, 235, 262, 280, 290, 332, 340, 409, 436

bulbuls (Pycnonotidae), 55–56, 73, 241, 243–44, 290

bushmeat hunting: of frugivores, 445–46, 469–72

Cactaceae, 37, 194, 196, 230, 234–35, 257, 262, 289, 293, 332, 336, 448, 476

capsaicin, 78–79

cebids (Cebidae), 52, 371–72, 378, 407

cercopithecines (Cercopithecidae), 52, 55, 144, 180, 204, 253–55, 348, 371, 378, 400, 407

cheaters: in pollination mutualisms, 15; in seed dispersal mutualisms, 15, 77–78

climate change: and adaptive radiation, 430–31; and rates of vertebrate diversification, 473–75; and threat to biodiversity, 441, 472–77